NEUROGENIC *Inflammation*

Edited by

Pierangelo Geppetti, M.D.
Department of Preclinical and Clinical Pharmacology
University of Florence
Florence, Italy

Peter Holzer, Ph.D.
Department of Experimental and Clinical Pharmacology
University of Graz
Graz, Austria

CRC

CRC Press

Boca Raton New York London Tokyo

Library of Congress Cataloging-in-Publication Data

Neurogenic inflammation / edited by Pierangelo Geppetti, Peter Holzer.
p. cm.
Includes bibliographical references and index.
ISBN 0-8493-7646-7 (alk. paper)
1. Inflammation--Pathophysiology. 2. Inflammation--Molecular aspects. 3. Neuropeptides. I. Geppetti, Pierangelo. II. Holzer, Peter.
[DNLM: 1. Nervous System--physiology. 2. Inflammation--physiopathology. WL 102 N494553 1996]
RB131.N48 1996
616.8--dc20
DNLM/DLC
for Library of Congress 95-38900
CIP

International Standard Book Number 0-8493-7646-7
Library of Congress Card Number 95-38900
Printed in the United States of America 1 2 3 4 5 6 7 8 9 0
Printed on acid-free paper

DEDICATION

To Mary and Andy
To Ulrike, Veronika, and Judith

PREFACE

Neurogenic inflammation refers to the ability of a subpopulation of primary sensory neurons to evoke local inflammatory responses at the site of their peripheral endings. Although functional evidence showing that primary sensory neurons could cause inflammatory effects at the level of the tissue that they innervate was first obtained more than one hundred years ago, it was only in the last two decades that inflammation caused by the release of neuropeptides from afferent nerve endings was recognized as a physiologically and pathologically relevant process. The early findings that formed the basis for this concept were summarized in a volume on "Antidromic Vasodilatation and Neurogenic Inflammation" which was published by L.A. Chahl, J. Szolcsanyi and F. Lembeck through Akademiai Kiado, Budapest in 1984. Since then unprecedented progress in our understanding of the molecular biology, pharmacology and pathophysiology of neurogenic inflammation has been made. Taking all advances together, it was felt that a comprehensive and up-to-date volume was badly needed, and the present book was designed to cover all the major contemporary aspects of neurogenic inflammation in health and diseases, and to put them into perspective with new therapeutic approaches to inflammatory and related diseases.

Several chapters of the book focus on the anatomical, neurochemical and functional characteristics of those neurons which when stimulated give rise to neurogenic inflammation. These neurons are mostly C-fiber polymodal nociceptors whose small dark cell bodies reside in the trigeminal and vagal sensory ganglia and in the dorsal root ganglia; it is important to realize that they are central not only to inflammatory reactions but also to nociception, particularly inflammatory pain. For obvious reasons, much emphasis is laid on capsaicin, the drug which owing to its selective excitotoxic action on thin afferent neurons, has greatly contributed to our current knowledge of sensory nerve functions. The molecular biology and pharmacology of the peptide transmitters, calcitonin gene-related peptide (CGRP) and the tachykinins substance P and neurokinin A, and the receptors that mediate neurogenic inflammatory responses comprise another important section of the book.

A large number of endogenous substances and autacoids may stimulate or sensitize sensory nerve endings, thus simultaneously producing pain and nociceptive responses, as well as neurogenic inflammation. On the other hand, both prejunctional and postjunctional mechanisms to inhibit the inflammatory function of sensory neurons have been identified. The regulation of afferent nerve endings by factors which stimulate or inhibit the release of sensory neuropeptides or interfere with their actions holds great potential for therapeutic intervention with inflammatory and nociceptive processes.

Neurogenic inflammation can be regarded as a mechanism that activates protective responses, thus bringing about a first line of defense to maintain the integrity of the tissue. However, if severe and prolonged stimulation occurs the inflammatory response evoked locally by sensory nerves may produce injury, instead of strengthening repair processes. In chronic pathological conditions, inflammatory mediators may further enhance their proinflammatory potential by releasing sensory neuropeptides, thus promoting the establishment of a vicious circle that contributes to the exacerbation and perpetuation of the disease. The hypothesis of the involvement of neurogenic inflammation in various human diseases is discussed critically in various chapters of the book. Reports of the effectiveness of sensory nerve defunctionalization by capsaicin have supported this hypothesis and detailed description of the pertinent clinical studies is also given.

Our understanding of neurogenic inflammation under various pathophysiological conditions has been greatly advanced by the recent molecular cloning of the genes that encode the precursors for CGRP, tachykinins and tachykinin NK_1, NK_2 and NK_3 receptors, the development of capsaicin analogs and antagonists, and the design of potent and selective antagonists for all the three tachykinin receptors. This book offers the reader a comprehensive and up-to-date overview of all those findings that herald new therapeutical strategies for the various diseases in which neurogenic inflammatory processes are involved. The editors trust that the volume fills a need to integrate the current knowledge in this most important field of pathophysiology and hope that the current book may become the first in a series that is continuously updating the growing knowledge in this expanding field. The various chapters have been written by the leading authorities in their area of expertise, and we are exceedingly grateful to all the contributors for providing so freely of their time and knowledge to put together what we think has become a true state-of-the-art book on neurogenic inflammation. Special thanks go to Marsha Baker and Renee Taub of CRC Press for seeing this volume through the press.

Pierangelo Geppetti
Peter Holzer

THE EDITORS

Pierangelo Geppetti, M.D., is at the Department of Preclinical and Clinical Pharmacology, Faculty of Medicine of the University of Florence in Florence, Italy, where he is Assistant Professor of Clinical Pharmacology. Dr. Geppetti obtained his M.D. degree in 1978 at the University of Florence and he was a Visiting Scientist at the Cardiovascular Research Institute of the University of California, San Francisco in 1991—1993. He is a member of the British Pharmacological Society, the American Association of Clinical Pharmacology, and the Italian Pharmacological Society.

Dr. Geppetti has authored several book chapters as well as more than 130 research papers. His major research interest include the basic mechanism of sensory nerve activation and the involvement of sensory neuropeptides in migraine and inflammatory airway diseases.

Peter Holzer, Ph.D., is Professor of Neuropharmacology in the Department of Experimental and Clinical Pharmacology at the University of Graz, Austria. Dr. Holzer obtained his Ph.D. degree in 1978 and he was a Visiting Scientist at the University of California Medical School in Los Angeles in 1989. Dr. Holzer is Chairman of the European Neuropeptide Club, Board Member of the International Neuropeptide Society, and Secretary of the Gastrointestinal Pharmacology Section of IUPHAR. He is a member of the American Gastroenterological Association, the British Pharmacological Section and the German Society of Experimental and Clinical Pharmacology and Toxicology. He serves on the Editorial Board of Neuroscience, Regulatory Peptides and Naunyn-Schmiedeberg's Archives of Pharmacology, and British Journal of Pharmacology.

Dr. Holzer has authored several books and chapters and more than 115 research papers. He has a long-standing interest in the role of afferent nerves and sensory neuropeptides in inflammatory processes within the skin and gastrointestinal tract.

CONTRIBUTORS

Rainer Amann
Department of Experimental and Clinical Pharmacology
University of Graz
Graz, Austria

Peter J. Barnes
Department of Thoracic Medicine
National Heart and Lung Institute
London, England

Lorand Barthó
Department of Pharmacology
University Medical School
Pecs, Hungary

Stuart Bevan
Sandoz Institute of Medical Research
London, England

Susan D. Brain
Pharmacology Group
Biomedical Sciences Division
King's College
London, England

Girolamo Caló
Institute of Pharmacology
University of Ferrara
Ferrara, Italy

F. Michael Cutrer
Stroke Research Laboratory
Department of Neurosurgery and Neurology
Massachusetts General Hospital
Harvard Medical School
Boston, Massachusetts

Ronald J. Docherty
Sandoz Institute for Medical Research
London, England

Andy Dray
Sandoz Institute for Medical Research
London, England

Ronald B. Emeson
Departments of Pharmacology and Molecular Physiology and Biophysics
Vanderbilt University
Nashville, Tennessee

William R. Ferrell
Institute of Physiology
University of Glasgow
Glasgow, Scotland

Tung M. Fong
Department of Molecular Pharmacology and Biochemistry
Merck Research Laboratories
Rahway, New Jersey

Pierangelo Geppetti
Department of Preclinical and Clinical Pharmacology
University of Florence
Florence, Italy

Rolf Håkanson
Department of Pharmacology
University of Lund
Lund, Sweden

Judith M. Hall
School of Biological Sciences
University of Surrey
Guildford, England

Peter Holzer
Department of Experimental and Clinical Pharmacology
University of Graz
Graz, Austria

Guy F. Joos
Department of Respiratory Diseases
University Hospital
Ghent, Belgium

Francis Y. Lam
Department of Pharmacology
Chinese University of Hong Kong
Shatin, Hong Kong

Alessandro Lecci
Pharmacology Research
A. Menarini Pharmaceuticals
Florence, Italy

Won S. Lee
Stroke Research Laboratory
Departments of Neurosurgery and
Neurology
Massachusetts General Hospital
Harvard Medical School
Boston, Massachusetts

John A. Lowe, III
Central Research Division
Pfizer, Inc.
Groton, Connecticut

David B. MacLean
Central Research Division
Pfizer, Inc.
Groton, Connecticut

Carlo A. Maggi
Pharmacology Department
A. Menarini Pharmaceuticals
Florence, Italy

Michael A. Moskowitz
Stroke Research Laboratory
Department of Neurosurgery and
Neurology
Massachusetts General Hospital
Harvard Medical School
Boston, Massachusetts

Jay A. Nadel
Cardiovascular Research Institute
University of California – San Francisco
San Fransisco, California

Quang T. Nguyen
Department of Pharmacology
University of Sherbrooke
Faculty of Medicine
Sherbrooke, Quebec, Canada

Jose L. Ochoa
Department of Neurology
Good Samaritan Hospital
and Medical Center
Portland, Oregon

Domenico Regoli
Department of Pharmacology
Faculty of Medicine
University of Sherbrooke
Sherbrooke, Quebec, Canada

Jordi Serra
Neuropathic Pain Unit
Neuromuscular Unit
Hospital de Bellvitge
Barcelona, Spain

R. Michael Snider
Berlex Biosciences
Richmond, California

Arpad Szallasi
Department of Pharmacology
Karolinska Institute
Stockholm, Sweden

Janos Szolcsanyi
Department of Pharmacology
University Medical School
Pecs, Hungary

Zun-Yi Wang
Department of Pharmacology
University of Lund
Lund, Sweden

Marina Ziche
Department of Preclinical
and Clinical Pharmacology
University of Florence
Florence, Italy

TABLE OF CONTENTS

PART I. GENERAL

PART II. PHARMACOLOGY

PART III. PATHOPHYSIOLOGY

Part I. General

Chapter 1

MOLECULAR BIOLOGY OF TACHYKININS

Tung M. Fong

CONTENTS

I. DISCOVERY OF TACHYKININS

In 1931, von Euler and Gaddum[1] discovered a hypotensive, spasmogenic, and non-cholinergic activity in the acid ethanol extract of horse brain and intestine. The active component in the dried powder of the extract was referred to as substance P, which is free of suggested meaning, and it appeared to be a protein or peptide. However, the exact structure of substance P (SP) remained unknown for another 40 years while other related peptides were characterized.

Since the initial discovery of SP in the horse brain, substances with similar activities were found in other species and tissues (Figure 1). The most notable was eledoisin from the salivary glands of the octopus *Eledone moschata*. Because of its high concentration in the salivary gland, eledoisin became the first tachykinin to be purified and sequenced.[2] Erspamer and coworkers subsequently sequenced several amphibian peptides possessing similar biological activity and structure. It was shown that this group of peptides elicits a fast contractile response from intestinal tissues compared to the slower action of bradykinin, and hence the term tachykinin was proposed. In addition, Erspamer noticed the common C-terminal sequence FXGLM-NH_2, which would explain that this motif was required to confer the characteristic biological activity upon synthetic analogs of eledoisin. These studies led to the prediction that SP should also contain the conserved C-terminal sequence,[3] which was subsequently confirmed by Chang and colleagues.[4]

The discovery of SP in the mammalian nervous system quickly led to intensive research to understand the *in vivo* functions of tachykinins. In the 1950s, Lembeck postulated that SP may be a sensory transmitter.[5] This hypothesis was vigorously tested in the 1970s as soon as synthetic SP became available. The convergence of anatomic, biochemical, pharmacological, and physiological studies demonstrated that SP is synthesized and released from primary afferent terminals, is bound to specific receptors, and is degraded by peptidases, thereby fulfilling the criteria of neurotransmitter identification.[6-8] The vasodilation or intestinal contraction in response to SP and the identification of SP as a transmitter in the central nervous system confirmed the suggestion made by Dale in 1935 that the chemical transmitter of the axon-reflex vasodilation might be identical to the sensory transmitter at the spinal cord synapses because both are released from the same neuron, one from peripheral terminals and the other from central terminals.[9] More importantly, the release of SP from peripheral

0-8493-7646-7/96/$0.00+$.50

	Mammals
RPKPQQ**FFGLM-NH2**	Substance P
HKTDS**FVGLM-NH2**	Neurokinin A
DMHDF**FVGLM-NH2**	Neurokinin B
DADSSIEKQVALLKALYGHGQISHKRHKTDS**FVGLM-NH2**	Neuropeptide K
DAGHGQISHKRHKTDS**FVGLM-NH2**	Neuropeptide γ
	Amphibians
pGlu-ADPNK**FYGLM-NH2**	Physalaemin
DVPKSDQ**FVGLM-NH2**	Kassinin
	Fish
AKFDK**FYGLM-NH2**	Scyliorhinin
	Mollusc
pGlu-PSKDA**FIGLM-NH2**	Eledoisin
	Insects
NTGDK**FYGLM-NH2**	Sialokinin I
FXGLM-NH2	***Consensus sequence***

FIGURE 1. Amino acid sequence of selected tachykinins.

terminals (in the skin, visceral organs, and blood vessel) is the first step in the phenomenon of neurogenic inflammation, which will be discussed in the other chapters.

Besides SP, two related peptides have been found in mammals by several groups.[10-13] Both neurokinin A (also called substance K, neuromedin L, or neurokinin α) and neurokinin B (also called neurokinin β or neuromedin K) were identified in the spinal cord first, and then in the brain and other tissues. The different and confusing names adopted by different groups prompted committees to recommend guidelines to standardize the nomenclature.[14] Thus far, it is generally accepted that tachykinins refer to all peptides from either vertebrates or invertebrates having the conserved C-terminal sequence FXGLM-NH_2. The term neurokinin is more appropriate for mammalian tachykinins (i.e., substance P, neurokinin A, and neurokinin B) because of their functions as neurotransmitters or neuromodulators.

II. BIOSYNTHESIS AND REGULATED RELEASE OF NEUROKININS

Mammalian tachykinins are encoded by two distinct genes, the preprotachykinin (PPT) I gene and the PPT II gene. SP, neurokinin A (NKA), and neurokinin B (NKB) are generated from these genes through elaborate posttranscriptional and posttranslational processing. The translated products are designated preprotachykinin because they contain a signal sequence for endoplasmic reticulum targeting and other amino acid sequences that will be removed through posttranslational processing.

Among the 7 exons of the PPT I gene, exon 3 contains the SP-coding sequence and exon 6 contains the NKA coding sequence (Figure 2). Three mRNA species can be generated through alternative splicing.[11,15] The α-PPT I mRNA encodes only substance P, while the β-PPT I mRNA and γ-PPT I mRNA encode both substance P and neurokinin A. In all three cases, precursor peptides are synthesized from each mRNA species, followed by posttranslational processing to generate substance P and neurokinin A. Due to the involvement of multiple proteolytic cleavages, three variants of neurokinin A may also be produced, namely neurokinin A(3–10), neuropeptide K (from exons 3–6 in β-PPT mRNA) and neuropeptide γ (from exons 3, 5, and 6 in γ-PPT mRNA). NKA(3–10) is pharmacologically similar to NKA.[16]

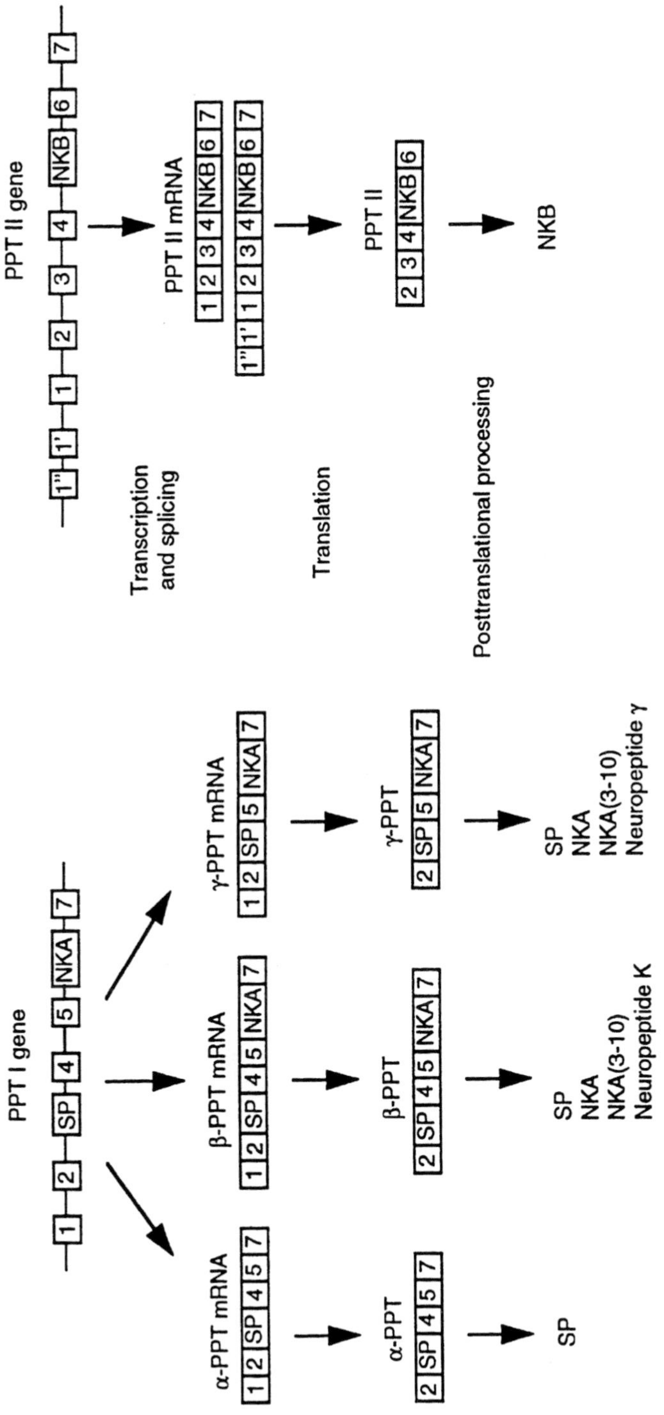

FIGURE 2. Biosynthesis of neurokinins.

On the other hand, both neuropeptide K and neuropeptide γ are slightly more potent than NKA.[17-20] Whether any of these variants of NKA is biologically significant will depend on their *in vivo* concentrations.

While the PPT I gene encodes SP and NKA, the PPT II gene encodes only NKB.[21] Two mRNA species are transcribed from this gene consisting of 9 exons. The two mRNA species differ only in the 5′ untranslated region. The translated product, PPT II, is processed to generate a single peptide NKB. Despite the fact that transcripts from the PPT II gene are processed differently compared to those of the PPT I gene, the overall resemblance in the structure of the PPT I and PPT II genes suggests that these two genes evolved from a common primordial gene via gene duplication. The two ancient PPT genes probably encoded NKA and NKB, respectively, or their ancestral versions. Later intragenic duplication in the ancient PPT I gene may have generated SP. Such an evolutionary relationship among the three neurokinins would not parallel that of the neurokinin receptors, among which the NK_1 receptor and the NK_3 receptor are more closely related.[22] It is likely that SP and the NK_1 receptor evolved more recently and independently from NKA and the NK_3 receptor, respectively.

In all preprotachykinins, the amino end of the SP, NKA, or NKB sequence is preceded by basic residues (Arg preceding the SP sequence, Lys-Arg preceding the NKA or NKB sequence), and the carboxy end is followed by Gly-Lys-Arg. Several endopeptidases are known to cleave at the C-terminal side of basic residues. However, the first residue in SP (i.e., Arg) is probably protected from proteolytic cleavage. After the proteolytic cleavage, precursors of SP, NKA, and NKB are generated, all containing Gly-Lys-Arg following the C-terminal consensus sequence. A carboxypeptidase is involved in removing the basic residues and the enzyme peptidyl-glycine-α-amidating monooxygenase will act to remove Gly and leave the amide group at the C-terminus.[23,24]

One of the most important events in peptide biosynthesis is the targeting and packaging of peptides in regulated secretory vesicles. Many neuropeptides, neurokinins included, are coreleased with other peptides or nonpeptide transmitters under certain physiological conditions. How each specific peptide is sorted is unclear. There is no obvious primary sequence feature that would segregate one peptide from another in the dense core vesicles. In addition, transfection studies suggested that foreign hormones are sorted correctly in a heterologous environment, implying that the mechanism of segregation is not cell specific.[25] There is evidence suggesting that peptides are usually aggregated in the Golgi network, and this aggregation may be linked to sorting in the presence of other Golgi proteins.[26]

Following sorting in the Golgi network, dense core vesicles containing peptides move via anterograde axonal transport to the release site at nerve terminals. Unlike the small, clear synaptic vesicles which are organized as active zone arrays and contain classical transmitters, the dense core vesicles are usually dispersed in the nerve endings and contain both peptides and classical transmitters. Furthermore, peptide release often requires multiple action potentials firing at high frequency as opposed to the release of classical transmitters following a single action potential.[27] This differential release mechanism is consistent with the suggestion that SP is only involved in nociceptive response evoked by noxious cutaneous stimulation or high-frequency electrical stimulation of sensory nerves, and therefore SP antagonists do not affect the response to innocuous input or low-frequency stimuli to C fibers.[28]

III. MOLECULAR BIOLOGY OF NEUROKININ RECEPTORS

For the three mammalian tachykinins, there exist three corresponding receptors in mammals.[29] Each receptor subtype is defined pharmacologically by the rank order of potency of three neurokinins. The NK_1 receptor/(NK_{1R}) receptor is characterized by a rank order of SP > NKA > NKB, while the NK_2 receptor/(NK_{2R}) is characterized by NKA > NKB > SP and the NK_3 receptor/(NK_{3R}) is characterized by NKB > NKA > SP. For each receptor subtype,

TABLE 1
Approximate Binding Affinities of Three Natural Peptides for the Cloned Neurokinin Receptors

	K_d, n*M*		
Receptor subtype	**SP**	**NKA**	**NKB**
NK_1	1	30	80
NK_2	1000	2	100
NK_3	1000	200	2

Note: The actual affinity will vary slightly, depending on the species origin and cell type. The affinities listed represent those for the high-affinity state of each receptor.

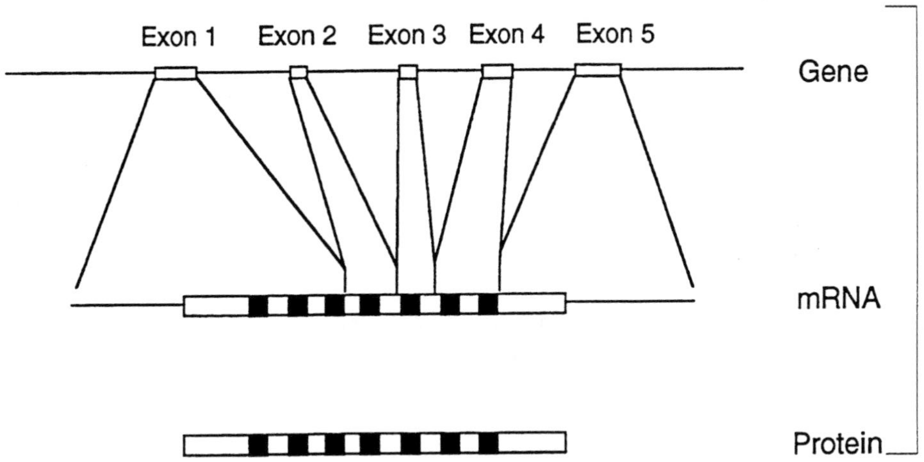

FIGURE 3. Gene structure of neurokinin receptors. The dark boxes represent transmembrane helices.

the most potent natural peptide has a binding affinity of about 1 n*M* for the high-affinity state of the receptor (see below). The pharmacological definition of receptor subtypes has been confirmed by recent molecular cloning and heterologous expression of the genes encoding each receptor subtype (Table 1).[30,31]

The overall genomic structure of all three neurokinin receptors is quite similar, extending over 60, 20, and 45 kb for the human NK_{1R}, NK_{2R}, and NK_{3R}, respectively.[22,32] All three genes consist of five exons, and the exon-intron boundaries are located at similar positions with respect to the protein sequence in the transmembrane region. The N-terminal region and transmembrane segments 1–3 reside in exon 1, transmembrane segment 4 resides in exon 2, transmembrane 5 resides in exon 3, transmembrane segments 6–7 reside in exon 4, and the C-terminal region resides in exon 5 (Figure 3). Both amino acid and nucleic acid sequence comparison revealed that the NK_{1R} and NK_{3R} are more similar to each other than to the NK_{2R}, suggesting that gene duplication may have generated the NK_{2R} and another primordial NK receptor gene, and the NK_{1R} and NK_{3R} genes evolved from another gene duplication event.[22] Unlike the production of multiple peptides from the preprotachykinin genes, the expression of neurokinin receptor genes generally leads to one final protein. Thus far, only one alternative splicing mRNA product has been detected for the human NK_{1R}, and the biological significance of this product is unknown.[33]

In general, the relative pattern of tissue distribution of neurokinin receptors as determined by radioligand binding correlates with that of mRNAs encoding each receptor as determined by Northern blot or RNase protection assay.[34-36] The NK_{1R} is widely expressed in the brain,

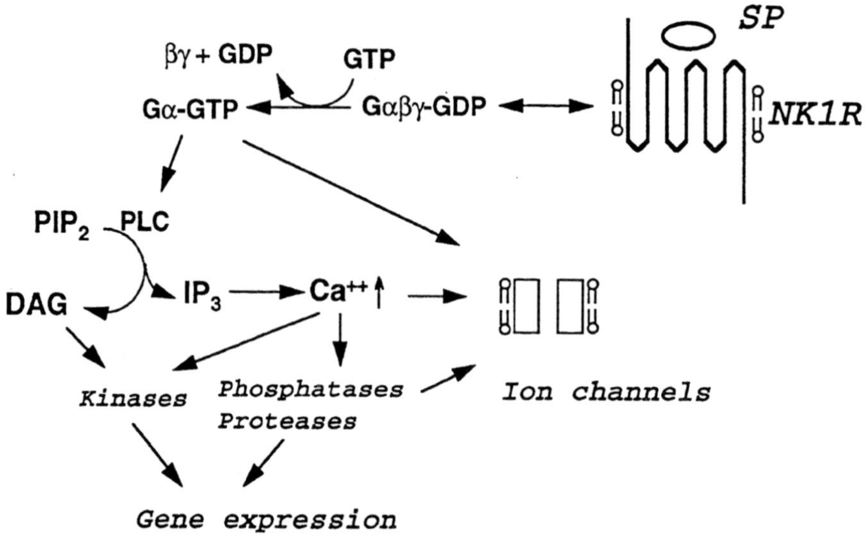

FIGURE 4. Intracellular signaling pathways evoked by substance P. Gα, G protein α-subunit; Gαβγ, heterotrimeric G protein; PLC phospholipase C; PIP_2, phosphatidylinositol-4,5-bisphosphate; DAG, 1,2-diacylglycerol; IP_3, inositol-1,4,5-trisphosphate.

spinal cord, and peripheral tissues such as intestine, lung, and lymphocytes. The NK_{2R} is expressed predominantly in peripheral tissues, while the NK_{3R} is expressed predominantly in the brain. While it is obvious that a particular receptor plays a crucial function in tissues where its expression level is high, the physiological significance of other receptors present at low abundance should not be overlooked. For example, specific NK_2 antagonists can influence certain behavioral aspects of mice even though the NK_{2R} expression level is low in the brain.[37] These studies point to the importance of *in situ* localization and the shortcomings of whole organ analysis such as Northern blot.

The molecular cloning of neurokinin receptors first suggested that these receptors are similar to rhodopsin in that they are characterized by seven putative transmembrane segments. Heterologous expression further demonstrated that these receptors function through the activation of G proteins,[38] which in turn activate other intracellular effectors, including ion channels and enzymes (Figure 4). Thus, neurokinin receptors are not enzymes themselves. Rather, they are conditional catalysts which, in combination with receptor agonists will catalyze the nucleotide exchange reaction of G proteins. Because of this prospect of protein–protein interaction, neurokinin receptors exist in at least two states, the free receptor and the G protein-associated receptor. The two different conformational states of the receptors are manifested as having different affinities for agonists, low affinity for the free receptor and high affinity for the G protein-associated receptor. A rigorous thermodynamic consideration will further require that both the free receptor and the G protein-associated receptor exist either in an active or an inactive conformation, although it is difficult to detect the active conformation directly. In contrast to agonists, most antagonists have the same affinity for both the free receptor and the G protein-associated receptor. The complexity of the signal transduction mechanism gives rise to different agonist binding affinities as determined in various assays depending on the cell type, the identity of G proteins, the receptor to G protein ratio, and the radioligand.

IV. PEPTIDE–RECEPTOR INTERACTIONS AND DOWNSTREAM CELLULAR EFFECTS

Although each of the three mammalian tachykinins bind to one receptor subtype with high affinity, each of the three receptor subtypes can be activated by all three neurokinins. Thus, there must be a minimal unit that would be recognized by all receptor subtypes, and yet each peptide would be unique in its molecular properties such that each peptide has a unique binding affinity for any particular receptor. Early studies using truncated SP have demonstrated that while the N-terminal residues can be deleted, the C-terminal residues are absolutely required.[39,40] The smallest units possessing agonist activity are hexapeptides such as SP(6–11), NKA(5–10), or NKB(5–10).[41] Furthermore, all these hexapeptides are less selective than their respective full-length counterparts.[41,42] It is clear, therefore, that the C-terminal portion of neurokinins is the essential part conferring agonist activity, while the N-terminal part somehow modulates the binding affinity and specificity.

Early studies of the structure–activity relationship of SP relied on using peptides with selected amino acid substitutions. For both SP and NKA, substitution of the conserved Phe (Phe^7 in SP or Phe^6 in NKA) with Ala is detrimental to the biological activity.[39] All other residues in SP or NKA can be substituted with Ala, accompanied by varying degrees of reduction in affinity. Seven of the 11 residues in SP can be substituted without affecting its activity or affinity, which can be interpreted as an indication that these side chains do not contribute significantly to direct receptor interactions. Substitution of Phe^8, Leu^{10}, or Met^{11} in SP with Ala results in a two- to fivefold reduction in apparent affinity for the NK_1 receptor. In contrast, substitution of Asp^4, Val^7, Leu^9, or Met^{10} in NKA with Ala leads to a 20- to 300-fold reduction in apparent affinity for the NK_2 receptor.[16] Thus, it appears that side chains in NKA play a more direct role than those in SP in either receptor binding or maintaining the peptide conformation.

Since the molecular cloning of neurokinin receptors, it has become clear that the transmembrane regions of different receptor subtypes are more homologous to each other than other regions. The availability of cloned receptors provides a second route from which to study the molecular recognition of neurokinin binding. For instance, recent mutagenesis studies on the SP binding site of the NK_1 receptor indicated that the C-terminal amide of SP binds at or near the conserved Asn^{85} of the second transmembrane segment.[43] However, the simplistic view that divergent peptide residues interact with divergent receptor residues does not explain much of the mutagenesis data,[44-47] and is not consistent with the uncorrelated evolutionary paths of neurokinins and neurokinin receptors. Based on extensive mutagenesis studies, a peptide–receptor interaction model has been proposed, predicting that the peptide selectivity is determined by receptor conformation, peptide conformation, and specific interactions with conserved receptor residues.[43,44] This model is consistent with structural studies based on drug–hemoglobin interactions or steroid–antibody interactions, in which the stereochemistry of binding is determined by the available van der Waals space, and within that space, by all possible electrostatic interactions.[48,49] The contribution of divergent and conserved residues of the NK_1 receptor to ligand binding can be illustrated by the differential binding affinity of an antagonist for the NK_1 receptors from various species. In this case, interspecies substitution of two divergent residues of the receptor can reverse the species selectivity of CP-96,345 and RP 67580.[50,51] Nonetheless, these divergent residues do not appear to interact directly with the ligand, and all the available data suggest that the divergent residues affect the local receptor conformation or the precise position of other conserved residues involved in direct ligand binding. Indeed, further investigation of the NK_{1R} led to the identification of several conserved residues that interact directly with antagonists.[52-56]

Upon binding to the receptor, SP elicits various physiological effects mainly through intracellular second messengers. In neurons, the effect is usually a depolarization lasting tens of seconds.[6] Both the slow onset and the long duration of these effects are consistent with the notion that SP modulates the activities of other fast transmitters such as glutamate acting on ligand-gated ion channels.[28] Such a modulatory role is perfectly suited for receptors that activate second messengers upon neurokinin binding.[57] The modulatory activity is also consistent with the localization of the majority of the NK_1 receptors on the somatic and dendritic surfaces of neurons instead of directly opposing the synaptic terminals.[58] Furthermore, neurokinin receptors are expressed in glial cells, and SP has been found to regulate transmitter transport and ion channels.[33,59-61] These effects of SP on glial cells are also expected to modulate the neuronal excitability indirectly.[62,63]

The most significant second messenger evoked by neurokinins is calcium (Figure 4). In neurons, calcium is released from intracellular stores, and indirectly affects various ion channels through calcium-dependent enzymes such as kinases or phosphatases. In addition, the G proteins activated by neurokinin receptors can inhibit calcium channel and potassium channel directly without the involvement of second messengers.[64-66] The combined effect of SP is depolarization and increased excitability as a result of reduced K^+ conductance, which is probably the major physiological function of neurokinins in pain transmission in the spinal cord.[28] Calcium is also responsible for the contraction of intestinal smooth muscle, where the activator calcium can come from both intracellular stores and extracellular space.[67] In blood vessels, SP mainly acts on endothelium, and the dilator action of SP depends on the formation of endothelium-derived relaxing factor (nitric oxide).[68,69]

V. REGULATION OF NEUROKININ GENE EXPRESSION

Neurogenic inflammation refers to the observation that stimulation of small-diameter afferent neurons often generates symptoms analogous to those of inflammation, including vasodilation, plasma extravasation, and recruitment of immune cells.[70] While the effects of vasodilation and plasma extravasation are acute, the effects of neurokinins on immune cells can be long-lasting due to the feedback regulation of neurokinin gene expression by cytokines. It is clear that the NK_1 receptor is expressed in B-lymphocyte[71,72] and probably other immune cells,[73,74] and the activation of NK_1 receptor results in a variety of reactions from immune cells. For example, SP augments immunoglobin secretion from B cells,[71] contributes to the migration and activation of polymorphonuclear leukocytes,[75,76] and enhances the secretion of interleukin-1, interleukin-6, tumor necrosis factor-α, or interferon-γ from various immune cells.[77-79] However, the effect of SP on immune cells other than B cells may not be mediated through neurokinin receptors. First, the potency of neurokinins in eliciting some responses is much lower than would be expected for neurokinin receptors.[80,81] Second, peptides inactive at the NK_1 receptor, e.g., SP(1–7), produce some biological responses.[82,83] Finally, SP can act on a non-neurokinin receptor in monocytes,[80] and on G proteins in mast cells.[84]

While SP plays a critical role in activating various immune cells, certain cytokines, in turn, regulate the expression and release of SP. For example, it has been shown that interleukin-1, acting through cholinergic differentiation factor (also called leukemia inhibitory factor), increases the expression of SP in injured sympathetic ganglia or cultured sympathetic neurons.[85,86] Such an increase in SP expression could be blocked by immunosuppressants.[85] The effect of cytokines on SP expression is apparently mediated by various transcriptional regulatory elements in the PPT I gene.[87] The mutually stimulatory influence of SP and cytokines on each other obviously completes a positive feedback cycle (Figure 5), which is one of the characteristics of neurogenic inflammation.

Another important regulator of neurokinin gene expression is nerve growth factor (NGF). It has been shown that NGF increases the transcription and translation of substance P.[88,89] Because NGF is synthesized by target tissues, this effect is also consistent with the reduction in SP content of sensory neuron upon axotomy.[90] These results suggest that the transcription

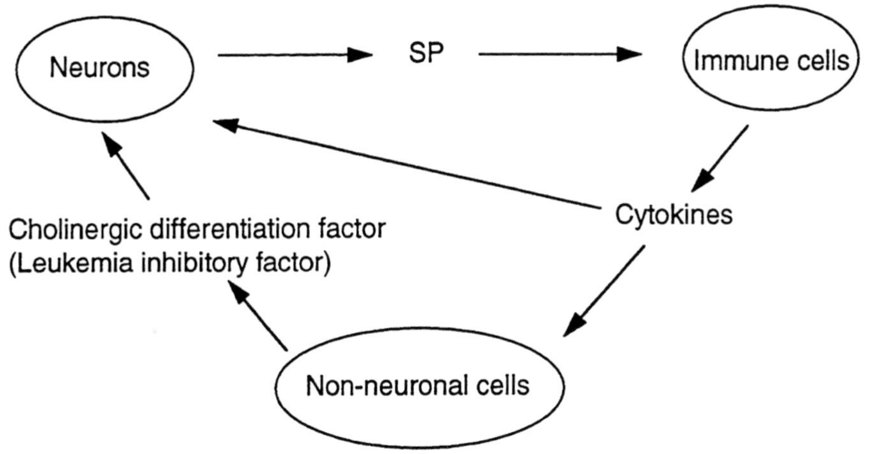

FIGURE 5. Mutual regulation of SP and cytokine expression.

of PPT I gene is regulated by the target tissues of innervation. It remains to be determined how NGF receptor activation upregulates the mRNA synthesis.

VI. CONCLUDING REMARKS

Tachykinins, originally identified as neurotransmitters, are important regulators of the immune system as well. Recent advances have demonstrated that neurokinin receptors are expressed in immune cells, cytokine receptors are expressed in neurons,[91] and the expression of neurokinins and cytokines are positively regulated by each other. Such a positive feedback cycle at the molecular level explains many of the phenomena related to neurogenic inflammation, and pharmacological intervention at this cycle is expected to be beneficiary for treating inflammatory diseases.[92] Besides SP, other neuropeptides can also regulate immune response. It has been suggested that SP and calcitonin gene-related peptide are coreleased, and both peptides contribute to inflammatory response through separate receptors.[93-95] The pleiotropic effects of neuropeptides and the involvement of multiple neuropeptides seem to necessitate the use of multiple pharmacological agents to break this vicious cycle. Further investigation of the cross-talk among the signaling pathways of various neuropeptides should, therefore, provide a rational basis for formulating an effective approach to control neurogenic inflammation.

REFERENCES

1. von Euler, U.S. and Gaddum, J.H., An unidentified depressor substance in certain tissue extracts, *J. Physiol.*, 72, 74, 1931.
2. Erspamer, V. and Anastasi, A., Structure and pharmacological actions of eledoisin, the active endecapeptide of the posterior salivary glands of Eledone, *Experientia*, 18, 58, 1962.
3. Erspamer, V., Biogenic amines and active polypeptides of the amphibian skin, *Annu. Rev. Pharmacol.*, 11, 327, 1971.
4. Chang, M.M., Leeman, S.E., and Niall, H.D., Amino acid sequence of substance P, *Nature New Biol.*, 232, 86, 1971.
5. Lambeck, F., Zur frage der zentralen ubertrangung afferenter impulse, *Naunyn-Schmiedebergs Arch. Exp. Pathol. Pharmakol.*, 219, 197, 1953.
6. Otsuka, M. and Konishi, S., Substance P, the first peptide neurotransmitter, *Trends Neurosci.*, 6, 317, 1983.

7. Nicoll, R.A., Schenker, C., and Leeman, S.E., Substance P as a neurotransmitter, *Annu. Rev. Neurosci.*, 3, 227, 1980.
8. Hokfelt, T., Kellerth, J.O., Nilsson, G., and Pernow, B., Experimental immunohistochemical studies on the localization and distribution of substance P in cat primary sensory neurons, *Brain Res.*, 100, 235, 1975.
9. Dale, H.H., Pharmacology and nerve-endings, *Proc. R. Soc. Med.*, 28, 319, 1935.
10. Maggio, J.E., Tachykinins, *Annu. Rev. Neurosci.*, 11, 13, 1988.
11. Nawa, H., Kotani, H., and Nakanishi, S., Tissue-specific generation of two preprotachykinin mRNAs from one gene by alternative RNA splicing, *Nature*, 312, 729, 1984.
12. Kimura, S., Okada, M., Sugita, Y., Kanazawa, I., and Munekata, E., Novel neuropeptides, neurokinin α and β, isolated from porcine spinal cord, *Proc. Jpn. Acad. Ser. B*, 59, 101, 1983.
13. Kangawa, K., Minamino, N., Fukuda, A., and Matsuo, H., Neuromedin K, a novel mammalian tachykinin identified in porcine spinal cord, *Biochem. Biophys. Res. Commun.*, 114, 533, 1983.
14. Henry, J.L., Discussions of nomenclature for tachykinins and tachykinin receptors, in *Substance P and Neurokinins*, Henry, J.L., Couture, R., Cuello, A.C., Pelletier, G., Quirion, R., and Regoli, D., Ed., Springer-Verlag, New York, 1987, p. xvii.
15. Krause, J.E., Churchwin, J.M., Carter, M.S., Xu, Z.S., and Hershey, A.D., Three rat preprotachykinin mRNAs encode the neuropeptides substance P and neurokinin A, *Proc. Natl. Acad. Sci. U.S.A.*, 84, 881, 1987.
16. Regoli, D., Rhaleb, N., Dion, S., Tousignant, C., Rouissi, N., Jukic, D., and Drapeau, G., Neurokinin A. A pharmacological study, *Pharmacol. Res.*, 22, 1, 1990.
17. Tatemoto, K., Lunderberg, J.M., Jornvall, H., and Mutt, V., Neuropeptide K: isolation, structure and biological activities of a novel brain tachykinin, *Biochem. Biophys. Res. Commun.*, 128, 947, 1985.
18. Beaujouan, J.C., Saffroy, M., Pelitet, F., Torrens, Y., and Glowinski, J., Neuropeptide K, scyliorhinin I and II: new tools in the tachykinin receptor field, *Eur. J. Pharmacol.*, 151, 353, 1988.
19. Dam, T.V., Takeda, Y., Krause, J.E., Escher, E., and Quirion, R., Gamma-preprotachykinin(72-92)-peptide amide: an endogenous preprotachykinin I gene derived peptide which preferentially binds to neurokinin-2 receptor, *Proc. Natl. Acad. Sci. U.S.A.*, 87, 246, 1990.
20. Krause, J.E., MacDanald, M.R., and Takeda, Y., The polyprotein nature of substance P precursors, *BioEssays*, 10, 62, 1989.
21. Nakanishi, S., Substance P precursor and kininogen: their structure, gene organization and regulation, *Physiol. Rev.*, 67, 1117, 1987.
22. Takahashi, K., Tanaka, A., Hara, M., and Nakanishi, S., The primary structure and gene organization of human substance P and neuromedin K receptors, *Eur. J. Biochem.*, 204, 1025, 1992.
23. Eipper, B.A. and Mains, R.E., Peptide α-amidation, *Annu. Rev. Physiol.*, 50, 333, 1988.
24. Sossin, W.S., Fisher, J.M., and Scheller, R.H., Cellular and molecular biology of neuropeptide processing and packaging, *Neuron*, 2, 1407, 1989.
25. Moore, H.P.H., Walker, M.D., Lee, F., and Kelly, R.B., Expressing a human proinsulin cDNA in a mouse ACTH-secreting cell, *Cell*, 35, 531, 1983.
26. Fumagalli, G. and Zanini, A., In cow anterior pituitary, growth hormone and prolactin can be packaged into separate granules of the same cell, *J. Cell Biol.*, 100, 2019, 1985.
27. Lundberg, J.M. and Holkeft, T., Coexistence of peptides and classical neurotransmitters, *Trends Neurosci.*, 6, 325, 1983.
28. De Koninck, Y. and Henry, J.L., Substance P-mediated slow excitatory postsynaptic potential elicited in dorsal horn neurons in vivo by noxious stimulation, *Proc. Natl. Acad. Sci. U.S.A.*, 88, 11344, 1991.
29. Iversen, L.L., Watling, K.J., McKnight, A.T., Williams, B.J., and Lee, C.M., Multiple receptors for substance P and related tachykinins, in *Topics in Medicinal Chemistry*, Vol. 65, Leeming, P.R., Ed., Royal Society of Chemists, London, 1987, 1.
30. Masu, Y., Nakayama, K., Tamaki, H., Harada, Y., Kuno, M., and Nakanishi, S., cDNA cloning of bovine substance K receptor through oocyte expression system, *Nature*, 329, 836, 1987.
31. Nakanishi, S., Mammalian tachykinin receptors, *Annu. Rev. Neurosci.*, 14, 123, 1991.
32. Gerard, N., Eddy, R.L., Shows, T.B., and Gerard, C., The human neurokinin A receptor, *J. Biol. Chem.*, 265, 20455, 1990.
33. Fong, T.M., Anderson, S.A., Yu, H., Huang, R.R.C., and Strader, C.D., Differential activation of intracellular effector by two isoforms of the human neurokinin-1 receptor, *Mol. Pharmacol.*, 41, 24, 1992.
34. Hershey, A.D. and Krause, J.E., Molecular characterization of a functional cDNA encoding the rat substance P receptor, *Science*, 247, 958, 1990.
35. Sasai, Y. and Nakanishi, S., Molecular characterization of rat substance K receptor and its mRNAs, *Biochem. Biophys. Res. Commun.*, 165, 695, 1989.
36. Shigemoto, R., Yokota, Y., Tsuchida, K., and Nakanishi, S., Cloning and expression of a rat neuromedin K receptor cDNA, *J. Biol. Chem.*, 265, 623, 1990.
37. Stratton, S.C., Beresford, I.J., Harvey, F.J., Turpin, M.P., Hagan, R.M., and Tyers, M.B., Anxiolytic activity of tachykinin NK2 receptor antagonists in the mouse light-dark box, *Eur. J. Pharmacol.*, 250, R11, 1993.

38. Gilman, A.G., G proteins: tranducers of receptor-generated signals, *Annu. Rev. Biochem.*, 56, 615, 1987.
39. Regoli, D., Escher, E., and Mizrahi, J., Substance P — structure-activity studies and the development of antagonists, *Pharmacology*, 28, 301, 1984.
40. Cascieri, M.A., Huang, R.R.C., Fong, T.M., Cheung, A.H., Sadowski, S., Ber, E., and Strader, C.D., Determination of the amino acid residues in substance P conferring selectivity and specificity for the rat neurokinin receptors, *Mol. Pharmacol.*, 41, 1096, 1992.
41. Regoli, D., Drapeau, G., Dion, S., and D'Orlean-Juste, P., Pharmacological receptors for substance P and neurokinins, *Life Sci.*, 40, 109, 1987.
42. Wormser, U., Laufer, R., Hart, Y., Chorev, M., Gilon, C., and Selinger, Z., Highly selective agonists for substance P receptor subtypes, *EMBO J.*, 5, 2805, 1986.
43. Huang, R.R.C., Yu, H., Strader, C.D., and Fong, T.M., Interaction of substance P with the second and seventh transmembrane domains of the neurokinin-1 receptor, *Biochemistry*, 33, 3007, 1994.
44. Fong, T.M. and Strader, C.D., Functional mapping of the ligand binding sites of G-protein coupled receptors, *Med. Res. Rev.*, 14, 387, 1994.
45. Strader, C.D., Fong, T.M., Tota, M.R., Underwood, D., and Dixon, R.A.F., Structure and function of G protein coupled receptors, *Annu. Rev. Biochem.*, 63, 101, 1994.
46. Yokota, Y., Akazawa, C., Ohkubo, H., and Nakanishi, S., Delineation of structural domains involved in the subtype specificity of tachykinin receptors through chimeric formation of substance P/substance K receptors, *EMBO J.*, 11, 3585, 1992.
47. Gether, U., Johansen, T.E., and Schwartz, T.W., Chimeric NK1/NK3 receptors: identification of domains determining the binding specificity of tachykinin agonists, *J. Biol. Chem.*, 268, 7893, 1993.
48. Perutz, M.F., Fermi, G., Abraham, D.J., Poyart, C., and Bursaux, E., Hemoglobin as a receptor of drugs and peptides: X-ray studies of the stereochemistry of binding, *J. Am. Chem. Soc.*, 108, 1064, 1986.
49. Arevalo, J.H., Taussig, M.J., and Wilson, I.A., Molecular basis of crossreactivity of antibody-antigen complementarity, *Nature*, 365, 859, 1993.
50. Fong, T.M., Yu, H., and Strader, C.D., Molecular basis for the species selectivity of the neurokinin-1 receptor antagonist CP-96,345 and RP67580, *J. Biol. Chem.*, 267, 25668, 1992.
51. Sachais, B.S., Snider, R.M., Lowe, I.J.A., and Krause, J.E., Molecular basis for the species selectivity of the substance P antagonist CP-96,345, *J. Biol. Chem.*, 268, 2319, 1993.
52. Fong, T.M., Yu, H., Cascieri, M.A., Underwood, D., Swain, C.J., and Strader, C.D., Interaction of glutamine-165 in the fourth transmembrane segment of the human neurokinin-1 receptor with quinuclidine antagonists, *J. Biol. Chem.*, 269, 14957, 1994.
53. Fong, T.M., Yu, H., Cascieri, M.A., Underwood, D., Swain, C.J., and Strader, C.D., The role of histidine-265 in antagonist binding to the neurokinin-1 receptor, *J. Biol. Chem.*, 269, 2728, 1994.
54. Fong, T.M., Cascieri, M.A., Yu, H., Bansal, A., Swain, C., and Strader, C.D., Amino-aromatic interaction between histidine-197 of the human neurokinin-1 receptor and CP-96,345, *Nature*, 362, 350, 1993.
55. Gether, U., Nilsson, L., Lowe, J.A. III, and Schwartz, T.W., Two specific residues at the top of the transmembrane segment V and VI of the neurokinin-1 receptor involved in binding of the nonpeptide antagonist CP96,345, *J. Biol. Chem.*, 269, 23959, 1994.
56. Huang, R.R.C., Yu, H., Strader, C.D., and Fong, T.M., Localization of the ligand binding site of the neurokinin-1 receptor: interpretation of chimeric mutants and single residue substitutions, *Mol. Pharmacol.*, 45, 690, 1994.
57. Hille, B., G protein-coupled mechanisms and nervous signaling, *Neuron*, 9, 187, 1992.
58. Liu, H., Brown, J.L., Jasmin, L., Maggio, J.E., Vigna, S.R., Mantyh, P.W., and Basbaum, A.L., Synaptic relationship between substance P and the substance P receptor, *Proc. Natl. Acad. Sci. U.S.A.*, 91, 1009, 1994.
59. Lee, C.M., Kum, W., Cockram, C.S., Teoh, R., and Young, J.D., Functional substance P receptors on a human astrocytoma cell line (U373MG), *Brain Res.*, 488, 328, 1989.
60. Johnson, C.L. and Johnson, C.G., Substance P regulation of glutamate and cystine transport in human astrocytoma cells, *Receptors Channels*, 1, 53, 1993.
61. Pradier, L., Heuillet, E., Hibert, J.P., Laville, M., Le Guern, S., and Doble, A., Substance P-evoked calcium mobilization and ionic current activation in the human astrocytoma cell line U373MG, *J. Neurochem.*, 61, 1850, 1993.
62. Wilkin, G.P. and Cholewinski, A., Peptides receptors on astrocytes, in *Glial Cell Receptors*, Kimelberg, H.K., Ed., Raven Press, New York, 1988, 223.
63. Kimelberg, H.K. and Norenberg, M.D., Astrocytes, *Sci. Am.*, 260, 66, 1989.
64. Bley, K.R. and Tsien, R.W., Inhibition of calcium and potassium channels in sympathetic neurons by neuropeptides and other ganglionic transmitters, *Neuron*, 4, 379, 1990.
65. Brown, A.M., A cellular logic for G protein coupled ion channel pathways, *FASEB J.*, 5, 2175, 1991.
66. Shapiro, M.S. and Hille, B., Substance P and somatostatin inhibit calcium channels in rat sympathetic neurons via different G protein pathways, *Neuron*, 10, 11, 1993.

67. Mayer, E.A., Loo, D.D.F., Kodner, A., and Reddy, S.N., Differential modulation of Ca^{++}-activated K^+ channels by substance P, *Am. J. Physiol.*, 257, G887, 1989.
68. D'Orleans-Juste, P., Dion, S., Mizrahi, J., and Regoli, D., Effects of peptides and non-peptides on isolated arterial smooth muscle, *Eur. J. Pharmacol.*, 114, 9, 1985.
69. Lippe, I.T., Stabentheiner, A., and Holzer, P., Role of nitric oxide in the vasodilator but not exudative component of mustard oil-induced inflammation in rat skin, *Agents Actions*, 38, C22, 1993.
70. Lembeck, F. and Holzer, P., Substance P as a neurogenic mediator of antidromic vasodilation and neurogenic plasma extravasation, *Naunyn-Schmiedeberg's Arch. Pharmacol.*, 310, 175, 1979.
71. Bost, K.L. and Pascual, D.W., Substance P: a late-acting B lymphocyte differentiation cofactor, *Am. J. Physiol.*, 262, C537, 1992.
72. Takeda, Y., Chou, K.B., Takeda, J., Sachais, B.S., and Krause, J.E., Molecular cloning, structural characterization and functional expression of the human substance P receptor, *Biochem. Biophys. Res. Commun.*, 179, 1232, 1991.
73. Roberts, A.I., Taunk, J., and Ebert, E.C., Human lymphocytes lack substance P receptors, *Cell. Immunol.*, 141, 457, 1992.
74. Calvo, C., Chavanel, G., and Senik, A., Substance P enhances IL-2 expression in activated human T cells, *J. Immunol.*, 148, 3498, 1992.
75. Perretti, M., Ahluwalia, A., Flower, R.J., and Manzini, S., Endogenous tachykinins play a role in IL-1 induced neutrophil accumulation: involvement of NK1 receptors, *Immunology*, 80, 73, 1993.
76. Serra, M.C., Bazzoni, F., Della Bianca, V., Greskowiak, M., and Rossi, F., Activation of human neutrophils by substance P, *J. Immunol.*, 141, 2118, 1988.
77. Okamoto, Y., Shirotori, K., Kudo, K., Ishikawa, K., Ito, E., Togawa, K., and Saito, I., Cytokine expression after the topical administration of substance P to human nasal mucosa, *J. Immunol.*, 151, 4391, 1993.
78. Lotz, M., Vaughan, J.H., and Carson, D.A., Effect of neuropeptides on production of inflammatory cytokines by human monocytes, *Science*, 241, 1218, 1988.
79. Janiszewski, J., Bienenstock, J., and Blennerhassett, M.G., Picomolar doses of substance P trigger electrical responses in mast cells without degranulation, *Am. J. Physiol.*, 36, C138, 1994.
80. Jeurissen, F., Kavelaars, A., Korstjens, M., Broeke, D., Franklin, R.A., Gelfand, E.W., and Heijnen, C.J., Monocytes express a non-neurokinin substance P receptor that is functionally coupled to MAP kinase, *J. Immunol.*, 152, 2987, 1994.
81. Schumann, M.A. and Gardner, P., Modulation of membrane K^+ conductance in T-lymphocytes by substance P via a GTP-binding protein, *J. Membr. Biol.*, 111, 133, 1989.
82. Maggi, C.A., Patacchini, R., Rovero, P., and Giachetti, A., Tachykinin receptors and tachykinin receptor antagonists, *J. Auton. Pharmacol.*, 13, 23, 1993.
83. Kreeger, J.S. and Larson, A.A., Substance P(1-7), a substance P metabolite, inhibits withdrawal jumping in morphine-dependent mice, *Eur. J. Pharmacol.*, 238, 111, 1993.
84. Mousli, M., Bueb, J.-L., Bronner, C., Rouot, B., and Landry, Y., G protein activation: a receptor-independent mode of action for cationic amphiphilic neuropeptides and venom peptides, *Trends Pharmacol. Sci.*, 11, 358, 1990.
85. Freidin, M. and Kessler, J.A., Cytokine regulation of substance P expression in sympathetic neurons, *Proc. Natl. Acad. Sci. U.S.A.*, 88, 3200, 1991.
86. Jonakait, G.M., Neural-immune interactions in sympathetic ganglia, *Trends Neurosci.*, 16, 419, 1993.
87. Kageyama, R., Sasai, Y., and Nakanishi, S., Molecular characterization of transcription factors that bind to the cAMP responsive region of the substance P precursor gene, *J. Biol. Chem.*, 1991, 15525, 1991.
88. Lindsay, R.M. and Harmar, A.J., Nerve growth factor regulates expression of neuropeptide genes in adult sensory neurons, *Nature*, 337, 362, 1989.
89. MacLean, D.B., Lewis, S.F., and Wheeler, F.B., Substance P content in cultured neonatal rat vagal sensory neurons: the effect of nerve growth factor, *Brain Res.*, 457, 53, 1988.
90. Hokfelt, T., Zhang, X., and Wiesenfeld-Hallin, Z., Messenger plasticity in primary sensory neurons following axotomy and its functional implications, *Trends Neurosci.*, 17, 22, 1994.
91. Rothwell, N.J., Functions and mechanisms of interleukin-1 in the brain, *Trends Pharmacol. Sci.*, 12, 430, 1991.
92. Lembeck, F., Donnerer, J., Tsuchiya, M., and Nagahisa, A., The nonpeptide tachykinin antagonist CP-96,345 is a potent inhibitor of neurogenic inflammation, *Br. J. Pharmacol.*, 105, 527, 1992.
93. White, D.M., Leah, J.D., and Zimmerman, M., The localization and release of substance P and CGRP at nerve fibre endings in rat cutaneous nerve neuroma, *Brain Res.*, 503, 198, 1989.
94. Goodman, E.C. and Iversen, L.L., Calcitonin gene-related peptide: novel neuropeptide, *Life Sci.*, 38, 2169, 1986.
95. Donnerer, J. and Amann, R., The inhibition of neurogenic inflammation, *Gen. Pharmacol.*, 24, 519, 1993.

Chapter 2

POSTTRANSCRIPTIONAL REGULATION OF CALCITONIN GENE-RELATED PEPTIDE (CGRP) mRNA PRODUCTION

Ronald B. Emeson

CONTENTS

I. INTRODUCTION

The calcitonin/calcitonin gene-related peptide (CGRP) gene is responsible for the generation of a number of distinct peptide hormones and peptide neurotransmitters mediating various physiological processes ranging from proliferation of bone osteoblasts[1-3] to regulation of acetylcholine receptor synthesis.[4-6] The starting point for studies of this gene began with the isolation and characterization of clones complementary to calcitonin mRNA from a rat medullary thyroid carcinoma (MTC) cDNA library.[7] Subsequent analyses of calcitonin gene expression demonstrated the production of a novel calcitonin-related RNA species, referred to as calcitonin gene-related peptide mRNA.[8] Isolation and sequence analyses of the calcitonin genomic DNA, as well as calcitonin and CGRP cDNAs, confirmed that both calcitonin and CGRP are generated by differential RNA splicing from a single genomic locus.[8-12] CGRP was one of the first examples of a neuropeptide that was initially predicted, and later identified, as a result of molecular cloning, without the biological basis for chemical characterization that existed for many other bioactive peptides. The discovery of a second gene encoding a CGRP-like molecule[13] further complicated the biological role of CGRP, as both of these peptides were shown to be widely expressed in the central and peripheral nervous systems, to act at multiple CGRP receptors in widespread effector systems, and to possess potent bioactivities in the regulation of diverse physiological processes. The calcitonin/CGRP gene and its resulting protein products have been studied extensively and used as an important model for examinations of tissue-specific gene expression, regulation of cell-specific RNA processing events, post-translational processing of peptide prohormones and the biological activity of multiple peptide products.

II. CALCITONIN

The existence of a calcium-regulating hormone with a biological activity antagonistic to that of parathyroid hormone was first demonstrated by Copp et al. (1962). This hormone, referred to as calcitonin for its ability to control the level or "tone" of calcium in the blood, was shown to be a 32-amino acid amidated peptide synthesized by the parafollicular (C) cells of the thyroid gland and the ultimobranchial gland in mammals and nonmammalian

0-8493-7646-7/96/$0.00+$.50

vertebrates, respectively.[14-16] The primary activity of calcitonin is to decrease circulating levels of calcium and phosphate by inhibiting bone resorption.[15,17,18] Calcitonin inhibits such resorption by a cAMP-dependent osteoclast inactivation, by decreasing tubular reabsorption of both calcium and phosphate in the proximal straight tubule of the kidney and by inhibiting the actions of vitamin D 1-α-hydroxylase, thus favoring formation of the less active 24,25(OH)-derivative of vitamin D.[19,20] Although the precise physiological role(s) of calcitonin remain controversial, the pharmacological actions of calcitonin regarding inhibition of bone resorption have demonstrated significant therapeutic value in pathophysiological disorders such as Paget's disease and severe osteoporosis.

Analyses of DNA complementary to rat calcitonin mRNA predicted the structure of a 17.5-kDa polypeptide precursor for calcitonin, from which this 32-amino acid peptide is posttranslationally excised and modified (Figure 1). Calcitonin messenger RNA is approximately 1 kb in length, encoding a predicted protein of 136 or 141 amino acids in rat and human, respectively.[7,21,22] The amino terminus of this prohormone contains a 25-amino acid long hydrophobic signal peptide characteristic of secreted proteins, the calcitonin peptide itself, as well as amino- and carboxyl-terminal flanking peptides. These latter peptides are separated from calcitonin by a pair of basic amino acid residues (Lys-Arg) at the amino terminus and Gly-Lys-Arg-Arg at the carboxyl terminus. These basic residues represent predicted sites of posttranslational proteolytic processing,[23-25] while the presence of a glycine residue in the prohormone allows the formation of an α-amide group at the calcitonin carboxyl terminus.[26,27]

The 57-amino acid amino-terminal peptide derived from the calcitonin prohormone (N-proCT) has been identified in extracts of rat and human C cells and this peptide has been shown to be secreted in equimolar amounts with calcitonin.[3] Recent studies have demonstrated that N-proCT acts as a cellular mitogen, causing the proliferation of bone osteoblasts and their progenitors.[1,3,28] The carboxyl-terminal flanking peptide of the calcitonin prohormone has been referred to as calcitonin carboxyl-terminal cleavage peptide (CCP) or katacalcin and ranges from 16 (rat) to 21 (human, bovine, chicken, salmon) amino acids in length.[7,29-32] While the existence of CCP was originally based upon cDNA sequence analyses alone, specific antisera prepared against a synthetic hexadecapeptide corresponding to the predicted amino acid sequence has allowed radioimmunochemical detection of CCP in both normal rat thyroid glands and medullary thyroid carcinomas.[33] In the human thyroid, CCP has also been shown to be cosecreted with calcitonin,[29] yet the biological role of CCP remains unknown.

III. CALCITONIN GENE-RELATED PEPTIDE (α-CGRP)

The production of multiple mRNA transcripts from the single rat calcitonin gene was first described during studies of calcitonin gene expression in serially passaged medullary thyroid carcinomas.[9] Thyroid C cell hyperplasia and the ultimate development of MTC are common physiological events in rodents, occurring in 10 to 40% of older animals.[34] Production of calcitonin by these tumors decreases spontaneously during a single transplantation, or gradually during successive transplantations, from the usual 1.5 to 4.5% of total cellular protein to less than 0.3% of total protein.[35] While originally considered to be a consequence of altered transcriptional regulation,[9] the "switching" of MTCs from states of high to low calcitonin production was shown to result from changes in RNA processing events generating a novel 1.2-kb RNA species, referred to as calcitonin gene-related peptide mRNA.[9] Sequence analyses of the rat CGRP cDNA predicted a 16-kDa prohormone precursor, which could be posttranslationally modified to produce a 37-amino acid amidated peptide hormone (CGRP), as well as amino- and carboxyl-terminal flanking peptides analogous to those derived from the calcitonin prohormone (Figure 1).

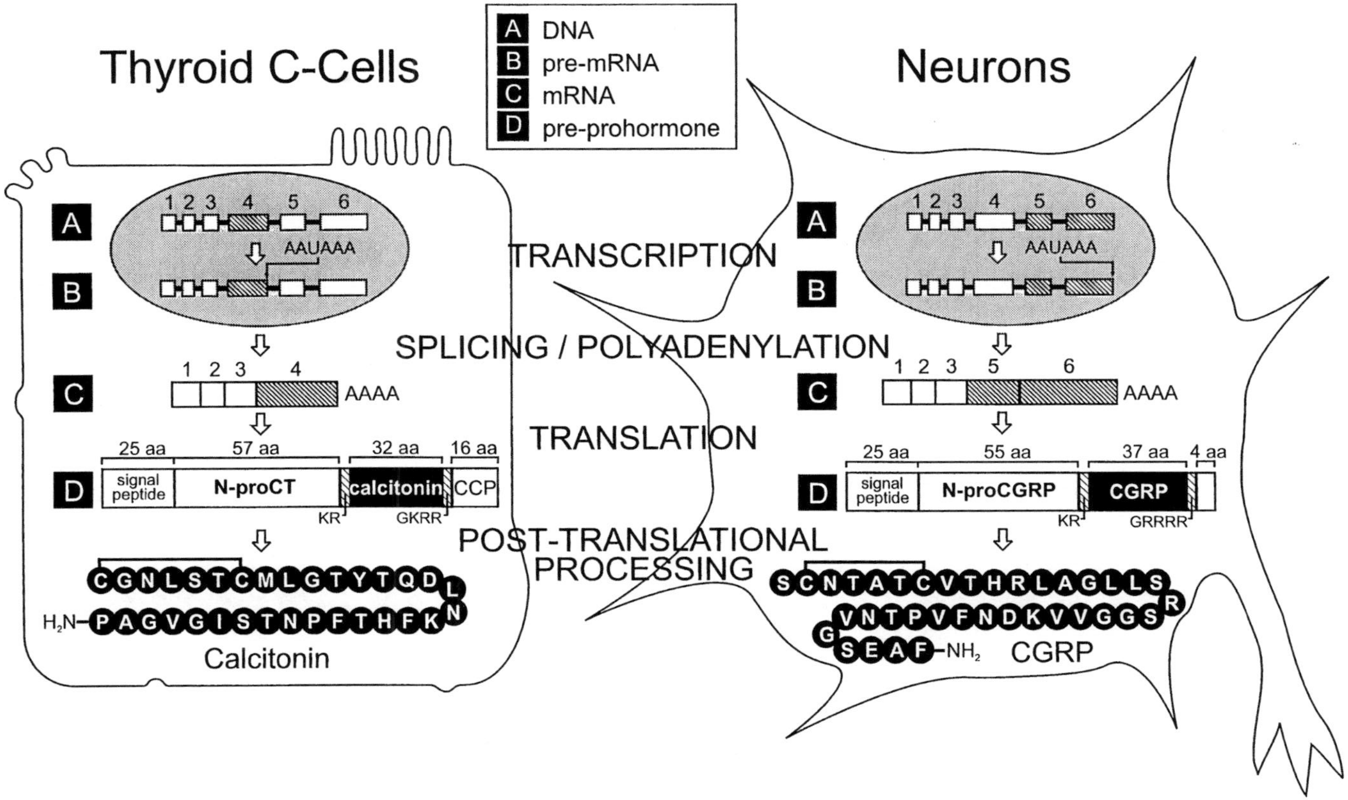

FIGURE 1. Cellular processes involved in the tissue-specific production of calcitonin and calcitonin gene-related peptide. Schematic representations of the rat calcitonin/CGRP gene, primary RNA transcript, mature mRNA species, peptide prohormone, and biologically active peptide products are presented, demonstrating differential patterns of messenger RNA and peptide expression in thyroid C cells and neurons. The location of polyadenylation signals (AAUAAA) and the sizes of predicted peptide products are indicated; sites of posttranslational proteolytic processing are designated with the one-letter amino acid code.

Isolation and sequence analyses of the calcitonin genomic DNA, as well as calcitonin and CGRP cDNAs, confirmed that both calcitonin and CGRP are generated by differential RNA splicing from a single genomic locus.[8-10,12] Examinations of the sequence for the two gene products revealed that CGRP and calcitonin share sequence identity through nucleotide 227 of the coding region, predicting that the initial 76 amino acids of each prohormone are identical. The nucleotide sequences then diverge completely, encoding unique C-terminal domains.[8] Protein-processing signals within the C-terminal region of CGRP predict the posttranslational excision of a 37-amino acid peptide containing a phenylalanyl-amide at its carboxyl terminus.[8] Based upon the structure of the calcitonin/CGRP gene and analyses of calcitonin and CGRP cDNAs, production of mature calcitonin transcripts involves splicing of the first three exons, which are common to both mRNAs, to the calcitonin-specific exon (exon 4) encoding the entire sequence for calcitonin, CCP and 3′ untranslated information. Alternative splicing of the first three exons to the fifth and sixth exons, which contain the entire CGRP coding sequence and 3′-noncoding sequences, respectively, results in the production of CGRP mRNA. The 55-residue amino-terminal flanking peptide (N-proCGRP) is 90% homologous to the 57-residue peptide derived from the calcitonin prohormone. Their first 51 amino acids are identical, reflecting the fact that the first three exons in the calcitonin/CGRP gene are utilized in both calcitonin and CGRP mRNA transcripts. The predicted peptide flanking the carboxyl terminus of CGRP is only four amino acids in length.

Using CGRP-specific antisera, CGRP-related immunoreactive material was shown to have a widespread anatomical distribution in the central nervous system, distinct from that of any known neuropeptide (Figure 2).[36-39] CGRP-immunoreactive cell bodies are present in several regions of the hypothalamus, including the preoptic and medial preoptic areas, the periventricular and anterior hypothalamic nuclei, the perifornical area, and in the lateral hypothalamus/medial forebrain bundle area. CGRP-positive cells are also found in the medial amygdaloid nucleus, the hippocampus, the ventromedial nucleus of the thalamus, in the periventricular gray, and in the fasciculus retroflexus extending laterally over the lemniscus medialis. In the mesencephalon, CGRP is found in the superior colliculus, parabigeminal nucleus, and in the peripeduncular area ventral to the medial geniculate body. In the hindbrain, cells are found in the parabrachial nucleus, ventral tegmental nucleus, lateral lemniscus, superior olive, lateral cuneate nucleus, and the nucleus of the solitary tract. CGRP-immunoreactive cell bodies are present in all cranial nuclei, including the trochlear nerve (IV), the trigeminal nerve (V), the facial nerve (VII), the hypoglossal nerve (XII), the oculomotor nerve (III), and the nucleus ambiguus. Central CGRP-stained pathways include projections from the thalamus to insular cortex,[40] from the hypothalamus to the lateral septal area,[41] and from the parabrachial nucleus to the visual sensory thalamus and cortex.[42] Immunoreactive afferents to the amygdala, bed nucleus of the stria terminalis, and certain hypothalamic nuclei also arise from the parabrachial nucleus.[43-45]

Although the physiological function(s) of CGRP have not been precisely defined, based upon its anatomical distribution, CGRP is thought to play an important role in autonomic, somatosensory, integrative, and motor functions.[36-38,46] CGRP is found in the C-fibers of sensory nerves associated with the smooth muscle of blood vessels and as free nerve endings in the skin, often colocalized with substance P.[47,48] CGRP can be released from these fibers both centrally and peripherally; when released peripherally it can produce a long-lasting vasodilation that is responsible for many of the proinflammatory actions of CGRP.[49] Although the majority of CGRP's actions as an inflammatory mediator are secondary to its actions as a vasodilator, both T lymphocytes and macrophages have CGRP receptors, which may be involved in the inhibition of interleukin-2 and superoxide production as well as stimulation of Na^+/H^+ exchange.[50-52] More recent studies have identified CGRP throughout the mammalian gastrointestinal tract, arising from a set of neurons in the myenteric and submucosal plexuses, and originating from sensory neurons whose cell bodies are located in the nodose and dorsal

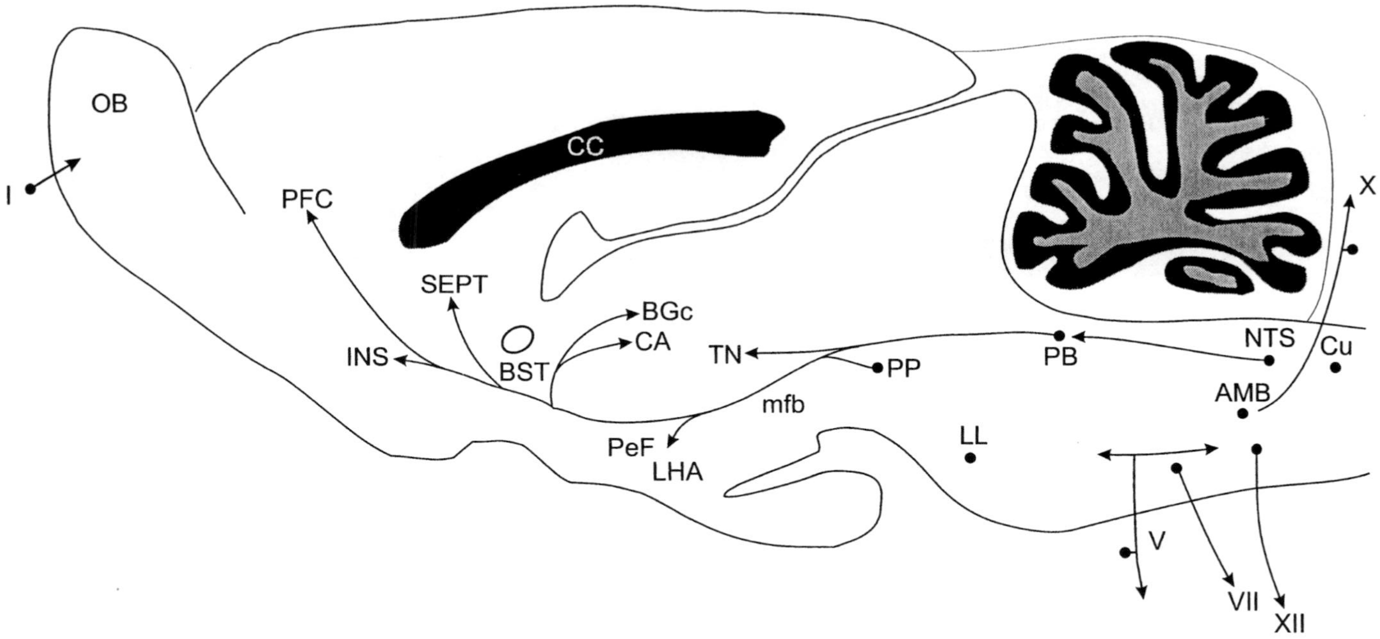

FIGURE 2. Schematic representation of a sagittal view of the adult rat brain demonstrating a summary of the major CGRP-stained cell groups (•) and pathways (arrows). For a more complete description of CGRP-containing pathways in the central nervous system, see accompanying text or Skofitsch and Jacobowitz[36,37] and Kawai et al.[39] Anatomical locations are abbreviated as follows: nucleus of the solitary tract (NTS); cuneate nucleus (Cu); nucleus ambiguus (AMB); parabrachial nucleus (PB); lateral lemniscus (LL); peripeduncular nucleus (PP); thalamic taste nucleus (TN); medial forebrain bundle (mfb); perifornical hypothalamic nucleus (PeF); lateral hypothalamic area (LHA); central amygdaloid nucleus (CA); caudal regions of the caudate/putamen and globis pallidus (BGc); bed nucleus of the stria terminalis (BST); lateral septal nucleus (SEPT); infralimbic prefrontal area (PFC); insular area of the cerebral cortex (INS); olfactory bulb (OB); primary olfactory fibers (I); corpus callosum (CC); trigeminal nerve (V); facial nerve (VII); vagus nerve (X); hypoglossal nerve (XII). (Adapted from Rosenfeld et al.[11].)

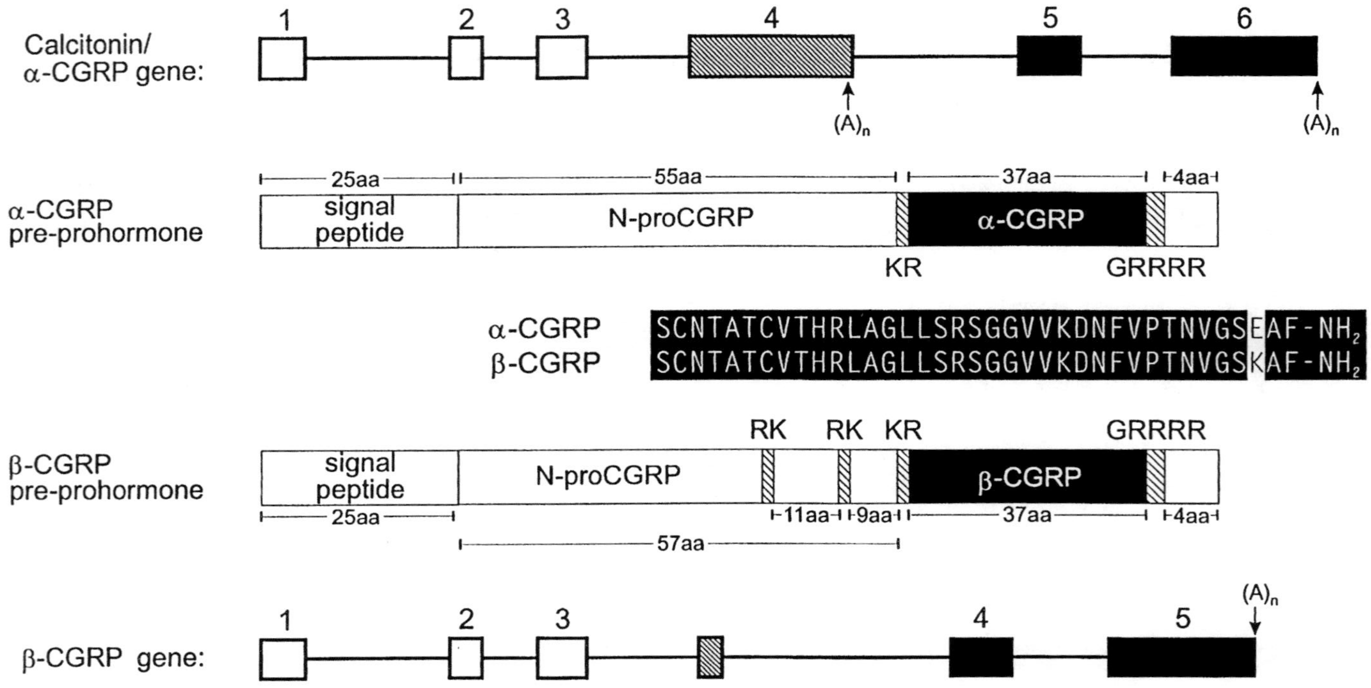

FIGURE 3. Schematic diagram of the rat α- and β-CGRP genes and their encoded peptide pre-prohormones. The genomic structure of the rat α- and β-CGRP genes are presented indicating the common (open rectangle), calcitonin-specific (hatched rectangle), and CGRP-specific (black rectangle) exons. The amino acid sequences of α- and β-CGRP are presented showing a single amino acid difference ($K^{35} \rightarrow E^{35}$) with the one-letter amino acid code. The location of polyadenylation signals $(A)_n$, the sites of predicted posttranslational proteolytic cleavage at pairs of basic amino acids (hatched), and the sizes of predicted peptide products are indicated.[13,65]

root ganglia.[53,54] Many studies have suggested that CGRP may serve not only as a neurotransmitter or neuromodulator in the enteric nervous system, but may also participate in local neurohumoral (paracrine) regulation of gastrointestinal effector systems.[55-57] CGRP has also been identified in the endocrine pancreas, where this peptide may play an important role in the regulation of carbohydrate metabolism,[58,59] and in the mouse olfactory lobe, where it has been postulated to serve as a trophic factor in neuronal development.[60,61] Additionally, CGRP has been colocalized with acetylcholine in presynaptic terminals of the neuromuscular junction, where it stimulates both acetylcholine receptor synthesis and the rate of receptor desensitization.[4,5,62]

IV. CALCITONIN GENE-RELATED PEPTIDE II (β-CGRP)

In addition to the gene encoding both calcitonin and CGRP, referred to as the calcitonin/α-CGRP gene, subsequent studies have uncovered the expression of a second CGRP-containing gene that encodes a peptide differing from α-CGRP by a single amino acid;[13,63,64] this peptide is referred to as β-CGRP (Figure 3). The nucleotide sequence of the rat β-CGRP cDNA predicts a 394-nucleotide open reading frame in which the first 256 nucleotides encode a 25-amino acid signal peptide and a 57-amino acid amino-terminal flanking peptide; this flanking peptide shares a 70% amino acid sequence similarity with the amino-terminal peptides derived from the calcitonin and α-CGRP prohormones. There are two sets of dibasic amino acid pairs (Arg-Lys) within this region, suggesting that three distinct peptide products could be posttranslationally generated from this region of the prohormone. In contrast, there are fewer than 5% base substitutions in the next 114 nucleotides of the β-CGRP mRNA, constituting the β-CGRP-encoding domain of the RNA transcript. This sequence predicts the excision of a 37-amino acid peptide containing a carboxyl-terminal phenylalanyl-amide residue, differing by only a single amino acid (lysine[35] for glutamic acid[35]) from the primary sequence of α-CGRP. The β-CGRP gene is very similar in overall structure to the α-CGRP gene (Figure 3), but the former does not contain an exon encoding a β-calcitonin-like peptide. Rather, the β-CGRP gene contains a nonfunctional exon which has significantly diverged from its α-calcitonin-specific (exon 4) counterpart and lacks a detectable polyadenylation signal.[65]

Analyses of RNA from various regions of rat brain have demonstrated an overlapping but nonidentical pattern of expression for these genes throughout the central nervous system. While the relative distribution of α- and β-CGRP is quite similar, the level of expression of β-CGRP is less than 20% of that found for α-CGRP in many brain regions, including the peripeduncular and parabrachial nuclei, superior olive, and trigeminal ganglion. The expression of β-CGRP exceeds that for α-CGRP in the oculomotor (III), trochlear (IV), and trigeminal (V) cranial nerves and is roughly equivalent in the hypoglossal nerve (XII) and the nucleus ambiguus.[13] The primary amino acid sequences of human α- and β-CGRP have also been determined and demonstrate highly conserved structures, differing from the rat peptides by only five amino acids.[63,66] Pharmacological comparisons of human α- and β-CGRP have indicated that these peptides exhibit qualitatively similar biological properties but are not necessarily equipotent.[67,68]

The close structural homology of α- and β-CGRP raises the possibility that not all of the CGRP-like immunoreactive material detected by antisera raised against α-CGRP is necessarily the product of α-CGRP gene expression. Preliminary studies have suggested that sensory neurons preferentially express α-CGRP, while enteric neurons express β-CGRP.[69] In organs where CGRP occurs predominantly in sensory fibers, the concentration of α-CGRP was found to be three to six times higher than that of β-CGRP. In the intestine, the concentration of β-CGRP was up to seven times higher than the product of the α-gene. The bulk of α-CGRP-containing fibers within the gastrointestinal tract have been shown to be eliminated by either extrinsic denervation or by degeneration of primary afferents utilizing capsaicin.[69-71]

V. MECHANISMS OF CELL-SPECIFIC ALTERNATIVE RNA PROCESSING

While calcitonin messenger RNA production is largely limited to the C cells of the thyroid gland, CGRP mRNA is widely expressed in discrete cell types throughout the central and peripheral nervous systems. S1 nuclease analyses have indicated that both calcitonin and CGRP mRNA are produced in the thyroid gland in a ratio of approximately 95–98:1.[11,72-74] Immunohistochemical and radioimmunoassay studies have further confirmed the presence of both calcitonin- and CGRP-immunoreactive material in parafollicular C cells and have demonstrated that most of these cells, in both rat and human tissues, react with both calcitonin and CGRP antisera.[72] These results have indicated that the ability to coproduce both calcitonin and CGRP peptide products is a feature of normal as well as transformed cells.[72] By contrast, little or no detectable calcitonin mRNA has been observed in a number of neuronal structures throughout the brain utilizing S1 nuclease protection, RNA blotting, and immunohistochemical analyses.[11,72,74] This differential production of calcitonin and CGRP mRNAs has raised a number of fundamental questions regarding the molecular mechanisms by which a single messenger RNA precursor can be posttranscriptionally processed, in a tissue-specific manner, to yield two distinct messenger RNA products.

A. CIS-ACTIVE REGULATORY ELEMENTS

For many genes, alternative splicing may simply reflect a kinetic process in which the ratio of splice site selection is solely determined by a hierarchy of relative "intrinsic strengths" among cis-active elements at competing splice sites.[75-78] A number of features in pre-mRNA transcripts have been implicated in alternative splice-site selection, including intron size,[79] the pyrimidine content of the 3′-splice site,[78,80,81] the location and sequence composition of the branchpoint,[82] RNA secondary structure,[83,84] exon sequences,[85-88] as well as proximity to other cis-active elements. The observation that alternative splicing of calcitonin/CGRP transcripts occurs in a tissue-specific manner suggests that cell-specific trans-acting factors must interact with the pre-mRNA; cis-active elements within the transcript itself, however, must ultimately determine the response to trans-acting factors in specifying the nature of the final spliced product.

Tissue culture model systems mimicking the RNA processing choices of thyroid C cells and neurons have been extensively utilized to identify potential cis-active regulatory elements involved in calcitonin/CGRP posttranscriptional processing. The RNA processing choice made by cell lines transfected with mutant calcitonin/CGRP transcription units represents a common experimental strategy for determining the effects of such mutations upon alternative 3′-splice site selection. When transfected with the wild-type calcitonin/CGRP gene, lymphoid (A20), human embryonic kidney (293), and human epithelial (HeLa) cell lines produce almost exclusively mature calcitonin transcripts, while rat pheochromocytoma (PC12), rat glioma (C6), and mouse teratocarcinoma (F9) cell lines produce >90% CGRP mRNA.[65,89-93]

To examine putative cis-active elements and/or RNA secondary structure involved in differential mRNA production, preliminary experiments have been directed at mutating sequences surrounding the calcitonin- and CGRP-specific splice junctions.[89,90] Results from these studies have indicated that while deletion of the calcitonin-specific splice acceptor (Δ4Ac, Figure 4) results in the efficient production of CGRP transcripts in a calcitonin-producing cell line (HeLa), inactivation of the CGRP 3′-splice site (Δ5Ac) does not generate the corresponding calcitonin transcripts in a CGRP-producing (F9) model system.[89,90] These observations have suggested that (1) there is no absolute requirement for a positive, cell-specific machinery to promote the splicing of exon 3 to exon 5 and (2) in the absence of a functional CGRP-specific splice acceptor, CGRP-producing cells are incapable of altering their differential splice-site selection to produce calcitonin mRNA. Additional mutational analyses of intronic sequence information near the calcitonin 3′-splice junction have failed to identify unique sequences mediating exon 4 inclusion or exclusion in calcitonin- and CGRP-producing cells, respectively (Figure 4);[91] these results have suggested that the relative strength of the calcitonin acceptor itself might represent a critical determinant of cell-specific RNA processing.

In vitro splicing analyses of the human calcitonin/CGRP gene have indicated that the branchpoint consensus sequence (YNYUR<u>A</u>Y) preceding the calcitonin-specific exon contains a unique uridine residue as the branchpoint nucleotide rather than the adenine moiety which normally occurs in this position.[94] In the rat gene, a cytidine nucleotide at the same position (–23 relative to the beginning of the fourth exon) is surrounded by an otherwise perfect match for the loosely defined metazoan branchpoint consensus. To examine the role of a nonstandard branchpoint nucleotide in the regulation of alternative RNA processing, the branchpoint uridine or cytidine residues were mutated to the canonical adenine nucleotide (C→A or U→A) in the human and rat calcitonin/CGRP genes, respectively.[65,91,92] Transfection of the mutant transcription unit into embryonic kidney, glioma, pheochromocytoma, and teratocarcinoma cells resulted in the predominant or exclusive production of mature calcitonin transcripts, regardless of the RNA processing phenotype observed with the wild-type calcitonin/CGRP gene.[65,91-93] These results have been interpreted to suggest that the branchpoint nucleotide is itself a specific cis-active regulatory sequence element mediating cell-specific mRNA production. Deletion or mutation of the branchpoint consensus sequence, however, had no effect upon cell-specific patterns of RNA processing, suggesting that the presence of an exon 4 suboptimal branchpoint nucleotide accounts for inefficient splicing, but is not in itself a regulatory sequence.[91] Additional mutant transcription units have shown that when the C→A branchpoint mutation, increasing acceptor strength, is paired with polypyrimidine mutations that decrease acceptor strength [Py→G(2)/C→A], tissue-specific splicing could be titrated with the overall strength of the calcitonin 3′-splice site.[91]

Extensive mutational analyses covering the entire alternatively spliced region from exon 3 through exon 5 have indicated that almost all of the mutations introduced had no effect on alternative splice-site selection in permanently transfected HeLa and F9 cells (Figure 4).[91] Exceptions were found when essential constitutive sites were removed (as in Δ4Ac) or when F9 cells were forced to use the calcitonin 3′-splice junction (Δ5Ac). Each of the mutations that were capable of activating the calcitonin-specific 3′-splice site in F9 cells apparently did so by increasing the strength of the exon 4 acceptor, overcoming the existing competition between exons 4 and 5. While these analyses did not eliminate the possibility of multiple, redundant regulatory elements, no evidence was found for a single cis-active inhibitory sequence required for the neuronal splicing pattern in CGRP-producing (F9) cells. In HeLa cells, the only mutations capable of altering RNA processing patterns were those in which the calcitonin-specific 3′-splice site was impaired. These results have suggested that the tissue-specificity of calcitonin and CGRP mRNA production occurs as a result of cell-type differences in the recognition of constitutive elements by the splicing apparatus.[91]

B. TRANS-ACTING FACTORS

A number of trans-acting factors regulating the alternative splice-site selection of *Drosophila* P-transposase[95] and genes involved in the *Drosophila* sex-determination pathway have been identified.[96-99] These studies have further suggested that such trans-acting factors directly interact with cis-active regulatory sequences present in the pre-mRNA. Such regulatory factors may affect differential RNA processing by binding to specific cis-active pre-mRNA sequences, thereby promoting or repressing the selection of alternative splice sites.[97,100] In

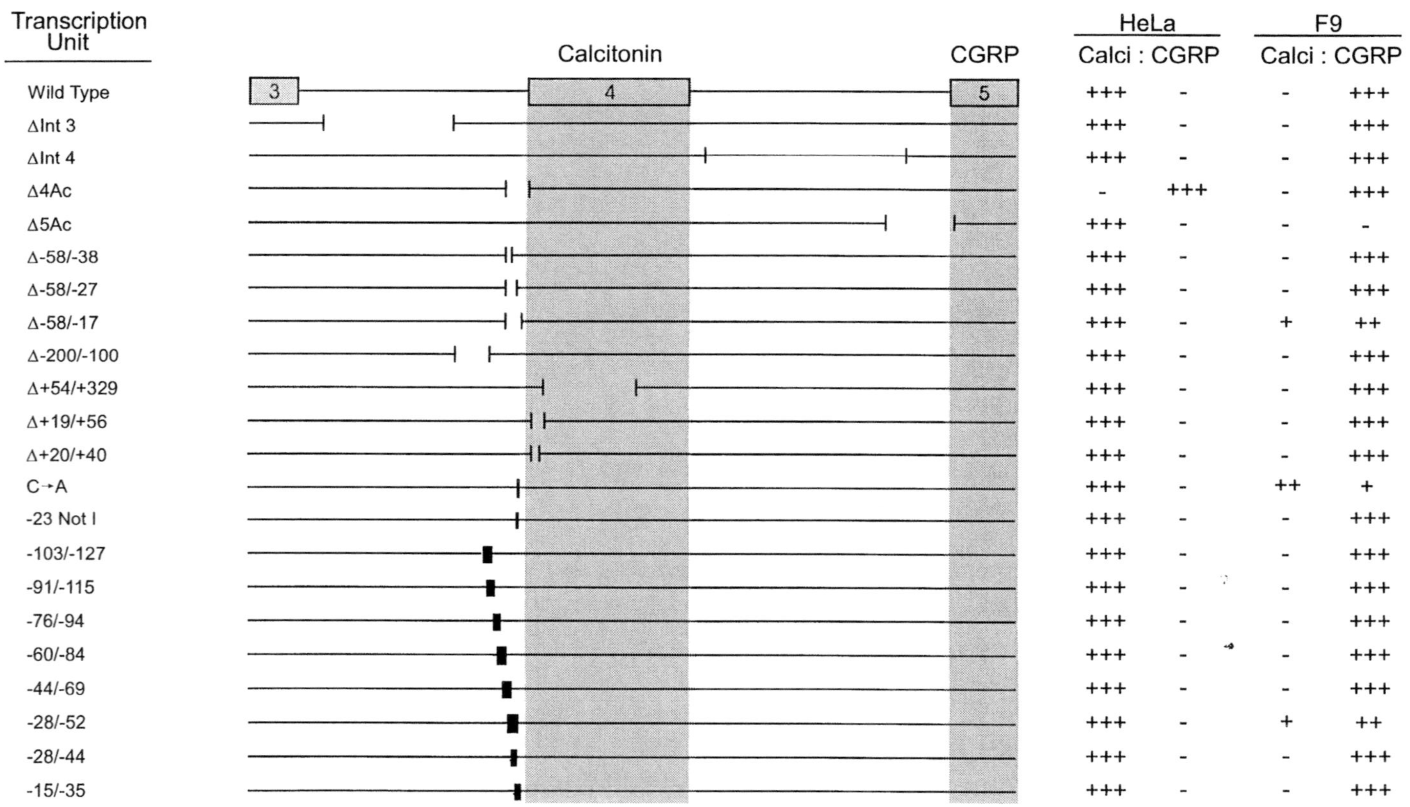

FIGURE 4. Mutational analyses of the rat calcitonin/CGRP gene revealed no unique sequence element(s) required for cell-specific RNA processing. This figure shows a partial listing of mutant calcitonin/CGRP transcription units spanning the entire alternatively spliced region; for a more complete description of each mutation see Yeakley et al. (1993).[91] A graphic representation of exons 3 through 5 is presented in which deletions are shown as missing portions of the solid line and sequence alterations are indicated with filled boxes. Alternative splicing patterns demonstrated by permanently transfected F9 and HeLa cells are represented symbolically as the proportion of each spliced product in the two cell types (+++, 75–100%; ++, 50–75%; +, 25–50%; –, 0–25%). (Adapted from Yeakley, J. M. et al., *Mol. Cell. Biol.*, 13, 5999, 1993.)

mammalian systems, at least one cellular factor, termed alternative splicing factor (ASF) or splicing factor 2 (SF2), has been found to alter 5′-splice site selection using either model pre-mRNA substrates derived from the human β-globin gene[101] or a pre-mRNA transcript derived from the early region of SV40 encoding large T and small t antigens.[102] This protein factor is required for early steps in spliceosome assembly and for the subsequent first cleavage-ligation reaction.[101,103] Although SF2 is required for the general splicing process, differences in the relative concentrations or activities of factors involved in constitutive splicing could play a role in mediating alternative RNA processing events. Indeed, the relative concentration of SF2 and the heterogeneous ribonucleoprotein (hnRNP) A1 has been shown to modulate alternative 5′-splice site selection of model β-globin RNA substrates spliced *in vivo* and *in vitro.*[104-106]

To determine whether the generation of mature calcitonin and CGRP transcripts requires the actions of a cell-specific machinery, the mouse metallothionein-I promoter was used to direct the expression of the rat calcitonin/CGRP gene to both neuronal and non-neuronal tissues in transgenic mice. Results from these studies indicated that in most tissues (liver. kidney, muscle, stomach, spleen, lung, and salivary gland), calcitonin mRNA represented >90% of the mature transcripts derived from the transgene. While Northern blotting analyses of whole brain mRNA indicated approximately equal levels of calcitonin and CGRP transcripts, detailed *in situ* hybridization analyses revealed that calcitonin mRNA was primarily expressed in non-neuronal tissues such as the pia mater and choroid plexus. In contrast, CGRP mRNA was largely associated with neuronal structures, including the hippocampus (CA3 pyramidal cells and dentate gyrus), reticular nucleus, ventrobasal complex of the thalamus, and layer V of the neocortex. Since mature calcitonin mRNA transcripts represented the predominant RNA species in almost all tissues examined, even in tissues in which the endogenous calcitonin/CGRP gene was not normally expressed, the simplest interpretation of these results suggested that calcitonin mRNA production is the unregulated or "default" RNA processing pattern, relying solely upon constitutive splicing components for its synthesis.[107] The ability to splice the calcitonin/CGRP primary transcript to produce mature CGRP mRNA must therefore be dependent upon a neuron-specific splicing machinery. Alternatively, production of calcitonin-specific transcripts could require a tissue-specific factor which is widely distributed, but absent from neuronal tissues.

To identify trans-acting factors controlling the cell-specific production of calcitonin and CGRP mRNAs, Roesser et al.[108] developed an *in vitro* splicing system using RNA transcripts containing the first two exons of human β-globin gene (βg1-2) or a chimeric molecule containing the first exon of the human β-globin gene and the calcitonin-specific exon (βg1-cal4).[108] RNA transcripts from the βg1-2 and βg1-cal4 transcription units were accurately spliced in an ATP-dependent manner when incubated with HeLa cell nuclear extracts. The addition of nuclear extracts from rat brain or F9 cells, but not spleen, specifically inhibited βg1-cal4 splicing, having no effect upon βg1-2 mRNA production. Thus, use of the calcitonin splice acceptor was preferentially inhibited *in vitro* by the addition of nuclear extracts from CGRP-producing cells, consistent with the presence of a regulator inhibiting calcitonin-specific 3′-splice site selection. Partially purified extracts from rat brain were then used to assess the RNA binding specificity of this inhibitory activity by examining the effect of adding unlabeled competitor RNAs to the βg1-cal4 splicing reaction. Addition of a 40-fold molar excess of a small RNA transcript extending from –16 to –36 (relative to the beginning of the fourth exon), but not tRNA or an unrelated sequence derived from the pBluescript polylinker, almost completely protected βg1-cal4 splicing from inhibition. The ability of added target RNA to specifically protect βg1-cal4 splicing *in vitro* implies that the rat brain activity inhibiting calcitonin splice acceptor use acts by binding directly to specific RNA sequences. Further support in favor of this hypothesis came from mobility-shift and ultraviolet (UV) cross-linking analyses in which partially purified extracts from rat brain were shown to interact directly with the 21-nucleotide intronic target sequence revealing the presence of two brain-specific polypeptides (approximately 43 and 41 kDa) which preferentially bind RNA containing the calcitonin splice acceptor.[108]

Using similar analytical strategies, Cote et al.[86] performed UV cross-linking studies to determine if RNA-binding proteins could be identified that interact with calcitonin-specific exonic sequences. UV cross-linking to short RNA probes, corresponding to the first 45 nucleotides of exon 4, revealed the presence of a 66-kDa protein which could be cross-linked to wild-type, but not to the mutant, exon sequences. Significant cross-linking of this 66-kDa protein to either RNA substrate was not observed in nuclear extracts derived from F9 cells. Furthermore, cross-linking experiments using a mixture of HeLa and F9 nuclear extracts also resulted in the absence of a 66-kDa protein binding, suggesting that a dominant neuronal (F9) factor inhibits recognition of exon 4 by inhibiting the binding of a 66-kDa polypeptide to calcitonin-specific exon sequences.[86] The transgenic analyses of Crenshaw et al.[107] and the UV cross-linking data obtained by Roesser et al.[108] and Cote et al.[86] have suggested a model in which a widely expressed exonic RNA-binding protein (66-kDa) is required for exon 4 inclusion in the mature calcitonin mRNA transcript. The production of CGRP transcripts would therefore be dependent upon the actions of a negative regulatory factor which binds to intronic sequence information and serves to inhibit an interaction between this 66-kDa calcitonin mRNA-requiring factor and the pre-mRNA. Additional studies will be required to confirm this model and to isolate the cellular factors mediating cell-specific alternative mRNA production from this complex neuroendocrine gene.

REFERENCES

1. Burns, D. M., Forstrom, J. M., Friday, K. E., Howard, G. A., and Roos, B. A., Procalcitonin's amino-terminal cleavage peptide is a bone-cell mitogen, *Proc. Natl. Acad. Sci. U.S.A.*, 86, 9519, 1989.
2. Burns, D. M., Howard, G. A., and Roos, B. A., An assessment of the anabolic skeletal actions of the common-region peptides derived from the CGRP and calcitonin prohormones, *Ann, NY Acad. Sci.*, 657, 50, 1992.
3. Burns, D. M., Birnbaum, R. S., and Roos, B. A., A neuroendocrine peptide derived from the amino-terminal half of rat procalcitonin, *Mol. Endocrinol.*, 3, 140, 1989.
4. New, H. V. and Mudge, A. W., Calcitonin gene-related peptide regulates muscle acetylcholine receptor synthesis, *Nature*, 323, 809, 1986.
5. Miles, K., Greengard, P., and Huganir, R. L., Calcitonin gene-related peptide regulates phosphorylation of the nicotinic acetylcholine receptor in rat myotubes, *Neuron*, 2, 1517, 1989.
6. Fontaine, B., Klarsfeld, A., and Changeux, J. P., Calcitonin gene-related peptide and muscle activity regulate acetylcholine receptor alpha-subunit mRNA levels by distinct intracellular pathways, *J. Cell Biol.*, 105, 1337, 1987.
7. Amara, S. G., David, D. N., Rosenfeld, M. G., Roos, B. A., and Evans, R. M., Characterization of rat calcitonin mRNA, *Proc. Natl. Acad. Sci. U.S.A.*, 77, 4444, 1980.
8. Amara, S. G., Jonas, V., Rosenfeld, M. G., Ong, E. S., and Evans, R. M., Alternative RNA processing in calcitonin gene expression generates mRNAs encoding different polypeptide products, *Nature*, 298, 240, 1982.
9. Rosenfeld, M. G., Amara, S. G., Roos, B. A., Ong, E. S., and Evans, R. M., Altered expression of the calcitonin gene associated with RNA polymorphism, *Nature*, 290, 63, 1981.
10. Rosenfeld, M. G., Lin, C. R., Amara, S. G., Stolarsky, L., Roos, B. A., Ong, E. S., and Evans, R. M., Calcitonin mRNA polymorphism: peptide switching associated with alternative RNA splicing events, *Proc. Natl. Acad. Sci. U.S.A.*, 79, 1717, 1982.
11. Rosenfeld, M. G., Mermod, J. J., Amara, S. G., Swanson, L. W., Sawchenko, P. E., Rivier, J., Vale, W. W., and Evans, R. M., Production of a novel neuropeptide encoded by the calcitonin gene via tissue-specific RNA processing, *Nature*, 304, 129, 1983.
12. Amara, S. G., Evans, R. M., and Rosenfeld, M. G., Calcitonin/calcitonin gene-related peptide transcription unit: tissue-specific expression involves selective use of alternative polyadenylation sites, *Mol. Cell. Biol.*, 4, 2151, 1984.

13. Amara, S. G., Arriza, J. L., Leff, S. E., Swanson, L. W., Evans, R. M., and Rosenfeld, M. G., Expression in brain of a messenger RNA encoding a novel neuropeptide homologous to calcitonin gene-related peptide, *Science,* 229, 1094, 1985.
14. Kumar, M. A., Foster, G. V., and MacIntyre, I., Further evidence for calcitonin—a rapid acting hormone which lowers plasma calcium, *Lancet,* 2, 480, 1963.
15. Hirsch, P. F., Voelkel, E. F., and Munson, P. L., Thyrocalcitonin hypocalcemic hypophosphatemic principle of the thyroid gland, *Science,* 146, 412, 1964.
16. Minvielle, S., Cressent, M., Delehaye, M. C., Segond, N., Milhaud, G., Jullienne, A., Moukhtar, M. S., and Lasmoles, F., Sequence and expression of the chicken calcitonin gene, *FEBS Lett.,* 223, 63, 1987.
17. Reynolds, J. J., Inhibition by calcitonin of bone resorption induced *in vitro* by vitamin A, *Proc. R. Soc. London (Biol.),* 170, 61, 1968.
18. Lenz, H. J., Rivier, J. E., and Brown, M. R., Biological actions of human and rat calcitonin and calcitonin gene-related peptide, *Regul. Pept.,* 12, 81, 1985.
19. Breimer, L. H., MacIntyre, I., and Zaidi, M., Peptides from the calcitonin genes: molecular genetics, structure and function, *Biochem. J.,* 255, 377, 1988.
20. Boden, S. D. and Kaplan, F. S., Calcium homeostasis, *Orthop. Clin. North Am.,* 21, 31, 1990.
21. Allison, J., Hall, L., MacIntyre, I., and Craig, R. K., The construction and partial characterization of plasmids containing complementary DNA sequences to human calcitonin precursor polyprotein, *Biochem. J.,* 199, 725, 1981.
22. Jacobs, J. W., Goodman, R. H., Chin, W. W., Dee, P. C., Habener, J. F., Bell, N. H., and Potts, J. T., Jr., Calcitonin messenger RNA encodes multiple polypeptides in a single precursor, *Science,* 213, 457, 1981.
23. Neurath, H., Proteolytic processing and regulation, *Enzyme,* 45, 239, 1991.
24. Andrews, P. C., Brayton, K. A., and Dixon, J. E., Post-translational proteolytic processing of precursors to regulatory peptides, *Exs,* 56, 192, 1989.
25. Darby, N. J. and Smyth, D. G., Endopeptidases and prohormone processing, *Biosci. Rep.,* 10, 1, 1990.
26. Eipper, B. A., Stoffers, D. A., and Mains, R. E., The biosynthesis of neuropeptides: peptide alpha-amidation, *Annu. Rev. Neurosci.,* 15, 57, 1992.
27. Bradbury, A. F. and Smyth, D. G., Peptide amidation, *Trends Biochem. Sci.,* 16, 112, 1991.
28. Guenther, H. L. and Fleisch, H., The procalcitonin amino-terminal cleavage peptide (N-proCT) lacks biological activity on normal clonal rat osteoblastic and preosteoblastic cells in vitro, *Calcif. Tissue Int.,* 49, 138, 1991.
29. Hillyard, C. J., Myers, C., Abeyasekera, G., Stevvensvenson, J. C., Craig, R. K., and MacIntyre, I., Katacalcin: a new plasma calcium-lowering hormone, *Lancet,* 1, 846, 1983.
30. Poschl, E., Lindley, I., Hofer, E., Seifert, J. M., Brunowsky, W., and Besemer, J., The structure of procalcitonin of the salmon as deduced from its cDNA sequence, *FEBS Lett.,* 226, 96, 1987.
31. Collyear, K., Girgis, S. I., Saunders, G., MacIntyre, I., and Holt, G., Predicted structure of the bovine calcitonin gene-related peptide and the carboxy-terminal flanking peptide of bovine calcitonin precursor, *J. Mol. Endocrinol.,* 6, 147, 1991.
32. Lasmoles, F., Jullienne, A., Day, F., Minvielle, S., Milhaud, G., and Moukhtar, M. S., Elucidation of the nucleotide sequence of chicken calcitonin mRNA: direct evidence for the expression of a lower vertebrate calcitonin-like gene in man and rat, *EMBO J.,* 4, 2603, 1985.
33. Amara, S. G., Jonas, V., ONeil, J. A., Vale, W., Rivier, J., Roos, B. A., Evans, R. M., and Rosenfeld, M. G., Calcitonin COOH-terminal cleavage peptide as a model for identification of novel neuropeptides predicted by recombinant DNA analysis, *J. Biol. Chem.,* 257, 2129, 1982.
34. Boorman, G. A. and Noord, M. V., and Hollander, C. F., Naturally occurring medullary thyroid carcinoma in the rat, *Arch. Pathol.,* 94, 35, 1972.
35. Roos, B. A., Yoon, M. J., Frelinger, A. L., Pensky, A. E., Birnbaum, R. S., and Lambert, P. W., Tumor growth and calcitonin during serial transplantation of rat medullary thyroid carcinoma, *Endocrinology,* 105, 27, 1979.
36. Skofitsch, G. and Jacobowitz, D. M., Quantitative distribution of calcitonin gene-related peptide in the rat central nervous system, *Peptides,* 6, 1069, 1985.
37. Skofitsch, G. and Jacobowitz, D. M., Calcitonin gene-related peptide: detailed immunohistochemical distribution in the central nervous system, *Peptides,* 6, 721, 1985.
38. Kawai, Y., Takami, K., Shiosaka, S., Emson, P. C., Hillyard, C. J., Girgis, S., MacIntyre, I., and Tohyama, M., Topographic localization of calcitonin gene-related peptide in the rat brain: an immunohistochemical analysis, *Neuroscience,* 15, 747, 1985.
39. Kresse, A., Jacobowitz, D. M., and Skofitsch, G., Distribution of calcitonin gene-related peptide in the central nervous system of the rat by immunocytochemistry and in situ hybridization histochemistry, *Ann. NY Acad. Sci.,* 657, 455, 1992.
40. Shimada, S., Shiosaka, S., Hillyard, C. J., Girgis, S. I., MacIntyre, I., Emson, P. C., and Tohyama, M., Calcitonin gene-related peptide projection from the ventromedial thalamic nucleus to the insular cortex: a combined retrograde transport and immunocytochemical study, *Brain Res.,* 344, 200, 1985.

41. Sakanaka, M., Magari, S., Emson, P. C., Hillyard, C. J., Girgis, S. I., MacIntyre, I., and Tohyama, M., The calcitonin gene-related peptide-containing fiber projection from the hypothalamus to the lateral septal area including its fine structures, *Brain Res.*, 344, 196, 1985.
42. Yasui, Y., Saper, C. B., and Cechetto, D. F., Calcitonin gene-related peptide immunoreactivity in the visceral sensory cortex, thalamus, and related pathways in the rat, *J. Comp. Neurol.*, 290, 487, 1989.
43. Shimada, S., Shiosaka, S., Emson, P. C., Hillyard, C. J., Girgis, S., MacIntyre, I., and Tohyama, M., Calcitonin gene-related peptidergic projection from the parabrachial area to the forebrain and diencephalon in the rat: an immunohistochemical analysis, *Neuroscience*, 16, 607, 1985.
44. Schwaber, J. S., Sternini, C., Brecha, N. C., Rogers, W. T., and Card, J. P., Neurons containing calcitonin gene-related peptide in the parabrachial nucleus project to the central nucleus of the amygdala, *J. Comp. Neurol.*, 270, 416, 1988.
45. Yamano, M., Hillyard, C. J., Girgis, S., MacIntyre, I., Emson, P. C., and Tohyama, M., Presence of a substance P-like immunoreactive neurone system from the parabrachial area to the central amygdaloid nucleus of the rat with reference to coexistence with calcitonin gene-related peptide, *Brain Res.*, 451, 179, 1988.
46. Rosenfeld, M. G., Amara, S. G., Birnberg, N. C., Mermod, J. J., Murdoch, G. H., and Evans, R. M., Calcitonin, prolactin, and growth hormone gene expression as model systems for the characterization of neuroendocrine regulation, *Recent Prog. Horm. Res.*, 39, 305, 1983.
47. Fried, K., Brodin, E., and Theodorsson, E., Substance P-, CGRP- and NPY-immunoreactive nerve fibers in rat sciatic nerve-end neuromas, *Regul. Pept.*, 25, 11, 1989.
48. Skofitsch, G. and Jacobowitz, D. M., Calcitonin gene-related peptide coexists with substance P in capsaicin sensitive neurons and sensory ganglia of the rat, *Peptides*, 6, 747, 1985.
49. Brain, S. D., Williams, T. J., Tippins, J. R., Morris, H. R., and MacIntyre, I., Calcitonin gene-related peptide is a potent vasodilator, *Nature*, 313, 54, 1985.
50. Vignery, A., Wang, F., and Ganz, M. B., Macrophages express functional receptors for calcitonin-gene-related peptide, *J. Cell. Physiol.*, 149, 301, 1991.
51. Wang, F., Millet, I., Bottomly, K., and Vignery, A., Calcitonin gene-related peptide inhibits interleukin 2 production by murine T lymphocytes, *J. Biol. Chem.*, 267, 21052, 1992.
52. Nong, Y. H., Titus, R. G., Ribeiro, J. M., and Remold, H. G., Peptides encoded by the calcitonin gene inhibit macrophage function, *J. Immunol.*, 143, 45, 1989.
53. Mulderry, P. K., Ghatei, M. A., Bishop, A. E., Allen, Y. S., Polak, J. M., and Bloom, S. R., Distribution and chromatographic characterisation of CGRP-like immunoreactivity in the brain and gut of the rat, *Regul. Pept.*, 12, 133, 1985.
54. Feher, E., Burnstock, G., Varndell, I. M., and Polak, J. M., Calcitonin gene-related peptide-immunoreactive nerve fibres in the small intestine of the guinea-pig: electron-microscopic immunocytochemistry, *Cell Tissue Res.*, 245, 353, 1986.
55. Beglinger, C., Born, W., Hildebrand, P., Ensinck, J. W., Burkhardt, F., Fischer, J. A., and Gyr, K., Calcitonin gene-related peptides I and II and calcitonin: distinct effects on gastric acid secretion in humans, *Gastroenterology*, 95, 958, 1988.
56. Raybould, H. E., Inhibitory effects of calcitonin gene-related peptide on gastrointestinal motility, *Ann. NY Acad. Sci.*, 657, 248, 1992.
57. Holzer, P. and Guth, P. H., Neuropeptide control of rat gastric mucosal blood flow. Increase by calcitonin gene-related peptide and vasoactive intestinal polypeptide, but not substance P and neurokinin A, *Circ. Res.*, 68, 100, 1991.
58. Molina, J. M., Cooper, G. J., Leighton, B., and Olefsky, J. M., Induction of insulin resistance in vivo by amylin and calcitonin gene-related peptide, *Diabetes*, 39, 260, 1990.
59. Leighton, B., Foot, E. A., Cooper, G. G., and King, J. M., Calcitonin gene-related peptide-1 (CGRP-1) is a potent regulator of glycogen metabolism in rat skeletal muscle, *FEBS Lett.*, 249, 357, 1989.
60. Denis-Donini, S., Calcitonin gene-related peptide influence on central nervous system differentiation, *Ann. NY Acad. Sci.*, 657, 344, 1992.
61. Denis-Donini, S., Expression of dopaminergic phenotypes in the mouse olfactory bulb induced by the calcitonin gene-related peptide, *Nature*, 339, 701, 1989.
62. Osterlund, M., Fontaine, B., Devillers-Thiery, A., Geoffroy, B., and Changeux, J. P., Acetylcholine receptor expression in primary cultures of embryonic chick myotubes. I. Discoordinate regulation of alpha-, gamma- and delta-subunit gene expression by calcitonin gene-related peptide and by muscle electrical activity, *Neuroscience*, 32, 279, 1989.
63. Steenbergh, P. H., Hoppener, J. W., Zandberg, J., Visser, A., Lips, C. J., and Jansz, H. S., Structure and expression of the human calcitonin/CGRP genes, *FEBS Lett.*, 209, 97, 1986.
64. Steenbergh, P. H., Hoppener, J. W., Zandberg, J., Lips, C. J., and Jansz, H. S., A second human calcitonin/CGRP gene, *FEBS Lett.*, 183, 403, 1985.
65. Bennett, M. M. and Amara, S. G., Molecular mechanisms of cell-specific and regulated expression of the calcitonin/alpha-CGRP and beta-CGRP genes, *Ann. NY Acad. Sci.*, 657, 36, 1992.

66. Petermann, J. B., Born, W., Chang, J. Y., and Fischer, J. A., Identification in the human central nervous system, pituitary, and thyroid of a novel calcitonin gene-related peptide, and partial amino acid sequence in the spinal cord, *J. Biol. Chem.*, 262, 542, 1987.
67. Holman, J. J., Craig, R. K., and Marshall, I., Human alpha- and beta-CGRP and rat alpha-CGRP are coronary vasodilators in the rat, *Peptides*, 7, 231, 1986.
68. Williams, G., Cardoso, H., Ball, J. A., Mulderry, P. K., Cooke, E., and Bloom, S. R., Potent and comparable vasodilator actions of A and B calcitonin gene-related peptides on the superficial subcutaneous vasculature of man, *Clin. Sci.*, 75, 309, 1988.
69. Mulderry, P. K. et al., Differential expression of alpha-CGRP and beta-CGRP by primary sensory neurons and enteric autonomic neurons of the rat, *Neuroscience*, 25, 195, 1988.
70. Sternini, C., Enteric and visceral afferent CGRP neurons. Targets of innervation and differential expression patterns, *Ann. NY Acad. Sci.*, 657, 170, 1992.
71. Lee, Y., Hayashi, N., Hillyard, C. J., Girgis, S. I., MacIntyre, I., Emson, P. C., and Tohyama, M., Calcitonin gene-related peptide-like immunoreactive sensory fibers form synaptic contact with sympathetic neurons in the rat celiac ganglion, *Brain Res.*, 407, 149, 1987.
72. Sabate, M. I., Stolarsky, L. S., Polak, J. M., Bloom, S. R., Varndell, I. M., Ghatei, M. A., Evans, R. M., and Rosenfeld, M. G., Regulation of neuroendocrine gene expression by alternative RNA processing. Colocalization of calcitonin and calcitonin gene-related peptide in thyroid C-cells, *J. Biol. Chem.*, 260, 2589, 1985.
73. Rosenfeld, M. G., Amara, S. G., and Evans, R. M., Alternative RNA processing: determining neuronal phenotype, *Science*, 225, 1315, 1984.
74. Tschopp, F. A., Tobler, P. H., and Fischer, J. A., Calcitonin gene-related peptide in the human thyroid, pituitary and brain, *Mol. Cell. Endocrinol.*, 36, 53, 1984.
75. Dominski, Z. and Kole, R., Selection of splice sites in pre-mRNAs with short internal exons, *Mol. Cell. Biol.*, 11, 6075, 1991.
76. Guo, W., Mulligan, G. J., Wormsley, S., and Helfman, D. M., Alternative splicing of β-tropomyosin pre-mRNA: cis-acting elements and cellular factors that block the use of a skeletal muscle exon in nonmuscle cells, *Genes Dev.*, 5, 2096, 1991.
77. Kuo, H. C., Nasim, F. H., and Grabowski, P. J., Control of alternative splicing by the differential binding of U1 small nuclear ribonucleoprotein particle, *Science*, 251, 1045, 1991.
78. Mullen, M. P., Smith, C. W. J., Patton, J. G., and Nidal-Ginard, B., α-tropomyosin mutually exclusive exon selection: competition between branchpoint/polypyrimidine tract determines default exon choice, *Genes Dev.*, 5, 642, 1991.
79. Fu, X. Y. and Manley, J. L., Factors influencing alternative splice site utilization in vivo, *Mol. Cell. Biol.*, 7, 738, 1987.
80. Freyer, G. A., O'Brien, J. P., and Hurwitz, J., Alterations in the polypyrimidine sequence affect the in vitro splicing reactions catalyzed by HeLa cell-free preparations, *J. Biol. Chem.*, 264, 14631, 1989.
81. Reed, R., The organization of 3′ splice-site sequences in mammalian introns, *Genes Dev.*, 3, 2113, 1989.
82. Reed, R. and Maniatis, T., The role of the mammalian branchpoint sequence in pre-mRNA splicing, *Genes Dev.*, 2, 1268, 1988.
83. Watakabe, A., Inoue, K., Sakamoto, H., and Shimura, Y. A., secondary structure at the 3′ splice site affects the in vitro splicing reaction of mouse immunoglobulin mu chain pre-mRNAs, *Nucleic Acids Res.*, 17, 8159, 1989.
84. Libri, D., Piseri, A., and Fiszman, M. Y., Tissue-specific splicing in vivo of the beta-tropomyosin gene: dependence on an RNA secondary structure, *Science*, 252, 1842, 1991.
85. Watakabe, A., Tanaka, K., and Shimura, Y., The role of exon sequences in splice site selection, *Genes Dev.*, 7, 407, 1993.
86. Cote, G. J., Stolow, D. T., Peleg, S., Berget, S. M., and Gagel, R. F., Identification of exon sequences and an exon binding protein involved in alternative RNA splicing of calcitonin/CGRP, *Nucleic Acids Res.*, 20, 2361, 1992.
87. Reed, R. and Maniatis, T. A., role for exon sequences and splice-site proximity in splice-site selection, *Cell*, 46, 681, 1986.
88. Black, D. L., Does steric interference between splice sites block the splicing of a short c-src neuron-specific exon in non-neuronal cells?, *Genes Dev.*, 5, 389, 1991.
89. Leff, S. E., Evans, R. M., and Rosenfeld, M. G., Splice commitment dictates neuron-specific alternative RNA processing in calcitonin/CGRP gene expression, *Cell*, 48, 517, 1987.
90. Emeson, R. B., Hedjran, F., Yeakley, J. M., Guise, J. W., and Rosenfeld, M. G., Alternative production of calcitonin and CGRP mRNA is regulated at the calcitonin-specific splice acceptor, *Nature*, 341, 76, 1989.
91. Yeakley, J. M., Hedjran, F., Morfin, J. P., Merillat, N., Rosenfeld, M. G., and Emeson, R. B., Control of calcitonin/calcitonin gene-related peptide pre-mRNA processing by constitutive intron and exon elements, *Mol. Cell. Biol.*, 13, 5999, 1993.
92. Adema, G. J., van-Hulst, K. L., and Baas, P. D., Uridine branch acceptor is a cis-acting element involved in regulation of the alternative processing of calcitonin/CGRP-I pre-mRNA, *Nucleic Acids Res.*, 18, 5365, 1990.

93. Adema, G. J. and Baas, P. D., Deregulation of alternative processing of calcitonin/CGRP-I pre-mRNA by a single point mutation, *Biochem. Biophys. Res. Commun.*, 178, 985, 1991.
94. Adema, G. J., Bovenberg, R. A., Jansz, H. S., and Baas, P. D., Unusual branch point selection involved in splicing of the alternatively processed calcitonin/CGRP-I pre-mRNA, *Nucleic Acids Res.*, 16, 9513, 1988.
95. Siebel, C. W., Fresco, L. D., and Rio, D. C., The mechanism of somatic inhibition of Drosophila P-element pre-mRNA splicing: multiprotein complexes at an exon pseudo-5′ splice site control U1 snRNP binding, *Genes Dev.*, 6, 1386, 1992.
96. Sosnowski, B. A., Belote, J. M., and McKeown, M., Sex-specific alternative splicing of RNA from the *transformer* gene results from sequence-dependent splice-site blockage, *Cell*, 58, 449, 1989.
97. Inoue, K., Hoshijima, K., Sakamoto, H., and Shimura, Y., Binding of the Drosophila *sex-lethal* gene to the alternative splice site of *transformer* primary transcript, *Nature*, 344, 461, 1990.
98. Ryner, L. C. and Baker, B. S., Regulation of *doublesex* pre-mRNA processing occurs by 3′-splice site activation, *Genes Dev.*, 5, 2071, 1991.
99. Tian, M. and Maniatis, T., Postive control of pre-mRNA splicing *in vitro*, *Science*, 256, 237, 1992.
100. Valcarcel, J., Singh, R., Zamore, P. D., and Green, M. R., The protein Sex-lethal antagonizes the splicing factor U2AF to regulate alternative splicing of transformer pre-mRNA, *Nature*, 362, 171, 1993.
101. Krainer, A. R., Conway, G. C., and Kozak, D., The essential pre-mRNA splicing factor SF2 influences 5′ splice site selection by activating proximal sites, *Cell*, 62, 35, 1990.
102. Ge, H. and Manley, J. L., A protein factor, ASF, controls cell-specific alternative splicing of SV40 early pre-mRNA in vitro, *Cell*, 62, 25, 1990.
103. Krainer, A. R., Conway, G. C., and Kozak, D., Purification and characterization of pre-mRNA splicing factor SF2 from HeLa cells, *Genes Dev.*, 4, 1158, 1990.
104. Mayeda, A., Helfman, D. M., and Krainer, A. R., Modulation of exon skipping and inclusion by heterogeneous nuclear ribonucleoprotein A1 and pre-mRNA splicing factor SF2/ASF, *Mol. Cell. Biol.*, 13, 2993, 1993.
105. Caceres, J. F., Stamm, S., Helfman, D. M., and Krainer, A. R., Regulation of alternative splicing *in vivo* by overexpression of antagonistic splicing factors, *Science*, 265, 1706, 1994.
106. Mayeda, A. and Krainer, A. R., Regulation of alternative pre-mRNA splicing by hnRNP A1 and splicing factor SF2, *Cell*, 68, 365, 1992.
107. Crenshaw, E. B., Russo, A. F., Swanson, L. W., and Rosenfeld, M. G., Neuron-specific alternative RNA processing in transgenic mice expressing a metallothionein-calcitonin fusion gene, *Cell*, 49, 389, 1987.
108. Roesser, J. R., Liittschwager, K., and Leff, S. E., Regulation of tissue-specific splicing of the calcitonin/calcitonin gene-related peptide gene by RNA-binding proteins, *J. Biol. Chem.*, 268, 8366, 1993.

Part II. Pharmacology

Chapter 3

NEUROGENIC INFLAMMATION: REEVALUATION OF AXON REFLEX THEORY

Janos Szolcsányi

CONTENTS

I. INTRODUCTION

The role of sensory nerves and axon reflexes in producing inflammation was recognized in the first decade of this century.[1] Nevertheless, neurogenic inflammation captured the interest of the international scientific community only during the 1980s. To the author's knowledge, the first book devoted to this topic was published in 1984.[2] In the 1970s one could find no references to neurogenic inflammation or its synonyms among the subject indexes of published works on inflammatory processes.[3,4] Axon reflex flare,[5] the phenomenon of "antidromic vasodilatation,"[6] and the recent revival of Lewis' concept of "nocifensor" system,[7,8] however, have often been related to neurogenic inflammation for many decades.[9] Therefore, in order to justify a reevaluation of the neurogenic inflammation concept, a clear description of the earlier theories and postulations put forward by the classical forerunners is necessary.

II. ANTIDROMIC VASODILATATION AND AXON REFLEX FLARE

It had been observed in the last century that stimulation of the peripheral end of cut dorsal roots elicited signs of cutaneous vasodilatation.[10] Bayliss described, for the first time, mediation of the response by "sensory afferent posterior root-fibres"[6] and on Langley's suggestion put the adjective "antidromic" to denote that under these conditions "impulses passing along sensory fibres in a direction contrary to what is regarded as the usual one".[11] Bruce[1] obtained evidence that undiluted mustard oil instilled into the eye of the rabbit did not induce inflammation after degeneration of the trigeminal nerve supply or after applying cocaine or alypine local anesthetic into the eye. For this reason the axon reflex arrangement proposed by Bayliss[6] for antidromic vasodilatation was adopted also as a possible ("wahrscheinlich") mechanism for neurogenic inflammation. The results obtained by Bruce with local anesthetics were, however, rather questionable on technical grounds.[12] According to Bayliss:[11] "In my first paper on the subject...(Ref. 6)...I pointed out that one possibility might be that the *dorsal root fibres divide near their peripheral terminations, one branch supplying the sensory end-organ in skin, muscle, etc., while the other ends as an efferent inhibitory end-organ on the muscular*

0-8493-7646-7/96/$0.00+$.50

coat of the arterioles...which is similar in some ways to that involved in Langley's 'axon-reflex'. Ninian Bruce (1910) advocated a similar view, producing experimental evidence in its favour... It seems desirable, nevertheless, that *the experiments should be repeated.*"

The blocking action of local anesthetics on neurogenic inflammation evoked by irritants was not substantiated in further animal experiments. One year later, however, strong support for the above description of the axon reflex theory was obtained in the human skin by Lewis and Grant.[13] In this and subsequent studies[5] local skin anesthesia prevented the cutaneous flare reaction surrounding local injuries. The wheal of the triple response was, however, not abolished in this way. Lewis[5] postulated a cutaneous network of sensory fibers, the size of which determines the area of flare. Again, according to his often-cited views, in this arrangement nerve terminals with vasodilator function are situated around the arterioles, and are distinct from the cutaneous receptors, which serve for initiation of the axon reflex.

An implication of the axon reflex flare to inflammation and trophic reactions was formulated within the frame of a new concept on "nocifensor nerves".[7,8] "For according to our present evidence distal stimulation of cutaneous nerves does provoke some of the phenomena of inflammation — namely, hyperalgesia and reddening — and may conceivably provoke more".[7] Lewis, in these late publications, postulated that axon reflex flare and spreading hyperalgesia is mediated by a "posterior root system" in which the fibers are "not sensory nerves" but "hitherto unrecognized...nocifensor nerves" that "are associated with local defense against injury". Although the nonsensory nature of the system rightly won little credence, the grain of truth in his foresight in respect to the functional significance of terminal arborizations of sensory nerves in local tissue defense and inflammation has been later verified and extended.[2,14-26]

III. NEUROGENIC INFLAMMATION

The first direct evidence for neurogenic inflammation evoked by antidromic stimulation of sensory nerves was obtained by the author in N. Jancsó's laboratories. The cardinal signs were "arteriolar vasodilatation, enhancement of vascular permeability, protein exudation, fixation of injected colloidal silver onto the walls of venules and, later, their storage in histiocytes". Furthermore, "secretion was discharged from the nose" in response to antidromic stimulation of the trigeminal nerve.[14] The symptoms could not be attributed to a simple enhancement of microcirculation (antidromic vasodilatation) or edema;[27] instead, the term neurogenic inflammation was favored.[12,14]

Subsequently, plasma extravasation was elicited both in the skin and in visceral organs by antidromic stimulation of the dorsal roots.[17] Furthermore, clear evidence was obtained for venular endothelial gap formation, leukocyte and platelet adherence to the endothelium,[28,29] and mast cell degranulation[30] of the rat trachea in response to antidromic stimulation of vagal afferents.

The selective blocking action of capsaicin pretreatment and antidromic stimulation of single afferent fibers provided the tools to show that neurogenic inflammation in the skin is mediated by C-polymodal nociceptors[19,24,31] and in interoceptive areas by sensory nerve endings which are also sensitive to the stimulatory and blocking actions of capsaicin.[17-25] There is no experimental result that might indicate a mediating role of afferents which are insensitive to capsaicin. Single unit studies provided the evidence for the existence of a subpopulation of primary afferent neurons characterized by their sensitivity (pharmacological receptive site) to capsaicin-type agents.[19,21,24,32,33] The similar excitability spectrum of polymodal and chemosensitive nociceptors with thin fibers of the C- or A-delta type and the unique function in evoking various local peptidergic sensory-efferent responses formed the basis for introduction of a pharmacological classification — the "capsaicin-sensitive" afferents — for designation of this substantial portion of the sensory neuron population.[19,32,33]

IV. SENSORY RECEPTORS WITH EFFERENT FUNCTION

According to the classical concept of Bayliss,[6,11] Bruce,[1] and Lewis,[5] the mediator(s) of antidromic vasodilatation, neurogenic inflammation or axon reflex flare are released from specialized effector nerve terminals of the primary afferent neurons and not from the sensory receptors themselves. This view fits the Cartesian reflex principle in the still accepted sense that the peripheral nervous system is composed of separate afferent and efferent elements. If it were the case then axonal conduction would be the prerequisite for evoking neurogenic inflammation, as has been proved for the axon reflex flare in the skin of humans,[5] monkeys,[34] and pigs.[35]

The following experimental results and considerations contradict this classical theory and seem to favor a dual function for the same site of the sensory nerve terminal axon (see References 17, 19, 21, 24–26, and 32 for reviews).

The first evidence against the obligatory requirement of axonal conduction was obtained in the eye and paw of the rat, as well as in the human skin. No inhibition of plasma extravasation to irritants or reddening of the human skin at the site of contact was observed after local anesthesia.[36] One might suppose, however, that capsaicin, xylene, mustard oil, or omega-chloracetophenone are special irritants which stimulate the receptors and are also suitable to release the mediator(s) from another ending. After subcutaneous injection of tetrodotoxin, however, plasma extravasation to capsaicin-type agents was not inhibited, and in the presence of tetrodotoxin its magnitude was related to the sensory stimulant potency of the irritants.[19] There was also a strong correlation between the nociceptive and neurogenic tracheoconstrictor effects of capsaicin, piperine, homovanilloyl octylamide, or homovanilloyl dodecylamide in the presence of tetrodotoxin in the guinea pig isolated tracheal strip preparation.[37] Tetrodotoxin (3 μM) did not inhibit the effect of these agents applied in a wide dose range (e.g., capsaicin from 33 nM to 3.3 μM).

Topical application of capsaicin onto the human skin to induce sensory desensitization abolished the axon reflex flare for several days. The time course of recovery of the function of the "efferent" and "afferent" side of the axon reflex flare was identical after capsaicin desensitization.[17,24]

All these observations indicate a matching of pharmacological sensitivity of the two functions of capsaicin-sensitive nerve endings, and invalidate the indispensable involvement of an axon reflex for activation of the efferent function of sensory nerve terminals.

Capsaicin and bradykinin in nanomolar concentrations evoke depolarization and discharges in a subset of neurons in dorsal root and trigeminal ganglia *in vitro.* Depolarization is accompanied by release of the putative neuropeptide transmitters (substance P and calcitonin gene-related peptide [CGRP]) from these cell bodies (see References 24 and 38). Intraarterial infusion of capsaicin into the rat excites, at similar doses, the central and peripheral endings of primary afferent neurons[39] and substance P is released by capsaicin from both endings (see Reference 20). Thus, the cell body and central terminals of capsaicin-sensitive primary afferent neurons can respond to noxious chemicals with the dual effect of spike initiation and release of sensory neuropeptides. Therefore, it is highly unlikely that in the periphery, the chemical excitability and the mediator releasing function might be confined to different nerve terminals.

In the rat, antidromic stimulation of single fibers of C-polymodal nociceptors elicits plasma extravasation localized to the spotlike receptive field.[31,40] In the rabbit[41] and human[42,43] skin C-polymodal nociceptors have multiple receptive fields; and the size of the receptive field corresponds to that of the flare evoked in the same cutaneous area.[24,44] Spatially restricted desensitization of a unit can be achieved by capsaicin.[43]

Histological evidence for the classical axon reflex arrangement has never been achieved. Instead, sensory nerve terminals or terminal varicosities of the cornea contain microvesicles

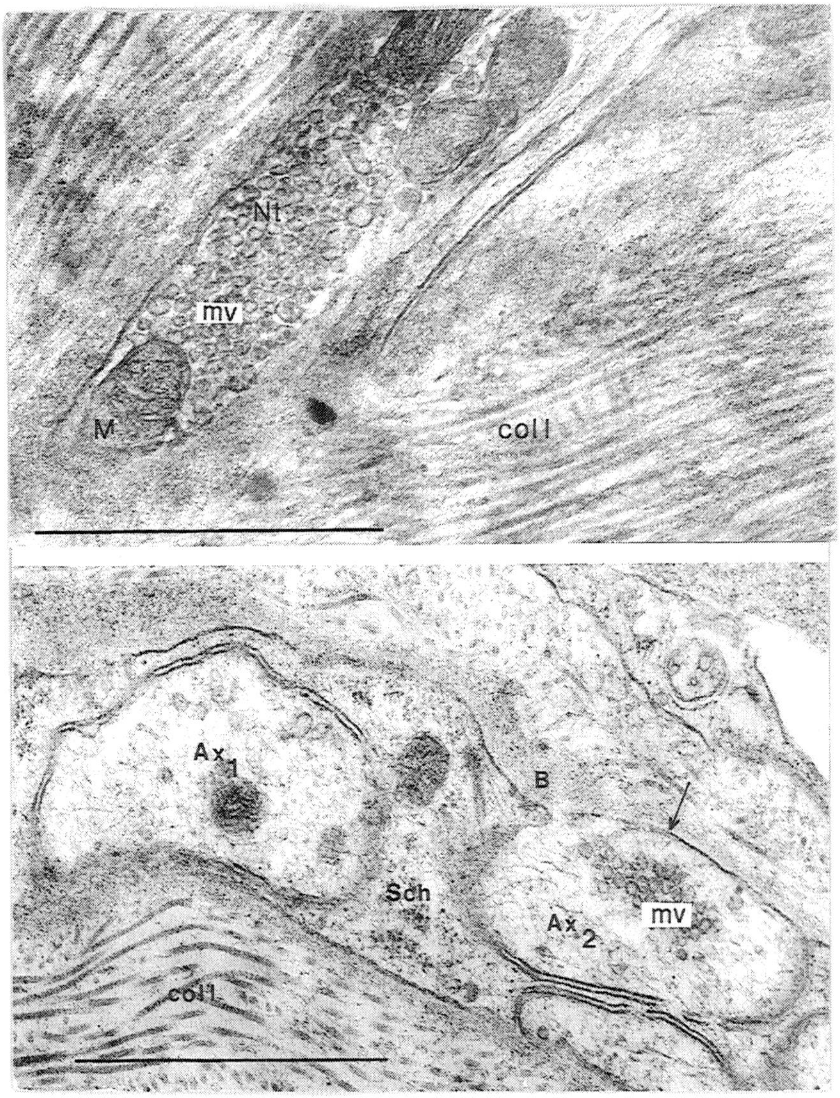

FIGURE 1. Ultrastructure of nervous elements in the cornea of the rat. The nerve terminal (Nt) in the upper part of the figure is filled with microvesicles (mv) and contains mitochondria (M). The nerve ending is not covered by a Schwann cell envelope. Two preterminal axon profiles (Ax1, Ax2) in the lower part of the figure are arranged in bundles within the Schwann cell (Sch) processes. The arrow indicates the free, uncovered part of the axon. Note the accumulated microvesicles at the site of Ax2. Coll, collagen fibers. Calibration line: 1 μm. Unpublished details from work reported in Reference 45.

and accumulated mitochondria like the neuroeffector varicosities (Figure 1; Reference 45). The chemosensitive fibers branch extensively in the cornea and form a very dense intraepithelial network of axons that also have axonal varicosities.[46,47] After local capsaicin pretreatment sensory desensitization of the eye took place and the number of vesicles in these chemoceptive corneal varicosities was decreased.[45]

The long latency and slow time course of the vasodilatation to antidromic nerve stimulation indicate that the mediator reaches the vessels by diffusion and not through a neuroeffector junction (see References 19 and 24).

The latency of axon reflex flare evoked in the volar surface of the forearm by transcutaneous electrical stimulation[48] or by noxious heat stimuli[49] was 4.9 ± 0.6 s and 3.17 ± 0.14 s at a distance of 10 and 8 mm, respectively. Antidromic stimulation of the superficial peroneal nerve at the ankle of healthy subjects elicited cutaneous vasodilatation on the dorsal surface of the foot after a delay of 13.6 ± 11 s.[50] The latency of vasodilatation in the plantar skin of the rat to dorsal root stimulation was 5.3 ± 0.2 s;[17] and in the pig's skin to transcutaneous electrical stimulation was 13.7 ± 1.2 s, in contrast to the much shorter latency of the 1.9 ± 0.5 s of the sympathetic vasoconstrictor response.[51] All these data were obtained with the aid of a highly sensitive laser Doppler flowmeter. The latency of plasma extravasation seems to be even longer.[52]

Histochemical and electron microscopic data also favor the notion that peptides released from terminal varicosities of capsaicin-sensitive afferents need to diffuse through a considerable distance of a few micrometers to reach the respective vascular binding sites,[53] or to reach the postcapillary venules where plasma extravasation takes place.[29,54] The sparse innervation of venules[29,53] speaks also against the existence of nerve endings specialized for efferent function. It is worth mentioning that at the sympathetic neuroeffector junction the diffusion delay of noradrenaline to reach the maximal electrochemical response is 200 ms in a distance of about 5 μm.[55]

Inhibitory junction potentials (IJP) elicited in the guinea pig mesenteric artery by electrical stimulation of capsaicin-sensitive fibers or by capsaicin itself appear after a few seconds, but obviously later than the adrenergic excitatory junction potential (EJP) of the same preparation.[56] After cessation of the electrical stimulus the evoked IJP lasts for several minutes, in contrast to the EJP which was over within a second.

V. REEVALUATION OF THE AXON REFLEX THEORY

Many key issues of the present reevaluation of axon reflex theory[24,25,32,33] were put forward by the author more than 10 years ago.[12,19,57] The proposed revision resides on the following suggestions.

1. A substantial portion of primary afferent neurons with B-type cell bodies and thin C- or A-delta fibers forms a separate subset of the peripheral nervous system, the sensory-efferent neural system. These chemoceptive neurons and their terminals are sensitive to the stimulatory and blocking actions of capsaicin *(capsaicin-sensitive afferents)*. Major representatives of this system are the cutaneous C-polymodal nociceptors. Their peripheral terminals and axonal varicosities (Figure 2) subserve a dual sensory-efferent function. These *sensory-effectors* form a new type of neural functional units which are suitable both for impulse initiation to the central nervous system and for release of neuropeptide mediator(s) to regulate tissue functions including arteriolar vasodilatation, neurogenic inflammation, and smooth muscle responses of various organs.
2. The release process in the generator region is not inhibited by tetrodotoxin or local anesthetics. Therefore, neurogenic inflammation or other local effector responses of sensory nerve terminals can develop *without axon reflexes.*
3. Axonal conduction is necessary within the terminal arborization of capsaicin-sensitive fibers to evoke axon reflex flare in the human skin or to elicit other tissue responses under conditions when the mediator(s) released from the activated sensory-effectors cannot reach the target cells (Figure 2E, site 1 for vasodilatation).
4. Axon reflexes can operate without axonal arborization either between two varicosities of a single fiber (Figure 2B, E), or by ephaptic transmission (coupling) between two C-afferent fibers at the periphery (Figure 2D).

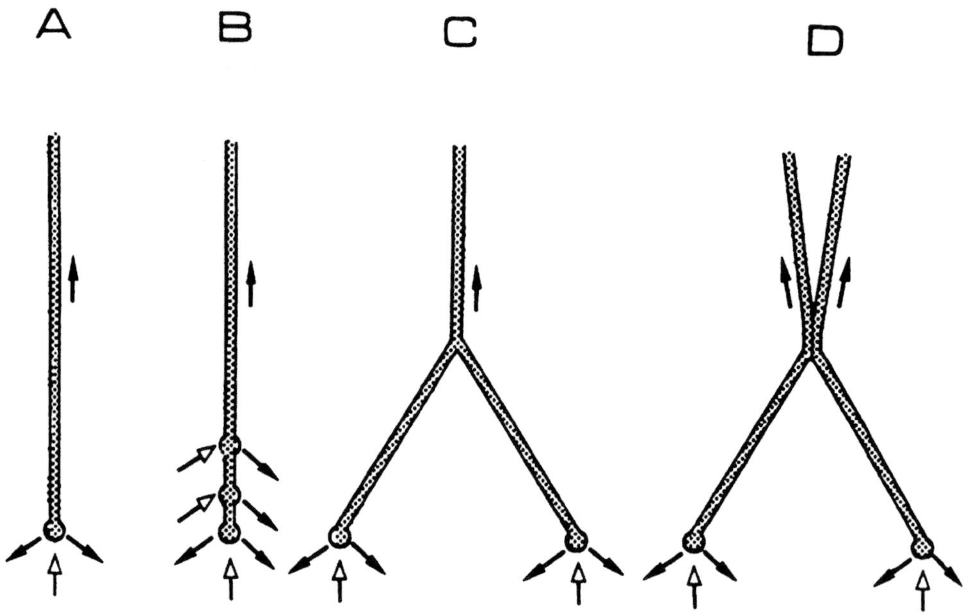

FIGURE 2. Schematic representation of various arrangements of axon reflexes according to the revised theory. (A) Nerve terminal with dual sensory-efferent function; white arrow, stimulus; black arrow, the evoked response. (B) Terminal axon with multiple sensory-effector varicosities forms the structural basis for axon reflex without axonal arborization. (C) Axonal arborization of a single fiber with two sensory-effector terminals. (D) Coupling between two fibers. (Reproduced from Reference 25. With permission.) (E) Multiple sensory-effectors of an arborizing axon: stimuli to site 2 or 3 can evoke vascular efferent responses without an axon reflex. Axon reflex is needed when the stimulus excites 1 type intraepithelial endings. NTS, nerve terminal spike; TTX, tetrodotoxin; SP, substance P; CGRP, calcitonin gene-related peptide (for more details see text).

According to this revised theory the neuroregulatory substrate of neurogenic inflammation and related phenomena resides *in sensu stricto* not in the axon reflex, but in the sensory-efferent nature of the operating neural elements. This neural mechanism seems to form a substantial part of the peripheral nervous system with the large number of capsaicin-sensitive sensory neurons suitable to evoke various local tissue responses throughout the body.[19-24] In the skin, where quantitative data of single unit studies with capsaicin and microneurostimulation are available, the following approximate calculation can be made. C-polymodal nociceptors comprise 60 to 70% of the unmyelinated afferents in various cutaneous nerves (see Reference 32). By close arterial injection in the rat and rabbit all C-polymodal nociceptors, with few exceptions, are activated by capsaicin.[58,59] Antidromic microneurostimulation of C-polymodal nociceptors of the rat saphenous nerve elicited in 66% of the units visible Evans blue accumulation at the receptive field.[31] Other types of unmyelinated or myelinated units gave no detectable signs of plasma extravasation.[31,40] Some A-delta fibers also appear to contribute to vasodilatation[60] and are sensitive to capsaicin.[59] In cutaneous nerves and spinal dorsal roots of mammals approximately 75% of the axons are unmyelinated.[61] Taking into account all these figures it turns out that at least 30% of the cutaneous afferents belong to that portion of the sensory-efferent neural system which mediates the neurogenic inflammation.

The revised axon reflex theory provides an explanation for some recent contradictory results in respect to the mode of action of capsaicin on the respiratory airways of the guinea pig.[37,62-66] Bronchoconstriction evoked by capsaicin-type agents is resistant to tetrodotoxin or omega conotoxin in isolated tissue preparations. Hence an axon reflex is not needed for the effector response under conditions where the released mediator could directly gain access to the target cells (smooth muscle, mast cell).[37,66] Under *in vivo* conditions bronchoconstriction evoked by capsaicin-type agents is only partially resistant to tetrodotoxin.[62,63] This state of

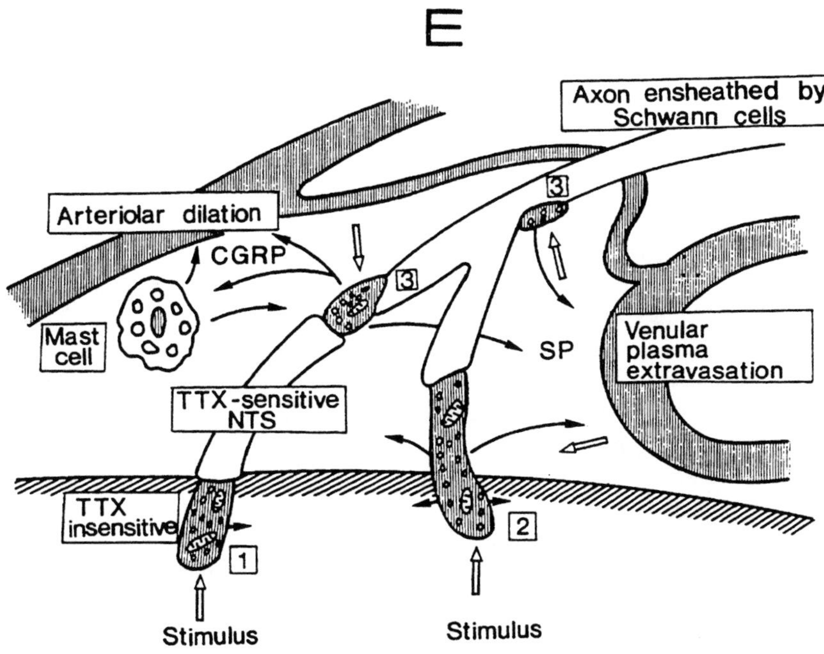

FIGURE 2. (Continued).

affairs indicates that sensory-effector sites preferentially reached from the pulmonary circulation served in respect to the evoked bronchoconstrictor response both as mediator releasing sites and as spike generators. In the latter case arborizations, terminals, or axonal varicosities of the fibers adjacent to the smooth muscle cells were activated by axonal conduction before capsaicin in the interstitial space could reach them by diffusion in a concentration sufficient for a direct activation. In an *in vitro* perfused lung preparation diffusion of capsaicin to these sites is probably hindered by an expanded interstitial space, causing in this way a more favorable experimental condition for provoking bronchoconstriction through axon reflexes.[64,65]

Part of the species differences in respect to the development of neurogenic inflammation and bronchoconstriction evoked by capsaicin in nonrodent mammals[67] might be attributed to similar differences in conditions favorable or not for diffusion of the released mediator from the sensory-effectors to the target cells. In this context it is interesting to mention a case report which shows that inhalation of capsaicin could produce in humans severe "pulmonary edema masquerading as croup" but disappearing within a day.[68]

VI. MAST CELLS, AXON REFLEX, AND NEUROGENIC INFLAMMATION

Antidromic stimulation of sensory nerves induced degranulation of the mast cells in the trachea and bronchi,[30] but not in the skin of the rat during the early stage of inflammation.[69,70] In the human skin substance P releases histamine, but the flare induced by capsaicin is not affected by antihistaminics.[16] There is no evidence that antidromic stimulation of dorsal roots or sensory nerves might induce discharge of histamine or other stored mediators from the mast cells in sufficient quantities for activation of sensory receptors and to induce in this way a spread of flare reaction. Mast cells seem to be involved in neurogenic inflammation as target cells for the released sensory neuropeptide mediators (Figure 2E), but their role as an essential step in a cascade of an axon reflex or "axon response" arrangement[9,15,16,20] has not yet been substantiated.[24,69,70] In fact, as mentioned earlier, the supposed mismatch between the size of the receptive field of C-polymodal nociceptors and the size of flare reaction in a given skin area is not supported by recent findings.[24,44]

VII. CONCLUSIONS

Antidromic stimulation of dorsal roots evokes not only vasodilatation but all characteristic responses of inflammation in the skin and in various internal organs. Electrical stimulation of capsaicin-insensitive afferents or sympathetic efferent fibers elicits no plasma extravasation under physiological conditions.[25] It is suggested that the capsaicin-sensitive primary afferents form a new type of neural system characterized by sensory-effector nerve terminals suitable for the dual function of impulse generation to the central nervous system and of release of neuropeptide mediators (tachykinins and CGRP) which elicit inflammation and other efferent tissue responses. These sensory-effector functional units form multiple sites as varicosities in the terminal arborization of a single afferent fiber. Neurogenic inflammation can be evoked without the intervention of an axon reflex. Axon reflex arrangements between varicosities of a single fiber, between terminals of axonal arborization, and through coupling between two afferent fibers are proposed. The powerful efferent function of nociceptive (e.g., C-polymodal nociceptive) fibers underlines their role in conditions in which stimulation is subliminal to evoke pain sensation, but is still sufficient to increase microcirculation in the skin.[17,32,33]

ACKNOWLEDGMENTS

This work was supported by research grants OTKA 016945 and ETT-735/1993. The author is indebted to his coworkers listed as coauthors in the references.

REFERENCES

1. Bruce, A. N., Über die Beziehungen der sensiblen Nervenendigungen zum Entzündungsvorgang, *Arch. Exp. Pathol. Pharmakol.*, 63, 424, 1910.
2. Chahl, L. A., Szolcsányi, J., and Lembeck, F., Eds., *Antidromic Vasodilatation and Neurogenic Inflammation*, Akadémiai Kiadó, Budapest, 1984.
3. Zweifach, B. W., Grant, L., and McCluskey, R. T., Eds., *The Inflammatory Process*, Vol. 1–3, Academic Press, New York, 1974.
4. Vane, J. R. and Ferreira, S. H., Eds., Part I, *Inflammation;* Part II, *Anti-Inflammatory Drugs, Handbook of Experimental Pharmacology*, Vol. 50, Springer-Verlag, Berlin, 1978.
5. Lewis, T., *The Blood Vessels of the Human Skin and Their Responses*, Shaw, London, 1927.
6. Bayliss, W. M., On the origin from the spinal cord of the vaso-dilator fibres of the hindlimb, and on the nature of these fibres, *J. Physiol.*, 26, 173, 1901.
7. Lewis, T., The nocifensor system of nerves and its reaction *Br. Med. J.*, 194, 431, 1937.
8. Lewis, T., The nocifensor system of nerves and its reaction, *Br. Med. J.*, 194, 491, 1937.
9. Lembeck, F., Sir Thomas Lewis's nocifensor system, histamine and substance-P-containing primary afferent nerves, *Trends Neuro. Sci.*, 6, 106, 1983.
10. Stricker, S., Untersuchungen über die Gefasswurzeln des Ischiadicus, *Sitzungsber. Kaiserl. Akad. Wiss. Wien.*, 3, 173, 1876.
11. Bayliss, W. M., The vaso-motor system, Longmans, Green and Co., London, 1923.
12. Szolcsányi, J., Capsaicin and neurogenic inflammation: history and early findings, in *Antidromic Vasodilatation and Neurogenic Inflammation*, Chahl, L. A., Szolcsányi, J., and Lembeck, F., Eds., Akadémiai Kiadó, Budapest, 1984, 7–26.
13. Lewis, T. and Grant, R. T., Vascular reactions of the skin to injury. II. The liberation of a histamine-like substance in injured skin; the underlying cause of factitious urticaria and of wheals produced by burning; and observations upon the nervous control of certain skin reactions, *Heart*, 11, 209, 1924.
14. Jancsó, N., Jancsó-Gábor, A., and Szolcsányi, J., Direct evidence for neurogenic inflammation and its prevention by denervation and by pretreatment with capsaicin, *Br. J. Pharmacol.*, 31, 138, 1967.

15. Holzer, P., Peptidergic sensory neurons in the control of vascular functions: mechanisms and significance in the cutaneous and splanchnic vascular beds, *Rev. Physiol. Biochem. Pharmacol.*, 121, 49, 1992.
16. Chahl, L., Antidromic vasodilatation and neurogenic inflammation, *Pharmacol. Ther.*, 37, 275, 1988.
17. Szolcsányi, J., Antidromic vasodilatation and neurogenic inflammation, *Agents Action*, 23, 4, 1988.
18. Pintér, E. and Szolcsányi, J., Plasma extravasation in the skin and pelvic organs evoked by antidromic stimulation of the lumbosacral dorsal roots of the rat, *Neuroscience*, 68, 603, 1995.
19. Szolcsányi, J., Capsaicin-sensitive chemoceptive neural system with dual sensory-efferent function, in Chahl, L. A., Szolcsányi, J., and Lembeck, F., Eds., *Antidromic Vasodilatation and Neurogenic Inflammation*, Akadémiai Kiadó, Budapest, 1984, 27–56.
20. Holzer, P., Local effector functions of capsaicin-sensitive sensory nerve endings: involvement of tachykinins and other neuropeptides, *Neuroscience*, 24, 739, 1988.
21. Maggi, C. A. and Meli, A., The sensory-efferent function of capsaicin-sensitive sensory neurons, *Gen. Pharmacol.*, 19, 1, 1988.
22. Barnes, P. J., Belvisi, M. G., and Rogers, D. F., Modulation of neurogenic inflammation: novel approaches to inflammatory disease, *Trends Pharmacol. Sci.*, 11, 185, 1990.
23. Lundberg, J. M., Capsaicin-sensitive sensory nerves in the airways — implications for protective reflexes and disease, in *Capsaicin in the Study of Pain*, Wood, J., Ed., Academic Press, London, 1993, 219.
24. Szolcsányi, J., Perspectives of capsaicin-type agents in pain therapy and research, in *Contemporary Issues in Pain Management*, Parris, W. C. V., Ed., Kluwer Academic, Norwell, 1991, 97.
25. Szolcsányi, J., Pintér, E., and Pethö, G., Role of unmyelinated afferents in regulation of microcirculation and its chronic distortion after trauma and damage, in *Reflex Sympathetic Dystrophy Pathophysiological Mechanisms and Clinical Implications*, Janig, W. and Schmidt, R. F., Eds., VCH Verlaggesellschaft, Weinheim, 1992, 245.
26. Lisney, S. J. W. and Bharali, L. A. M., The axon reflex: an outdated idea or a valid hypothesis? *News Physiol. Sci.*, 4, 45, 1989.
27. Van Arman, C. G., Oedema and increased vascular permeability, in *Anti-Inflammatory Drugs, Handbook of Experimental Pharmacology*, Vol. 50, Part II, Vane, J. R. and Ferreira, S. H., Eds., Springer-Verlag, Berlin, 1978, 75.
28. McDonald, D. M., Neurogenic inflammation in the rat trachea. I. Changes in venules, leucocytes and epithelial cells, *J. Neurocytol.*, 17, 583, 1988.
29. McDonald, D. M., Mitchell, R. A., Gabella, G., and Haskell, A., Neurogenic inflammation. II. Identity and distribution of nerves mediating the increase in vascular permeability, *J. Neurocytol.*, 17, 605, 1988.
30. Kiernan, J. A., Degranulation of mast cells in the trachea, bronchi of the rat following stimulation of the vagus nerve, *Int. Arch. Allergy Appl. Immunol.*, 91, 398, 1990.
31. Bharali, L. A. M. and Lisney, S. J. W., The relationship between unmyelinated afferent type and neurogenic plasma extravasation in normal and reinnervated skin, *Neuroscience*, 47, 703, 1992.
32. Szolcsányi, J., Actions of capsaicin on sensory receptors, in *Capsaicin in the Study of Pain*, Wood, J., Ed., Academic Press, London, 1993, 1.
33. Szolcsányi, J., Capsaicin-sensitive chemoceptive B-afferents: a neural system with dual sensory-efferent function, *Behav. Brain Sci.*, 13, 316, 1990.
34. Treede, R.-D., Meyer, R. A., Davis, K. D., and Champbell, J. N., Intradermal injections of bradykinin or histamine cause a flare-like vasodilatation in monkey. Evidence from laser Doppler studies, *Neurosci. Lett.*, 115, 201, 1990.
35. Pierau, Fr.-K. and Szolcsányi, J., Neurogenic inflammation: axon reflex in pigs, *Agents Actions*, 26, 231, 1989.
36. Jancsó, N., Jancsó-Gábor, A., and Szolcsányi, J., The role of sensory nerve endings in neurogenic inflammation induced in human skin and in the eye and paw of the rat, *Br. J. Pharmacol.*, 33, 32, 1968.
37. Szolcsányi, J., Tetrodotoxin-resistant non-cholinergic neurogenic contraction evoked by capsaicinoids and piperine on the guinea-pig trachea, *Neurosci. Lett.*, 42, 83, 1983.
38. Dymshitz, J. and Vasko, M. R., Endothelin-1 enhances capsaicin-induced peptide release and cGMP accumulation in cultures of rat sensory neurons, *Neurosci. Lett.*, 167, 128, 1994.
39. Pethö, G. and Szolcsányi, J., Analysis of the site of action of capsaicin on primary afferent neurones, *Neuropeptides*, 22, 51, 1992.
40. Kenins, P., Identification of the unmyelinated sensory nerves which evoke plasma extravasation in response to antidromic stimulation, *Neurosci. Lett.*, 25, 137, 1981.
41. Kenins, P., The functional anatomy of the receptive fields of rabbit C polymodal nociceptors, *J. Neurophysiol.*, 59, 1098, 1988.
42. Nordin, M., Low-threshold mechanoreceptive nociceptive units with unmyelinated (C) fibres in the human supraorbital nerve, *J. Physiol.*, 624, 229, 1990.
43. La Motte, R. H., Lundberg, L. E. R., and Torebjörk, H. E., Pain, hyperalgesia and activity in nociceptive C units in humans after intradermal injection of capsaicin, *J. Physiol.*, 48, 749, 1992.
44. Wardell, K., Naver, H. K., Nilsson, G. E., and Wallin, B. G., The cutaneous vascular axon reflex in humans characterized by laser Doppler perfusion imaging, *J. Physiol.*, 460, 185, 1993.

45. Szolcsányi, J., Jancsó-Gábor, A., and Joó, F., Functional and fine structural characteristics of the sensory neuron blocking effects of capsaicin, *Naunyn-Schmiedebergs Arch. Pharmakol. Exp. Pathol.*, 287, 157, 1975.
46. Rózsa, A. J. and Beuerman, R. W., Density and organization of free nerve endings in the corneal epithelium of the rabbit, *Pain*, 14, 105, 1982.
47. Tanelian, D. L. and MacIver, M. B., Simultaneous visualization and electrophysiology of corneal A-delta and C fiber afferents, *J. Neurosci. Meth.*, 32, 213, 1990.
48. Magerl, W., Szolcsányi, J., Westerman, R. A., and Handwerker, H. O., Laser Doppler measurements of skin vasodilation elicited by percutaneous electrical stimulation of nociceptors in human skin, *Neurosci. Lett.*, 82, 349, 1987.
49. Lynn, B. and Cotsell, B., The delay in onset of vasodilator flare in human skin at increasing distances from a localized noxious stimulus, *Microvasc. Res.*, 41, 197, 1991.
50. Blumberg, H. and Wallin, G., Direct evidence of neurally mediated vasodilation in hairy skin of the human foot, *J. Physiol.*, 382, 105, 1987.
51. Barthó, L., Ernst, R., Pierau, Fr.-K., Sann, H., Faulstroh, K., and Pethö, G., An opioid peptide inhibits capsaicin-sensitive vasodilation in the pig's skin, *Neuropeptides*, 23, 227, 1992.
52. Brokaw, J. J. and McDonald, D. M., Neurally mediated increase in vascular permeability in the rat trachea: onset, duration, and tachyphylaxis, *Exp. Lung Res.*, 14, 757, 1988.
53. Kruger, L., Morphological features of thin sensory afferent fibers: a new interpretation of "nociceptor" function, in *Progress in Brain Research*, Vol. 74, Hamann, W. and Iggo, A., Eds., Elsevier, Amsterdam, 1988, 253.
54. Baluk, P., Nadel, J. A., and McDonald, D. M., Substance P-immunoreactive sensory axons in the rat respiratory tract: a quantitative study of their distribution and role in neurogenic inflammation, *J. Comp. Neurol.*, 319, 586, 1992.
55. Gonon, F., Msghina, M., and Stjarne, L., Kinetics of noradrenaline released by sympathetic nerves, *Neuroscience*, 56, 535, 1993.
56. Meehan, A. G., Hottenstein, O. D., and Kreulen, D. L., Capsaicin-sensitive nerves mediate inhibitory junction potentials and dilatation in guinea-pig mesenteric artery, *J. Physiol.*, 443, 161, 1991.
57. Szolcsányi, J., Capsaicin type pungent agents producing pyrexia, in *Handbook of Experimental Pharmacology*, Vol. 60, Milton, A. S., Ed., Springer, Berlin, 1982, 437.
58. Szolcsányi, J., Selective responsiveness of polymodal nociceptors of the rabbit ear to capsaicin, bradykinin and ultra violet irradiation, *J. Physiol.*, 388, 9, 1987.
59. Szolcsányi, J., Anton, F., Reeh, P. W., and Handwerker, H. O., Selective excitation by capsaicin of mechano-heat sensitive nociceptors in rat skin, *Brain. Res.*, 446, 262, 1988.
60. Jänig, W. and Lisney, S. J. W., Small diameter myelinated afferents produce vasodilatation but not plasma extravasation in rat skin, *J. Physiol.*, 415, 477, 1989.
61. Berthold, C.-H., Morphology of normal peripheral axons, in *Physiology and Pathobiology of Axons*, Waxman, S. G., Ed., Raven Press, New York, 1987, 3.
62. Lai, Y.-L., Role of the axon reflex in capsaicin-induced bronchoconstriction in guinea pigs, *Respir. Physiol.*, 83, 35, 1991.
63. Zhang, H.-Q. and Lai, Y.-L., Axon reflex in resiniferatoxin-induced bronchoconstriction in guinea pigs, *Respir. Physiol.*, 92, 13, 1993.
64. Kröll, F., Karlsson, J.-A., Lundberg, J. M., and Persson, C. G. A., Capsaicin-induced bronchoconstriction and neuropeptide release in guinea pig perfused lung, *J. Appl. Physiol.*, 68, 1679, 1990.
65. Lou, Y. P., Franco-Cereceda, A., and Lundberg, J. M., Omega-conotoxin inhibits CGRP release and bronchoconstriction by a low concentrations of capsaicin, *Acta Physiol. Scand.*, 141, 135, 1991.
66. Maggi, C. A., The pharmacology of the efferent function of sensory nerves, *J. Auton. Pharmacol.*, 11, 173, 1991.
67. Manzini, S., Goso, C., and Szállási, Á., Sensory nerves and tachykinins, in *Neuropeptides in Respiratory Medicine*, Kaliner, M. A., Barnes, P. J., Kunkel, G. H., and Baraniuk, J. N., Eds., Marcel Dekker, New York, 1994, 173.
68. Winograd, H. L., Acute croup in an older child, *Clin. Pediatr.*, 16, 884, 1977.
69. Kowalski, M. L. and Kaliner, M. A., Neurogenic inflammation, vascular permeability, and mast cells, *J. Immunol.*, 140, 3905, 1988.
70. Baraniuk, J. N., Kowalski, M. L., and Kaliner, M. A., Relationship between permeable vessels, nerves, and mast cells in rat cutaneous neurogenic inflammation, *J. Appl. Physiol.*, 68, 2305, 1990.

Chapter 4

THE VANILLOID RECEPTOR

Arpad Szallasi

CONTENTS

I. INTRODUCTION

A. CAPSAICIN RECEPTOR: ARGUMENTS FOR AND AGAINST

Primary sensory neurons are activated by a great variety of noxious chemical stimuli.[1-3] Upon activation, these nerves release proinflammatory mediators which, in turn, initiate the cascade of inflammatory events collectively referred to as neurogenic inflammation.[1-3] Among the activators of primary sensory neurons, capsaicin (Figure 1), the pungent principle in hot peppers, is unique in that it renders the excited nerves insensitive to further stimuli.[4,5] This peculiar dual response (initial excitation followed by long-lasting desensitization) to capsaicin is so characteristic of these neurons that they are often called "capsaicin-sensitive neurons".[1-5]

The neural specificity of capsaicin actions, along with the fairly strict structural requirements for capsaicin-like activity,[6] and the striking species-related differences[2,7] (only mammals respond to capsaicin and even some mammalian species, such as the rabbit or the

0-8493-7646-7/96/$0.00+$.50

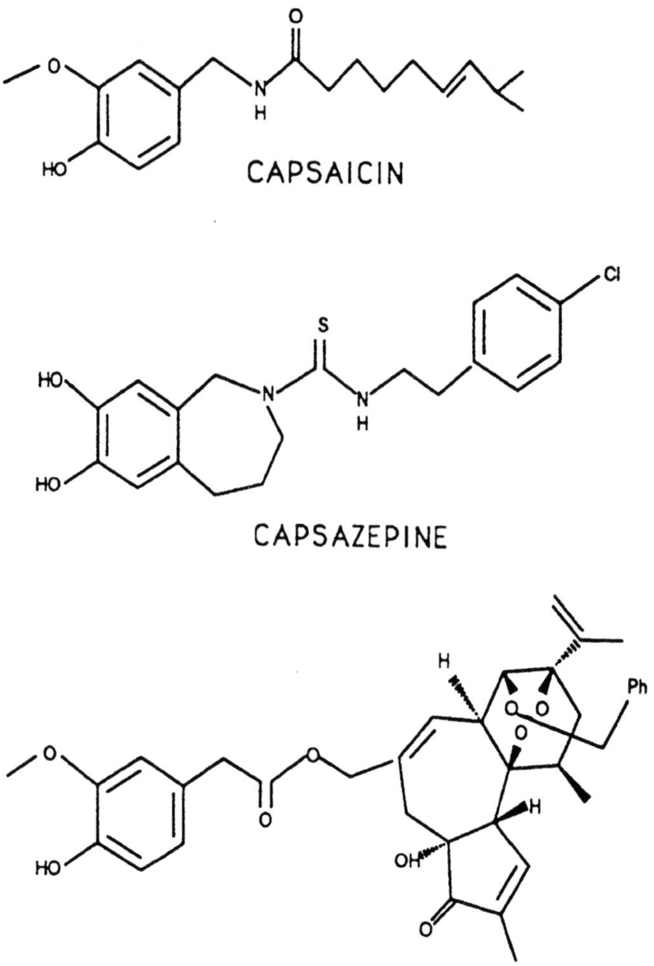

FIGURE 1. Structures of capsaicin, capsazepine, and resiniferatoxin.

hamster, show marginal sensitivity) strongly suggested that capsaicin interacts at a specific recognition site (receptor) expressed predominantly, if not exclusively, by a well-defined subdivision of sensory nerves in some mammals, including man.[8] When high doses of capsaicin are used, the tissue- and species-selectivity of capsaicin actions is, however, no longer observed.[2,7] This observation, along with the demonstrated ability of capsaicin to influence membrane fluidity, to inhibit a variety of enzymes, and to regulate a number of channels, until recently has clouded the arguments for the existence of the putative capsaicin receptor.[2]

B. FAILURE TO IDENTIFY THE CAPSAICIN RECEPTOR

Given the therapeutic potential of capsaicin to mitigate pain[9] and to improve disease states in which neurogenic inflammation plays a major role,[1,10] the identification of the capsaicin receptor is of utmost practical importance. Several attempts were made, therefore, to demonstrate specific binding of radiolabeled dihydrocapsaicin[11] or capsaicin-like photoaffinity probes.[12] A combination of relatively low affinity and high lipophilicity of capsaicin, however, destined these efforts to fail.

II. RESINIFERATOXIN: AN ULTRAPOTENT CAPSAICIN ANALOG (VANILLOID) WITH A UNIQUE SPECTRUM OF ACTIONS

The search for a "capsaicin receptor" has been given a major impetus by our recent discovery that resiniferatoxin (RTX) (Figure 1), a phorbol-related diterpene, mimics most of the capsaicin actions, such as hypothermia and induction of neurogenic inflammation, with several thousandfold higher potency than capsaicin itself.[13] As RTX structural analogs and capsaicinoids differ dramatically in the rest of the molecule but share a homovanillyl group as a structural motif essential for biological activity (Figure 1), these bioactive compounds now appear to be best termed vanilloids.[13] In keeping with this terminology, the receptor at which vanilloids interact may be referred to as the vanilloid receptor.

The peculiar spectrum of actions of RTX relative to capsaicin has been reviewed extensively,[13-15] as has its therapeutic potential.[10] In general, RTX and capsaicin induce a similar pattern of responses; however, they differ strikingly in their relative potencies for different responses.[13,14] For example, whereas for most of the responses characteristic of capsaicin, RTX is 100- to 10,000-fold more potent, it is only equal in potency to capsaicin to cause acute ocular pain.[13] The basis for these striking differences in relative potencies remains to be established; the most likely explanation is, however, a combination of pharmacokinetic differences and receptor heterogeneity. Importantly, at least in the rat, RTX also has unique actions, such as the desensitization without prior excitation of pulmonary chemoreceptors.[16] Such differences between RTX and capsaicin actions may play a pivotal role in the much broader "therapeutic range" of RTX.[10]

Since the ultrapotency of RTX implies that this compound occupies vanilloid receptors at much lower concentrations than does capsaicin, which, in turn, means a more favorable ratio of nonspecific binding at the K_d, the demonstration of the long-sought "capsaicin receptor" by using radiolabeled (tritiated) RTX seemed feasible. In fact, in 1990 we were the first to demonstrate high-affinity, saturable specific binding sites for RTX shared by capsaicin.[17] This initial vanilloid receptor assay, along with its limitations, and the discovery of a vanilloid-binding plasma protein which provided the methodological means to improve the vanilloid receptor assay (reduced nonspecific binding), have been detailed elsewhere.[14,15] This chapter focuses on our current understanding of the pharmacology of vanilloid receptors in the rat, the species in which the pattern of capsaicin- and RTX-induced biological responses is best explored, and on the emerging differences between rat and human vanilloid receptors.

III. SPECIFIC BINDING OF RESINIFERATOXIN

Generally speaking, specific binding of RTX represents the vanilloid receptor.[15] It should be noted, however, that RTX, although with much lower affinity, also interacts at the phorbol ester receptor, protein kinase C (PKC),[14] binds to the well-known drug-binding domain of α_1-acid glycoprotein (AGP),[14,15] and is recognized by a cytoplasmic, non-tissue-specific target.[18] Whereas the several thousandfold difference between the affinity of RTX to vanilloid receptors (for example in rat sensory ganglion membranes, K_d = 20 p*M*) and to partially purified PKC (K_i = 400 n*M*) makes the biological relevance of this latter interaction extremely unlikely,[14] RTX binding to AGP may have important pharmacokinetic consequences.[14,15]

A. PHARMACOLOGICAL SPECIFICITY

Specific binding of RTX to tissues representing the cell bodies, central as well as peripheral terminals of capsaicin-sensitive neurons, is fully displaced by capsaicin but is not inhibited at all by biologically inactive capsaicin and RTX congeners;[15] moreover, relative binding affinities of vanilloids are consistent with their relative *in vivo* potencies.[13-15]

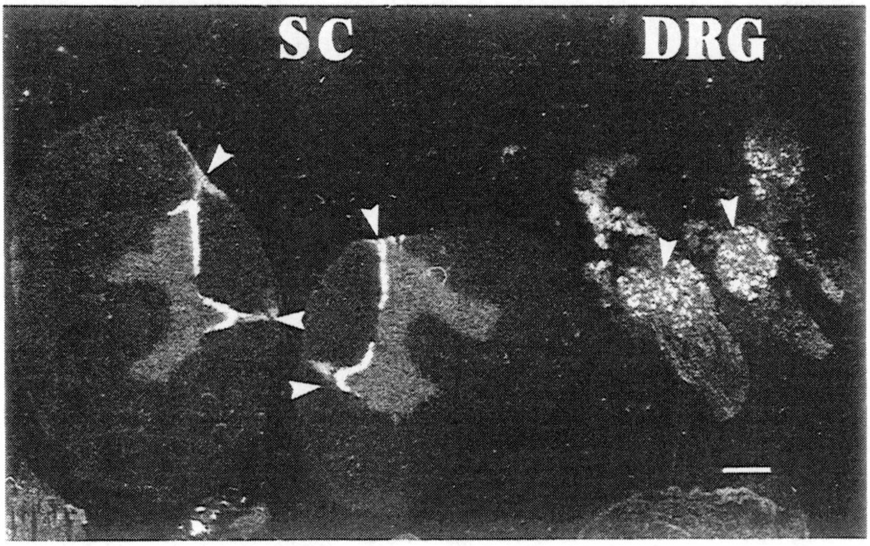

FIGURE 2. Autoradiograph of a cryostat section of two pig spinal cords (SC) and a dorsal root ganglion (DRG) that has been incubated with 1 n*M* [^{3}H]resiniferatoxin. Arrows indicate high-density labeling over the known termination area of vanilloid-sensitive nerves in the superficial dorsal horn (marginal zone and substantia gelatinosa); in the DRG, the high-density labeling is likely to represent the perikarya of vanilloid-sensitive neurons. Marker indicates 1 mm. (From Szallasi, A. et al., *Eur. J. Pharmacol.*, 264, 217, 1994. With permission.)

Capsazepine (Figure 1), a compound shown to inhibit vanilloid actions with Schild plots consistent with a competitive mechanism,[19] is also a competitive inhibitor of RTX binding.[15] By contrast, the so-called "functional antagonist" ruthenium red[20] does not compete for specific RTX binding sites.[17]

B. TISSUE SPECIFICITY AND LOCALIZATION

Specific RTX binding can be detected in tissues representing the perikarya (dorsal root ganglia and trigeminal ganglia), the central (dorsal horn of the spinal cord), as well as the peripheral (urinary tract: bladder and urethra; airways: nasal mucosa, windpipe, bronchi; colon) terminals of primary sensory neurons, but not in tissues, such as cerebellum, liver, skeletal or heart muscle, which do not show responses to capsaicin.[15] Autoradiographic mapping revealed high densities of [^{3}H]RTX binding sites exclusively over areas (Rexed laminae I and II) in the dorsal horn of the spinal cord known to be rich in central terminals of capsaicin-sensitive neurons[21] (Figure 2).

C. SPECIES SPECIFICITY

It is generally accepted that only mammalian species respond to capsaicin.[2,7] In accord, no specific RTX binding could be detected in the sensory ganglia of reptiles (alligators) or birds (chickens).[15] Only a limited correlation was found between the capsaicin sensitivity of mammals and the level of RTX binding sites they express (see below), implying that the presence of vanilloid receptors is a prerequisite, but their actual density is not the sole determinant, of *in vivo* capsaicin sensitivity.

D. ABLATION OF CAPSAICIN-SENSITIVE NEURONS IS ACCOMPANIED BY A LOSS OF RESINIFERATOXIN BINDING SITES

Whereas it is debated whether vanilloids may cause gross neurotoxicity via specific, vanilloid receptor-mediated mechanisms in adult animals, it is clear that neonatal vanilloid

TABLE 1
Vanilloid Receptors in the Rat

Tissue	Resiniferatoxin (K_d, p*M*)	Capsaicin (K_i, μ*M*)	Capsazepine (K_i, μ*M*)
Positive Cooperative Binding			
Dorsal root ganglia	18–24	0.6–2.0	3.5
Trigeminal ganglia	20	1.4	4.0
Spinal cord	13–16	0.3	4.0
Noncooperative Binding			
Urinary bladder (?)	30–87	0.5	5.0
Urethra	105	N.D.	N.D.
Airways	250	0.1	0.1
Colon	3000	3.0	0.1

Note: N.D., not determined; ?, although we reported noncooperative binding in the urinary bladder (Szallasi, A. et al., *Life Sci.*, 52, PL221, 1993), others, in contrast, observed positive binding cooperative (Acs et al., *Life Sci.*, 55, 1017, 1994).

treatment ablates capsaicin-sensitive nerves with remarkable selectivity.[7] Following neonatal RTX or capsaicin administration, as expected, a dramatic, though incomplete, loss of specific RTX binding sites was observed in sensory ganglia, spinal cord, and airways of the rat.[15] Whether the remaining RTX binding sites simply reflect submaximal dosage or indicate neural heterogeneity (some neurons are less sensitive to toxicity than others) is yet to be determined.

When RTX was given to adult rats, a dose-dependent loss of vanilloid receptors was observed.[15] This receptor loss occurred later than the loss of the biological responses, but the recovery of RTX binding, at least in the urinary bladder, accompanied the recovery of the neurogenic inflammatory response.[15] It is likely, therefore, that a ligand-induced receptor loss may contribute to the long-term desensitization to vanilloids.

IV. THE VANILLOID RECEPTOR IN THE RAT

A. RECEPTOR TYPES

Given the peculiar pattern of RTX actions in the rat,[13,14] the emerging concept of a heterologous vanilloid receptor comprising several types and even subtypes is hardly unexpected (Table 1). At present, the criteria to typify vanilloid receptor types and subtypes are ill defined; nevertheless, the existence of at least two basic receptor types can be postulated:[15]

1. A "cooperative type", such as found in sensory ganglia and spinal cord, which binds RTX with high (approximately 20 p*M*) affinity in a positive cooperative fashion (with a cooperativity index close to 2)
2. A "noncooperative type" which binds RTX with 5- (urinary tract) to 150-fold (colon) lower affinity in a noncooperative manner, and is usually detected in peripheral tissues

Given the contrasting findings as to the cooperativity of RTX binding in the urinary bladder,[15] this terminology appears to be less controversial than the previously suggested "central" and "peripheral" types.[15] It may be speculated that the different affinities for RTX (for example, 250 p*M* in the airways and 3000 p*M* in the colon) as well as the contrasting affinity for capsaicin relative to capsazepine (for example, capsaicin is approximately 10-fold more potent than capsazepine in the bladder, 30-fold less potent in the colon, and equipotent to capsazepine in the airways) dissect further vanilloid receptor types/subtypes.[15]

TABLE 2
Inhibitors and Modulators of Resiniferatoxin Binding to the Vanilloid Receptor in the Rat

1. Competitive binding inhibitors
   Agonists: resiniferanoids, capsaicinoids
   Antagonist: capsazepine
2. Noncompetitive or mixed binding inhibitors
   Low pH (protons)
   Heavy metal cations (e.g., nickel, cobalt, cadmium)
   Sulfhydryl-reactive agents (e.g., *N*-ethylmaleimide, *p*-chloromercuribenzoate)
3. Modulators
   Buffer composition (e.g., concentration of calcium)
   Temperature
   Reducing/oxidizing agents (e.g., reduced glutathione, dithiothreitol)

B. STRUCTURE-AFFINITY RELATIONS

In 1975, Szolcsányi and Jancsó-Gábor proposed a model to describe the interactions between capsaicin and its receptor.[6] This model, based on the analysis of structure–activity relations of capsaicinoids, is clearly inadequate to explain the ultrapotency of RTX. Although data furnished so far by the structure–activity analysis of resiniferanoids are insufficient to create an improved vanilloid receptor model, it is clear that both the homovanillyl moiety and the orthopenylacetate group (Figure 1) are essential for the ultrapotency of RTX, as is a precise steric relationship between these two chemical features. In fact, all the chemical modifications of RTX resulted in a loss of activity which ranged from severalfold (e.g., tinyatoxin, which differs from RTX only in that it is lacking the methoxy group of the homovanillyl ester) through several thousandfold (e.g., replacement of the orthophenylacetate moiety by a 13-phenylacetate substituent), to a dramatic, complete loss of capsaicin-like activity (e.g., resiniferonol 9,13,14-orthophenylacetate, the C20 deesterified parent diterpene of RTX).[13-15] Interestingly, capsaicinoids and resiniferanoids appear to display somewhat different structure–activity relations. A dramatic example is the change of the linkage of the homovanillyl moiety from ester to amide which increases the potency of capsaicin congeners,[6] but diminishes that of the resiniferanoids.[22] Although the role of the homovanillyl moiety in receptor binding is firmly established, it remains to be determined whether the orthophenylacetate group actively participates in receptor binding or just helps to anchor the rest of the molecule in the lipid bilayer. Some of these RTX congeners display an intriguing pattern of biological activity. For example, 12-deoxyphorbol 13-phenylacetate 20-homovanillate is comparable in potency to capsaicin for provoking acute ocular pain and desensitizing the neurogenic inflammatory pathway, but is totally inactive for inducing hypothermia.[13,14] This finding may be interpreted as further indication of vanilloid receptor heterogeneity.

C. MODULATORS OF RESINIFERATOXIN BINDING

Several agents are known to inhibit RTX binding by a noncompetitive mechanism[23,24] (Table 2). These agents are of interest because they also activate (e.g., protons) or, alternatively, block (e.g., *N*-ethylmaleimide, heavy metal cations) vanilloid-sensitive nerves[1-3] and thus a link between their interaction at vanilloid receptors and their action on vanilloid-sensitive nerves can be postulated. The mechanism(s) by which these agents inhibit RTX binding is (are) unknown. Inhibition of RTX binding by, for example, protons and reducing agents (e.g., reduced glutathione), or by protons and heavy metal cations appear to be additive, indicating separate targets, as is the inhibition of RTX binding by the above agents and vanilloids.[25]

V. SPECIES-RELATED DIFFERENCES IN RECEPTOR BINDING

The guinea pig, rat, and hamster represent species highly sensitive, sensitive, and marginally sensitive, respectively, to capsaicin.[2,7] A simple mechanism to explain the contrasting susceptibilities might have been a species-related difference in vanilloid receptor expression. This hypothesis is, however, not supported by the available findings. Guinea pig tissues, in which the *in vivo* susceptibility might imply the existence of high-affinity/high-density vanilloid receptors, in fact, bind RTX with relatively low affinity and the density of RTX binding sites in guinea pig urinary bladder and airways is also lower than in the rat.[15] It is very likely, therefore, that the higher *in vivo* vanilloid sensitivity of the guinea pig reflects differences in signal transduction and/or effector mechanisms (e.g., guinea pig nerves are known to contain more substance P-like immunoreactivity than rat nerves[26]), rather than differences in vanilloid receptor expression. The lack of detectable specific RTX binding sites in the peripheral tissues (urinary bladder as well as airways) of the hamster[15] appears to be in accord with the marginal sensitivity of this species to vanilloids. On the other hand, this finding, though it is in accord with the biology, is surprising, since hamster and rat dorsal root ganglion (DRG) neurons express specific RTX binding sites in comparable densities.[15] It is likely, therefore, that for some obscure reason hamster DRG neurons either do not transport the vanilloid receptors they synthesize to the periphery or, alternatively, the receptors are modified during the transport and thus lose their ability to bind vanilloids.

In this context it may be of relevance that RTX was shown to contract the iris sphincter muscle of the rabbit, another species marginally susceptible to vanilloids,[7] with a biphasic dose-response curve displaying EC_{50} values of a 10,000-fold difference.[27] It is not infeasible that the second phase of the dose-response curve represents a second, low (submicromolar to micromolar) affinity vanilloid receptor, undetected by the present binding methodology, which may be very closely related to the high-affinity receptor. In other words, RTX (or capsaicin) is not necessarily a suitable tool to detect vanilloid receptors if this receptor, like the tachykinin receptors[28] or the excitatory amino acid receptors,[29] shows substantial heterogeneity in ligand binding properties. This assumption implies that the expression of vanilloid receptors may be, in fact, more ubiquitous than predicted by capsaicin sensitivity.

VI. HUMAN VANILLOID RECEPTORS

Although the pharmacology of vanilloid-sensitive nerves is less studied in human beings than in animals, by and large capsaicin appears to have a similar pattern of actions in humans and rodents.[8]

A. POST-MORTEM NEURAL TISSUES

Methodologically, the demonstration of specific RTX binding sites in human neural tissues (in the dorsal horn of the spinal cord,[30,31] in the cuneate and gracile nuclei, and in the spinal nucleus of the trigeminal nerve[31]) dissected post-mortem seems to be an important advance, since such tissues may be readily obtained; nevertheless, the *in vivo* relevance of vanilloid receptor assays utilizing post-mortem tissues is yet to be confirmed. In contrast to human bronchi specimens removed during operation (see below), in human airways obtained post-mortem no, or very marginal, specific RTX binding could be detected;[32] these findings imply that, probably due to autolysis, vanilloid receptors are quickly destroyed in peripheral tissues. Vanilloid receptors in the central nervous system appear to be better protected: rat spinal cord preparations obtained freshly or 24 h post-mortem bind RTX with almost identical parameters.[33]

There are four major differences between the binding properties of human and rat spinal vanilloid receptors:[30,31] (1) both the association and dissociation of [^{3}H]RTX is significantly faster in human spinal cord; (2) capsazepine is recognized by human vanilloid receptors with an approximately 70-fold higher affinity; (3) human vanilloid receptors bind RTX with a 40- to 500-fold lower affinity; and (4) the lower affinity of RTX binding in humans is associated with receptor densities 3- to 15-fold higher than in the rat. With regard to binding cooperativity, we have obtained contrasting results (cooperativity indices in the range of 1.0 to 1.7) using spinal cord specimens obtained from different morgues.[30,34] The basis for this variance is unknown; possibilities include the time elapsed between clinical death and sample removal, the age of the deceased, as well as the diseases from which they suffered. Peter M. Blumberg's laboratory[31] reported cooperative RTX binding in all three major central endings of primary afferent neurons using tissues obtained within 2 h after death from victims of fatal traffic accidents.

A. *EX VIVO* BRONCHI

In human and guinea pig bronchi comparable levels of specific RTX binding could be detected.[32] This result is surprising since (1) human bronchi show no, or minimal, constriction in response to capsaicin,[35] and (2) the density of substance P-like immunoreactivity (SP-LI; generally considered to be a marker of vanilloid-sensitive nerves[7]) is much lower in human than in guinea pig airways.[36] Both capsaicin and RTX aerosols, however, provoke severe coughing in human volunteers as well as in guinea pig experiments.[37] These findings imply (1) that, as compared to mediating protective reflexes, the vanilloid-sensitive innervation of human airways plays a minor role in regulating the contractile tone and (2) that SP-LI may not be a suitable marker to identify vanilloid-sensitive nerves in human airways.

VII. CONCLUDING REMARKS

A. THE VANILLOID RECEPTOR: THERE BUT WHY?

The vanilloid receptor is thought to be a novel, nonselective cation channel or at least to be closely coupled to one.[38] Its radiation inactivation size[39] (270 kDa in porcine sensory ganglia and 290 kDa in spinal cord) is predictive of a receptor complex; interactions among the members of this receptor complex represent a possible mechanism to explain positive binding cooperativity.[15] If this assumption holds true, noncooperative binding, such as observed in peripheral tissues,[15] may indicate the existence of monomeric receptor form(s). Binding experiments suggest that the vanilloid-binding domain is regulated by a redox site[15] and a protonation site,[24] and that sulfhydryl groups may play an important role in ligand binding.[23] The physiological function(s) of this receptor is (are) unknown. Postulating the existence of endogenous vanilloid(s), it is easy to visualize that by maintaining a tonic release of sensory neuropeptides the vanilloid receptor may mediate a variety of tonic, vaso-, and immunoregulatory functions, whereas by a mass release of the same neuropeptides endogenous vanilloids produced under pathobiological conditions may initiate a number of defense mechanisms, such as neurogenic inflammation, the core issue of this book.

B. DO ENDOGENOUS VANILLOIDS EXIST?

Although considerable indirect evidence implies the existence of endogenous vanilloids, as yet no direct proof is available. In general, all endogenous substances that activate vanilloid-sensitive nerves are suspect endogenous vanilloids. Most of these substances (e.g., prostanoids, lipoxins, etc.) do not interact with RTX binding,[15] thus they are very likely to act on primary targets other than the vanilloid receptor. Others, such as protons,[24] turned out to be noncompetitive inhibitors of RTX binding; nonetheless, the relation between biological actions of these agents and their noncompetitive interaction at vanilloid receptors is yet to be determined. Positive cooperativity of binding is thought to serve as an amplification mechanism to enhance the activity of endogenous ligands present at low concentrations.[40] This hypothesis predicts the existence of endogenous vanilloids in tissues, such as the spinal cord, where the vanilloid receptor displays a clearly cooperative binding behavior.[15]

C. THE VANILLOID RECEPTOR: A LONE RECEPTOR OR MEMBER OF A RECEPTOR SUPERFAMILY ?

Although the vanilloid receptor in its ligand binding properties or electrophysiological features does not seem to resemble any other known receptor (it is often described as a "novel" cation channel[38]), there are intriguing parallels between the vanilloid receptor and receptors of similar molecular target sizes. For example, both the vanilloid (270 kDa[39]) and the nicotinic acetylcholine receptor (300 kDa[41]) show positive cooperative binding behavior,[15,42] are regulated by nerve growth factor,[43,44] and are blocked by ruthenium red.[45] Whether these similarities are purely coincidental or reflect intimate relations remains to be determined.

ACKNOWLEDGMENTS

I thank Drs. Peter M. Blumberg (National Cancer Institute, Bethesda, MD) and Jan M. Lundberg (Karolinska Institute, Stockholm, Sweden) for reading of the manuscript.

REFERENCES

1. Maggi, C. A. and Meli, A., The sensory-efferent function of capsaicin-sensitive neurons, *Gen. Pharmacol.*, 19, 1, 1988.
2. Holzer, P., Capsaicin: cellular targets, mechanisms of action, and selectivity for thin sensory neurons, *Pharmacol. Rev.*, 43, 144, 1991.
3. Maggi, C. A., The pharmacology of the efferent function of sensory nerves, *J. Auton. Pharmacol.*, 11, 173, 1991.
4. Jancsó, N., Desensitization with capsaicin and related acylamides as a tool for studying the function of pain receptors, in *Pharmacology of Pain*, Lin, K., Armstrong, D., and Pardo, E. G., Eds., Pergamon Press, Oxford, 1968, 33–55.
5. Szolcsányi, J., Capsaicin-sensitive chemoceptive neural system with dual sensory-efferent function, in *Antidromic Vasodilatation and Neurogenic Inflammation*, Chahl, L. A., Szolcsányi, J., and Lembeck, F., Eds., Akadémiai Kiadó, Budapest, 1984, 27–52.
6. Szolcsányi, J. and Jancsó-Gábor, A., Sensory effects of capsaicin congeners. I. Relationship between chemical structure and pain-producing potency, *Drug Res.*, 25, 1877, 1975.
7. Buck, S. H. and Burks, T. F., The neuropharmacology of capsaicin: review of some recent observations, *Pharmacol. Rev.*, 38, 179, 1986.
8. Fuller, R. W., The human pharmacology of capsaicin, *Arch. Int. Pharmacodyn.*, 303, 147, 1990.
9. Szolcsányi, J., Perspectives of capsaicin-type agents in pain therapy and research, in *Contemporary Issues in Chronic Pain Management*, Parris, W. C. V., Ed., Kluwer, Boston, 1991, 97–124.
10. Szallasi, A. and Blumberg, P. M., Mechanisms and therapeutic potential of vanilloids (capsaicin-like molecules), *Adv. Pharmacol.*, 24, 123, 1993.
11. Miller, M. S., Buck, S. H., Sipes, I. G., and Burks, T. F., Capsaicinoid-induced local and systemic antinociception without substance P depletion, *Brain Res.*, 244, 193, 1982.
12. James, I. F., Walpole, C. S. J., Hixon, J., Wood, J. N., and Wrigglesworth, R., Long-lasting activity by a capsaicin-like photoaffinity probe, *Mol. Pharmacol.*, 33, 643, 1988.
13. Szallasi, A. and Blumberg, P. M., Resiniferatoxin and analogs provide novel insights into the pharmacology of the vanilloid (capsaicin) receptor, *Life Sci.*, 47, 1399, 1990.
14. Blumberg, P. M., Szallasi, A., and Ács, G., Resiniferatoxin — an ultrapotent capsaicin analogue, in *Capsaicin in the Study of Pain*, Wood, J. N., Ed., Academic Press, London, 1993, 45–62.
15. Szallasi, A., The vanilloid (capsaicin) receptor: receptor types and species differences, *Gen. Pharmacol.*, 25, 223, 1994.
16. Szolcsányi, J., Szallasi, A., Szallasi, Z., Joó, F., and Blumberg, P. M., Resiniferatoxin, an ultrapotent selective modulator of capsaicin-sensitive primary afferent neurons, *J. Pharmacol. Exp. Ther.*, 255, 923, 1990.

17. Szallasi, A. and Blumberg, P. M., Specific binding of resiniferatoxin, an ultrapotent capsaicin analog, by dorsal root ganglion membranes, *Brain Res.*, 524, 106, 1990.
18. Ninkina, N. N., Willoughby, J. J., Beech, M. M., Coote, P. R., and Wood, J. N., Molecular cloning of a resiniferatoxin-binding protein, *Mol. Brain Res.*, 22, 39, 1994.
19. Bevan, S., Hothi, S., Hughes, G., James, I. F., Rang, H. P., Shah, K., Walpole, C. S. J., and Yeats, J. C., Capsazepine: a competitive antagonist of the sensory neurone excitant capsaicin, *Br. J. Pharmacol.*, 107, 544, 1992.
20. Amann, R. and Maggi, C. A., Ruthenium red as a capsaicin antagonist, *Life Sci.*, 49, 849, 1991.
21. Szallasi, A., Nilsson, S., Hökfelt, T., and Lundberg, J. M., Visualizing vanilloid (capsaicin) receptors in pig spinal cord by [^{3}H]resiniferatoxin autoradiography, *Brain Res.*, in press.
22. Walpole, C. S. J., Bevan, S., Bloomfield, G., Breckenridge, R., James, I. F., Ritchie, T., Szallasi, A., Winter, J., and Wrigglesworth, R., Similarities and differences in the structure-activity relationships of capsaicin and resiniferatoxin analogues, submitted.
23. Szallasi, A. and Blumberg, P. M., [^{3}H]resiniferatoxin binding by the vanilloid receptor: species-related differences, effects of temperature and sulfhydryl reagents, *Naunyn-Schmiedeberg's Arch. Pharmacol.*, 347, 84, 1993.
24. Szallasi, A., Blumberg, P. M., and Lundberg, J. M., Proton inhibition of [^{3}H]resiniferatoxin binding to vanilloid (capsaicin) receptors in rat spinal cord, *Eur. J. Pharmacol. Mol. Pharmacol. Sect.*, 289, 181, 1995.
25. Szallasi, A., unpublished data, 1994.
26. Maggi, C. A., Giuliani, S., Santicioli, P., Abelli, L., Geppetti, P., Somma, V., Renzi, D., and Meli, A., Species-related variations in the effects of capsaicin on urinary bladder functions: relation to bladder content of substance P-like immunoreactivity, *Naunyn-Schmiedeberg's Arch. Pharmacol.*, 336, 546, 1987.
27. Wang, Z.-Y. and Håkanson, R., Effect of resiniferatoxin on the isolated rabbit iris sphincter muscle: comparison with capsaicin and bradykinin, *Eur. J. Pharmacol.*, 213, 235, 1992.
28. Watling, K. J., Nonpeptide antagonists herald new era in tachykinin research, *Trends Pharmacol. Sci.*, 13, 266, 1992.
29. Iversen, L. L. and Kemp, J. A., *The NMDA Receptor*, IRL Press, Oxford, 1994.
30. Szallasi, A. and Goso, C., Characterization by [^{3}H]resiniferatoxin binding of a human vanilloid (capsaicin) receptor in post-mortem spinal cord, *Neurosci. Lett.*, 165, 101, 1994.
31. Ács, G., Palkovits, M., and Blumberg, P. M., [^{3}H]resiniferatoxin binding by the human vanilloid (capsaicin) receptor, *Mol. Brain Res.*, 23, 185, 1994.
32. Szallasi, A., Goso, C., and Manzini, S., Resiniferatoxin binding to vanilloid receptors in guinea pig and human airways, *Am. J. Respir. Crit. Care Med.*, 152, 59, 1995.
33. Szallasi, A., unpublished data, 1994.
34. Szallasi, A., unpublished data, 1994.
35. Lundberg, J. M., Martling, C.-R., and Saria, A., Substance P- and capsaicin-induced contractions of human bronchi, *Acta Physiol. Scand.*, 119, 49, 1983.
36. Lundberg, J. M., Hökfelt, T., Martling, C.-R., Saria, A., and Cuello, C., Substance P-immunoreactive sensory nerves in the lower respiratory tract of various mammals including man, *Cell Tissue Res.*, 235, 251, 1984.
37. Laude, E. A., Higgins, K. S., and Morice, A. H., A comparative study of the effects of citric acid, capsaicin and resiniferatoxin on the cough challenge in guinea pig and man, *Pulm. Pharmacol.*, 6, 171, 1993.
38. Bevan, S. and Szolcsanyi, J., Sensory neuron-specific actions of capsaicin: mechanisms and applications, *Trends Pharmacol. Sci.*, 11, 330, 1990.
39. Szallasi, A. and Blumberg, P. M., Molecular target size of the vanilloid (capsaicin) receptor in pig dorsal root ganglia, *Life Sci.*, 48, 1863, 1991.
40. Maderspach, K. and Fajszi, Cs., Beta-adrenergic receptors of brain cells. Membrane integrity implies apparent positive cooperativity and higher affinity, *Biochim. Biophys. Acta*, 692, 469, 1982.
41. Lo, M. M. S., Barnard, E. A., and Dolly, J. O., Size of the acetylcholine receptors in the membrane. An improved version of the radiation inactivation method, *Biochemistry*, 21, 2210, 1982.
42. Fels, G., Wolff, E. K., and Maelicke, A., Equilibrium binding of acetylcholine to the membrane-bound acetylcholine receptor, *Eur. J. Biochem.*, 127, 31, 1982.
43. Winter, J., Walpole, C. S. J., Bevan, S., and James, I. F., Characterization of resiniferatoxin binding sites on sensory neurons: co-regulation of resiniferatoxin binding and capsaicin sensitivity in adult rat dorsal root ganglia, *Neuroscience*, 57, 747, 1993.
44. Mandelzys, A., Cooper, E., Verge, V. M., and Richardson, P. M., NGF induces functional nicotinic acetylcholine receptors on rat sensory neurons, *Neuroscience*, 37, 523, 1990.
45. Franco-Cereceda, A., Rydh, M., and Dalsgaard, C. -J., Nicotine and capsaicin but not potassium-evoked CGRP release from cultured guinea pig spinal ganglia is inhibited with ruthenium red, *Neurosci. Lett.*, 137, 72, 1992.

Chapter 5

THE IONIC BASIS OF CAPSAICIN-EVOKED RESPONSES

Stuart Bevan and Ronald J. Docherty

CONTENTS

I. INTRODUCTION

Capsaicin, the pungent ingredient from peppers of the *Capsicum* family, has been widely exploited as a tool in sensory neuron biology and the study of neurogenic inflammation. Its excitatory actions have been well documented at the level of single nerve fibers and have led to the conclusion that capsaicin excites some, but not all, primary afferent neurons. The axonal conduction velocities of the capsaicin-sensitive sensory neurons are in the C- and Aδ-fiber range, and faster conducting fibers are usually not excited by capsaicin.[1] Application of capsaicin to nerves that innervate areas (e.g., skin) that are commonly subjected to stimuli from the external environment (exteroceptors) leads to the sensation of pain. For example, intradermal injection of capsaicin causes a painful burning sensation.[2] Capsaicin also excites afferent neurons innervating internal organs (interoceptors) and such a stimulation elicits autonomic responses. There is no precise relationship between capsaicin-sensitive nerves and any neuronal type when classified on the basis of their biochemical, biophysical, or morphological characteristics. This has led to the proposal that sensitivity to capsaicin be used as the diagnostic feature of this group of nerves.[3] Many, although not all, capsaicin-sensitive neurons contain neuropeptides, notably substance P and calcitonin gene-related peptide (CGRP), and exposure to capsaicin evokes the release of these neuropeptides. The local release of peptides can also be elicited by electrical stimulation, which suggests that any adequate stimulus that activates capsaicin-sensitive nerves will release neuropeptides and evoke neurogenic inflammation.[4]

Capsaicin has been employed as a cytotoxin, as high systemic concentrations selectively damage the stimulated neurons. The degree of damage depends on the age of the animals. Neurons in neonatal animals can be killed,[5,6] but adult neurons are more resistant and, although the axons are destroyed, many of the cell bodies remain intact.[7,8] Despite the widespread use of capsaicin, the mechanisms by which it exerts its unique effects have been poorly understood. However, investigations over the last 7 years have begun to unravel the ways in which capsaicin acts.

0-8493-7646-7/96/$0.00+$.50

II. CAPSAICIN DEPOLARIZES AND EXCITES SENSORY NEURONS

The depolarization of sensory neurons by capsaicin has been shown by voltage recording from whole nerve trunks[9,10] and from single neurons either in intact ganglia or after enzymatic dissociation of the ganglia.[11-15] The depolarization is concentration dependent and is accompanied by an increase in membrane conductance (Figure 1).[13-15] The change in membrane conductance suggests that capsaicin acts in some way to open ion channels in the plasma membrane. Studies on single dorsal root ganglion (DRG) and nodose ganglion neurons have shown that the depolarization wanes despite the continued presence of capsaicin, so that the membrane potential declines over several minutes to a value close to the original resting potential.[15,16] This repolarization is associated with a decrease in membrane conductance from the peak level, which suggests that it is due, at least in part, to the closure of ion channels. In some neurons, activation of another current will also contribute to the repolarization and may result in a final hyperpolarization. The available data suggest that the late hyperpolarization is a secondary phenomenon which depends on the presence of extracellular calcium during the initial capsaicin-induced depolarization.[16] The likely sequence of events is that calcium enters during the depolarization and raises the intracellular free calcium concentration,[17,18] which in turn activates a calcium-activated potassium current.

III. PERMEABILITY CHANGES EVOKED BY CAPSAICIN

The ionic basis of the capsaicin-evoked depolarization has been studied by voltage clamp of isolated DRG and trigeminal neurons.[15,19-22] There appears to be little or no difference in the responses in neurons from neonatal and adult rats. Capsaicin evokes a concentration-dependent inward current associated with an increase in membrane conductance in neurons voltage clamped at negative membrane potentials (Figure 2A). One striking feature of the response to capsaicin is the slow onset of the current.[20,21,23] This slow response is seen even with rapid drug application systems that change the extracellular solution bathing the cell in less than 100 msec[15,20,23] and the peak current in these conditions is often not achieved until about 20 sec after the beginning of the drug application. The time constant of the capsaicin-evoked current has been reported to be concentration dependent, with values of 3.8 and 2 sec for 0.3 and 3 μM capsaicin.[20] Such slow responses are in marked contrast to the rapid currents evoked by other agonists such as gamma-aminobutyric acid, adenosine triphosphate, and 5-hydroxytryptamine in the same cells with the same drug application systems[10,20] and suggest that access of capsaicin to its "receptor" or a subsequent molecular reaction is the rate-limiting step for the activation of the current. A recent study by Lui and Simon[21] reported that capsaicin evoked two types of inward current in rat trigeminal neurons: a rapid current with a time to peak of 1.7 sec and a slower current with a time to peak of 24.5 sec. The relationship between these two currents and the single type of current reported to date in DRG neurons[10,20] is unclear.

The amplitude of the capsaicin-evoked current is concentration dependent and EC_{50} values close to 0.3 μM have been reported for the initial responses to capsaicin.[15,20] Figure 2C shows a log concentration-response curve for the normalized currents with an estimated EC_{50} of about 290 nM, which is similar to the values calculated from membrane potential changes (see above) and radioactive ion flux experiments (see below). The Hill coefficient measured from these experiments was 1.95 ± 0.30, which is similar to the mean value of 1.8 ± 0.4 reported by Vlachová and Vyklický[20] and suggests that at least two molecules of capsaicin must bind to the receptor to open the associated ion channel.

The amplitude of the capsaicin-evoked inward current depends on the holding membrane potential and is reduced as the potential approaches about 0 mV. The current reverses direction at about this membrane potential and an outward current is evoked at positive membrane

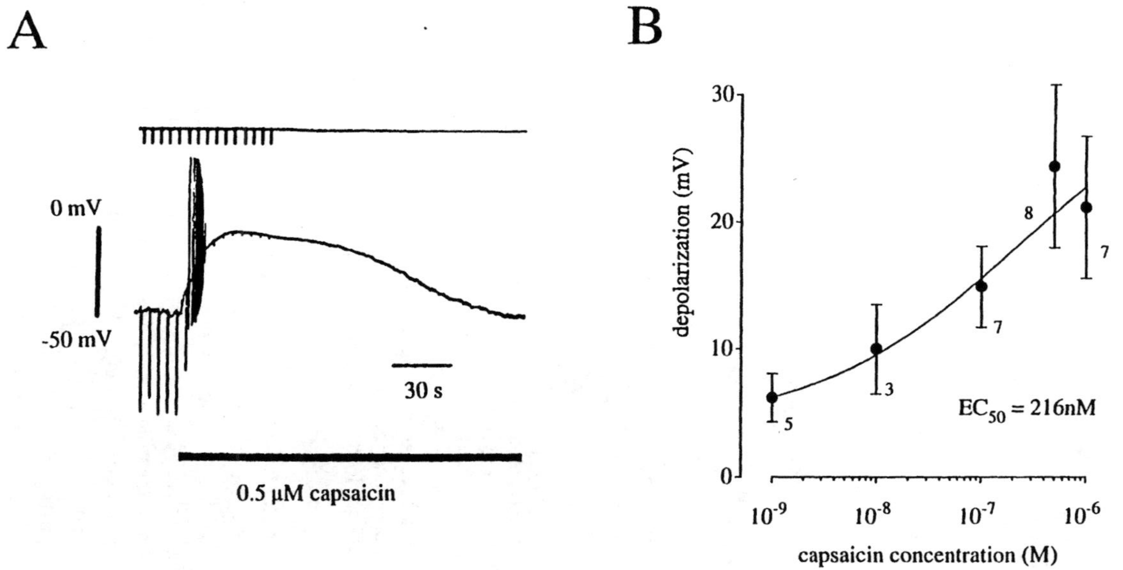

FIGURE 1. (A) Membrane depolarization evoked by 0.5 μ*M* capsaicin in an adult rat cultured DRG neuron. Bar shows duration of capsaicin application. Upper trace shows timing of brief 0.1 nA hyperpolarizing current pulses used to monitor membrane conductance. Note increase in conductance evoked by capsaicin and the repolarization in the continued presence of the agonist. (B) Log concentration-depolarization curve for capsaicin in DRG neurons. The numbers adjacent to each point represent number of cells studied. (Data reproduced with permission from Bevan, S. and Docherty, R. J., in *Capsaicin in the Study of Pain*, Wood, J. N., Ed., Academic Press, London, 1993, chap. 2.)

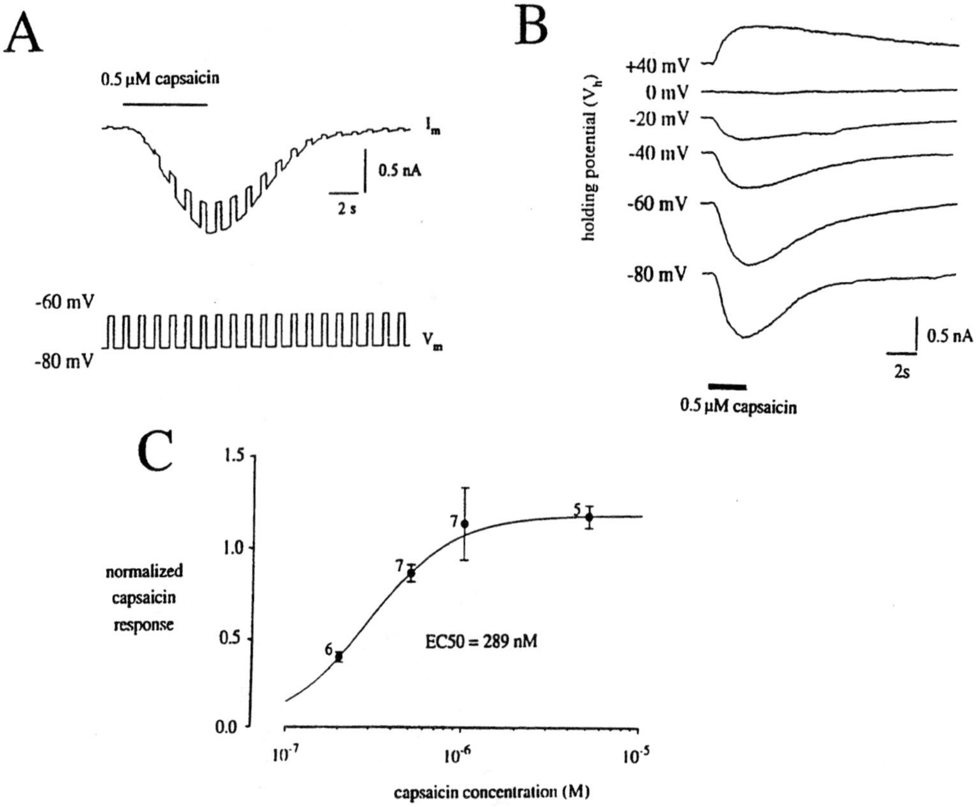

FIGURE 2. (A) Upper trace: inward current evoked by 0.5 μM capsaicin in voltage clamped rat DRG neuron. The increase in membrane conductance is indicated by the increased amplitude of the current steps. Lower trace: Command holding potential stepped between –80 and –60 mV. (B) Membrane currents evoked by 0.5 μM capsaicin at different holding potentials to show a reversal potential of about 0 mV in this example. (C) Log concentration-response curve for normalized capsaicin-evoked current. Currents recorded at +40 mV and normalized to the response to 1 μM capsaicin in each cell. The numbers adjacent to each point represent number of cells studied. (Data reproduced with permission from Bevan, S. and Docherty, R. J., in *Capsaicin in the Study of Pain*, Wood, J. N., Ed., Academic Press, London, 1993, chap. 2.)

potentials (see Figure 2B). The potential at which the current reverses direction, the zero current or reversal potential, is determined by the ionic permeability of the conductance pathway. The reversal potential for the capsaicin-evoked current is close to 0 mV when the cell is dialyzed with either CsCl- or KCl-based solutions and bathed in an external medium composed mostly of NaCl. Ion replacement experiments have been used to show that the conductance pathway has little or no permeability to Cl^- and yielded estimates of the relative permeabilities for a number of cations. The ionic permeabilities for monovalent cations relative to Na^+ can be estimated, using the Goldman-Hodgkin-Katz constant field equation, from the change in the reversal potential that occurs when external Na^+ is replaced with another monovalent cation. The relative permeabilities for Ca^{2+} and Mg^{2+} are estimated in a similar way, although a modified equation is necessary for divalent cations. These experiments yielded the following permeability sequence:[22,24]

$$Ca^{2+} > Mg^{2+} > \text{guanidinium} > K^+ > Cs > Na > \text{choline}$$

The relative permeabilities are greater for divalent than for monovalent cations (e.g., $P_{Ca}/P_{Na} = 5$), but in general the conductance pathway discriminates poorly between the naturally occurring cations and allows relatively bulky cations such as choline ($P_{choline}/P_{Na}$ = 0.35) to cross the membrane. A less direct measure of membrane permeability can be obtained from ion replacement experiments in which the cytotoxic effects of capsaicin are measured. Results from such a study suggest that cations as large as arginine are permeable.[23] Cytochemical studies, which use a stain for accumulated cobalt ions as an index of capsaicin sensitivity, also show that the capsaicin-activated channel is permeable to cobalt.[25,26] Other electrophysiological estimates of the reversal potential (+10 mV) of the capsaicin-evoked current were also explained by a conductance increase to monovalent cations and calcium.[20] These conclusions are broadly consistent with the earlier findings of Marsh et al.,[16] who examined the capsaicin-induced conductance increase in rat nodose ganglion neurons bathed in physiological medium and impaled with sharp KCl-filled microelectrodes. They concluded that capsaicin increases the permeability to cations but their estimates of reversal potential were complicated by the superimposition of other conductances (e.g., chloride) that were activated by Ca^{2+} that entered the neurons through the capsaicin-operated channel.

A capsaicin-evoked increase in Ca^{2+} permeability has also been demonstrated with intracellular dyes (e.g., fura-2) that monitor the Ca^{2+} concentration.[17,18] Application of capsaicin elevates intracellular Ca^{2+} in rat DRG neurons in a concentration-dependent manner, with an EC_{50} concentration of 70 n*M*.[18]

An increased cation permeability has also been shown with radioactive ion flux methods. Capsaicin stimulates the uptake of $^{45}Ca^{2+}$, which appears to accumulate in intracellular compartments by energy-dependent mechanisms.[26] The idea that the additional Ca^{2+} is sequestered in intracellular organelles is consistent with the observations that capsaicin induces changes in mitochondrial structure that are typical of Ca^{2+} overloading.[27-29] $^{45}Ca^{2+}$ uptake experiments yielded EC_{50} estimates for capsaicin of about 0.2 μ*M*, which is higher than the value obtained with direct measurement of the evoked changes in intracellular calcium concentration (70 n*M*).[18] Similar lower EC values (60 to 70 n*M*) were found in other ion flux assays which monitored Na^+ or^{14}[C]-guanidinium influx and the efflux of $^{86}Rb^+$ from rat cultured DRG neurons.[26]

IV. CAPSAICIN-ACTIVATED ION CHANNELS

There is now strong evidence that capsaicin acts at a specific membrane receptor found only in the sensitive subpopulation of sensory neurons. The existence of a receptor was originally proposed by Szolcsányi and Jancsó-Gábor[30,31] on the basis of the stucture–activity relationships of a range of synthetic analogs of capsaicin. The receptor hypothesis has been strengthened by the synthesis of further analogs of capsaicin,[32] all of which act like capsaicin. Furthermore, recent studies with resiniferatoxin (RTX), which acts as a highly potent capsaicin-like agonist,[23,33] and the development of capsazepine, a competitive antagonist of both capsaicin and RTX,[10] argue strongly for the existence of a specific capsaicin receptor.

The activity of capsaicin-operated ion channels has been studied in isolated membrane patches from DRG neurons[19,20,34] (see Figure 3). The activity of this ion channel is blocked by ruthenium red,[35] which is known to block both capsaicin-induced ion fluxes in sensory neurons[26] and the action of capsaicin in multicellular preparations.[36,37] Single channel recordings such as those shown in Figure 3B have shown that the ion channels activated by RTX and capsaicin have very similar, if not identical, properties,[24,38] which strengthens the argument that both agonists have a common mode of action. The single channel conductance appears to depend on membrane potential; a conductance of 20 to 30 pS has been estimated at holding potentials of –50 to –80 mV,[20,22] while the conductance is about 70 pS at +40 mV.[22] It is unclear if the ion channel is part of an integral receptor/ion channel complex. The finding that single channel activity can be recorded in isolated membrane patches shows that the binding site for capsaicin must be closely associated with the ion channel in the membrane, although it is possible that the binding site resides on a distinct molecule.

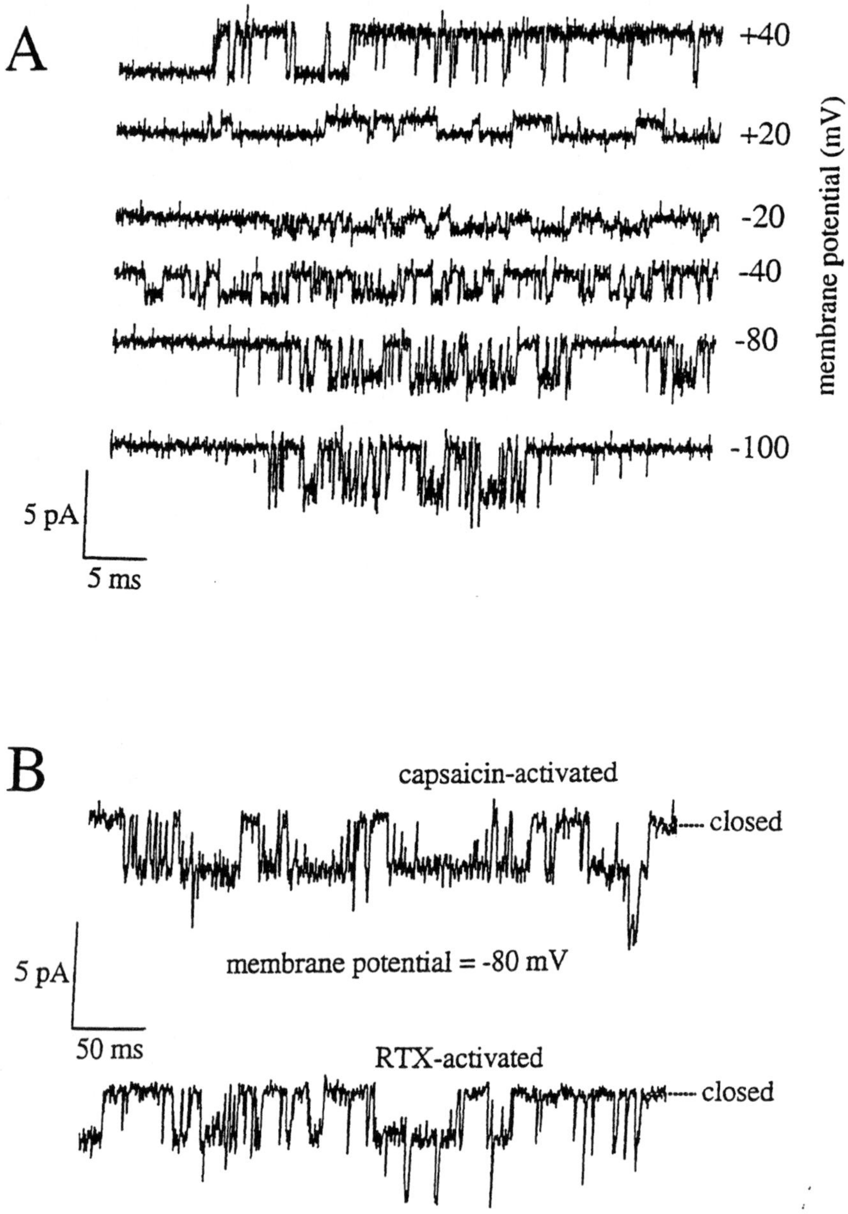

FIGURE 3. (A) Single channel activity evoked by 0.1 μM capsaicin in an outside-out membrane patch from a DRG neuron at different holding potentials. Data reproduced with permission from Bevan, S. and Docherty, R. J., in *Capsaicin in the Study of Pain*, Wood, J. N., Ed., Academic Press, London, 1993, chap. 2.) (B) Single channel currents evoked by capsaicin and RTX in a membrane patch at –80 mV. (Data reproduced with permission from Bevan, S. and Szolcsányi, J., *Trends Pharmacol. Sci.*, 11, 330, 1990.)

V. PROTONS ACTIVATE CAPSAICIN-OPERATED CHANNELS

What is the normal physiological role of the capsaicin receptor? The available evidence does not readily support the existence of endogenous capsaicinoids as natural ligands for the receptor. No structurally related endogenous compound with similar biological activity has been described.[39] Furthermore the capsaicin antagonist, capsazepine, blocks the acute nociceptive and chronic antinociceptive and anti-inflammatory effects of capsaicin but does not by itself appear to have any significant effect on nociceptive mechanisms.[40] Although the possibility that the receptor/ion channel is normally activated by a capsaicin-like molecule has not been ruled out, it seems likely that capsaicin stimulates an "unphysiological" activation of the channel. The endogenous activator may be structurally unrelated to capsaicin, and hydrogen ions (protons) are candidates for this role. Recent studies have shown that protons can activate the same ion channels as capsaicin and would be expected to stimulate these sensory nerves and evoke neurogenic inflammation by liberating neuropeptides from the peripheral nerve terminals. This may be (patho)physiologically relevant as reductions in extracellular pH occur in a variety of inflammatory and traumatic conditions (e.g., ischemia). In these situations, stimulation of capsaicin-sensitive nerves by protons will have an immediate beneficial effect. The local release of peptides will induce a vasodilatation and the firing of the afferent nerves will evoke a reflex rise in blood pressure.

A step reduction in the pH of the extracellular medium activates two types of depolarizing current in rat voltage clamped DRG neurons (see Figure 4A). The first is a transient current that has a short duration (2 to 5 sec) and is due to an increase in Na^+ conductance,[41-43] while the second current can persist for minutes and has a different ionic basis. The persistent current is due to a relatively nonselective increase in cation permeability and is found only in capsaicin-sensitive DRG neurons.[44] The characteristics of the capsaicin- and proton-evoked currents are remarkably similar: (1) the evoked currents have a similar time course (Figure 4A), (2) the amplitudes of the two responses are well correlated in any given neuron (Figure 4B), (3) the properties of the single channel currents evoked by either protons or capsaicin are very similar, if not identical,[19] and (4) the chemosensitivities to capsaicin and protons (persistent current) are regulated to the same degree by the presence or absence of nerve growth factor.[45] Protons therefore represent one physiologically relevant mechanism of activating the capsaicin-operated channels.

The capsaicin antagonists, ruthenium red and capsazepine, have been used as tools to investigate how protons can act. Ruthenium red acts as a noncompetitive capsaicin antagonist[10,35,46] and inhibits proton-evoked nerve depolarization[47] and neuropeptide release,[48a-50] all at low micromolar concentrations. Unfortunately the utility of ruthenium red as an antagonist is limited as, at slightly higher concentrations, it can also affect the responses to other stimuli and inhibits neuropeptide release evoked by electrical field stimulation and potassium depolarization.[51] Nevertheless the results obtained with ruthenium red are consistent with the hypothesis that protons activate capsaicin-sensitive nerves and that both agents activate the same ion channels. Capsazepine has also been shown to block the effects of protons on sensory nerve preparations, although the mechanism of blockage is unclear. One obvious site of action is at the level of the receptor/ion channel. Here the published data are contradictory. Proton-evoked depolarizations of rat and guinea pig vagus nerve are not blocked by capsazepine.[47,52] Furthermore, the sustained proton-evoked current in rat DRG neurons is not blocked by capsazepine at concentrations that inhibit the actions of capsaicin.[47] In contrast, Lui and Simon[21] reported that capsazepine is able to block both the transient and persistent proton-evoked currents in rat trigeminal ganglion neurons. Capsazepine also failed to antagonize the low pH-evoked activation of nerves in the tail of a neonatal rat tail-cord preparation,[48] although it can inhibit proton-induced firing of single afferent fibers innervating guinea pig airways.[52] Several reports have also shown that capsazepine blocks proton-evoked neuropeptide release from sensory nerves in various tissue preparations,[48a,50,53] although the mechanism of inhibition has not been resolved. It is possible that low pH solutions act, in part, to liberate a capsaicin-like molecule from tissues and capsazepine blocks the action of this second agent. In this context it is of interest that the ability of the prostanoid PGI_2 to release CGRP from isolated guinea pig heart or to contract guinea pig tracheal muscle is inhibited by capsazepine.[54,55] An alternative explanation is that capsazepine can directly inhibit the interaction between protons and the receptor/ion channel under some circumstances. Perhaps protons activate

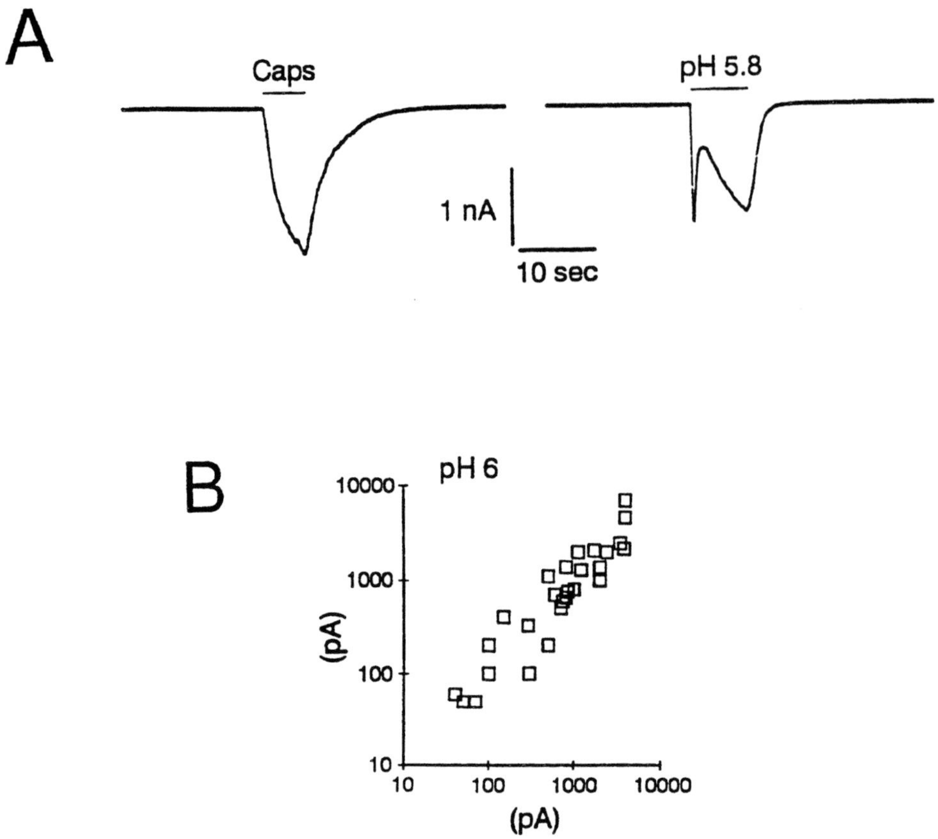

FIGURE 4. (A) Currents evoked by capsaicin (left) and pH 5.8 medium (right) in a single DRG neuron. Note the similarity in size and time course of the capsaicin-evoked current and the slower proton-evoked current. (B) Plot to show the good correlation between the amplitude of the capsaicin-evoked current (ordinate) and the slower proton-evoked (pH 6) current (abscissa). (Data reproduced with permission from Bevan, S. and Docherty, R. J., in *Capsaicin in the Study of Pain*, Wood, J. N., Ed., Academic Press, London, 1993, chap. 2.)

multiple types of receptor/ion channel which vary in their distribution and sensitivity to capsazepine.

VI. EVOKED RELEASE OF NEUROPEPTIDES

Capsaicin and other excitatory agents stimulate the release of neuropeptides (CGRP, neurokinin A, and substance P) from sensory neurons. Neuropeptide release has been used extensively to investigate many facets of the actions of capsaicin, both on cultured sensory neurons[56] and on tissues.[46,57-60] Two distinct mechanisms are responsible for the release of the peptides. First, the influx of calcium through capsaicin-activated ion channels will raise the concentration of free, intracellular calcium. This promotes a local release of peptides which is independent of action potential generation and propagation. Second, the depolarization evoked by capsaicin stimulates action potentials in the neuron which propagate and thus depolarize more distant regions of the nerve, such as nerve collaterals. In this way, neuropeptides are released at regions distant from the site at which capsaicin acts. The pharmacological properties of capsaicin-evoked neuropeptide release will depend on the involvement of propagated action potentials in the preparation under study. Blockade of voltage gated sodium and calcium channels with tetrodotoxin (TTX) and ω-conotoxin (CTX), which do not inhibit the capsaicin-induced calcium flux,[26] will have little or no effect if the release is stimulated by the local action of capsaicin. In contrast, release may be inhibited by these agents if the measured neuropeptide release is dependent on nerve activity and calcium entry through voltage gated calcium channels. Such a difference probably accounts for the findings of Lundberg and co-workers, who reported that release of CGRP from guinea pig lung was blocked by TTX and CTX when low concentrations of capsaicin were used, but not inhibited when release was stimulated by high concentrations of capsaicin.[61-63] A caveat associated with this argument is that some sodium currents in sensory nerves are insensitive to TTX[64] and the ability of CTX to inhibit the neuropeptide release evoked by depolarization varies between species and even between different sensory nerve preparations from a single species.[4]

Both the "local" and "propagated action potential" modes of evoked neuropeptide release are dependent on extracellular calcium,[36,56,57] although there are quantitative differences in the calcium dependence which reflect the different routes of calcium entry into the neuron. Simple removal of extracellular calcium, which lowers the concentration to micromolar levels, will block the release triggered by calcium influx through voltage gated calcium channels.[36,57] However, a significant calcium flux occurs via capsaicin-activated channels in the same ionic conditions[26] and this influx is sufficient to stimulate a significant release of neuropeptides.[36] It is necessary to lower the calcium concentration even further by the addition of chelators such as EDTA in order to abolish the capsaicin-evoked local release of neuropeptides.[36,56,57]

Although capsaicin initially stimulates neuropeptide release it can also, in the longer term, have an inhibitory effect such that a variety of stimuli no longer evoke peptide release and neurogenic inflammation despite the presence of near normal levels of peptides in the nerves.[65-68] This phenomenon is the basis for the paradoxical anti-inflammatory effects of capsaicin.[69] One explanation for the long-term inhibitory effect of capsaicin is inhibition of voltage gated calcium channels (see below), which would reduce neuropeptide release evoked by nerve depolarization.

VII. DESENSITIZATION

The ability of capsaicin to desensitize neurons has been well documented.[70] However, reports in the literature describe two distinct phenomena under the general heading of desensitization. The first is a pharmacological desensitization, where prolonged or repeated applications of capsaicin lead to a progressive decline in the size of subsequent responses to capsaicin; this phenomenon is referred to here as "desensitization". The second phenomenon is a "functional desensitization", in which a challenge with capsaicin leads to a reduction or loss of responsiveness to other stimuli. Although the two phenomena are often observed concomitantly in the same tissue they can be separated when low concentrations of capsaicin are employed.[71] In these conditions, responses to capsaicin are reduced or lost selectively and responses to other noxious or innocuous stimuli are unchanged.

The process of capsaicin desensitization has been studied in single cells.[18,71,72] The degree of desensitization can be minimized by the use of low concentrations (0.5 μM or less) of capsaicin and infrequent applications of agonist, but the phenomenon is rarely eliminated by such precautions. Typically a single application of capsaicin causes a marked attenuation of responses to subsequent challenges with the drug. There is good evidence that desensitization is a calcium-dependent process. Little or no desensitization is seen when calcium is removed from the extracellular medium[18] and desensitization is greatly reduced when cells are loaded with BAPTA, a fast calcium buffer, or when the extracellular calcium is replaced with barium.[73] In contrast no desensitization is seen when neurons are challenged with capsaicin in calcium-free medium while simultaneously raising the intracellular calcium concentration by application of the metabolic inhibitor FCCP, which will liberate Ca^{2+} from intracellular stores.[18] A possible explanation for these results is that desensitization occurs when calcium acts locally either near or within the capsaicin-activated ion channel and that the local increase in calcium concentration is not matched when calcium is released from internal stores.

Recent evidence indicates that the increase in intracellular Ca^{2+} concentration promotes desensitization by stimulating protein phosphatase 2B (calcineurin), which is a calcium- and calmodulin-dependent cytosolic enzyme. Calcineurin is inhibited by a complex of cyclosporin A and and its cytoplasmic binding protein cyclophilin.[74] Capsaicin desensitization is almost abolished when the cyclosporin-cyclophilin complex is introduced into the cytoplasm of rat DRG neurons.[72] Conversely, increased responsiveness to capsaicin was reported when DRG neurons were exposed to treatments that raised the intracellular levels of cyclic AMP.[75] These findings suggest that the activity of the capsaicin-activated ion channels is in some way regulated by the level of phosphorylation of either the receptor/ion channel or an associated protein. It is possible that activation of calcineurin is also responsible for the functional desensitization of sensory neurons which may occur via a calcium-dependent dephosphorylation of other substrates such as ion channels or enzymes. This hypothesis is consistent with the observations that the functional desensitization induced by capsaicin is dependent on the presence of extracellular calcium[76] and is blocked by the antagonist ruthenium red.[36,77]

VIII. IONIC BASIS OF NEUROTOXICITY

The ability of capsaicin to kill or damage a subpopulation of mammalian sensory neurons, particularly in neonatal rats,[5] has been exploited in a wide range of studies. The mechanisms of capsaicin- and RTX-induced neurotoxicity have been studied *in vitro*, where experiments have shown that the neurotoxic effects of the compounds can be explained by the evoked influx of Na^+ and Ca^{2+} (see Reference 23). Neurotoxicity is greatly reduced when the external sodium chloride is replaced with sucrose and is abolished if calcium is also removed from the medium. Cell damage is still evident in calcium-free medium when sodium chloride is replaced with either choline chloride or arginine hydrochloride, which argues that both choline and arginine can permeate the activated ion channels. However, the neurotoxic effects of the agonists are abolished in calcium-free sodium glutamate solutions, which suggests that the Na^+-mediated damage requires the influx of chloride via a separate anion conductance pathway. In such conditions, the net effect is an accumulation of sodium chloride, which is followed by an influx of water, osmotic swelling, damage, and lysis. The neurotoxic effects of calcium can be seen in the absence of external sodium chloride and may be mediated by activation of calcium-dependent proteolytic enzymes[24] such as calpain.[78]

IX. EFFECTS ON OTHER MEMBRANE CURRENTS

Capsaicin can also affect other membrane currents by mechanisms that are distinct from its specific actions on sensory neurons. In all reported cases the "nonspecific" effects of capsaicin result in an inhibition of a resting or voltage-activated conductance. Molluscan neurons are depolarized by capsaicin in a concentration-dependent manner with a decrease in membrane conductance due to inhibition of voltage activated sodium, potassium, and calcium currents.[79,80] The concentrations of capsaicin required to inhibit these currents are about 50- to 100-fold higher than those required to activate the specific membrane current in mammalian sensory neurons. Similar, high concentrations of capsaicin inhibit potassium currents recorded from frog nodes of Ranvier,[81,82] rat demyelinated internodes,[83] crayfish giant axons,[84] guinea pig and chicken DRG neurons,[85] rat ventricular myocytes,[86] rat Schwann cells[87] and rat pituitary melanotrophs.[88] Detailed kinetic studies of currents in rat Schwann cells and melanotrophs have shown that capsaicin blocks open voltage-activated potassium channels with K_d values of 8.7 and 17.4 μM.[87,88] Capsaicin also blocks sodium currents in crayfish giant axons,[84] frog nodes of Ranvier,[82] and guinea pig and chicken DRG neurons,[85] but once again these effects are only seen with high concentrations of capsaicin.

There have been several reports that capsaicin inhibits voltage-activated calcium currents but the characteristics of the effects indicate that two mechanisms operate. In guinea pig DRG neurons capsaicin was reported to inhibit an inactivating calcium current found in about 50% of cells.[89] The inhibition is associated with a negative shift in the voltage dependence of the current, and an acceleration in its rates of activation and inactivation. No obvious sensory neuron agonist-like effects were reported in this study and for this reason the inhibition is unlikely to be secondary to an evoked calcium influx. In contrast, capsaicin causes a relatively long-lasting inhibition of voltage-activated calcium currents in rat DRG neurons which is mediated by the agonist-evoked influx of calcium and the resultant increase in intracellular free calcium concentration.[17,90,91] The basis for this inhibition has not been fully elucidated but presumably involves activation of an unidentified calcium-dependent process. In this way capsaicin inhibits voltage-activated calcium currents in a cell-specific manner and at the low concentrations that are required for agonism. One likely consequence of this calcium current inhibition is that neurotransmitter release from the nerve endings will be impaired and this phenomenon could underlie the longer-term peripheral anti-inflammatory effects of systemically administered capsaicin.

X. SUMMARY

In the last 10 years, we have made considerable advances in understanding the ionic mechanisms that are responsible for the excitatory and neurotoxic actions of capsaicin and related compounds.

- Capsaicin depolarizes and excites nociceptive sensory neurons by opening a cation-selective ion channel. Calcium entry regulates the activity of these channels, probably by altering the level of phosphorylation.
- Protons activate the same ion channels as capsaicin and can be considered as endogenous activators. Activation will occur in conditions where the pH falls, such as inflammation and ischemia. It is possible that endogenous agonists directly activate the capsaicin receptor: if so, they may or may not bind to the same site as capsaicin.
- Capsaicin promotes the the influx of Ca^{2+} in two ways: (1) via the capsaicin-activated ion channels and (2) via voltage gated Ca channels that are opened by the local depolarization and propagated action potentials. Protons will also depolarize the neuron by opening the same ion channels. The Ca^{2+} influx elicited by these agents stimulates the release of neuropeptides and evokes neurogenic inflammation.
- Functional desensitization of the neurons can be induced by capsaicin. This is calcium mediated and may involve inactivation of voltage-activated calcium channels.
- Capsaicin and RTX neurotoxicity results from the influx of Na^+ and Ca^{2+} through the agonist-operated ion channels.

REFERENCES

1. Szolcsányi, J., Actions of capsaicin on sensory receptors., in *Capsaicin in the Study of Pain*, Wood, J. N., Ed., Academic Press, London, 1993, chap. 1.
2. Lamotte, R. H., Shain, C. N., Simone, D. A., and Tsai, E. F. P., Neurogenic hyperalgesia: psychophysical studies of underlying mechanisms, *J. Neurophysiol.*, 66, 190, 1991.
3. Szolcsányi, J., Capsaicin-sensitive chemoceptive neural system with dual sensory-efferent function, in *Antidromic Vasodilatation and Neurogenic Inflammation*, Chahl, L. A., Szolcsányi, J., and Lembeck., F., Eds., Akadémiai Kiadó, Budapest, 1984, 27.

4. Maggi, C. A., The pharmacological modulation of neurotransmitter release, in *Capsaicin in the Study of Pain*, Wood, J. N., Ed., Academic Press, London, 1993, chap. 8.
5. Jancsó, G., Király, E., and Jancsó-Gábor, A., Pharmacologically induced selective degeneration of chemosensitive primary sensory neurons, *Nature*, 270, 741, 1977.
6. Nagy, J. I., Iversen, L. L., Goedert, M., Chapman, D., and Hunt, S. P., Dose-dependent effects of capsaicin on primary sensory neurons in the neonatal rat, *J. Neurosci.*, 3, 399, 1983.
7. Chung, K., Schwen, R. J., and Coggeshall, R. E., Ureteral axon damage following subcutaneous administration of capsaicin in adult rats, *Neurosci. Lett.*, 53, 221, 1985.
8. Chung, K., Klein, C. M., and Coggeshall, R. E., The receptive part of the primary afferent axon is most vulnerable to systemic capsaicin in adult rats, *Brain Res.*, 511, 222, 1990.
9. Hayes, A. G., Hawcock, A. B., and Hill, R. G., The depolarising action of capsaicin on rat isolated sciatic nerve, *Life Sci.*, 35, 1561, 1984.
10. Bevan, S., Hothi, S. K., Hughes, G., James, I. F., Rang, H. P., Shah, K., Walpole, C. S. J., and Yeats, J. C., Capsazepine: a competitive antagonist of the sensory neurone excitant capsaicin, *Br. J. Pharmacol.*, 107, 544, 1992.
11. Baccaglini, P. I. and Hogan, P. G., Some rat sensory neurons in culture express characteristics of differentiated pain sensory cells, *Proc. Natl. Acad. Sci. U.S.A.*, 80, 594, 1982.
12. Williams, J. T. and Zieglgansberger, W., The acute effects of capsaicin on rat primary afferents and spinal neurons, *Brain Res.*, 253, 125, 1982.
13. Taylor, D. C. M., Pierau, Fr.-K., Szolcsányi, J., Krishtal, O., and Petersen, M., Effect of capsaicin on rat sensory neurons, in *Thermal Physiology*, Hales, J. R. S., Ed., Raven Press, New York, 1984, 23.
14. Heyman, I. and Rang, H. P., Depolarizing responses to capsaicin in a subpopulation of rat dorsal root ganglion cells, *Neurosci. Lett.*, 56, 69, 1985.
15. Bevan, S. and Docherty, R. J., Cellular mechanisms of the action of capsaicin, in *Capsaicin in the Study of Pain*, Wood, J. N., Ed., Academic Press, London, 1993, chap. 2.
16. Marsh, S. J., Stansfeld, C. E., Brown, D. A., Davey, R., and McCarthey, D., The mechanism of action of capsaicin on sensory C-type neurons and their axons in vitro, *Neuroscience*, 23, 275, 1987.
17. Bleakman, D., Brorson, J. R., and Miller, R. J., The effects of capsaicin on voltage-gated calcium currents and calcium signals in cultured dorsal root ganglion cells, *Br. J. Pharmacol.*, 101, 423, 1990.
18. Cholewinski, A., Burgess, G. M., and Bevan, S., The role of calcium in capsaicin-induced desensitization in rat cultured dorsal root ganglion neurons, *Neuroscience*, 55, 1015, 1993.
19. Bevan, S., Forbes, C. A., and Winter, J., Protons and capsaicin activate the same ion channels in rat isolated dorsal root ganglion neurons, *J. Physiol.*, 459, 401P, 1993.
20. Vlachová, V. and Vyklický, L., Capsaicin-induced membrane currents in cultured sensory neurons of the rat, *Physiol. Res.*, 42, 301, 1993.
21. Liu, L. and Simon, S. A., A rapid capsaicin-activated current in rat trigeminal ganglion neurons, *Proc. Natl. Acad. Sci. U.S.A.*, 91, 738, 1994.
22. Bevan, S. and Forbes, C. A., unpublished data, 1988.
23. Winter, J., Dray, A., Wood, J. N., Yeats, J. C., and Bevan, S., Cellular mechanism of action of resiniferatoxin: a potent sensory neuron excitotoxin, *Brain Res.*, 520, 131, 1990.
24. Bevan, S. and Szolcsányi, J., Sensory neuron-specific actions of capsaicin: mechanisms and applications, *Trends Pharmacol. Sci.*, 11, 330, 1990.
25. Winter, J., Characterization of capsaicin-sensitive neurons in adult rat dorsal root ganglion cultures, *Neurosci. Lett.*, 80, 134, 1987.
26. Wood, J. N., Winter, J., James, I. F., Rang, H. P., Yeats, J., and Bevan, S., Capsaicin-induced ion fluxes in dorsal root ganglion cells in culture, *J. Neurosci.*, 8, 3208, 1988.
27. Joó, F., Szolcsányi, J., and Jancsó-Gábor, A., Mitochondrial alterations in the spinal ganglion cells of the rat accompanying the long-lasting sensory disturbance induced by capsaicin, *Life Sci.*, 8, 621, 1969.
28. Szolcsányi, J., Joó, F., and Jancsó-Gábor, A., Mitochondrial changes in preoptic neurons after capsaicin desensitization of the hypothalamic thermodetectors in rats, *Nature*, 229, 116, 1971.
29. Jancsó, G., Karcsu, S., Király, E., Szebeni, A., Toth, L., Bacsy, E., and Joó, F., and Parducz, A., Neurotoxin induced nerve cell degeneration: possible involvement of calcium, *Brain Res.*, 295, 211, 1984.
30. Szolcsányi, J. and Jancsó-Gábor, A., Sensory effects of capsaicin congeners. I. Relationship between chemical structure and pain-producing potency of pungent agents, *Drug Res.*, 25, 1877, 1975.
31. Szolcsányi, J. and Jancsó-Gábor, A., Sensory effects of capsaicin congeners. II. Importance of chemical structure and pungency in desensitizing activity of capsaicin, *Drug Res.*, 26, 33, 1976.
32. Walpole, C. S., Wrigglesworth, R., Bevan, S., Campbell, E. A., Dray, A., James, I. F., Masdin, K. J., Perkins, M. N., and Winter, J., Analogs of capsaicin with agonist activity as novel analgesic agents; structure-activity studies, *J. Med. Chem.*, 36, 2381, 1993.
33. Szállási, A. and Blumberg, P. M., Resiniferatoxin, a phorbol-related diterpene, acts as an ultrapotent analog of capsaicin, the irritant constituent in red pepper, *Neuroscience*, 30, 515, 1989.

34. Forbes, C. A. and Bevan, S., Properties of single capsaicin-activated channels, *Soc. Neurosci. Abstr.*, 14, 260.20, 1988.
35. Dray, A., Forbes, C. A., and Burgess, G. M., Ruthenium red blocks the capsaicin-induced increase in intracellular calcium and activation of membrane currents in sensory neurons as well as the activation of peripheral nociceptors in vitro, *Neurosci. Lett.*, 110, 52, 1990.
36. Maggi, C. A., Patacchini, R., Santicioli, P., Giuliani, S., Geppetti, S., Geppetti, P., and Meli, A., Protective action of ruthenium red toward capsaicin desensitization of sensory fibers, *Neurosci. Lett.*, 88, 201, 1988.
37. Amann, R. and Lembeck, F., Ruthenium red selectively prevents capsaicin-induced nociceptor stimulation, *Eur. J. Pharmacol.*, 161, 227, 1989.
38. Bevan, S., unpublished data, 1989.
39. Wood, J. N., Walpole, C., James, I. F., Dray, A., and Coote, P. R., Immunochemical detection of photoaffinity-labelled capsaicin-binding proteins from sensory neurons, *FEBS Lett.*, 269, 381, 1990.
40. Perkins, M. N. and Campbell, E. A., Capsazepine reversal of the antinociceptive action of capsaicin in vivo, *Br. J. Pharmacol.*, 107, 329, 1992.
41. Krishtal, O. A. and Pidoplichko, V. I., A receptor for protons in the nerve cell membrane, *Neuroscience*, 5, 2325, 1980.
42. Krishtal, O. A. and Pidoplichko, V. I., A receptor for protons in the membrane of sensory neurons may participate in nociception, *Neuroscience*, 6, 2599, 1981.
43. Konnerth, A. and Lux, H. D., and Morad, M., Proton-induced transformation of calcium channels in chick dorsal root ganglion cells, *J. Physiol.*, 386, 603, 1987.
44. Bevan, S. and Yeats, J. C., Protons activate a cation conductance in a sub-population of rat dorsal root ganglion neurons, *J. Physiol.*, 433, 145, 1991.
45. Bevan, S. and Winter, J., Nerve growth factor (NGF) modulates the chemosensitivity of adult rat dorsal root ganglion neurons in culture, *J. Physiol.*, 476, 39*P*, 1994.
46. Maggi, C. A., Bevan, S., Walpole, C. S. J., Rang, H. P., and Guiliani, S., A comparison of capsazepine and ruthenium red as capsaicin antagonists in the rat isolated urinary bladder and vas deferens, *Br. J. Pharmacol.*, 108, 801, 1993.
47. Bevan, S., Rang, H. P., and Shah, K., Capsazepine does not block the proton-induced activation of rat sensory neurons, *Br. J. Pharmacol.*, 107, 235P, 1992.
48. Dray, A., Patel, I., Naeem, S., Rueff, A., and Urban, L., Studies with capsazepine on peripheral nociceptor activation by capsaicin and low pH: evidence for a dual effect of capsaicin, *Br. J. Pharmacol.*, 107, 236P, 1992.

48a. Lou, Y. P. and Lundberg, J. M., Inhibition of low pH evoked activation of airway sensory nerves by capsazepine, a novel capsaicin-receptor antagonist, *Biochem. Biophys. Res. Commun.*, 189, 537, 1992.

49. Geppetti, P., Del Bianco, E., Patacchini, R., Santicioli, P., Maggi, C. A., and Tramontana, M., Low pH-induced release of CGRP from capsaicin-sensitive sensory nerves: mechanism of action and biological response, *Neuroscience*, 41, 295, 1991.
50. Santicioli, P., Del Bianco, E., Figini, M., Bevan, S., and Maggi, C. A., Effect of capsazepine on the release of calcitonin gene-related peptide-like immunoreactivity (CGRP-LI) induced by low pH, capsaicin and potassium in rat soleus muscle, *Br. J. Pharmacol.*, 110, 609, 1993.
51. Amann, R. and Maggi, C. A., Ruthenium red as a capsaicin antagonist, *Life Sci.*, 49, 849, 1991.
52. Fox, A. J., Urban, L., Barnes, P. J., and Dray, A., Capsazepine selectively inhibits capsaicin and low pH evoked firing of single C-fibers in the guinea pig airways in vitro, submitted, 1994.
53. Franco-Cereceda, A. and Lundberg, J. M., Capsazepine inhibits pH- and lactic acid-evoked release of calcitonin gene-related peptide from sensory nerves in guinea pig heart, *Eur. J. Pharmacol.*, 221, 183, 1992.
54. Franco-Cereceda, A., Källner, G., and Lundberg, J. M., Cyclo-oxygenase products released by low pH have capsaicin-like actions on sensory nerves in the isolated guinea pig heart, *Cardiovasc. Res.*, 28, 365, 1994.
55. Mapp, C. E., Bevan, S., Boniotti, A., Papi, A., Chitano, P., Coser, E., Saetta, M., Maestrelli, P., Ciaccia, A., and Fabbri, L., unpublished data, 1993.
56. Vedder, H. and Otten, U., Biosynthesis and release of tachykinins from rat sensory neurons in culture, *J. Neurosci. Res.*, 30, 288, 1991.
57. Gamse, R., Molnar, A., and Lembeck, F., Substance P release from spinal cord slices by capsaicin, *Life Sci.*, 25B, 629, 1979
58. Del Bianco, E., Santicioli, P., Tramontana, M., Maggi, C. A., Cecconi, R., and Geppetti, P., Different pathways by which extracellular Ca^{2+} promotes calcitonin gene-related peptide release from central terminals of capsaicin-sensitive afferents of guinea pigs: effects of capsaicin, high K^+ and low pH media, *Brain Res.*, 566, 46, 1991.
59. Lou, Y. P., Franco-Cereceda, A., and Lundberg, J. M., Different ion channel mechanisms between low concentrations of capsaicin and high concentrations of capsaicin and nicotine regarding peptide release from pulmonary afferents, *Acta Physiol. Scand.*, 146, 119, 1992.

60. Santicioli, P., Del Bianco, E., Geppetti, P., and Maggi, C. A., Release of calcitonin gene-related peptide-like (CGRP-LI) immunoreactivity from rat isolated soleus muscle by low pH, capsaicin and potassium, *Neurosci. Lett.*, 143, 19, 1992.
61. Lou, Y. P., Franco-Cereceda, A., and Lundberg, J. M., Omega-conotoxin inhibits CGRP release and bronchoconstriction evoked by a low concentration of capsaicin, *Acta Physiol. Scand.*, 141, 135, 1991.
62. Lou, Y. P., Karlsson, J. A., Franco-Cereceda, A., and Lundberg, J. M., Selectivity of ruthenium red in inhibiting bronchoconstriction and CGRP release induced by afferent C-fibre activation in the guinea-pig lung, *Acta Physiol. Scand.*, 142, 191, 1991.
63. Kroll, F., Karlsson, J. A., Lundberg, J. M., and Persson, C. G., Capsaicin-induced bronchoconstriction and neuropeptide release in guinea pig perfused lungs, *J. Appl. Physiol.*, 68, 1679, 1990.
64. Campbell, D. T., Large and small vertebrate sensory neurons express different Na and K channel subtypes, *Proc. Natl. Acad. Sci. U.S.A.*, 89, 9569, 1992.
65. McMahon, S. B., Lewin, G., and Bloom, S. R., The consequences of long-term topical capsaicin application in the rat, *Pain*, 44, 301, 1991.
66. Anand, P., Bloom, S. R., and McGregor, G. P., Topical capsaicin pretreatment inhibits axon reflex vasodilation caused by somatostatin and vasoactive intestinal polypeptide in human skin, *Br. J. Pharmacol.*, 78, 665, 1983.
67. Tóth-Kásá, I., Jancsó, G., Bognar, A., Husz, S., and Obal, F., Capsaicin prevents histamine-induced itching, *Int. J. Clin. Pharmacol. Res.*, 6, 163, 1986.
68. Bjerring, P. and Arendt-Nielsen, L., Inhibition of histamine skin flare reaction following repeated topical applications of capsaicin, *Allergy*, 45, 121, 1990.
69. Campbell, E. A., Bevan, S., and Dray, A., Clinical applications of capsaicin and its analogs, in *Capsaicin in the Study of Pain*, Wood, J. N., Ed., Academic Press, London, 1993, chap. 12.
70. Holzer, P., Capsaicin: cellular targets, mechanisms of action, and selectivity for thin sensory neurons, *Pharmacol. Rev.*, 43, 143, 1991.
71. Dray, A., Bettany, J., and Forster, P., Actions of capsaicin on peripheral nociceptors of the neonatal rat spinal cord-tail *in vitro*: dependence of extracellular and independence of second messengers, *Br. J. Pharmacol.*, 101, 727, 1990.
72. Yeats, J. C., Boddeke, H. W. G. M., and Docherty, R. J., Capsaicin desensitization in rat dorsal root ganglion neurons is due to activation of calcineurin, *Br. J. Pharmacol.*, 107, 238P, 1992.
73. Yeats, J. C. and Docherty, R. J., and Bevan, S., Calcium-dependent and independent desensitization of capsaicin-evoked responses in voltage-clamped adult rat dorsal root ganglion (DRG) neurons in culture, *J. Physiol.*, 446, 390P, 1992.
74. Liu, J., Farmer, J. D., Lane, W. S., Friedman, J., Weissman, I., and Schreiber, S. L., Calcineurin is a common target of cyclophilin-cyclosporin A and FKBP-FK506 complexes, *Cell*, 66, 807, 1991.
75. Pitchford, S. and Levine, J. D., Prostaglandins sensitize nociceptors in cell culture, *Neurosci. Lett.*, 132, 105, 1991.
76. Santicioli, P., Patacchini, R., Maggi, C. A., and Meli, A., Exposure to calcium-free medium protects sensory fibers by capsaicin desensitization, *Neurosci. Lett.*, 80, 167, 1987.
77. Amann, R., Donnerer, J., Maggi, C. A., Giuliani, S., DelBianco, E., Weihe, E., and Lembeck, F., Capsaicin desensitization in vivo is inhibited by ruthenium red, *Eur. J. Pharmacol.*, 186, 169, 1990.
78. Chard, P. S., Savidge, J. R., Bleakman, D., and Miller, R. J., Calpain inhibitors prevent capsaicin-dependent neurotoxicity in dorsal root ganglion neurons, *Soc. Neurosci. Abstr.*, 18, 465.9, 1992.
79. Erdelyi, L. and Such, G., Effects of capsaicin on molluscan neurons: a voltage clamp study, *Comp. Biochem. Physiol.*, 85C, 319, 1986.
80. Erdelyi, L., Such, G., and Nedeljkovic, M., Effects of capsaicin on molluscan neurons: an intracellular study, *Comp. Biochem. Physiol.*, 85C, 313, 1986.
81. Dubois, J. M., Capsaicin blocks one class of K^+ channels in the frog node of Ranvier, *Brain Res.*, 245, 372, 1982.
82. Bethge, E. W., Bohuslavizki, K. H., Hansel, W., Kneip, A., and Koppenhofer, E., Effects of some potassium channel blockers on the ionic currents in myelinated nerve, *Gen. Physiol. Biophys.*, 10, 225, 1991.
83. Röper, J. and Schwartz, J. R., Heterogeneous distribution of fast and slow potassium channels in myelinated rat nerve fibers, *J. Physiol.*, 416, 93, 1989.
84. Yamanaka, K., Kigoshi, S., and Muramatsu, I., Conduction-block induced by capsaicin in crayfish giant axon, *Brain Res.*, 300, 113, 1984.
85. Petersen, M., Pierau, F.-K., and Weyrich, M., The influence of capsaicin on membrane currents in dorsal root ganglion neurons of guinea pig and chicken, *Pflugers Arch.*, 409, 403, 1987.
86. Castle, N. A., Differential inhibition of potassium currents in rat ventricular myocytes by capsaicin, *Cardiovasc. Res.*, 26, 1137, 1992.
87. Baker, M. and Ritchie, J. M., The action of capsaicin on type I delayed rectifier K^+ currents in rabbit Schwann cells, *Proc. R. Soc. London Ser. B*, 255, 259, 1994.

88. Kehl, S., Block by capsaicin of voltage-gated K^+ currents in melanotrophs of the rat pituitary, *Br. J. Pharmacol.*, 112, 616, 1994.
89. Petersen, M., Wagner, G., and Pierau, F.-K., Modulation of calcium-currents by capsaicin in a subpopulation of sensory neurons of guinea pig, *Naunyn-Schmiedeberg's Arch. Pharmacol.*, 339, 184, 1989.
90. Robertson, B., Docherty, R. J., and Bevan, S., Capsaicin inhibits voltage-activated calcium currents in a subpopulation of adult rat dorsal root ganglion neurons, *Soc. Neurosci. Abstr.*, 15, 354, 1989.
91. Docherty, R. J., Robertson, B., and Bevan, S., Capsaicin causes prolonged inhibition of voltage-activated calcium currents in adult rat dorsal root ganglion neurons in culture, *Neuroscience*, 40, 513, 1991.

Chapter 6

AFFERENT RESPONSES OF SENSORY NEURONS

Andy Dray

CONTENTS

I. INTRODUCTION

All tissues are innervated by afferent neurons, but the properties of these fibers may differ markedly depending on whether they are somatic afferents, innervating the skin, joints, and muscles or visceral afferents, innervating the cardiovascular system, the gastrointestinal tract, or renal and reproductive systems. In the context of the present discussion, reference will be made only to fine, slowly conducting afferent fibers, rather than rapidly conducting proprioceptive or mechanoreceptive somatic afferents.

Fine afferents (C- and Aδ-fibers) respond to a range of physiological stimuli, including heat, cold, and low-intensity mechanical stimulation or distension. A proportion of somatic afferent fibers may be nociceptive and may be specialized for this purpose. However, it is by no means clear that this is the case in visceral systems, where there is only sparse evidence for afferents that serve specific nociceptive function under normal physiological conditions (for discussion see Reference 1). Nociceptive fibers are activated by a variety of high-intensity stimuli, which are potentially tissue damaging, including heat and pressure, and the majority of nociceptors, the polymodal nociceptors, also respond to chemical stimulation. Indeed, chemical signaling is likely to be the most common and diverse form of signal generation in all types of afferent fibers, though this would not necessarily be accompanied by specific sensations as with nociceptors. In addition the mechanisms by which chemical stimuli rather than physical stimuli generate nerve signals are perhaps best understood.

Substances that are made during tissue damage and inflammation produce manifold and complex changes, ranging from overt activation and sensitization of afferent fibers to alterations of chemical phenotype and structure. The actions of these substances on fine afferents and the physiological and pathophysiological significance of these events will form the focus of this chapter.

0-8493-7646-7/96/$0.00+$.50

II. SENSORY FIBER SPECIALIZATION

Strict physiological criteria are needed to identify afferent fibers in order to study the function and properties of the nerve terminals where sensory transduction occurs. It is more convenient, however, from an experimental point of view, to study the cell body, located in spinal ganglia, and a number of distinct properties of these cells have been noted which distinguish different types of neuron.[2] Reliance can also be placed on identifying fine primary afferent neurons by the fact that many, but not all, fibers can be stimulated by capsaicin. Since many studies are made *in vitro,* particularly with preparations of cells maintained in culture, the particular experimental conditions must be born in mind; especially when attempting to extrapolate the data obtained in this manner to the interactions of chemicals with afferent nerves *in vivo.* However, cultured neurons are convenient for characterizing fundamental molecular properties of sensory nerve membranes without interference from heterogeneous surrounding tissues.

Morphological specializations of fine afferent nerve terminals (e.g., articular and corneal nerve endings) have been described. In the knee joint capsule these fibers branch at their endings and branches may be spread along blood vessels (chemosensitive?) or extend into dense connective tissue (mechanosensitive?). Characteristic axonal beads occur along their length, together with an end bulb structure; features characteristic of receptive sites.[3] In the cornea, C-fiber endings also form branching clusters which run perpendicular to the corneal surface while Aδ-fibers are distinctive by running parallel to the corneal surface.[4] Further characterization of these features is required, as well as more studies of specializations of other afferent endings. There is also *molecular specialization* of sensory neurons, as chemical interactions are likely to occur at specific receptors on the nerve terminal (peripheral or central). Under pathophysiological conditions these processes will be in a state of continuous change as the nerve terminals may be undergoing simultaneous structural modifications such as sprouting or repair following an injury or inflammation. Little is known, however, of any differences in the sensory properties of nerve fibers undergoing morphological alterations. However, since many mediators may be present around sensory nerve terminals following an injury or inflammation, it is conceivable that many chemical interactions would be occurring simultaneously; perhaps at different sites on the same terminal or at several nerve branches. This would create an opportunity for signal integration once signal transduction had occurred.

Normally, sensory neurons would be responsive to trophic factors in their environment (e.g., nerve growth factor), as this would be important for maintaining normal physiological processes.[5] Significantly the chemical products of inflammation allow sensory fibers to increase their sensitivity for detecting and responding to adverse conditions and these effects are related to the production of hyperalgesia.[6] An extreme example of the range of sensitivity that sensory fibers are capable of occurs with "silent" or "sleeping" nociceptors. These normally appear unresponsive, even to intense physiological stimuli, and may therefore represent the extreme range of sensory neuron sensitivity. However, when activated by inflammatory mediators, or exogenously administered irritants, they become responsive to mechanical stimulation or exhibit spontaneous activity.[7,8] The mechanisms involved are not yet clear but it is likely that they are similar to other mechanisms of activation and sensitization described below. Additionally it is becoming evident that inflammatory mediators induce changes in the neurochemistry of sensory neurons to enhance neurogenic interactions with surrounding tissues, which may be important also for facilitating tissue repair. These changes are also necessary for boosting nociceptive signal transmission by the enhanced release of sensory neuropeptide transmitters in the spinal cord. Indeed activation of fine afferents is necessary for the transmission of normal, physiologically important nociceptive signals,[2,6,9] but have additional critical function in inducing hyperexcitability in the spinal cord following peripheral injury. Under these pathological circumstances activation of large-diameter, low-threshold, mechanoreceptive Aβ-fibers is also capable of eliciting painful sensations.[9,10]

III. SIGNAL TRANSDUCTION

Chemical mediators derived from damaged tissue, microvessels, immune cells and surrounding tissues, and other sensory and sympathetic nerves usually interact with a multiplicity of receptors on sensory nerve terminals. These receptors, for the most part, are coupled with a limited repertoire of cellular regulatory intermediates (G-proteins, second messengers) which alter membrane permeability and cellular ion concentration. Thus, signal generation by unrelated chemicals and, therefore, the output of sensory neurons can be amplified or attenuated by adjustments of common pathways. Transient or long-lasting alterations in membrane excitability may follow, as well as changes in cellular biochemistry and structural modification of cytoarchitecture and connectivity.

As in other excitable tissues, cellular excitability changes in sensory neurons occur mainly by permeability changes to potassium, sodium, and calcium ions. These ion channels can be distinguished by a number of neurotoxins.[11] However, it is unusual that many fine sensory neurons also express sodium channels that are insensitive to blockade by tetrodotoxin and it is notable that the expression of this type of sodium channel is regulated by nerve growth factor (NGF),[12] which occurs in greater abundance during infalmmation.[13] It is tempting to speculate that these features are more than casually related and may be important for peripheral mechanisms of hyperalgesia and allodynia. Indeed sodium channel density may increase at sites of injury of large peripheral nerve fibers[14] and cause pain. Under these conditions abnormal sodium channel activation may be associated with the production of hyperalgesia, allodynia, or spontaneous pain.[2,15]

Calcium entry through voltage-gated calcium channels affects many cellular processes, including the regulation of other membrane channels, the release of transmitters, and the regulation of many enzymes (especially kinases and phosphatases), which cause a variety of cellular effects. The inward calcium current contributes directly to membrane depolarization, and mainly accounts for the characteristic "hump" on the falling phase of the C cell action potential. Voltage gated calcium channels are sensitive to many agents (e.g., baclofen, noradrenaline, dopamine, somatostatin, opiates), which inhibit high-threshold (L and/or N) channels in sensory neurons (see References 2 and 11). Calcium-mediated T-currents carry a substantial depolarizing current, and may be responsible for repetitive firing of sensory neurons following a single stimulus. T channels are unaffected by most of the inhibitory modulators that act on L and/or N channels, although opiates inhibited T-currents as well as high-threshold currents. In contrast, the entry of calcium through high-threshold (L or N) channels, which depends greatly on the amplitude and duration of the action potential, is modulated by chemical agents. In general, calcium entry through T channels is mainly involved in the control of neuronal firing. Calcium entry through high-threshold N channels controls transmitter release, which can be blocked by ω-conotoxin, but is relatively insensitive to dihydropyridines.[2,16] Unfortunately, there are no selective T channel blockers, so the contribution of these channels to the release of neurochemicals from sensory neurons cannot be assessed as directly.

Other sources of intracellular calcium are derived from activation of receptor-coupled second messengers, particularly 1,4,5-inositol-trisphosphate (IP_3), which mobilizes intracellular calcium stores. However, the complex spatial and temporal relationships of these intracellular sources of calcium are not well understood;[17] mobilization of intracellular calcium is likely to affect the activity of a number of systems highlighted earlier.

In some sensory neurons, several inflammatory mediators, including bradykinin, prostaglandins, and serotonin, increase excitability, and thus contribute to hyperalgesia, by blocking specific calcium-activated K channels.[18,19] These are responsible for the later phases of the after-hyperpolarization (AHP) that follows an action potential, in some types of visceral sensory neurons. Calcium-activated K channels impose a period of reduced excitability after each action potential, and are thus important in regulating the firing pattern of the neuron.[2,11,18,19]

IV. FACTORS AFFECTING THE EXCITABILITY OF AFFERENTS

A variety of exogenous substances such as irritants (capsaicin, mustard oil) as well as endogenously produced substances (inflammatory mediators, neurogenic mediators) can directly excite fine afferents. Other chemicals administered exogenously or generated after injury induce sensitization of these afferents so that their threshold for activation by other stimuli (chemical or physical) is lowered. The features of these effects, activation or sensitization, can be distinguished for different substances.

A. ACTIVATORS OF SENSORY NEURONS

1. Kinins

Bradykinin and kallidin (lysyl-bradykinin) are liberated from precursor kininogen in the vasculature and from damaged tissues, respectively, by the action of proteolytic enzymes. Kinins can also be formed by immune cells (mast cells, basophils) associated with acute inflammatory reactions following the release of cellular proteases. Bradykinin is one of the most potent stimulants of nociceptors (skin, joint, muscle) and sensory neurons. Kallidin also activates nociceptors and the mechanism of action is presumed to be identical to that of bradykinin, though fewer studies have directly addressed this. Both bradykinin and kallidin are rapidly degraded by kininases to generate the active metabolites desArg^{9}bradykinin or des-Arg^{10}-kallidin, respectively.[20] To date, however, the pathophysiological significance of these metabolites, in changing the activity of sensory neurons, has received little attention.

Though there are two main classes of kinin receptor, B_1 and B_2, bradykinin and kallidin act mainly on the B_2 receptor to activate sensory neurons.[21] The B_2 receptor belongs to the family of receptors with seven transmembrane spanning domains and which is coupled with G proteins (possibly Gq/11). Several specific antagonists have been made for the B_2 receptor, including peptides (e.g., Hoe 140) and nonpeptides (e.g., WIN 64338).[22] The B_1 receptor is also G protein-coupled[23] and is preferentially activated by des-Arg^9-bradykinin or des-Arg^{10}-kallidin, while Leu^8-des-Arg^9-bradykinin has been used for many years as the prototypic antagonist of this receptor. It is unclear at present whether B_1 receptors occur on sensory neurons.

Bradykinin-induced activation of sensory neurons occurs through activation of phospholipase C. This generates two intracellular second messengers, IP_3 and diacylglycerol (DAG), following cleavage of membrane phospholipids. IP_3 stimulates the release of calcium from intracellular stores, producing a rise in free cytosolic calcium concentration within the cell. The effect of DAG is to activate protein kinase C (PKC) and to phosphorylate cellular proteins, including membrane receptors and ion channels. PKC plays a key role in the excitation of sensory neurons by bradykinin,[24,25] which is associated with an inward ionic current and an increase in membrane conductance, mainly to sodium ions.[24] B_2 receptor activation is also coupled with a phospholipase A_2-mediated increase in arachidonic acid production. The role of arachidonic acid is not clear but, as in other cells, this acts purely as a precursor for the elaboration of other prostanoids, particularly prostaglandins.[26] Since bradykinin can induce calcium influx[27] in sensory neurons, this may cause a number of secondary effects, such as the release of neuropeptides, e.g., substance P and calcitonin gene-related peptide (CGRP), and contribute towards arachidonic acid production via the activation of phospholipase C and subsequent metabolism of DAG.

2. 5-Hydroxytryptamine (5-HT)

Mild and transient pain is produced by 5-HT when applied to a blister base.[28] This suggests that the release of 5-HT, from platelets and mast cells, can directly excite sensory neurons. In keeping with this, 5-HT-induced pain and sensory neuron excitation can be blocked by the 5-HT_3/5-HT_4 receptor antagonist tropisetron (ICS 205,930).[28,29] Although the 5-HT_3 binding site is part of a cation (Na^+)-selective ion channel,[30] which may open to depolarize and activate sensory neurons, other data indicate that the effects of 5-HT may also be mediated in part by other mechanisms involving cellular second messengers. Thus, 5-HT-induced activation of sensory neurons was attenuated by staurosporine and by ICS 205,930,[31] suggesting that this effect was likely due to PKC activation and in part to adenylate cyclase, which is coupled with the 5-HT_4 receptor. 5-HT also appears to activate sensory neurons via 5-HT_2 receptors which are linked through G proteins to close potassium channels. This also results in membrane depolarization and repetitive neuronal firing.[19,32]

3. Protons

Proton activation of unmyelinated fibers is likely to contribute to the sensation of aching and discomfort following tissue acidosis after muscle exercise. Indeed direct activation of nociceptors accounts for the sharp stinging pain produced by local injections of acidic solutions. Such observations suggest that protons may have a pathophysiological role in hypoxia/anoxia, as well as in inflammation,[33] in which the pH of the extracellular environment is known to fall. Pain thus results from direct nociceptor activation and by enhancing the effects of other inflammatory mediators.[34]

Proton-induced activation of sensory neurons involves two types of membrane depolarization. One type is associated with a rapid increase in membrane cation permeability evoked by pH changes in the normal physiological range, whereas the second type is associated with a more prolonged increase in membrane permeability and is evoked by lower extracellular pH. The latter is produced by concentrations of protons that can give rise to sustained nerve activation[35] and which have been shown to induce an enhancement of mechanosensitivity in afferent fibers.[36]

It is likely that protons activate nociceptors by acting on the external membrane surface, although activation could also occur following the intracellular generation of protons.[36] In support of this is the observation that CO_2-induced activation of nociceptors, presumably via the generation of carbonic acid, can be abolished by inhibition of intracellular carbonic anhydrase.

It is curious that activation of sensory neurons by protons has features that are strikingly similar to those shown by capsaicin.[37] The significance of this is not entirely clear, but capsaicin, a pungent principle in *Capsicum* peppers, has highly specific actions on polymodal nociceptors[37,38] to cause membrane depolarization by increasing the permeability to sodium and calcium ions. This is achieved by the opening of ion channels with unique properties which, for example, differ from those of voltage gated ion channels described earlier, but which are coupled with a capsaicin receptor. These channels are not blocked by a number of conventional ion channel blockers but may be blocked by ruthenium red.[39]

The capsaicin receptor on sensory neurons can be distinguished by the competitive antagonist, capsazepine,[38,40] suggesting that there may be an endogenous ligand that recognizes the receptor under physiological or pathophysiological circumstances. However, although capsaicin and protons share similar features of sensory neuronal activation, it seems that protons are not ubiquitous candidates for the capsaicin receptor, as capsazepine has no effect on proton-induced activation of sensory neurons.[41] Other studies indicate, however, that proton-induced activation of trigeminal neurons[42] or tracheal afferent fibers,[43] as well as proton-induced neuropeptide release from visceral sensory neurons, can be inhibited by capsazepine.[44] These observations raise the possibility that proton-induced activation of some types of sensory neurons may occur via a capsaicin receptor. Interestingly, the responsiveness of sensory neurons to both capsaicin and protons is controlled by NGF, normally released from tissues such as fibroblasts and keratinocytes, and whose production is stimulated during inflammation.[13]

4. Nitric Oxide

There is abundant evidence that this reactive molecule is associated with a variety of pain conditions and is involved in regulating the activity of sensory neurons.[45] Nitric oxide can be made by many neural and non-neural cells. It is formed from L-arginine following the activation of constitutive nitric oxide synthase (cNOS) by calcium and other cofactors and alters cellular processes mainly via the activation of guanylate cyclase and the production of cyclic GMP.[46] Small and medium-sized sensory neurons are able to make NO[47] and though it is possible that NO acts intracellularly in the cell of origin, current observations favor the view that it is released to act nearby since an increase in cyclic GMP occurs in satellite cells of sensory ganglia upon administration of NO donors.[48] However, because of its extreme reactivity it seems unlikely that NO acts unchanged but is likely to interact with other chemical species soon after its formation.

Although intradermal injection of NO can cause pain,[49] there is little evidence for direct sensory neuron activation by NO.[50] However, NO donors have been postulated to activate cerebral sensory fibers directly to cause the release of the vasodilator, CGRP.[51] In keeping with this, NO has been suggested to contribute to migraine and other types of head pain.[52] There is also evidence that NO can change the excitability of sensory neurons indirectly by affecting their sensitivity to bradykinin. A cyclic GMP mechanism is involved in this process and may work by inducing receptor desensitization.[50,53,54]

It is important to note that a calcium-independent form of NOS (i-NOS) can be induced in a number of immune cell types, including macrophages, and in microglia cells in the central nervous system during inflammation. In addition, following a peripheral nerve injury, NOS is also induced in many small sensory neurons which are likely to be involved in nociception.[55] The i-NOS can produce high concentrations of NO, sufficient to induce a number of systemic pathological effects (e.g., hypotension, vascular leakage) including cytotoxicity, especially when combined with other reactive oxygen species to form peroxynitrite. i-NOS has also been shown to have an important role in the regulation of an inducible form of cyclooxygenase (COX)-2 activity[56] and thereby the production of prostanoids, which also act as powerful regulators of sensory neuron excitability.

B. SUBSTANCES THAT MAINLY SENSITIZE SENSORY FIBERS

Prostaglandins rarely activate cutaneous nociceptive afferents directly and do not evoke pain when injected intradermally into human skin.[57] Other studies have indicated, however, that both PGE_2 and prostacyclin (PGI_2) can increase the activity of nociceptors.[58,59] These findings raise the possibility that prostanoids could contribute directly to the activation of afferent neurons in inflammatory conditions such as arthritis. Indeed studies of rat isolated sensory (DRG) neurons have shown that application of PGE_2 can evoke excitatory, depolarizing currents in some neurons[60] and induce the release of the sensory neuropeptides, substance P and CGRP.[26] This effect could be the basis for any direct neuronal activation.

The major effect of prostaglandins, however, is to sensitize afferent neurons to heat and mechanical stimulation, as well as to noxious chemical agents. Cyclic AMP stimulation is involved since agents that elevate cyclic AMP produce hyperalgesia by a direct action of cyclic AMP on the afferent neuron.[61] The ionic mechanism that underlies this sensitization has not been fully elucidated but in visceral sensory neurons (nodose ganglion cells) increased excitability is often associated with the inhibition of a long-lasting spike after-hyperpolarization (slow AHP) which is regulated by a cyclic AMP-dependent, calcium-activated potassium conductance mechanism. The slow AHP, following a single action potential, produces a state of inexcitability which limits the number of action potentials that can be evoked upon stimulation. Prostaglandins D_2 and E_2 increase excitability via blockade of this mechanism.[18] In somatic sensory neurons, however, there is little or no post-spike slow AHP and few studies have explored the mechanism of activation and sensitization by prostaglandins. It is likely, however, that other types of potassium conductance mechanism may be blocked by prostanoids to induce repetitive neuronal discharges after neural activation.

As mentioned earlier, bradykinin is a potent endogenous algogen that can induce pain by direct stimulation of nociceptors; it can also sensitize nociceptors. Indeed there is a strong synergism between the actions of bradykinin and other algogenic substances, e.g., prostaglandins and 5-HT. Sympathetic neurons may also be involved in bradykinin-mediated sensitization and mechanical hyperalgesia,[62] possibly through the release of other mediators such as prostaglandins. In the skin, however, sympathectomy did not alter heat hyperalgesia induced by bradykinin.[63,64]

Bradykinin, through prostanoid formation, inhibits the post-spike slow AHP in visceral neurons by stimulating cyclic AMP formation. This allows the cell to fire repetitively following depolarization, a mechanism that can account for sensitization of sensory neurons and the induction of hyperalgesia. On the other hand cyclic GMP-depedent mechanisms have also been implicated in the action of bradykinin at B_2 receptors on sensory neurons. Thus, bradykinin-induced IP_3 production and activation of sensory neurons is reduced in the presence of cyclic GMP, possibly via desensitization of this receptor.[53] Nitric oxide appears to be involved in this process, since inhibitors of NOS attenuated bradykinin-induced desensitization in cultured sensory neurons.[50,53]

B_1 receptors also contribute to the production of pain and hyperalgesia and the expression of B_1 receptor activity is increased remarkably by inflammatory mediators, particularly by cytokines. In addition B_1 receptor antagonists such Leu^8-des-Arg^9-bradykinin, are analgesic under these circumstances.[20] There is no evidence, however, that B_1 receptor ligands activate or sensitize sensory neurons directly, so that the hyperalgesic actions are likely to be secondary to the actions of other mediators such as prostanoids or neuropeptides released from nearby cells.

5-HT also sensitizes nociceptors and lowers their threshold to heat and pressure stimuli.[65] Sensitization and 5-HT-induced hyperalgesia involve both $5\text{-}HT_1$ and $5\text{-}HT_2$ receptors.[66] Activation of a cyclic AMP-dependent kinase system is required for 5-HT-mediated sensitization of nociceptors since hyperalgesia was blocked by Rp-cAMPS (an inhibitor of cyclic AMP), and augmented by a phosphodiesterase inhibitor.[62] Furthermore it has been proposed that 5-HT increases sensory neural excitability by a cyclic AMP-mediated reduction in K^+ permeability. This mechanism attenuates the slow AHP that follows the action potential in visceral neurons and provokes repetitive firing of these cells.[19] In addition to these observations, the 5-HT-induced enhancement of bradykinin-evoked pain in the human blister base was blocked by ICS 205,930, a $5\text{-}HT_3/5\text{-}HT_4$ receptor antagonist.

It is important to mention that sensory neuronal excitability and afferent activity is also influenced by peripheral inhibitory factors. Perhaps the most important of these and the best studied are the opioid peptides, enkephalin and endorphin. Thus, during inflammation there is an induction of opioid peptide synthesis by immune cells coupled with the expression of opioid receptors in peripheral cutaneous nerve fibers.[67] Stimulation of immune cells would allow opioids to be released close to afferent nerve terminals. In keeping with this view, exogenous opioids may induce analgesia by directly depressing activity in peripheral sensory neurons[68,69] and by decreasing sympathetic fiber activity.[70] The mechanism of action of opioids on sensory neuron membranes is complex. Depression of neural excitability is mediated by μ, δ, and κ-receptors via receptor-coupled activation of a G protein and inhibition of adenylate cyclase, and by a reduction of membrane calcium conductance and/or increased potassium ion conductance.[71,72]

C. EXCITABILITY CHANGES BY TROPHIC FACTORS AND THE GENERATION OF ECTOPIC ACTIVITY

Nerve growth factor (NGF) is produced by a number of peripheral target tissues, including fibroblasts and Schwann cells, and acts as a trophic factor during sensory neuronal development.

NGF acts on a neural membrane receptor, trk A, one of several tyrosine kinase-coupled receptors. It is bound and transported to the cell soma where it acts to promote gene transcription,[73] and thereby regulate gene expression. This leads to an increase in the synthesis of several important proteins and neuropeptides.[74] For example, NGF increases the synthesis, axoplasmic transport, and neuronal content of substance P and CGRP[13,74] and regulates the levels of at least two types of ion channel in DRG neurons, the capsaicin receptor/ion channel[75] and the tetrodotoxin-resistant Na^+ channel.[12] In both cases, exposure to NGF promotes channel expression while removal of the growth factor results in a large reduction of complete loss of the channel. NGF may also modify sensory neuron properties by promotion of axonal sprouting, thereby increasing the peripheral receptive field.[76] A similar effect may occur at the central terminals of nociceptive neurons exposed to abnormally high levels of NGF. If so, it would explain the increased strength of synaptic connections between sensory and dorsal horn neurons after experimental elevation of NGF *in vivo*,[5] although other mechanisms, such as greater release of sensory neuropeptides, could also play an important role. These changes may explain how NGF induces hyperalgesia in a number of inflammatory conditions. Thus, NGF levels are elevated by experimental inflammation induced by Freund's adjuvant injection or exposure to ultraviolet irradiation.[3,77] It is also increased in the synovial fluid taken from patients with rheumatoid arthritis.[78] Injection of NGF produces increased sensitivity to noxious stimuli, while administration of antibodies to NGF to animals exposed to NGF or inflammatory mediators reduces the responses to painful and inflammatory stimuli.[5,77]

Peripheral axotomy or nerve lesions can induce ectopic activity in the resulting neuroma tissue, which is formed at the site of injury but also in undamaged sensory ganglia that are remote from the injury. Ectopic activity is thought to account for spontaneous pain that occurs in certain types of painful neuropathies. Several compounds (amitryptiline, carbamazepine, lidocaine) produce analgesia by reducing ectopic activity without blocking nerve conduction.[79] The mechanism underlying this effect is likely due to the block of activated sodium ion channels. This suggests that abnormal ion channel activity is associated with the nerve damage and the related pain. There is little information about how ectopic activity is generated but accumulation of sodium channels in large diameter peripheral nerves at the sites of nerve injury has been noted.[14,15]

Disturbances of peripheral nerve excitability in neuropathic pain may also occur through alterations in potassium or calcium ion function as well as by altered local blood flow. But there is relatively little evidence in support of these alternatives. Inhibition of potassium ion conduction increases nerve excitability and disturbed potassium channel function may be worth further investigation. On the other hand calcium channel blockers such as nimodipine improve experimental diabetic and cisplatin-induced neuropathy (see Reference 22), possibly through improving local blood flow or by altering the excitability of peripheral afferent and sympathetic nerves.

V. SUMMARY

Sensory neurons respond to their environment in a number of complex ways: by activation, with the generation of nerve impulses to signal painful as well as nonpainful events; by sensitization, so that exogenous stimuli (thermal, mechanical, and chemical) may be more effectively detected and transmitted; by changes in phenotype, which prepares the sensory neurons for prolonged environmental changes and which facilitates the processes of tissue repair.

Sensory neurons show particular sensitivity to chemical factors in the environment which vary from trophic elements required for normal physiological function, to products of tissue damage and inflammation. As far as we know, most of these substances operate through membrane receptors coupled with ion channels to regulate cellular biochemical processes and genomic processes which provide opportunities for a vast range of cellular modifications. An appreciation of these events is emerging but progress will only be enhanced by the development of new and specific tools to characterize them. In addition careful and painstaking studies are required to monitor the rapid and long-lasting phenotypic changes that sensory neurons undergo and to identify the appearance and function of novel cell proteins. Traditionally the approach to studying these factors has been piecemeal, one element being studied without particular reference to the rest. However, the outcome of complex changes in sensory cell function, which will be occurring simultaneously, may only be appreciated by using simulation methods and interactive cellular models.

REFERENCES

1. Cervero, F., Sensory innervation of the viscera: peripheral basis of visceral pain, *Physiol. Rev.*, 74, 95, 1994.
2. Rang, H. P., Bevan, S. J., and Dray, A., Nociceptive peripheral neurones: cellular properties, in *Textbook of Pain*, Wall, P. D. and Melzack, R., Eds., Churchill-Livingstone, Edinburgh, 1994, 57.
3. Heppelmann, B., Messlinger, K., Neiss, W. F., and Schmidt, R. F., Ultrastructural three-dimensional reconstruction of group III and group IV sensory nerve endings ("free nerve endings") in the knee joint capsule of the cat: evidence for multiple receptive sites, *J. Comp. Neurol.*, 292, 103, 1990.
4. Maciver, M. B. and Tanelian, D. L., Structural and functional specialization of Aδ and C fiber free nerve endings innervating rabbit corneal epithelium, *J. Neurosci.*, 13, 4511, 1993.
5. Lewin, G. R. and Mendell, L. M., Nerve growth factor and nociception, *Trends Neurosci.*, 16, 353, 1993.
6. Treede, R. D., Meyer, R. A., Raja, S. N., and Campbell, J. N., Peripheral and central mechanisms of cutaneous hyperalgesia, *Prog. Neurobiol.*, 38, 397, 1992.
7. Häbler, H. J., Jänig, W., and Koltzenburg, M., Activation of unmyelinated afferent fibers by mechanical stimuli and inflammation of the urinary bladder in the cat, *J. Physiol.*, 425, 545, 1990.
8. Handwerker, H. O., Kilo, S., and Reeh, P. W., Unresponsive afferent nerve fibres in the sural nerve of the rat, *J. Physiol.*, 435, 229, 1991.
9. Gracely, R. H., Lynch, S. A., and Bennet, G. J., Painful neuropathy: altered central processing maintained dynamically in peripheral input, *Pain*, 51, 175, 1992.
10. Woolf, C. J. and Doubell, T. P., The pathophysiology of chronic pain — increased sensitivity of low threshold Aβ fibre inputs, *Curr. Biol.*, 4, 525, 1994.
11. Nowycky, M., Voltage gated ion channels in dorsal root ganglion neurons, in *Sensory Neurons: Diversity, Development and Plasticity*, Scott, S. A., Ed., Oxford University Press, New York, 1992, 97.
12. Aguayo, L. G. and White, G., Effects of nerve growth factor o TTX- and capsaicin-sensitivity in adult rat sensory neurones, *Brain Res.*, 570, 61, 1992.
13. Bonnerer, J., Schuligoi, R., and Stein, C., Increased content and transport of substance P and calcitonin gene-related peptide in sensory nerves innervating inflamed tissue: evidence for a regulatory function of nerve growth factor *in vivo*, *Neuroscience*, 49, 693, 1992.
14. Devor, M., Govrin-Lippman, R., and Angelides, K., Na^+ channel immunolocalization in peripheral mammalian axons and changes following nerve injury and neuroma formation, *J. Neurosci.*, 13, 1976, 1992.
15. Matzner, O. and Devor, M., Hyperexcitability at sites of nerve injury depends on voltage sensitive Na^+ channels, *J. Neurophysiol.*, 72, 349, 1994.
16. Perney, T. M., Hirning, L. D., Leeman, S. E., and Miller, R. J., Multiple calcium channels mediate transmitter release from peripheral neurons, *Proc. Natl. Acad. Sci. U.S.A.*, 83, 6656, 1986.
17. Taylor, C. W., Ca^+ sparks a wave of excitement, *Trends Pharmacol. Sci.*, 15, 271, 1994.
18. Weinreich, D. and Wonderlin, W. F., Inhibition of calcium-dependent spike after-hyperpolarization increases excitability of rabbit visceral sensory neurons, *J. Physiol.*, 394, 415, 1987.
19. Christian, E. P., Taylor, G. E., and Weinreich, D., Serotonin increases excitability of rabbit C-fibre neurons by two distinct mechanisms, *J. Appl. Physiol.*, 67, 584, 1989.
20. Dray, A. and Perkins, M., Bradykinin and inflammatory pain, *Trends Neurosci.*, 16, 99, 1993.
21. Hall, J. M., Bradykinin receptors: pharmacological properties and biological roles, *Pharmacol. Ther.*, 56, 131, 1992.
22. Dray, A., Urban, L., and Dickenson, A., Pharmacology of chronic pain, *Trends Pharmacol. Sci.*, 15, 190, 1994.
23. Menke, J. G., Borkowski, J. A., Bierilo, K. K., MacNeil, T., Derrick, A. W., Schneck, K. A., Ransom, R. W., Strader, C. D., Linemeyer, D. L., and Hess, J. F., Expression cloning of a human B_1 bradykinin receptor, *J. Biol. Chem.*, 269, 21583, 1994.

24. Burgess, G. M., Mullaney, J., McNeil, M., Dunn, P., and Rang, H. P., Second messengers involved in the action of bradykinin on cultured sensory neurons, *J. Neurosci.*, 9, 3314, 1989.
25. Dray, A., Patel, I. A., Perkins, M. N., and Rueff, A., Bradykinin-induced activation of nociceptors: receptor and mechanistic studies on the neonatal rat spinal cord-tail preparation in vitro, *Br. J. Pharmacol.*, 107, 1129, 1992.
26. Vasko, M. R., Campbell, W. B., and Waite, K. J., Prostaglandin E2 enhances bradykinin-stimulated release of neuropeptides from rat sensory neurons in culture, *J. Neurosci.*, 14, 4987, 1994.
27. Thayer, S. T., Ewald, D. A., Perney, T. M., and Miller, R. J., Regulation of calcium homeostasis in sensory neurons by bradykinin, *J. Neurosci.*, 8, 4089, 1988.
28. Richardson, B. P., Engel, G., Donatsch, P., and Stadler, P. A., Identification of serotonin M-receptor subtypes and their specific blockade by a new class of drugs, *Nature*, 316, 126, 1985.
29. Dumuis, A., Gozlan, H., Sebben, M., Ansanay, H., Rizzi, C. A., Turconi, M., Monferini, K., Schiantarelli, P., Ladensky, H., and Bockaert, J., Characterisation of a novel serotonin ($5HT_4$) receptor antagonist of the azabicyclokylbenzimidazoline class: DAU6285, *Naunyn-Schmiedeberg's Arch. Pharmacol.*, 345, 264, 1993.
30. Maricq, A. V., Peterson, A. S., Brake, A. J., Meyers, R. M., and Julius, D., Primary structure and functional expression of the 5HT3 receptor, a serotonin-gated ion channel, *Science*, 254, 432, 1991.
31. Robertson, B. and Bevan, S., Properties of 5-hydroxytryptamine$_3$ receptor-gated currents in adult dorsal root ganglion neurones, *Br. J. Pharmacol.*, 102, 272, 1991.
32. Todorovic, S. and Anderson, E. G., 5-HT2 and 5-HT3 receptors mediate two distinct depolarizing responses in rat dorsal root ganglion neurons, *Brain Res.*, 511, 71, 1990.
33. Corbe, S. M. and Poole-Wilson, P. A., The time of onset and severity of acidosis in myocardial ischaemia, *J. Mol. Cell. Cardiol.*, 12, 745, 1980.
34. Kessler, W., Kirchoff, H., Reeh, P. W., and Handwerker, H. O., Excitation of cutaneous afferent nerve endings in vitro by a combination of inflammatory mediators and conditioning effect of substance P, *Exp. Brain Res.*, 91, 467, 1992.
35. Bevan, S. and Yeats, J., Protons activate a cation conductance in a sub-population of rat dorsal root ganglion neurons, *J. Physiol.*, 433, 145, 1991.
36. Steen, K. H., Reeh, P. W., Anton, F., and Handwerker, H. O., Protons selectively induce lasting excitation and sensitization to mechanical stimuli of nociceptors in rat skin, in vivo, *J. Neurosci.*, 12, 86, 1992.
37. Bevan, S., Rang, H. P., and Shah, K., Capsazepine does not block the proton induced activation of rat sensory neurons, *Br. J. Pharmacol.*, 107, 235P, 1992.
38. Dray, A., Neuropharmacological mechanisms of capsaicin and related substances, *Biochem. Pharmacol.*, 44, 611, 1992.
39. Bevan, S. and Szolcsanyi, J., Sensory neuron-specific actions of capsaicin: mechanisms and applications, *Trends Pharmacol. Sci.*, 11, 330, 1990.
40. Bevan, S., Hothi, S., Hughes, G., James, I. F., Rang, H. P., Shah, K., Walpole, C. S. J., and Yeats, J. C., Capsazepine: a competitive antagonist of the sensory neurone excitant capsaicin, *Br. J. Pharmacol.*, 197, 44, 1992.
41. Liu, L. and Simon, S. A., A rapid capsaicin-activated current in rat trigeminal ganglion neurons, *Proc. Natl. Acad. Sci. U.S.A.*, 91, 738, 1994.
42. Fox, A. J., Urban, L., Barnes, P. J., and Dray, A., Capsazepine selectively inhibits capsaicin and low pH-evoked firing of single C-fibres in the guinea-pig airways in vitro through an epithelium-independent mechanism, *Br. J. Pharmacol.*, in press.
43. Lou, Y.-P. and Lundberg, J. M., Inhibition of low pH evoked activation of airway sensory nerves by capsazepine, a novel capsaicin-receptor antagonist, *Biochem. Biophys. Res. Commun.*, 189, 537, 1992.
44. Meller, S. T. and Gebhart, G. F., Nitric oxide (NO) and nociceptive processing in the spinal cord, *Pain*, 52, 127, 1993.
45. Moncada, S., Palmer, R. M., and Higgs, E. A., Nitric oxide: physiology, pathophysiology, and pharmacology, *Pharmacol. Rev.*, 43, 109, 1991.
46. Aimi, Y., Fujimura, M., Vincent, S. R., and Kimura, H., Localization of NADPH-diaphorase-containing neurons in sensory ganglia of the rat, *J. Comp. Neurol.*, 306, 382, 1991.
47. Morris, R., Southam, E., Braid, D. J., and Garthwaite, J., Nitric oxide may act as a messenger between dorsal root ganglion neurones and their satellite cells, *Neurosci. Lett.*, 137, 29, 1992.
48. Holthusen, H. and Arndt, J. O., Nitric oxide evokes pain in humans on intracutaneous injection, *Neurosci. Lett.*, 165, 71, 1994.
49. McGehee, D. S., Goy, M. F., and Oxford, G. S., Involvement of the nitric oxide-cyclic GMP pathway in the desensitization of bradykinin responses of cultured rat sensory neurons, *Neuron*, 9, 315, 1992.
50. Wei, P., Moskowitz, M. A., Boccalini, P., and Kontos, H. A., Calcitonin gene-related peptide mediates nitroglycerin and sodium nitroprusside-induced vasodilatation in feline cerebral arterioles, *Circ. Res.*, 70, 1313, 1992.
51. Olesen, J., Thomsen, L. L., and Iversen, H., Nitric oxide is a key molecule in migraine and other vascular headaches, *Trends Pharmacol. Sci.*, 15, 149, 1994.

52. Bradley, C. and Burgess, G. A., Nitric oxide synthase inhibitor reduces desensitisation of bradykinin-induced activation of phospholipase C in sensory neurones, *Trans. Biochem. Soc.*, 23, 4353, 1993.
53. Rueff, A., Patel, I. A., Urban, L., and Dray, A., Regulation of bradykinin sensitivity in peripheral sensory fibres of the neonatal rat by nitric oxide and cyclic GMP, *Neuropharmacology*, in press.
54. Verge, V. M. K., Xu, Z., Xu, X.-J., Wiesenfelt-Hallin, Z., and Hokfelt, T., Marked increase in nitric oxide synthase mRNA in rat dorsal root ganglia after peripheral axotomy: in situ hybridization and functional studies, *Proc. Natl. Acad. Sci. U.S.A.*, 89, 11617, 1992.
55. Salvemini, D., Misko, T. P., Masferrer, J. L., Seibert, K., Cuiire, M. G., and Needleman, P., Nitric oxide activates cyclooxygenase enzymes, *Proc. Natl. Acad. Sci. U.S.A.*, 90, 7240, 1993.
56. Handwerker, H. O., Influences of algogenic substances and prostaglandins on the discharges of unmyelinated cutaneous nerve fibres identified as nociceptors, in *Advances in Pain Research and Therapy*, Bonica, J. J. and Albe-Fessard, D., Eds., Raven Press, New York, 1976, 41.
57. Rueff, A. and Dray, A., Sensitization of peripheral afferent fibres in the in vitro neonatal rat spinal cord-tail by bradykinin and prostaglandins, *Neuroscience*, 54, 527, 1993.
58. Birrell, G. J., McQueen, D. S., Iggo, A., Coleman, R. A., and Grubb, B. D., PGI_2-induced activation and sensitization of articular mechanonociceptors, *Neurosci. Lett.*, 124, 5, 1991.
59. Mizumura, K., Sato, J., and Kumazawa, T., Effects of prostaglandins and other putative chemical intermediaries on the activity of canine testicular polymodal receptors studied in vitro, *Pflugers Arch.*, 408, 565, 1987.
60. Nicol, G. D., Klingberg, D. K., and Vasko, M. R., Prostaglandin E_2 increases calcium conductance and stimulates release of substance P in avian sensory neurons, *J. Neurosci.*, 12, 1917, 1992.
61. Taiwo, Y. O. and Levine, J. D., Further confirmation of the role of adenyl cyclase and of cAMP-dependent protein kinase in primary afferent hyperalgesia, *Neuroscience*, 44, 131, 1991.
62. Taiwo, Y. O. and Levine, J. D., Characterization of the arachidonic acid metabolite mediating bradykinin and norepinephrine hyperalgesia, *Brain Res.*, 492, 397, 1988.
63. Kolzenburg, M., Kress, M., and Reeh, P. W., The nociceptor sensitization by bradykinin does not depend on sympathetic neurons, *Neuroscience*, 46, 465, 1992.
64. Meyer, R. A., Davis, K. D., Raja, S. N., and Campbell, J. N., Sympathectomy does not abolish bradykinin induced cutaneous hyperalgesia in man, *Pain*, 51, 323, 1992.
65. Taiwo, Y. O., Heller, P. H., and Levine, J. D., Mediation of serotonin hyperalgesia by the cAMP second messenger system, *Neuroscience*, 48, 479, 1992.
66. Rueff, A. and Dray, A., 5-Hydroxytryptamine-induced sensitization and activation of peripheral fibres in the neonatal rat are mediated via different 5-hydroxytryptamine-receptors, *Neuroscience*, 50, 899, 1992.
67. Taiwo, Y. O. and Levine, J. D., Further confirmation of the role of adenyl cyclase and of cAMP-dependent protein kinase in primary afferent hyperalgesia, *Neuroscience*, 44, 131, 1991.
68. Stein, C., Peripheral analgesic actions of opioids, *Pain Symptom Manage.*, 6, 119, 1991.
69. Andreev, N. and Dray, A., Opioids suppress activity of polymodal nociceptors in rat paw skin induced by ultraviolet irradiation, *Neuroscience*, 58, 793, 1994.
70. Taiwo, Y. O. and Levine, J. D., κ and δ-opioids block sympathetically dependent hyperalgesia, *J. Neurosci.*, 11, 928, 1991.
71. Shen, K.-F. and Crain, S. M., Dynorphin prolongs the action potential of mouse sensory ganglion neurons by decreasing a potassium conductance whereas another specific kappa opioid does so by increasing a calcium conductance, *Neuropharmacology*, 29, 343, 1990.
72. Nomura, K., Reuveny, E., and Narahashi, T., Opioid inhibition and desensitization of calcium channel currents in rat dorsal root ganglion neurons, *J. Pharmacol. Exp. Ther.*, 270, 466, 1994.
73. Wood, J. N., Lillycrop, K. A., Dent, K. L., Ninkina, N. N., Beech, M. M., Willoughby, J. J., Winter, J., and Latcham, D. S., Regulation of expression of the neuronal POU protein Oct-2 by nerve growth factor, *J. Biol. Chem.*, 267, 17787, 1992.
74. Lindsay, R. M. and Harmar, A. J., Nerve growth factor regulates expression of neuropeptide genes in adult sensory neurons, *Nature*, 337, 362, 1989.
75. Winter, J., Forbes, C. A., Sterberg, J., and Lindsay, J. M., Nerve growth factor (NGF) regulates adult rat cultured dorsal root ganglion neuron responses to capsaicin, *Neuron*, 1, 973, 1988.
76. Diamon, J., Holmes, M., and Coughlin, M., Endogenous NGF and nerve impulses regulate the collateral sprouting of sensory axons in the skin of the adult rat, *J. Neurosci.*, 12, 1454, 1992.
77. Woolf, C. J., Safieh-Garabedian, B., Ma, Q.-P., Crilly, P., and Winter, J., Nerve growth factor contributes to the generation of inflammatory sensory hyperalgesia, *Neuroscience*, 62, 327, 1994.
78. Aloe, L., Tuveri, M. A., Carcassi, U., and Levi-Montalcini, R., Nerve growth factor in the synovial fluid of patients with chronic arthritis, *Arthritis Rheum.*, 35, 351, 1992.
79. Devor, M., Wall, P. D., and Catalan, N., Systemic lidocaine silences ectopic neuroma and DRG discharge without blocking nerve conduction, *Pain*, 48, 261, 1992.

Chapter 7

PHARMACOLOGY OF THE EFFERENT FUNCTION OF PRIMARY SENSORY NEURONS

Carlo A. Maggi

CONTENTS

I. CAPSAICIN AND THE EFFERENT FUNCTION OF SENSORY NERVES

The introduction of capsaicin for studying primary afferent neurons[1,2] has heralded the beginning of a new era in sensory pharmacology,[3-8] since for the first time a substance has become available for producing both a selective stimulation and a selective blockade of transmitter release from sensory neurons.

The actions of capsaicin on a subpopulation of neuropeptide-containing primary afferent neurons involve the activation of a specific receptor which binds capsaicin and other capsaicin-like agents (the "vanilloid" receptor). The activation of the vanilloid receptor leads to the opening of a peculiar type of receptor-operated cation channel. The consequent influx of sodium (Na) and calcium ions (Ca) leads to a transient excitation of primary afferents, followed by a more prolonged condition of refractoriness, during which the primary afferents are unresponsive to further applications of capsaicin-like agents and other stimuli as well. It is the latter property, or stage of action ("desensitization" of the primary afferent neurons) of vanilloid receptor agonists which makes capsaicin and some of its congeners such useful tools in sensory neuron research. By using one or more of the several variants of the capsaicin "desensitization" technique,[6] it has been possible to define the biological roles of these afferent neurons in various organs and systems and to precisely assess whether a given stimulus or drug acts, partly or totally, by stimulating or inhibiting this subpopulation of sensory nerves.[8]

The ability of the capsaicin-sensitive primary afferent neurons to release neurotransmitters from both their central and peripheral endings enables them to exert a dual, sensory, and "efferent", or "local effector", function.[3-5] The term "neurogenic inflammation" is used quite often to describe the "efferent" function of sensory nerves, although it does not account for the trophic or "nonpathogenic" role that the peripheral release of sensory neuropeptides seems to play in various regions of the body (e.g., skin, gastric mucosa, etc.).[9] The evidence for an

0-8493-7646-7/96/$0.00+$.50

TABLE 1
Pharmacological Preparations Amenable to Study of the Pharmacology of the Efferent Function of Capsaicin-Sensitive Nerves

Preparation	Stimulus	Response	Mediator	Ref.
Rabbit iris sphincter	EFS, BK	Contraction	TKs	10,11
Guinea pig left atrium	EFS, BK	Inotropism	CGRP	12,13
Guinea pig bronchus	EFS	Contraction	TKs	14,15
Guinea pig renal pelvis	EFS	Inotropism	TKs	16
Guinea pig ureter	EFS	Relaxation	CGRP	17
Guinea pig pulmonary artery	EFS	Relaxation	CGRP	18
Rat urinary bladder	EFS	Contraction	TKs	19
Rat ureter	EFS	Relaxation	CGRP	17

Abbreviations: EFS, electrical field stimulation; BK, bradykinin; TKs, tachykinins; CGRP, calcitonin gene-related peptide.

"efferent" release of transmitters from these sensory neurons is based on the following experimental approaches.

1. The demonstration, in preparations disconnected from the central nervous system, that nerve stimulation produces functional responses (e.g., changes in smooth muscle tone) which can be blocked by pretreatment with high doses of capsaicin (capsaicin desensitization); the acute application of capsaicin should reproduce this nerve-mediated effect(s); the acute response to capsaicin should undergo desensitization and should be prevented by chronic extrinsic denervation
2. The demonstration that the application of capsaicin (or nerve stimulation) produces an active, Ca-dependent, increase in the outflow of transmitters known to be actively synthesized by primary afferent neurons and transported to their peripheral endings
3. The demonstration of the existence, in the given preparation, of nerve profiles containing the transmitter(s) candidate responsible for the response(s) observed upon acute capsaicin application; the demonstration that the corresponding nerve profiles are sensitive to capsaicin desensitization
4. The demonstration that antagonists or blockers of the candidate transmitter(s) released by capsaicin block or modify the acute response produced by application of capsaicin

When combining these four sets of criteria it can be concluded that an efferent function of sensory nerves has now been demonstrated in a number of tissues/preparations that are amenable to a pharmacological analysis of this process (Table 1).[10-19] The models listed in Table 1 are especially suitable for this purpose, because multiple reproducible responses involving transmitter(s) release can be elicited over consecutive stimulation cycles. The responses measured in these test systems involve the measurement of changes in muscle tone in guinea pig or rat preparations. For obvious reasons, these two species have been extensively investigated under this point of view, but there is also evidence that the efferent function of sensory nerves is especially prominent in rats and guinea pigs, as compared to other rodent (e.g., hamster)[20] or nonrodent (e.g., pig)[21] species. The bulk of evidence collected in various preparations from various species indicates a major role for peptides of the tachykinin (TK) family (substance P and neurokinin A) and for calcitonin gene-related peptide (CGRP) in mediating the efferent function of capsaicin-sensitive nerves.[9] In the literature, examples can be found of paradigms in which other mediators, e.g., vasoactive intestinal polypeptide (VIP)[22] or adenosine triphosphate[23] have been implicated as mediators released from sensory nerves in the peripheral nervous system, yet, for various reasons, this evidence is incomplete under the items 1 to 4 described above.

TABLE 2
Agents that Directly or Indirectly Activate the Capsaicin-Sensitive Primary Afferent Neurons

1. Natural principles from plants
   Capsaicin, resiniferatoxin, piperine, eugenol, curcumin, mustard oil
2. Mediators of inflammation and transmitters
   Bradykinin, arachidonic acid metabolites, serotonin, histamine, PAF, acetylcholine, neurotensin, cholecystokinin octapeptide, GABA (via $GABA_A$ receptors), interleukin-1
3. Chemicals from the environment
   Vapor phase of smoke, toluene diisocyanate, allergens, bacterial chemotactic peptides, ether, formalin, xylene, chloroacetophenone, acrolein, ouabain
4. Others
   H^+, ischemia, heat, mechanical stimuli, electrical stimulation, high potassium

In this chapter, a brief outline is provided of the pharmacology of the efferent function of sensory nerves. A quite wide literature exists on this topic, which was reviewed a few years ago.[8] Because of the limitations of space, this chapter does not intend to provide a comprehensive review, rather it focuses on certain basic principles and outlines trends for future research in this field.

II. PHARMACOLOGICAL MODULATION OF THE RELEASE PROCESS

A. AGENTS THAT INDUCE THE RELEASE OF SENSORY NEUROPEPTIDES FROM PERIPHERAL ENDINGS OF SENSORY NEURONS

Various agents or stimuli, especially chemicals, are able to produce the release of stored transmitters from capsaicin-sensitive primary afferent neurons (Table 2). Some stimuli listed in Table 2 act indirectly to produce the release of sensory neuropeptides, through the generation of active metabolites (e.g., acrolein produced from cyclophosphamide)[24] or the release of other mediators (e.g., prostanoids produced by bradykinin).[25] In most instances it is not an easy task to dissect out whether a given agent acts, totally or partially, in an indirect manner to induce the peripheral release of sensory neuropeptides: the relevant experiments measuring the release of TKs/CGRP should involve the physiological sites of release, i.e., the peripheral terminals of primary afferents, and hence need to be performed in multicellular preparations. Therefore, several possibilities are open, and a given stimulus may act on other cell types to induce the release of mediator(s) which, in turn, acts on primary afferents. Studies on sensory neurons in culture can provide a hint as to whether a given mediator does or does not have a direct excitatory action on primary afferents, assuming that receptors expressed in the cell body of sensory neurons are also present on its peripheral terminals. A major caveat in this respect exists because cultured primary afferent neurons are, *de facto*, axotomized neurons and this kind of lesion is known to produce a number of phenotypic changes in the neuronal body.

Several chemical agents that excite the efferent function of sensory neurons originate from the environment: many of them are capable of producing tissue damage, pain, and inflammation or are anyway potentially harmful for tissues; in this respect, the term neurogenic inflammation appears fully justified for describing the efferent function of sensory nerves.[9] Several agents listed in Table 2 are mediators of inflammation (e.g., bradykinin) and also in this case the term neurogenic inflammation is fully justified to outline a neurogenic contribution or reinforcement to the overall inflammatory process. On the other hand, certain mediators listed in Table 2 are produced/released in different relative amounts during normal function and inflammation: prostanoids are a clear example of this case. In the rat urinary bladder, for example, prostanoids appear to act as a chemical link between physiological distension of the viscus, mimicking spontaneous bladder filling, and the excitation of capsaicin-sensitive afferents.[26] In this case, the mechanoreceptivity of afferent nerves seems to be amplified through the action of prostanoids produced in physiological conditions[27]. Two recent studies, performed in the guinea pig colon inferior mesenteric artery,[28] and in the ferret trachea,[29] have provided evidence that a moderate or physiological-like degree of mechanical stretch can produce a peripheral release of sensory neuropeptides from capsaicin-sensitive afferents. Although the possibility that prostanoids were acting as intermediate links between stretch/distension and the local release of sensory neuropeptides was not investigated in those studies,[28,29] these observations strongly suggest that the efferent function of sensory nerves can be activated, at least in certain viscera, by stimuli that are physiological rather than pathological. Another category can be established for certain chemicals that are present in body fluids but do not normally come into contact with primary afferent neurons. A good example is urine which, in normal conditions, contains several chemicals (KCl, bradykinin, protons) that are effective in producing the release of mediators from afferent nerves.[30-32] In this case, the barrier function of the urothelium prevents the contact with sensory nerve terminals present in the mucosa lining the urinary tract. The contact may happen in pathophysiological conditions, producing a breakdown of the barrier, e.g., damage of the urothelium because of the passage of a stone in the ureter.[33]

With regard to the mechanisms through which chemicals activate the efferent functions of sensory neurons, the results of several studies[30,34-41] enable the following conclusions to be drawn:

1. Whatever stimulus is applied, the influx of extracellular Ca is mandatory for sustaining the secretion of transmitter(s) from peripheral endings of primary afferents.
2. The influx of extracellular Ca produced by vanilloid receptor agonists chiefly occurs through the cation channel coupled with the vanilloid receptor. In the absence of extracellular Ca, Na ions flowing through the channel can produce enough depolarization to activate afferent discharge, while the efferent function is prevented. Protons (low pH media) appear to use the same secretory pathway, sensitive to blockade by ruthenium red (see below), possibly by direct activation of the cation channel.
3. The release process activated by agents that act through the cation channel coupled with the vanilloid receptor is blocked by relatively low concentrations of ruthenium red;[36,38,42] the molecular mechanism of this blocking action is unknown, but it may involve an interaction with sialic acid residues on the cell membrane.
4. A number of chemicals produce Ca influx and transmitter release from the peripheral endings of capsaicin-sensitive afferent neurons through voltage-sensitive Ca channels; in the guinea pig, N-type Ca channels sensitive to the blocking action of ω-conotoxin (CTX) play a relevant role in excitation-secretion coupling of sensory neuropeptides.[37] Under certain circumstances, especially those that favor the occurrence of axon reflexes, also low concentrations of vanilloid receptor agonists may use this pathway for producing transmitter release.[41]
5. The capsaicin receptor antagonist, capsazepine,[43] is a moderately potent, but selective ligand, to block the release of sensory neuropeptides from peripheral endings of primary afferent neurons induced by vanilloid receptor agonists. Capsazepine acts as a competitive antagonist in this respect, while the profile of action of ruthenium red is more like that of a noncompetitive antagonist.[44]

In a number of experimental paradigms, capsazepine acts as a quite selective antagonist of capsaicin or vanilloid agonists, without affecting the responses induced by a number of other stimuli. However, some studies have shown the ability of capsazepine to inhibit the release of sensory neuropeptides induced by low pH at capsazepine concentrations similar or even lower than those effective toward capsaicin.[45,46] In the rat soleus muscle[46] the threshold

TABLE 3
Agents that Directly or Indirectly Inhibit Transmitter Release from Capsaicin-Sensitive Primary Afferent Neurons

Opioids
Noradrenaline (via α_2 receptors)
Neuropeptide Y
Galanin
Somatostatin
GABA (via $GABA_B$ receptors)
Adenosine
Serotonin
Mast cell stabilizers (sodium cromoglycate)
K channel openers
Nifedipine and ω-conotoxin
Ruthenium red
Capsazepine
Diuretics

concentration of capsazepine for inhibiting CGRP release by low pH is 3 μ*M*, that inhibiting CGRP release by capsaicin is 10 μ*M*, and that inhibiting CGRP release by KCl is 30 μ*M* (the latter representing a putative nonspecific effect). It is unclear whether the inhibitory action exerted by capsazepine toward low pH may be a nonspecific effect; the main (and most attractive) alternative explanation is that, in certain conditions, low pH may induce the release/generation of an endogenous ligand(s) for the vanilloid receptor. Such a ligand, if it exists at all, has not been discovered thus far. The development of capsazepine congeners with much higher (n*M* rather than μ*M*) affinity for the vanilloid receptor, is needed to clarify the issue.

B. AGENTS THAT INHIBIT THE RELEASE OF SENSORY NEUROPEPTIDES FROM PERIPHERAL ENDINGS OF SENSORY NEURONS

Various agents (Table 3) have been reported to inhibit the "efferent" function of sensory nerves. The issue is of practical relevance in therapeutics since a prejunctional inhibition of the evoked transmitter release could be a rational approach for controlling neurogenic inflammation. Some of the agents that inhibit the release of sensory neuropeptides (e.g., noradrenaline and neuropeptide Y[13]) act as neurotransmitters in the peripheral nervous system, indicating a possible communication network between different types of effector nerves. Opioids have a well-characterized inhibitory action on the release of sensory neuropeptides:[10,17] the overexpression and putative release of opioids by cells of the immune system during chronic inflammation[47] suggest a neuroimmune link through which neurogenic inflammation could be dampened in pathophysiological conditions. It is interesting to note that some endogenous mediators (e.g., serotonin or GABA) appear to have a dual effect on the "efferent" release of sensory neuropeptides, being able to either activate or inhibit the process by acting via different receptor types (Table 3). The possible pathophysiological significance of this dual effect is a matter of speculation.

It is worth mentioning that various agents that produce a prejunctional inhibition of the release of sensory neuropeptides do not uniformly work against all applied stimuli. In particular, the concept has emerged that the release process induced by stimuli that produce a ruthenium red-sensitive and ω-conotoxin (CTX)-resistant release of sensory neuropeptides (e.g., capsaicin in high concentration) cannot be substantially inhibited by, e.g., opioids or α_2 adrenoceptor agonists, while stimuli that produce a ruthenium red-resistant but CTX-sensitive release of sensory neuropeptides are sensitive to prejunctional inhibition.[10,12,13,38,48]

III. PHARMACOLOGICAL MODULATION OF SYSTEMS THAT INACTIVATE RELEASED TRANSMITTERS

Various peptidases are able to cleave TKs and CGRP and may be important for regulating the intensity and duration of action of the released peptides. When considering whether a given peptidase may be important in regulating the physiological action of sensory neuropeptides, several criteria should be met.[8]

1. The peptidase should be present in the target tissue at appropriate locations with regard to nerve profiles containing the sensory neuropeptides and target cells of the released transmitters
2. The peptidase should be able to cleave the peptide transmitter(s) with appropriate kinetics and affinity in relation to the amounts released during nerve stimulation, yielding inactive or less potent fragments
3. Inhibitors of the given peptidase should be able to increase the recovery of the released transmitter
4. Inhibitors of a given peptidase should be able to enhance or prolong the action of the peptide transmitter, either administered exogenously or when released from sensory nerves

With regard to the peripheral nervous system and the efferent function of primary afferent neurons, the above-mentioned criteria are fulfilled by endopeptidase 24.11, also known as "enkephalinase" or "neutral endopeptidase".[8] Endopeptidase 24.11 has been shown to be involved in terminating the action of endogenous TKs and CGRP released from peripheral endings of sensory neurons in, e.g., the guinea pig urinary bladder[49] and ureter,[50] respectively. Other peptidases, such as aminopeptidases and the angiotensin-converting enzyme are able to degrade TKs in the peripheral nervous system, but evidence for their involvement in the control of the efferent function of sensory nerves is, for various reasons, incomplete.

IV. PHARMACOLOGICAL MODULATION OF RECEPTORS THAT MEDIATE THE ACTION OF RELEASED TRANSMITTERS

The actions of TKs and CGRP are mediated by specific receptors expressed on target cells; accordingly, drugs that block these receptors modulate neurogenic inflammation. The actions of TKs encoded by their common, C-terminal sequence, are mediated by three receptors, termed NK_1, NK_2, and NK_3, for which substance P (SP), neurokinin A (NKA), and neurokinin B possess the highest affinity, respectively.[51] The three receptors have been isolated and cloned,[52] and all belong to the superfamily of G protein-coupled receptors with seven putative transmembrane spanning segments (see Chapter 3 in this book). In recent years, a number of potent and selective antagonists of the three tachykinin receptors have been developed, especially for the NK_1 and NK_2, receptors (see Chapter 10 in this book). Some actions of SP such as mast cell degranulation are mediated by its N-terminal region and, by definition, do not involve NK_1, NK_2, or NK_3 receptors.[51] It is also doubtful whether this effect is receptor mediated at all. Mast cell degranulation by substance P could be relevant during neurogenic inflammation not only as a further brick of the "neuroimmune" connection,[9] leading to reinforcement of the inflammatory process via the release of mast cell-derived mediators, but also because some mast cell-derived peptidases could limit the intensity/extent of neurogenic inflammation through the cleavage of sensory neuropeptides.[53] By using receptor selective antagonists, both NK_1 and NK_2 receptors have been shown to be junctionally activated through the efferent function of sensory nerves.[15,16,19] In general, a remarkable specialization appears to exist with regard to the activation of different target cells by TKs in the same organ/system. A clear example supporting this statement originates from studies in the airways: at this level, endogenous TKs released by capsaicin-sensitive afferents produce plasma protein extravasation, stimulate airway secretion, and produce bronchoconstriction. Experiments with selective antagonists have clearly shown that the two former effects are mediated by the NK_1 receptor, while bronchoconstriction by endogenous TKs is chiefly produced by activation of the NK_2 receptor.[54,55]

The issue of CGRP receptors is less well defined, since these receptors have not yet been isolated. Indirect elements suggest that they should be G protein-coupled;[56] adenylate cyclase stimulation and cAMP accumulation have been shown repeatedly to be involved in CGRP actions in the peripheral nervous system.[56] The C-terminal fragment of human αCGRP, CGRP(8–37) acts as a competitive antagonist at certain CGRP receptors.[57,58] Chiefly based on a differential sensitivity to the blocking action of this ligand, it has been proposed that two types of CGRP receptors exist,[58] $CGRP_1$ receptors being blocked by CGRP(8–37) with micromolar affinity and $CGRP_2$ receptors being insensitive to the action of CGRP(8–37). By using CGRP(8–37), the release of endogenous CGRP from the peripheral endings of sensory nerves has been implicated in various aspects of neurogenic inflammation, such as antidromic vasodilation and increase in cutaneous blood flow induced by activation of primary afferents.[59,60] Furthermore, a mediator role of CGRP in the efferent function of sensory nerves has been demonstrated in various preparations, including the guinea pig ileum[61] and pulmonary artery[62] and in the rat vas deferens.[63]

V. CONCLUSIONS AND FUTURE RESEARCH TRENDS

The existence of an efferent function of sensory nerves has been known since the previous century as a phenomenon of uncertain physiological significance.[3] The introduction of capsaicin as a probe for studying sensory neurons[1-3] has enabled an accurate assessment of the neurobiology of this function, the definition of its anatomical and neurochemical substrates, and its overall physiological significance. Drawing a parallel, capsaicin has been revealed to have the same or similar value that, e.g., atropine or guanethidine have had for defining other, more "classical" examples of autonomic neurotransmission. The use of capsaicin "desensitization", with its multiple variants, has also enabled definition of substances or stimuli that act, directly or indirectly, by stimulating or inhibiting transmitter release from the peripheral endings of sensory nerves: thus, a new "sensory pharmacology" has developed,[7,8] of which a schematic outline has been provided in this chapter. Some aspects of this pharmacology appear to be peculiar to this subset of primary afferents, being linked to the expression of a specific receptor that recognizes capsaicin and other vanilloid receptor agonists. Whether this receptor can accommodate endogenous substances is a major challenge for future research in the field.

The receptor appears to be coupled (or perhaps forms part of) a peculiar type of ion channel which mediates the action of vanilloid receptor agonists. The demonstration that protons can activate this channel[64,65] and produce release of sensory neuropeptides[30,45,46] warrants a pathophysiological relevance of this mechanism in regulating sensory neuron function. It appears quite conceivable that selective blockers of the ion channel coupled with the vanilloid receptor will represent a novel class of analgesic and anti-inflammatory drugs; the development of such ligands is a further challenge in this line of research, and will be crucial to understand the true relevance of this particular mechanism leading to sensory neuron activation.

Additionally, the use of capsaicin has enabled the demonstration that a number of transmitters, mediators, and autacoids act on sensory neurons and modulate neurogenic inflammation; to this list of chemicals also some drugs currently used in human therapy can be added. It has been proposed that the therapeutic action of some drugs (e.g., the antimigraine activity of the serotonin receptor agonist, sumatriptan[66] or the antiasthmatic action of certain diuretics[67]) may be linked to an inhibition of neurogenic inflammation at discrete sites in the body. Should this hypothesis be correct, then a role for neurogenic inflammation in human diseases would be proven; yet, in purely pharmacological terms, the development of drugs that specifically interfere with sensory nerves and modulate neurogenic inflammation without influencing other cell types in the body appears possible and would represent a major advancement in both basic and applied research.

REFERENCES

1. Jancso, N., Jancso-Gabor, A., and Szolcsányi, J., Direct evidence for neurogenic inflammation and its prevention by denervation and by pretreatment with capsaicin, *Br. J. Pharmacol.*, 31, 138, 1967.
2. Jancso, N., Jancso-Gabor, A., and Szolcsányi, J., The role of sensory nerve endings in neurogenic inflammation induced in human skin and in the eye and paw of the rat, *Br. J. Pharmacol.*, 32, 32, 1968.
3. Szolcsányi, J., Capsaicin-sensitive chemoceptive neural system with dual sensory-efferent function, in *Antidromic Vasodilatation and Neurogenic Inflammation*, Chahl, L. A., Szolcsányi, J., and Lembeck, F., Eds., Akademiai Kiado, Budapest, Hungary, 1984, 27.
4. Maggi, C. A. and Meli, A., The sensory-efferent function of capsaicin-sensitive sensory neurons, *Gen. Pharmacol.*, 19, 1, 1988.
5. Holzer, P., Local effector functions of capsaicin-sensitive sensory nerve endings: involvement of tachykinins, CGRP and other neuropeptides, *Neuroscience*, 24, 739, 1988.
6. Holzer, P., Capsaicin: cellular targets, mechanisms of action and selectivity for thin sensory neurons, *Pharmacol. Rev.*, 43, 144, 1991.
7. Maggi, C. A., Capsaicin and primary afferent neurons: from basic science to human therapy? *J. Autonom. Nerv. Syst.*, 33, 1, 1991.
8. Maggi, C. A., The pharmacology of the efferent function of sensory nerves, *J. Autonom. Pharmacol.*, 11 173, 1991.
9. Maggi, C. A., Tachykinins and CGRP as co-transmitters released from peripheral endings of sensory nerves, *Prog. Neurobiol.*, 45, 1, 1995.
10. Ueda, N., Muramatsu, I., Hayashi, H., and Fujiwara, M., Effects of Met-enkephalin on the SP-ergic and cholinergic responses in the rabbit iris sphincter muscle, *J. Pharmacol. Exp. Ther.*, 226, 507, 1983.
11. Hakanson, R., Beding, B., Ekman, R., Heilig, M., Wahlestedt, C., and Sundler, F., Multiple tachykinin pools in sensory nerve fibres in the rabbit iris, *Neuroscience*, 21, 943, 1987.
12. Maggi, C. A., Giuliani, S., Manzini, S., and Meli, A., GABA-A receptor-mediated positive inotropism in guinea pig isolated left atria: evidence for the involvement of capsaicin-sensitive nerves, *Br. J. Pharmacol.*, 97, 103, 1989.
13. Giuliani, S., Maggi, C. A., and Meli, A., Prejunctional modulatory action of neuropeptide Y on peripheral terminals of capsaicin-sensitive sensory nerves, *Br. J. Pharmacol.*, 98, 407, 1989.
14. Grundstrom, N., Andersson, R. G. G., and Wikberg, J. E. S., Pharmacological characterization of the autonomous innervation of the guinea pig tracheobronchial smooth muscle, *Acta Pharmacol. Toxicol.*, 49, 150, 1981.
15. Maggi, C. A., Patacchini, R., Rovero, P., and Santicioli, P., Tachykinin receptors and noncholinergic bronchoconstriction in the guinea pig isolated bronchi, *Am. Rev. Respir. Dis.*, 144, 363, 1991.
16. Maggi, C. A., Patacchini, R., Eglezos, A., Quartara, L., Giuliani, S., and Giachetti, A., Tachykinin receptors in guinea pig renal pelvis: activation by exogenous and endogenous tachykinins, *Br. J. Pharmacol.*, 107, 27, 1992.
17. Maggi, C. A. and Giuliani, S., The neurotransmitter role of CGRP in the rat and guinea pig ureter: effect of a CGRP antagonist and species-related differences in the action of omega conotoxin on CGRP release from primary afferents, *Neuroscience*, 43, 261, 1991.
18. Maggi, C. A., Patacchini, R., Perretti, F., Tramontana, M., Manzini, S., Geppetti, P., and Santicioli, P., Sensory nerves, vascular endothelium and neurogenic relaxation of the guinea pig isolated pulmonary artery, *Naunyn-Schmiedeberg's Arch. Pharmacol.*, 342, 78, 1990.
19. Meini, S. and Maggi, C. A., Evidence for a capsaicin-sensitive, tachykinin-mediated, component in the nanc contraction of the rat urinary bladder to nerve stimulation, *Br. J. Pharmacol.*, 112, 1123, 1994.

20. Maggi, C. A., Giuliani, S., Santicioli, P., Abelli, L., Geppetti, P., Somma, V., Renzi, D., and Meli, A., Species-related variations in the effects of capsaicin on urinary bladder functions: relation to bladder content of SP-LI, *Naunyn-Schmiedeberg's Arch. Pharmacol.*, 336, 546, 1987.
21. Alving, K., Matran, R., and Lundberg, J. M., Capsaicin-induced local effector responses, autonomic reflexes and sensory neuropeptide depletion in the pig, *Naunyn-Schmiedeberg's Arch. Pharmacol.*, 343, 37, 1991.
22. Maggi, C. A., Theodorsson, E., Santicioli, P., Patacchini, R., Barbanti, G., Turini, D., Renzi, D., and Giachetti, A., Motor response of the human isolated colon to capsaicin and its relationship to release of VIP, *Neuroscience*, 39, 833, 1990.
23. Holton, F. A. and Holton, P., The capillary dilator substances in dry powders of spinal roots: a possible role of ATP in chemical transmission from nerve endings, *J. Physiol. (London)*, 126, 124, 1954.
24. Ahluwalia, A., Maggi, C. A., Santicioli, P., Lecci, A., and Giuliani, S., Characterization of the capsaicin-sensitive component of cyclophosphamide-induced inflammation in the rat urinary bladder, *Br. J. Pharmacol.*, 111, 1017, 1994.
25. Geppetti, P., Del Bianco, E., Tramontana, M., Vigano, M., Folco, G. C., Maggi, C. A., Manzini, S., and Fanciullacci, M., Arachidonic acid and bradykinin share a common pathway to release neuropeptides from capsaicin-sensitive sensory nerve fibers of the guinea pig heart, *J. Pharmacol. Exp. Ther.*, 259, 759, 1991.
26. Maggi, C. A., Giuliani, S., Conte, B., Furio, M., Santicioli, P., Meli, P., Gragnani, L., and Meli, A., Prostanoids modulate reflex micturition by acting through capsaicin-sensitive afferents, *Eur. J. Pharmacol.*, 145, 105, 1988.
27. Maggi, C. A., Prostanoids as local modulators of reflex micturition, *Pharmacol. Res.*, 25, 13, 1992.
28. Meehan, A. G. and Kreulen, D. L. A., Capsaicin-sensitive inhibitory reflex from the colon to mesenteric arteries in the guinea pig, *J. Physiol. (London)*, 448, 153, 1992.
29. Coburn, R. F., Mitchell, H., Dey, R. D., and Alkon, J., Capsaicin-sensitive stretch responses in ferret trachealis muscle, *J. Physiol. (London)*, 475, 293, 1994.
30. Geppetti, P., Tramontana, M., and Patacchini, R., Del Bianco, E., Santicioli, P., and Maggi, C. A., Neurochemical evidence for the activation of the efferent function of capsaicin-sensitive nerves by lowering of the pH in the guinea pig urinary bladder, *Neurosci. Lett.*, 114, 101, 1990.
31. Maggi, C. A., Patacchini, R., Santicioli, P., Geppetti, P., Cecconi, R., Giuliani, S., and Meli, A., Multiple mechanisms in the motor responses of the guinea pig isolated urinary bladder to bradykinin, *Br. J. Pharmacol.*, 98, 619, 1989.
32. Maggi, C. A., Santicioli, P., Del Bianco, E., and Giuliani, S., Local motor responses to bradykinin and bacterial chemotactic peptide FMLP in the guinea pig isolated renal pelvis and ureter, *J. Urol.*, 148, 1944, 1992.
33. Maggi, C. A., Giuliani, S., Del Bianco, E., Geppetti, P., Theodorsson, E., and Santicioli, P., CGRP in the regulation of urinary tract motility, *N.Y. Acad. Sci.*, 657, 328, 1992.
34. Gamse, R., Molnar, A., and Lembeck, F., Substance P release from spinal cord slices by capsaicin, *Life Sci.*, 25, 629, 1979.
35. Maggi, C. A., Santicioli, P., Geppetti, P., Parlani, M., Astolfi, M., Del Bianco, E., Patacchini, R., Giuliani, S., and Meli, A., The effect of calcium free medium and nifedipine on the release of SP-LI and contractions induced by capsaicin in the isolated guinea pig and rat bladder, *Gen. Pharmacol.*, 20, 445, 1989.
36. Maggi, C. A., Santicioli, P., Geppetti, P., Parlani, M., Astolfi, M., Pradelles, P., Patacchini, R., and Meli, A., The antagonism induced by ruthenium red of the actions of capsaicin on the peripheral terminbals of sensory neurons: further studies, *Eur. J. Pharmacol.*, 154, 1, 1988.
37. Maggi, C. A., Patacchini, R., Giuliani, S., Santicioli, P., and Meli, A., Evidence for two independent modes of activation of the efferent function of capsaicin-sensitive nerves, *Eur. J. Pharmacol.*, 156, 367, 1988.
38. Maggi, C. A., Patacchini, R., Santicioli, P., Giuliani, S., Del Bianco, E., Geppetti, P., and Meli, A., The 'efferent' function of capsaicin-sensitive nerves: ruthenium red discriminates between different mechanisms of activation, *Eur. J. Pharmacol.*, 170, 167, 1989.
39. Geppetti, P., Tramontana, M., Santicioli, P., Del Bianco, E., Giuliani, S., and Maggi, C. A., Bradykinin-induced release of CGRP from capsaicin-sensitive nerves in guinea pig atria: mechanism of action and calcium requirements, *Neuroscience*, 38, 687, 1990.
40. Maggi, C. A., Tramontana, M., Cecconi, R., and Santicioli, P., Neurochemical evidence for the involvement of N-type calcium channels in transmitter secretion from peripheral endings of sensory nerves in guinea pigs, *Neurosci. Lett.*, 114, 203, 1990.
41. Lou, Y. P., Franco-Cereceda, A., and Lundberg, J. M., Different ion channel mechanisms between low concentrations of capsaicin and high concentrations of capsaicin and nicotine regarding peptide release from pulmonary afferents, *Acta Physiol. Scand.*, 146, 119, 1992.
42. Amann, R. and Maggi, C. A., Ruthenium red as a capsaicin antagonist, *Life Sci.*, 49, 849, 1991.
43. Bevan, S., Hothi, S., Hughes, G., James, I. F., Rang, H. P., Shah, K., Walpole, C. S. J., and Yeats, J. C., Capsazepine: a competitive antagonist of the sensory neurone excitant capsaicin, *Br. J. Pharmacol.*, 107, 544, 1992.

44. Maggi, C. A., Bevan, S., Walpole, C. S. J., Rang, H. P., and Giuliani, S., A comparison of capsazepine and ruthenium red as capsaicin antagonists in the rat isolated urinary bladder and vas deferens, *Br. J. Pharmacol.*, 108, 801, 1993.
45. Lou, Y. P. and Lundberg, J. M., Inhibition of low pH evoked activation of airway sensory nerves by capsazepine, a novel capsaicin-receptor antagonist, *Biochem. Biophys. Res. Commun.*, 189, 537, 1992.
46. Santicioli, P., Del Bianco, E., Figini, M., Bevan, S., and Maggi, C. A., Effects of capsazepine on the release of CGRP-LI induced by low pH capsaicin and potassium in rat soleus muscle, *Br. J. Pharmacol.*, 110, 609, 1993.
47. Stein, C., Hassan, A. H. S., Przewlocki, R., Gramsch, C., Peter, K., and Herz, A., Opioids from immunocytes interact with receptors on sensory nerves to inhibit nociception in inflammation, *Proc. Natl. Acad. Sci. U.S.A.*, 87, 5935, 1990.
48. Lou, Y. P., Franco-Cereceda, A., and Lundberg, J. M., Variable $\alpha_2$adrenoceptor-mediated inhibition of bronchoconstriction and peptide release upon activation of pulmonary afferents, *Eur. J. Pharmacol.*, 210, 173, 1992.
49. Maggi, C. A., Astolfi, M., Santicioli, P., Tramontana, M., Leoncini, G., Geppetti, P., Giachetti, A., and Meli, A., Effect of thiorphan on the response of guinea-pig isolated urinary bladder to exogenous and endogenous tachykinins, *J. Urol.*, 144, 1546, 1990.
50. Maggi, C. A. and Giuliani, S. A., A thiorphan-sensitive mechanism regulates the action of both exogenous and endogenous calcitonin gene-related peptide (CGRP) in the guinea pig ureter, *Regul. Peptides*, 51, 263, 1994.
51. Maggi, C. A., Patacchini, R., Rovero, P., and Giachetti, A., Tachykinin receptors and receptor antagonists, *J. Autonom. Pharmacol.*, 13, 23, 1993.
52. Nakanishi, S., Mammalian tachykinin receptors, *Annu. Rev. Neurosci.*, 14, 123, 1991.
53. Brain, S. D. and Williams, T. J., SP regulates the vasodilator activity of CGRP, *Nature*, 335, 73, 1988.
54. Eglezos, A., Giuliani, S., Viti, G., and Maggi, C. A., Direct evidence that capsaicin-induced plasma protein extravasation is mediated through tachykinin NK-1 receptors, *Eur. J. Pharmacol.*, 209, 277, 1991.
55. Maggi, C. A., Giuliani, S., Ballati, L., Lecci, A., Manzini, S., Patacchini, R., Renzetti, A. R., Rovero, P., Quartara, L., and Giachetti, A., In vivo evidence for tachykininergic transmission using a new NK-2 receptor selective antagonist, MEN 10376, *J. Pharmacol. Exp. Ther.*, 257 1172, 1991.
56. Poyner, D. R., CGRP: multiple actions, multiple receptors, *Pharmacol. Ther.*, 56, 23, 1992.
57. Chiba, T., Yamaguchi, A., Yamatani, T., Nakamura, A., Morishita, T., Inui, T., Fukase, M., Noda, T., and Fujita, T., CGRP receptor antagonist hCGRP, 8-37, *Am. J. Physiol.*, 256, E331, 1989.
58. Dennis, T., Fournier, A., Cadieaux, A., Pomerlau, F., Jolicoeur, F. B., St. Pierre, S., and Quirion, R., hCGRP(8-37) a CGRP antagonist revealing CGRP receptor heterogeneity in brain and periphery, *J. Pharmacol. Exp. Ther.*, 254, 123, 1990.
59. Brain, S. D., Hughes, S. R., Cambridge, H., and O'Driscoll, G., The contribution of CGRP to neurogenic vasodilator responses, *Agents Actions*, 38, C19, 1993.
60. Escott, K. J. and Brain, S. D., Effect of CGRP antagonist, CGRP(8-37) on skin vasodilatation and oedema induced by stimulation of the rat saphenous nerve, *Br. J. Pharmacol.*, 110, 722, 1993.
61. Bartho, L., Petho, G., Antal, A., Holzer, P., and Szolcsányi, J., Two types of relaxation due to capsaicin in the guinea pig isolated ileum, *Neurosci. Lett.*, 81, 146, 1987.
62. Butler, A., Worton, S. P., O'Shaughnessy, C. T., and Connor, H. E., Sensory nerve-mediated relaxation of guinea pig isolated pulmonary artery: prejunctional modulation by α_2-adrenoceptor agonists but not sumatriptan, *Br. J. Pharmacol.*, 109, 126, 1993.
63. Maggi, C. A., Santicioli, P., Theodorsson-Norheim, E., and Meli, A., Immunoblockade of the response to capsaicin in the rat vas deferens: evidence for the involvement of endogenous CGRP, *Neurosci. Lett.*, 78, 63, 1987.
64. Bevan, S. and Yeats, J., Protons activate a cation conductance in a subpopulation of rat dorsal root ganglion neurones, *J. Physiol. (London)*, 433, 145, 1991.
65. Liu, L. and Simon, S. A., A rapid capsaicin-activated current in rat trigeminal ganglion neurons, *Proc. Natl. Acad. Sci. U.S.A.*, 91, 738, 1994.
66. Buzzi, M. G., Moskowitz, M. A., Peroutka, S. J., and Byun, B., Further characterization of the putative 5-HT receptor which mediates blockade of neurogenic plasma extravasation in rat dura mater, *Br. J. Pharmacol.*, 103, 1421, 1991.
67. Elwood, W., Lotvall, J. O., Barnes, P. J., and Chung, K. F., Loop diuretics inhibit cholinergic and noncholinergic nerves in guinea pig airways, *Am. Rev. Respir. Dis.*, 143, 1340, 1991.

Chapter 8

PHARMACOLOGY OF TACHYKININS

Domenico Regoli, Quang T. Nguyen, and Girolamo Caló

CONTENTS

I. INTRODUCTION

Molecular biological studies have demonstrated the existence of three tachykinin-like peptides in mammals: the undecapeptide substance P, and two decapeptides, neurokinin A and neurokinin B, whose primary structures are shown below:

SP: Arg-Pro-Lys-Pro-Gln-Gln-Phe-Phe-Gly-Leu-Met-NH_2
NKA: His-Lys-Thr-Asp-Ser-Phe-Val-Gly-Leu-Met-NH_2
NKB: Asp-Met-His-Asp-Phe-Phe-Val-Gly-Leu-Met-NH_2

The three peptides, encoded in the mammalian genes, however, are expressed almost exclusively in neurons and act as neurotransmitters in the central nervous system (CNS) and as essential components of the NANC (nonadrenergic, noncholinergic) system in the autonomic nerves. Neurokinins, particularly substance P, are released from sensory neurons by various stimuli, both in the spinal cord, where they contribute to pain transmission,[1] and in the peripheral tissues, where they produce neurogenic inflammation.[2] Biological activities of neurokinins are mediated by three receptor types, NK_1, NK_2, and NK_3,[3] which are expressed by a variety of target cells, which include neurons, smooth muscles, endothelium, endocrine and exocrine secretory epithelia, lymphocytes, and other blood cells.[4] Through receptor activation, neurokinins exert a variety of biological actions that extend from vasodilatation to vasoconstriction, increase of vascular permeability, neuronal and smooth muscle activation, stimulation of exocrine (salivary, intestinal, pancreatic, bronchial secretions) and endocrine (hypothalamus, adeno- and neurohypophysis) glands, as well as of lymphocytes, fibroblasts, and other cells involved in immunological and hematological events.[5] The recent identification of selective agonists and nonpeptide antagonists for the three neurokinin receptors have been instrumental not only in establishing the pharmacological profile of each receptor type[6] (and for revealing differences between receptor types among species[4,7,8]), but also for determining the roles of neurokinin receptors in physiology and physiopathology.[8] In the text that follows, we present a concise description of (1) the pharmacological profile of each neurokinin receptor type and subtype, (2) a comparison of the data obtained in pharmacological,

0-8493-7646-7/96/$0.00+$.50

TABLE 1
Parmacological Characterization of Neurokinin Receptors with Agonists and Antagonists

	NK_1		NK_2		NK_3	
Preparations	**RVC**	RUBt	**RPA**	HUB	**RPV**	GPIt
Agonists (pD_2)						
SP	**8.6**	7.1	**6.1**	5.6	**5.8**	7.4
[Sar^9,$Met(O)_2^{11}$]SP	**8.6**	8.0	**Inactive**	Inactive	**Inactive**	7.2
Ac[Arg^6Sar^9,$Met(O)_2^{11}$]$SP_{(6-11)}$	**8.7**	ND	**Inactive**	Inactive	**Inactive**	6.6
NKA	**7.3**	5.8	**8.2**	7.4	**6.4**	7.0
[βAla^8]$NKA_{(4-10)}$	**<5.0**	<5.0	**8.6**	7.2	**6.1**	5.8
[Nle^{10}]$NKA_{(4-10)}$	**5.4**	5.2	**8.0**	7.1	**6.5**	5.8
NKB	**7.2**	5.7	**7.4**	7.2	**7.7**	8.4
[$MePhe^7$]NKB	**5.6**	Inactive	**5.2**	6.1	**8.3**	9.1
Senktide	**Inactive**	6.3	**5.4**	5.7	**7.6**	9.2
Preparations	**RVC**	RUBt	**RPA**	HUB	**RPV**	GPIt
Agonist	**SP**	Ac[Arg^6]	**NKA**	[βAla^8] $NKA_{(4-10)}$	[$MePhe^7$]NKB	
Antagonists (pA_2)						
CP 99994	**8.9**	6.0	**Inactive**	Inactive	**Inactive**	5.0
SR 140333	**9.8**	8.0	**5.8**	5.8	**5.8**	5.8
SR 48968	**6.1**	5.9	**9.8**	7.6	**<5.0**	5.8
GR 94800	**6.9**	5.6	**8.2**	7.5	**6.0**	5.6
R 486	**Inactive**	5.6	**Inactive**	α 0.4	**7.5**	6.1
R 820	**5.4**	5.4	**<5.0**	ND	**7.6**	Inactive

RVC, rabbit vena cava; RUBt, rat urinary bladder treated with SR 48968 1.7 μM; RPA, rabbit pulmonary artery; HUB, hamster urinary bladder; RPV, rat portal vein; GPIt, guinea pig ileum treated with CP 99994 1 μM and SR 48968 1.7 μM.
pD_2, –log of the concentration of agonist producing 50% of the maximal response.
pA_2, –log of the concentration of antagonist that reduces the effect of a double dose of agonist to that of a single dose.

biochemical, and molecular biological studies on NK receptors, and (3) a description of the biological effects exerted by neurokinins.

II. BASIC PHARMACOLOGY OF NEUROKININ RECEPTORS

A fairly precise characterization of the three neurokinin receptor types has been carried out on isolated blood vessels by applying a classical pharmacological approach and the concepts and equations of the occupational theory.[3,4,6] This work has required the identification of monoreceptor systems, the use of natural and selective agonists, as well as the utilization of specific and selective antagonists.

Three isolated blood vessels have been identified and extensively used for this purpose, the rabbit vena cava (RVC), the rabbit pulmonary artery (RPA), and the rat portal vein (RPV)[9-11] (see Table 1, bold type). Three groups of selective agonists ([Sar^9,$Met(O)_2^{11}$]SP and Ac[Arg^6,Sar^9,$Met(O)_2^{11}$]$SP_{(6-11)}$ for the NK_1,[6,12] [βAla^8]$NKA_{(4-10)}$ and [Nle^{10}]$NKA_{(4-10)}$ for NK_2,[13,14] [$MePhe^7$]NKB[6,15] or senktide[16] for the NK_3), as well as the three naturally occurring neurokinins have been systematically tested, together with a couple of antagonists (generally of nonpeptide nature) for each receptor. The results, summarized in Table 1, bold type, indicate that: (1) the order of potency of the neurokinins for the three receptors differs in that it is SP > NKA > NKB in the NK_1; NKA > NKB > SP in the NK_2; and NKB > NKA > SP in the NK_3 functional site. (2) NK_1 selective agonists are active in the RVC and inactive in the other two tissues; NK_2 selective agonists show a much higher activity (by 1.5 to more than 3.0 log units) in the RPA than in other preparations; and NK_3 selective agonists show high affinity in the RPV and much less (by 2.0 to 3.0 log units) in the other tissues.

Therefore, results obtained with agonists clearly differentiate between three different receptor types. Such a differentiation is validated by the results obtained with antagonists, namely CP 99994[17] and SR 140333,[18] two NK_1 selective compounds; SR 48968[19] and GR 94800,[20] two NK_2 antagonists recently described; R 486[21] or R 820,[4,22] two small peptidic compounds that show fairly high affinities for the NK_3 site of the RPV. Indeed, the first two compounds (CP 99994 and SR 140333) are extremely active (pA2 ≥ 9) on the NK_1 receptor type, while being almost inactive on the other receptors. The two NK_2 antagonists, the nonpeptide SR 48968 and the peptide GR 94800, are active on the NK_2 preparation and show little activity for NK_1 and NK_3 sites. The same can be said for the last two compounds, which are inhibitors of the contraction of the RPV (NK_3) and almost inactive in the other tissues.

Taken together, the data obtained with agonists and with antagonists demonstrate the existence of three pharmacologically distinct entities which can be characterized with the basic criteria recommended by Schild,[23] namely, the order of potency of agonists and the estimation of the apparent affinity of antagonists, which are very selective for each respective receptor type. Furthermore, all antagonists used in the present analysis and in recent reports have been shown to be competitive, CP 99994 by McLean et al.,[17] RP 67580 by Garret et al.,[24] SR 48968 by Emonds-Alt et al.,[19] R 486 by Drapeau et al.,[21] and R 820 by Regoli et al.[22] This is essential for an appropriate use of these compounds for the characterization of neurokinin receptor types and subtypes.

III. NEUROKININ RECEPTOR SUBTYPES

This topic has been recently reviewed by Regoli et al.,[22] who have characterized pharmacologically two NK_1 receptor subtypes, the NK_{1rb} (rb = rabbit) and the NK_{1r} (r = rat); two NK_2 receptor subtypes, the NK_{2rb} (NK_{2A}, according to Maggi et al.[14]) and the NK_{2h} (h = hamster) (NK_{2B}, according to Maggi et al.[14]); and the NK_{3r} and NK_{3gp} (gp = guinea pig). With respect to agonists, NK_{1rb} and NK_{1r} receptors differ, the former being more sensitive to the naturally occurring neurokinins than the latter. Similarly, NK_{2rb} shows higher affinities for the natural peptides and the NK_2 selective agonists when compared with NK_{2h}. Conversely, the NK_{3r} of the RPV shows a significantly lower affinity for the naturally occurring neurokinins than the NK_{3gp}, a new receptor subtype identified with binding assays by Petitet et al.[25] and validated with bioassays by Nguyen et al.[26] Selective agonists are more discriminative for the NK_{3gp} than for NK_{3r}.

The distinction between subtypes is confirmed by data obtained with antagonists. Thus, CP 99994 is almost 3 log units less active on the NK_{1r} receptor than on the NK_{1rb} site;[17,22] conversely, RP 67580 is more active on the rat (pA_2 8.0) than on the rabbit receptor (pA_2 7.2). SR 48968 shows higher affinity (by 2.2 log units) for the NK_{2rb} site than for NK_{2h},[27] while R 396 is more active in blocking the effect of NK_2 agonists in the hamster (urinary bladder:[13] pA_2 7.5, or trachea:[14] pA_2 7.6) than in the rabbit (pA_2 5.6); R 820 is a fairly potent antagonist on the NK_{3r} site, but is inactive on the NK_{3gp}.[4,22] Other compounds under investigation show an opposite pharmacological profile, in that they are active on the NK_{3gp} receptor and inactive on the NK_{3r} functional site.[28]

TABLE 2
Affinities of Neurokinins Evaluated by Bioassay (Pharmacology), Binding (Biochemistry), and Binding to COS Cells (Molecolar Biology)

	NK_1	NK_2	NK_3
Pharmacology (pD_2)			
Preparation	RVC	RPA	RPV
SP	8.6	6.1	5.8
NKA	7.3	8.2	6.4
NKB	7.2	7.4	7.7
Biochemistry (–log IC_{50} *M*)			
Preparation	GPB	RDSM	GPB
Ligand	$[^3H][Sar^9,Met(O)_2^{11}]SP$	$[^3H]NKA$	$[^3H]$senkide
SP	9.4	6.7	6.9
NKA	6.3	8.1	5.9
NKB	6.8	7.5	8.1
Molecular Biology (–log IC_{50} *M*)			
Preparation	COS	COS	COS
Ligand	$[^{125}I]$BH-SP	$[^{125}I]$NKA	$[^{125}I]$BH-ELE
SP	9.8	7.0	7.0
NKA	7.7	9.6	7.7
NKB	7.3	8.3	9.5

RVC, rabbit vena cava; RPA, rabbit pulmonary artery; RPV, rat portal vein;
GPB, guinea pig brain membranes; RDSM, rat duodenum smooth muscle membranes;
COS, CV1 origin SV40 cells;
pD_2, –log of the concentration of agonist producing 50% of the maximal response.

IV. COMPARISON OF DATA OBTAINED IN PHARMACOLOGICAL, BIOCHEMICAL, AND MOLECULAR BIOLOGICAL STUDIES ON NEUROKININ RECEPTORS

In parallel with the pharmacological studies described above, binding assays have been performed in various laboratories, using the three groups of compounds analyzed in Table 1 as competitors of the binding of radiolabeled ligands specific for NK_1, NK_2, or NK_3 sites.[29-31]

In other experiments, the cDNA coding for the three neurokinin receptors have been identified,[32-34] isolated, and transfected in appropriate systems (COS, CHO cells), in order to obtain homogeneous pure NK_1, NK_2, or NK_3 receptor sites for measuring actual affinities by the binding to membranes or to intact cells[35,36] and to demonstrate the activation of second messengers (generally IP_3 accumulation).[37] The three sets of data available for the naturally occurring neurokinins are compared in Table 2. Thus, apparent affinities of neurokinins (pD_2 values 7.7 to 8.6), evaluated with biological assays, are comparable to the affinities measured by binding (at least for NK_2 and NK_3 receptor sites); they are, however, lower than the affinities measured by binding of radioactive ligands to the homogeneous populations of receptor sites, that are found in the membranes of COS cells. Factors to be considered for explaining data are the peptide degradation, the accessibility of the receptor sites, and the efficacy of the biological systems to translate the peptide–receptor interaction into the contractile responses. Worthy of notice, however, is the perfect correspondence of the order of

TABLE 3
Neurokinin Biological Activities: Implication of Receptor Types

Effect	Target Cell	Mediator	NK_1	NK_2	NK_3
Neurotransmitters in the CNS	CNS neurons	Several mediators	+	+	+
Decrease of blood pressure	Artery endothelium	Release of NO	+		
Plasma extravasation	Vein smooth muscle	—	+	+	
Release of histamine, PGs, etc.	Mastocytes	Histamine	nrm	nrm	nrm
	Macrophages	PGs	+	+	
	Lymphocytes	Superoxide anions	+		
Smooth muscle contraction	Smooth muscle	PGs?	+	+	+
Release of transmitters	Parasympathetic terminals	Acetylcholine			+
in the PNS	Sympathetic terminals	Noradrenaline		+	
Release of secretory products	Exocrine glands	—	+		

Note: CNS, central nervous system; PNS, peripheral nervous system; PGs, prostanoids; nrm, nonreceptor mechanism, direct interference with G proteins? (see Mousli et al.[57])

potency of agonists and the ratio of activity between the three neurokinins in the various tests. In recent studies, the accumulation of IP_3 in COS or CHO cells has been measured and shown to correlate well with the neurokinin binding, both in terms of IC_{50} and over a range of concentrations expressed by the concentration–response curves.[37,38] Nakanishi et al.[36] have also shown that the activation of NK_1, NK_2, and NK_3 receptors expressed in CHO cells is associated with an increase of IP_3 (obtained with low concentrations of agonists) and of cAMP (obtained with higher concentrations, by a factor of 10, of neurokinins).

V. NEUROKININ RECEPTOR TYPES IN TARGET CELLS: BIOLOGICAL EFFECTS

Neurokinins are primarily neuropeptides and neurotransmitters (Table 3) in the CNS. The three naturally occurring peptides (SP, NKA, NKB) and the three receptor types (NK_1, NK_2, NK_3) have been demonstrated in a variety of CNS nuclei and a certain complementarity between peptides and receptors appears to be present in some areas of the brain. The abundant literature in this field has been analyzed by Pernow[39] and more recently by Otsuka and Yoshioka.[40] The role and function of neurokinins on the CNS are largely unknown, and their involvement in pain transmission, regulation of movements (dopamine), and salt or water intake (vasopressin, angiotensin) is still hypothetical, as is the attribution of such effects to one or the other receptor type. Recent work with antagonists points to the involvement of NK_1 receptors in emesis[41] and of NK_2 receptors in cough;[42] however, this relies on antagonist specificity and selectivity.

Application of neurokinins and their selective agonists intracerebroventricularly has been found to be followed by marked changes of cardiovascular functions (increase of blood pressure and heart rate and, eventually, increase of peripheral blood flow)[43,44] and of animal behavior (increased locomotion, grooming, scratching, skin biting), which have been interpreted as "fulfilling the . . . criteria of a classic defense reaction".[43] These central effects are blocked by the respective NK_1, NK_2, and NK_3 receptor antagonists,[44] and appear in large part to be due to activation of the autonomic nerves and to be mediated by classical neurotransmitters, since (in the periphery) they are blocked by α or β adrenoreceptor antagonists.[43,45]

However, the effects of neurokinin B and NK_3 receptor activation may go through the neurohypophysis, being blocked by antagonists of the vasopressin V_1 receptor.[46] In some circumstances the central effects of neurokinins may contribute to neurogenic inflammation. Large amounts of neurokinins are also found in the periphery, especially in the myenteric plexus and in the sensory nerves. For instance, Brimijoin et al.[47] have shown that 90% of the SP immunoreactivity of the lung C-fibers acceding to the nodose ganglion is in the peripheral bronchi, and might be released upon sensory nerve stimulation[2] and generate neurogenic inflammation.

Sensory neuropeptides find a variety of targets in peripheral tissues, some of which are presented in Table 3. NK_1 receptors have been visualized by historadiography[48] and their functional role has been demonstrated by pharmacological experiments in isolated vessels (e.g., the dog carotid artery,[49] the rabbit aorta with intact endothelium[50]) and in isolated rat mesentery.[51] NK_1 receptors are localized in the endothelial membrane and activate the release of EDRF (endothelium-dependent relaxing factor; in fact, nitric oxide),[49,52,53] which diffuses to the smooth muscle, where it induces arterial relaxation via guanilyl-cyclase activation. This is the mechanism by which exogenous neurokinins, particularly SP, reduce blood pressure[4] and increase local blood flow in neurogenic inflammation. NK_1 receptors are present in the capillary endothelium and in the veins,[9] where they stimulate (contract) the smooth muscle, producing venoconstriction (with the resulting increase in intracapillary pressure) presumably by favoring the accumulation of IP_3 and Ca^{2+}. By this mechanism neurokinins, through NK_1 receptors, may favor plasma extravasation and accumulation of fluid in the extracellular space (edema).

Other targets for neurokinins are the mastocytes, macrophages, and fibroblasts, from which neurokinins may promote the release of other endogenous agents, such as histamine, prostanoids, superoxides, which contribute to neurogenic inflammation in various ways, including the stimulation of sensory fibers that produce pain or decrease pain threshold.[40,54] Some of these cells, for instance the lymphocytes, have receptors, presumably of the NK_1 type,[55] while others, such as the mast cells may be activated to release histamine and 5-hydroxytryptamine[56] by a nonspecific mechanism that does not require specific receptor sites. This mechanism has been discussed by Mousli et al.[57] as dependent on positive charges, which mediate the direct interaction of agents with G proteins.

Important for the overall proinflammatory role of the neurokinins is the stimulation of chemotaxis of blood cells, like neutrophils[5] or sedentary cells, such as alveolar macrophages[58] or monocytes,[59] as well as the participation of SP and congeners in the immune responses through activation of lymphocyte activity and proliferation. SP has been found to promote the release of cytokines in the brain[60] and the production of interleukin-6 from U373MG, a human astrocytoma cell line.[61] The majority of the above-mentioned effects is mediated by NK_1 receptors. The other neurokinin receptors are involved in smooth muscle contraction, especially the NK_2, which is responsible for the activation of intestinal, tracheobronchial, and other (e.g., urethral or vescical) smooth muscles in man.[8]

No neurokinin autoreceptors have been demonstrated clearly in the sensory fibers, while NK_2 and NK_3 functional sites have been found, respectively, in sympathetic (the rat vas deferens[62]) and parasympathetic (the guinea pig ileum[63]) nerve fibers. Exocrine glands, such as the salivary, bronchial, intestinal, and pancreatic glands, respond to neurokinins with an increase of secretory products that may contribute to the tissue reactions to noxious or irritant stimuli, as exemplified by the tracheobronchial system.[64]

VI. CONCLUSIONS

Three tachykinin-like peptides (SP, NKA, and NKB) have been identified and cloned in a variety of mammalian cells. Each peptide acts, as favorite agonist, on a respective receptor type (NK_1, NK_2, and NK_3). These three receptors have been fully characterized and cloned. The neurokinin receptor system has been studied using bioassay techniques, and, as pharmacological tools, both the naturally occurring agonists, some synthetic selective agonists, and competitive antagonists of peptidic and nonpeptidic nature. The recent studies carried out with potent and selective nonpeptidic antagonists not only have confirmed the classification of the neurokinin receptors into three distinct receptor types, but have also revealed differences between receptors among species, suggesting the existence of receptor subtypes. Indeed, NK_{1rb} and NK_{1r}, NK_{2rb} and NK_{2h}, NK_{3gp} and NK_{3r} have been pharmacologically described in this chapter.

Pharmacological data obtained in three monoreceptor systems (RVC, RPA, and RPV) have been compared with data of biochemical (binding) and molecular biological (cloned receptors expressed in CHO cells) studies. A very good correspondence has been demonstrated between the three sets of data.

Finally, a list of target cells showing biological responses to one or the other receptor type has been presented to describe the most important physiological implications of neurokinins and their receptors.

ACKNOWLEDGMENTS

The experimental work reported in this chapter was supported by grants from the Medical Research Council of Canada (MRCC) and the Quebec Heart Foundation. D.R. is a career investigator of the MRCC.

REFERENCES

1. Otsuka, M. and Yanagisawa, M., Pain and neurotransmitters, *Cell. Mol. Neurobiol.*, 10, 293, 1990.
2. Lembeck, F. and Holzer, P., Substance P as a neurogenic mediator of antidromic vasodilatation and neurogenic plasma extravasation, *Naunyn-Schmiedeberg's Arch. Pharmacol.*, 310, 175, 1979.
3. Regoli, D., Drapeau, G., and D'Orleans-Juste, P., Pharmacological receptors for substance P and neurokinins, *Life Sci.*, 40, 109, 1987.
4. Regoli, D., Boudon, A., and Fauchere, J. L., Receptors and antagonists for substance P and related peptides, *Pharmacol. Rev.*, in press.
5. Hartung, H. P. and Toyka, K. V., Substance P, the immune system and inflammation, *Int. Rev. Immunol.*, 4, 229, 1989.
6. Regoli, D., Drapeau, G., Dion, S., and Couture, R., New selective agonists for neurokinin receptors: pharmacological tools for receptor characterization, *Trends Pharmacol. Sci.*, 9, 290, 1988.
7. Gitter, B. D., Waters, D. C., Bruns, R. F., Mason, N. R., Nixon, J. A., and Howbert, J. J., Species differences in affinities of nonpeptide antagonists for substance P receptors, *Eur. J. Pharmacol.*, 197, 237, 1991.
8. Maggi, C. A., Patacchini, R., Rovero, P., and Giachetti, A., Tachykinin receptors and tachykinin receptor antagonists, *J. Autonom. Pharmacol.*, 13, 23, 1993.
9. Nantel, F., Rouissi, N., Rhaleb, N. E., Jukic, D., and Regoli, D., Pharmacological evaluation of the angiotensin, kinin and neurokinin receptors on the rabbit vena cava, *J. Cardiovasc. Pharmacol.*, 18, 398, 1991.
10. D'Orleans-Juste, P., Dion, S., Drapeau, G., and Regoli, D., Different receptors are involved in the endothelium-mediated relaxation and the smooth muscle contraction of the rabbit pulmonary artery in response to substance P and related neurokinins, *Eur. J. Pharmacol.*, 125, 27, 1986.
11. Mastrangelo, D., Mathison, R., Huggel, H. J., Dion, S., D'Orleans-Juste, P., Rhaleb, N. E., Drapeau, G., Rovero, P., and Regoli, D., The rat isolated portal vein: a preparation sensitive to neurokinins, particularly neurokinin B, *Eur. J. Pharmacol.*, 134, 321, 1987.
12. Drapeau, G., D'Orleans-Juste, P., Dion, S., Rhaleb, N. E., Rouissi, N., and Regoli, D., Selective agonists for substance P and neurokinin receptors, *Neuropeptides*, 10, 43, 1987.
13. Dion, S., Rouissi, N. E., Nantel, F., Jukic, D., Rhaleb, N. E., Tousignant, C., Telemaque, S., Drapeau, G., Regoli, D., Naline, E., Advenier, C., Rovero, P., and Maggi, C. A., Structure-activity study of neurokinins: antagonists for the NK-2 receptor, *Pharmacology*, 41, 184, 1990.

14. Maggi, C. A., Patacchini, R., Astolfi, M., Rovero, P., Giuliani, S., and Giachetti, A., NK-2 receptor agonists and antagonists, *Ann. N.Y. Acad. Sci.*, 632, 184, 1991.
15. Drapeau, G., D'Orleans-Juste, P., Dion, S., Rhaleb, N. E., and Regoli, D., Specific agonists for neurokinin B receptors, *Eur. J. Pharmacol.*, 136, 401, 1987.
16. Wörmser, U., Laüfer, R., Hart, Y., Chorev, M., Gilon, C., and Selinger, Z., Highly selective agonists for substance P receptor subtypes, *EMBO J.*, 5, 2805, 1986.
17. McLean, S., Ganong, A., Seymour, P. A., Snider, R. M., Desai, M. C., Rosen, T., Bryce, D. K., Longo, K. P., Reynolds, L. S., Robinson, G., Schmidt, A. W., Siok, C., and Heym, J., Pharmacology of CP 99994, a non-peptide antagonist of the tachykinin (NK-1) receptor, *Regul. Peptides*, 46, 120, 1993.
18. Emonds-Alt, X., Doutremepuich, J. D., Jung, M., Proietto, E., Santucci, V., Vanbroek, D., Vilain, P., Suobrie, P., Le Fur, G., and Breliere, J. C., SR 140333 a non-peptide antagonist of substance P (NK-1) receptor, *Neuropeptides*, 24, 231, 1993.
19. Emonds-Alt, X., Vilain, P., Goulaouic, P., Proietto, V., Van Broeck, D., Advenier, C., Naline, E., Neliat, G., Le Fur, G., and Breliere, J. C., A potent and selective non-peptide antagonist of the neurokinin A (NK-2) receptor, *Life Sci.*, 50, 101, 1992.
20. McElroy, A. B., Clegg, S. P., Deal, M. J., Ewan, G. B., Hagan, R. M., Ireland, S. J., Jordan, C. C., Porter, B., Ross, B. C., Ward, P., and Whittington, A. R., Highly potent and selective heptapeptide antagonists of the neurokinin NK-2 receptor, *J. Med. Chem.*, 35, 2582, 1992.
21. Drapeau, G., Rouissi, N., Nantel, F., Rhaleb, N. E., Tousignant, C., and Regoli, D., Antagonists for the neurokinin NK-3 receptor evaluated in selective receptor systems, *Regul. Peptides*, 31, 125, 1990.
22. Regoli, D., Nguyen, Q. T., and Jukic, D., Neurokinin receptor subtypes characterized by biological assays, *Life Sci.*, 54, 2035, 1994.
23. Schild, H. O., Receptor classification with special reference to β adrenergic receptor, in *Drug Receptors*, Rang, H. P., Ed., University Park Press, Baltimore, 1973, 29.
24. Garret, C., Carruette, A., Fardin, V., Moussaoui, S., Peyronel, J. F., Blandrard, J. C., and Laduron, P. M., Pharmacological properties of a potent and selective non-peptide substance P antagonist, *Proc. Natl. Acad. Sci. U.S.A.*, 88, 10208, 1991.
25. Petitet, F., Beaujouan, J. C., Saffroy, M., Torrens, Y., and Glowinski, J., The nonpeptide NK-2 antagonist SR 48968 is also a NK-3 antagonist in the guinea pig but not in the rat, *Biochem. Biophys. Res. Commun.*, 191, 180, 1993.
26. Nguyen, Q. T., Jukic, D., Chretien, L., Gobeil, F., Boussougou, M., and Regoli, D., Two NK-3 receptor subtypes: demonstration by biological and binding assays, *Neuropeptides*, 27, 157, 1994.
27. Advenier, C., Rouissi, N., Nguyen, Q. T., Emonds-Alt, X., Breliere, J. C., Neliat, G., Naline, E., and Regoli, D., Neurokinin A (NK-2) receptor revisited with SR 48968, a potent non-peptide antagonist, *Biochem. Biophys. Res. Commun.*, 184, 1418, 1992.
28. Nguyen, Q. T., Nguyen, X. K., Amonds-Alt, X., Breliere, J. C., Cellier, E., Couture, R., and Regoli, D., Pharmacological characterization of SR 142801, a new nonpeptide antagonist of the neurokinin NK-3 receptor. presented at Peptides in Tissue Injury, Montreal, July 31 to August 5, 1994.
29. Cascieri, M. A., Chicchi, G. G., and Liang, T., Demonstration of two distinct tachykinin receptors in rat brain cortex, *J. Biol. Chem.*, 260, 1501, 1985.
30. Lavielle, S., Chassaing, G., Ploux, D., Loeuillet, D., Besseyre, J., Julien, S., Marquet, A., Convert, O., Beaujouan, J. C., Torrens, Y., Bergstrom, L., Saffroy, M., and Glowinski, J., Analysis of tachykinin binding site interactions using constrained analogues of the tachykinins, *Biochem. Pharmacol.*, 37, 41, 1988.
31. Mussap, C. J., Geraghty, D. P., and Burcher, E., Tachykinin receptors: a radioligand binding perspective, *J. Neurochem.*, 60, 1987, 1993.
32. Ingi, T., Kitajima, Y., Minamitake, Y., and Nakanishi, S., Characterization of ligand binding properties and selectivities of three rat tachykinin receptors by transfection and functional expression of their coded cDNAs in mammalian cells, *J. Pharmacol. Exp. Ther.*, 259, 968, 1991.
33. Nakanishi, S., Mammalian tachykinin receptors, *Annu. Rev. Neurosci.*, 14, 123, 1991.
34. Gerard, N. P., Bao, L., Ping, H. X., and Gerard, C., Molecular aspects of the tachykinin receptors, *Regul. Peptides*, 43, 21, 1993.
35. Guard, S. and Watson, S. P., Tachykinin receptor types: classification and membrane signalling mechanism, *Neurochem. Int.*, 18, 149, 1991.
36. Nakanishi, S., Nakajima, Y., and Yokota, Y., Signal transmission and ligand binding domains of the tachykinin receptors, *Regul. Peptides*, 46, 37, 1993.
37. Nakajima, Y., Tsuchida, K., Negishi, M., Ito, S., and Nakanishi, S., Direct linkage of three tachykinin receptors to stimulation of both phosphatidyl-inositol hydrolysis and cyclic AMP cascade in transfected Chinese hamster ovary cells, *J. Biol. Chem.*, 267, 2437, 1992.
38. Hermans, E., Jeanjean, A. P., Fardin, V., Pradier, L., Garret, C., Laduron, P. M., Octave, J. M., and Maloteaux, J. M., Interaction of the substance P receptor antagonist RP 67580 with the rat brain NK-1 receptor expressed in transfected CHO cells, *Eur. J. Pharmacol.*, 245, 43, 1993.

39. Pernow, B., Substance, P, *Pharmacol. Rev.*, 35, 85, 1983.
40. Otsuka, M. and Yoshioka, K., Neurotransmitter functions of mammalian tachykinins, *Physiol. Rev.*, 73, 229, 1993.
41. Tattersall, F. D., Rycroft, W., Hargreaves, R. J., and Hill, R. G., The tachykinins NK-1 receptor antagonist CP 99994 alterates cisplatin-induced emesis in the ferret, *Eur. J. Pharmacol.*, 250, R5-6, 1993.
42. Advenier, C., Girard, V., Naline, E., and Emonds-Alt, X., Antitussive effect of SR 48968, a non-peptide tachykinin (NK-2) receptor antagonist, *Eur. J. Pharmacol.*, 250, 169, 1993.
43. Unger, T., Carolus, S., Demmert, G., Ganten, D., Lang, R. E., Maser-Gluth, C., Steinberg, H., and Veelken, R., Substance P induces a cardiovascular defence reaction in the rat. Pharmacological characterization, *Circ. Res.*, 63, 812, 1988.
44. Tschöppe, C., Picard, P., Culman, N. J., Prat, A., Itoi, K., Regoli, D., Unger, T., and Couture, R., Use of selective antagonists to dissociate the central cardiovascular and behavioural effects of tachykinins on NK-1 and NK-2 receptors in the rat, *Br. J. Pharmacol.*, 107, 750, 1992.
45. Couture, R., Hassessian, H., and Gupta, A., Studies on the cardiovascular effects produced by the spinal action of substance P in the rat, *J. Cardiovasc. Pharmacol.*, 11, 270, 1988.
46. Takano, Y. and Kamiya, H., Tachykinin receptor subtypes: cardiovascular roles of tachykinin peptides, *Asian Pacif. J. Pharmacol.*, 6, 341, 1991.
47. Brimijoin, S., Lundberg, J. M., Brodin, E., Hökfelt, T., and Nilsson, G., Axonal transport of substance P in the vagus and sciatic nerves of the guinea pig, *Brain Res.*, 191, 443, 1980.
48. Burcher, E., Mussap, C. J., Geraghty, D. P., McClure-Sharp, J. M., and Watkins, D. J., Concepts and characterization of tachykinin receptors, *Ann. N.Y. Acad. Sci.*, 632, 123, 1991.
49. D'Orleans-Juste, P., Dion, S., Mizrahi, J., and Regoli, D., Effects of peptides and non-peptides on isolated arterial smooth muscle: role of endothelium, *Eur. J. Pharmacol.*, 114, 9, 1985.
50. Snider, R. M., Constantine, J. W., and Lowe, J. A., III, Longo, K. P., Lebel, W. S., Woody, H. A., Drozda, S. E., Desai, M. C., Vinick, F. J., Spenser, R. W., and Hess, H. J., A potent nonpeptide antagonist of the substance P (NK-1) receptors, *Science*, 251, 435, 1991.
51. Claing, A., Telemaque, S., Cadieux, A., Fournier, A., Regoli, D., and D'Orleans-Juste, P., Non-adrenergic and non-cholinergic arterial dilatation and venoconstriction are mediated by CGRP and NK-1 receptor respectively in the mesenteric vasculature of the rat following perivascular nerve stimulation, *J. Pharmacol. Exp. Ther.*, 266, 1226, 1992.
52. Zawadzki, J. V., Furchgott, R. F., and Cherry, D., The obligatory role of endothelial cells in the relaxation of arterial smooth muscle by substance P, *Fed. Proc.*, 40, 689, 1981.
53. Palmer, R. M. J., Ferridge, A. G., and Moncada, S., Nitric oxide release accounts for the biological activity of endothelium-derived relaxing factor, *Nature*, 327, 524, 1987.
54. Leah, I. D., Cameron, A. A., and Snow, P. J., Neuropeptides in physiologically identified mammalian sensory neurons, *Neurosci. Lett.*, 56, 257, 1985.
55. Payan, D. G., Brewster, D. R., and Goetzl, E. J., Stereospecific receptors for substance P on cultured human IM-9 lynphoblasts, *J. Immunol.*, 133, 3260, 1984.
56. Devillier, P., Drapeau, G., Renoux, M., and Regoli, D., Role of the N-terminal arginine in the histamine releasing activity of substance P, bradykinin and related peptides, *Eur. J. Pharmacol.*, 168, 53, 1989.
57. Mousli, M., Bueb, J. L., Bronner, C., Rouot, B., and Landry, Y., G protein activation: a receptor-independent mode of action for cationic amphiphilic neuropeptides and venom peptides, *Trends Pharmacol. Sci.*, 11, 358, 1990.
58. Brunelleschi, S., Vanni, L., Ledda, F., Giotti, A., Maggi, C. A., and Fantozzi, R., Tachykinins activate guinea pig alveolar macrofages: involvement of NK-2 and NK-1 receptors, *Br. J. Pharmacol.*, 100, 417, 1990.
59. Lots, M., Vaughan, J. H., and Carson, D. A., Effect of neuropeptides on production of inflammatory cytokines by human monocytes, *Science*, 241, 1218, 1988.
60. Martin, F. C., Charles, A. C., Sanderson, M. J., and Merrill, J. E., Substance P stimulates IL-1 production by astrocytes via intracellular calcium, *Brain Res.*, 599, 13, 1992.
61. Gitter, B. D., Regoli, D., Howbert, J., Glasebrook, A. L., and Waters, D. C., Interleukin-6 secretion from human astrocytoma cells induced by substance P, *J. Neuroimmunol.*, in press.
62. Lee, C. M., Iversen, L. L., Hanley, M. R., and Sandberg, B. E. B., The possible existence of multiple receptors for substance P, *Naunyn-Schmiedeberg's Arch. Pharmacol.*, 318, 281, 1982.
63. Laüfer, R., Wörmser, U., Friedman, Z. Y., Gilon, C., Chorev, M., and Selinger, Z., Neurokinin B is a preferred agonist for a neuronal substance P receptor and its action is antagonized by enkephalin, *Proc. Natl. Acad. Sci. U.S.A.*, 82, 7444, 1985.
64. Barnes, P. J., Neurogenic inflammation and asthma, *J. Asthma*, 29, 165, 1992.

Chapter 9

PHARMACOLOGY OF CALCITONIN GENE-RELATED PEPTIDE

Judith M. Hall and Susan D. Brain

CONTENTS

I. INTRODUCTION

The 37-amino acid peptide, calcitonin gene-related peptide (CGRP), along with the tachykinins substance P and neurokinin A, are the major peptide transmitters present in primary afferent C-fiber neurons. CGRP-containing sensory nerve fibers have been shown to innervate tissues of diverse origins, including exocrine and endocrine glands, cardiac muscle, visceral smooth muscle, and epithelia, where they have a predominantly perivascular localization. Release of CGRP from the peripheral termini of primary afferent neurons contributes to various efferent components of the neurogenic inflammatory response and, in particular, CGRP may be considered to be the primary mediator of neurogenic vasodilatation. CGRP is also present in intrinsic enteric neurons innervating hollow organs, autonomic, central nervous system (CNS), and motor fibers, and also in some non-neuronal cell types. CGRP has numerous biological activities unrelated to neurogenic inflammation. This chapter summarizes the current status of our understanding of the pharmacology of CGRP. In addition, we consider the evidence for the proposed existence of CGRP receptors and briefly consider potential therapeutic applications arising from knowledge of the biological effects of CGRP.

0-8493-7646-7/96/$0.00+$.50

II. SYNTHESIS AND DISTRIBUTION

A. CGRP

The unique discovery in 1982[1] and the regulation of expression of the 37-amino acid peptide CGRP have been the subject of great interest, as discussed in detail by Emeson in Chapter 2 of this volume. CGRP has now been isolated from several species, and is known to exist in two forms (α and β) in species including the rat and human. αCGRP is encoded by the calcitonin gene, and the expression of CGRP or calcitonin is tissue specific, whereas the second form, βCGRP, is the only biologically active product of a separate gene (see Emson, Chapter 2). Figure 1 shows the amino acid sequences of some of the various identified forms of CGRP which share overall greater than 80% sequence homology, along with a number of other endogenous peptides that share some degree of sequence homology and some biological effects in common with CGRP (see Section II.B).

Soon after the discovery of CGRP, it was realized that both the α and β forms of CGRP are expressed primarily in neural tissues,[2] with CGRP-like immunoreactivity (CGRP-LI) localized to discrete pathways in both the central and peripheral nervous systems (see References 3 to 5). Peripherally, CGRP-containing sensory fibers innervate most vascular and nonvascular tissues, predominantly in perivascular primary afferent unmyelinated C-fibers, where it is often colocalized with other transmitters, notably the tachykinins substance P and neurokinin A. CGRP-LI has, however, also been demonstrated in afferent Aδ-fibers, in both sympathetic and parasympathetic nerves, intrinsic enteric neurons in hollow organs, notably the gastrointestinal tract, and CGRP-LI also coexists with acetylcholine (ACh) in motor neurons (see References 4 and 5). In the CNS, CGRP-LI is widespread and is particularly abundant in the hypothalamus and in some brainstem nuclei (see Reference 3). The distribution of the two forms of CGRP has not been examined extensively; however, in the nervous system of the gastrointestinal tract, α-CGRP is found predominantly in extrinsic sensory neurons, whereas the β form is preferentially expressed in enteric myenteric and submucosal intrinsic neurons.[6] Other cell sources of CGRP have also been suggested, including thyroid C cells, especially in older rats, and in medullary thyroid carcinoma in man, where CGRP coexists with calcitonin.[7]

B. AMYLIN AND ADRENOMEDULLIN

Amylin[8] (also known as insulinoma amyloid peptide and islet amyloid polypeptide) and adrenomedullin[9] are two recently described peptides that share limited sequence homology with CGRP (Figure 1). The homology resides in a six-residue ring structure formed by an intramolecular disulfide linkage (an area important in terms of CGRP receptor recognition, see Section IV.A), and there is also some homology in the C-terminal amide sequence. Amylin is colocalized with insulin in the β cells of the pancreas in many patients with non-insulin-dependent diabetes mellitus. In contrast, adrenomedullin is a hormone, originally isolated from human pheochromocytoma cells, which circulates in the blood in high concentrations (e.g., 3 fmol ml^{-1},[10]). This latter observation has led to speculation that adrenomedullin may be an endogenous circulating form of CGRP, an issue of interest in relation to receptor classification (Section IV.A). Both amylin and adrenomedullin have been shown to have biological activities in common with CGRP, including that of microvascular vasodilatation,[11-14] and recent studies have demonstrated that many of these effects are due to activation of CGRP receptors (Section IV.A). However, it is also suggested that amylin acts via separate receptors to inhibit carbohydrate metabolism (see Section X).

III. RELEASE AND DEGRADATION OF CGRP

Calcium-dependent release of CGRP-LI from sensory nerves has been demonstrated both centrally and peripherally following electrical stimulation, stimulation with high K^+ media or

Sequences of CGRP and related peptides

```
        1-  2-  3-  4-  5-  6-  7-  8- 9 - 10- 11- 12- 13- 14- 15- 16- 17- 18- 19- 20- 21- 22- 23- 24-

            ┌───────────────────┐
        Ala-Cys-Asp-Thr-Ala-Thr-Cys-Val-Thr-His-Arg-Leu-Ala-Gly-Leu-Leu-Ser-Arg-Ser-Gly-Gly-Val-Val-Lys-
        Asn-----Asn-----------------------------------------------------------------------Met---------
        Ser-----Asn-----------------------------------------------------------------------------------
        Ser-----Asn-----------------------------------------------------------------------------------
        Lys-----Asn-----------------Ala-----Gln-------------Asn-Phe-----Val-His-----Ser-Asn-Asn-Phe-Gly-
Ser-Phe-Gly-Cys-Arg-Phe-Gly-----Cys-Thr-Val-Gln-Lys---------His-Gln-Ile-Tyr-Gln-Phe-Thr-Asp-Lys-Asp-----
            └───────────────────┘

13- 14- 15- 16- 17- 18- 19- 20- 21- 22- 23- 24- 25- 26- 27- 28- 29- 30- 31- 32- 33- 34- 35- 36- 37- 38-


25- 26- 27- 28- 29- 30- 31- 32- 33- 34- 35- 36- 37
Asn-Asn-Phe-Val-Pro-Thr-Asn-Val-Gly-Ser-Lys-Ala-Phe-NH2      Human αCGRP
Ser--------------------------------------------------NH2      Human βCGRP
Asp-------------------------------------Glu----------NH2      Rat αCGRP
Asp--------------------------------------------------NH2      Rat βCGRP
Ala-Ile-Leu-Ser-Ser---------------------Asn-Thr-Tyr-NH2       Human amylin
Asp-----Val-Ala-----Arg-Ser-Lys-Ile-----Pro-Gln-Gly-Tyr-NH2   Human adrenomedullin13-52
39- 40- 41- 42- 43- 44- 45- 46- 47- 48- 49- 50- 51- 52
```

FIGURE 1. Amino acid residues common to alpha human CGRP (αCGRP*h*) are shown for some other partly homologous peptides. Non-identical homologous residues and the internal disulfide bonds are indicated.

acute capsaicin treatment in numerous tissues.[15,16] In addition, a range of inflammatory mediators have been shown to stimulate, or augment, the peripheral release of neuropeptides, including CGRP, from sensory nerves. These include bradykinin, histamine, serotonin, high K^+ and hydrogen ions, and, further, release may be dependent on prostaglandin synthesis (see References 16 to 18). Bradykinin-evoked CGRP release has been shown in functional studies to result from B_2 receptor stimulation, though direct neurochemical evidence for this is still lacking (see References 17 and 19). In the majority of tissues, the source of the released CGRP-LI is considered to be derived primarily from capsaicin-sensitive C-fibers. Thus, neonatal pretreatment of animals with capsaicin significantly depletes tissue levels of CGRP and prevents its subsequent evoked release.[20-22]

The means by which CGRP is degraded following its release are not well documented. CGRP has been shown to be metabolized by enzymes present in cerebrospinal fluid,[23] by tryptase and chymase released from activated mast cells,[24,25] and, albeit slowly, by neutral endopeptidase E.C.3.4.24.11.[26]

IV. RECEPTORS FOR CGRP

Despite extensive research over the last decade and the publication of numerous original papers on the effects and radioligand binding characteristics of CGRP, the pharmacological characteristics and possible existence of multiple receptors for CGRP has not, as yet, been established. In contrast to the situation with tachykinins, the CGRP receptor gene has not yet been cloned, nor has the receptor protein been fully isolated (see Reference 27). Importantly, for the functional characterization of CGRP receptors, the naturally occurring forms of CGRP exhibit very similar pharmacology and have not proved useful in the initial characterization of CGRP receptors; further, there are currently no selective, high-affinity antagonists or selective agonists available for classification purposes.

A. RECEPTOR CLASSIFICATION

The classification scheme that has been adopted by most workers was proposed by Dennis and colleagues.[28] Two CGRP subtypes of receptor were postulated; $CGRP_1$ receptors (found in guinea pig atria) were defined as being sensitive to the antagonist $CGRP_{8-37}$,[30] whereas $CGRP_2$ receptors (found in rat vas deferens) were less sensitive to this antagonist, but had previously been shown to be preferentially activated by the synthetic CGRP agonist analog [Cys(ACM)^{2,7}]-CGRP.[29] In practice, however, the presence of $CGRP_1$ or $CGRP_2$ receptors has, in general, been concluded solely based on the inhibitory potency of $CGRP_{8-37}$. This has led to the conclusion that most of the vascular effects of CGRP are mediated via $CGRP_1$ receptors (see Section V).

A consideration of the now extensive literature on the pharmacological effects of $CGRP_{8-37}$ does not permit any firm conclusions to be drawn regarding the reliability of the $CGRP_1$/$CGRP_2$ receptor classification scheme, or the existence of subtypes of these receptors. The reasons for this are multifold, and have been discussed extensively by Poyner.[5] In particular, use of the putative $CGRP_2$ receptor-selective agonist [Cys(ACM)^{2,7}]-CGRP has not been adopted to any great extent, so its potential usefulness is not established. Furthermore, the putative $CGRP_1$ receptor antagonist, $CGRP_{8-37}$, has not proved as valuable for receptor classification purposes as originally speculated, and other CGRP receptor antagonists of potential value, such as rat [Tyr^0]-$CGRP_{(28-37)}$[31] have not been incorporated into many studies. Notably, although differences in the affinity of $CGRP_{8-37}$ have been reported, no consistent pattern of affinity estimates has emerged. Thus, pK_B or pA_2 estimates ranging over three orders of magnitude (5.5 to 8.5) have been reported (see Reference 5). Although these observations may well indicate the existence of CGRP receptor heterogeneity, as proposed by several groups, such a conclusion may well be premature in view of the possible existence of species homologues of receptors[32] which could, in theory, account for some of these differences, and also, there may be complications resulting from differences in biodegradation. Furthermore, the reasons for the frequently reported noncorrespondence between antagonist affinities as determined in functional as compared to radioligand binding studies, and discrepancies in affinity estimates for $CGRP_{8-37}$ reported in the same tissue when determined by different groups, and indeed in some cases by the same group, need to be resolved (discussed in detail by Poyner[5]).

Of recent interest are the findings that several peptides, which have some degree of sequence homology with CGRP (see Section II.B), can interact and stimulate the same receptors. These peptides include amylin and adrenomedullin (and its 13–52 fragment). Thus, amylin evokes vasodilatation in the rabbit skin[11] and rat isolated perfused kidney,[12] and inhibits nerve and spasmogen-evoked contractions of nonvascular smooth muscle[33] via an interaction with $CGRP_{8-37}$ sensitive receptors. Adrenomedullin causes vasodilatation, in preparations including rat skin and the hamster cheek pouch microvasculature, via an interaction with $CGRP_{8-37}$-sensitive receptors.[13] The main structural similarity between these peptides is a six-residue ring structure formed by an intramolecular disulfide linkage and some further residues in common (Figure 1). Previous studies have shown that opening or modifying the disulfide bond leads to a considerable loss of agonist activity of CGRP.[34,35] This suggests that the disulfide bond may be important for agonist activity of amylin and adrenomedullin at CGRP receptors. Incorporation of such ligands into studies of CGRP receptors may well be a useful future approach to classification.

B. RECEPTOR SIGNAL TRANSDUCTION

Although the CGRP receptor genes have yet to be cloned, it is generally accepted that CGRP mediates the majority of its known biological effects through the stimulation of seven-transmembrane spanning G protein-linked receptors, and available biochemical evidence is compatible with this proposal. For example, in the neonatal rat heart, some of the effects of CGRP are pertussis toxin sensitive[36] and in some, though not all, radioligand binding studies GDP, GTP, and analogs have been shown to reduce CGRP receptor binding affinity (see Poyner,[5] for discussion). The most consistently reported receptor-effector coupling mechanism by CGRP is its stimulation of adenylyl cyclase and increase in cAMP levels,[37-40] although one report showed an increase in inositol monophosphate accumulation in response to CGRP.[41] The mechanism of vasodilatation for CGRP appears to be tissue specific. Vasodilator responses to CGRP can be endothelium-independent, especially in the microvasculature.[42,43] Alternatively, endothelium-dependent, nitric oxide-mediated relaxation vasodilatation is observed in some large vessels and this has been extensively studied in the rat aorta.[44,45] A novel mechanism of vasorelaxation was reported by Nelson and co-workers.[42] The ATP-gated potassium channel blocker, glibenclamide, was shown to inhibit CGRP-evoked vasodilatation in rabbit mesenteric arteries, thus suggesting that vasorelaxation results from hyperpolarization following receptor-gated K^+-channel opening. CGRP has also been shown to activate the muscarinic-gated K^+ channel in neonatal rat atrial cells.[36] In contrast, several studies, notably in the CNS, have reported that CGRP evokes K^+ channel closing and consequent membrane depolarization.[46]

V. CARDIOVASCULAR SYSTEM

A. MICROVASCULATURE AND NEUROGENIC INFLAMMATION

The best established action of CGRP is its potent ability to cause vasodilatation, especially in microvascular beds. CGRP, in femtomole to picomole doses, was shown to cause a long-lasting erythema when injected intradermally into the skin of several species, including human[47] (see Brain, Chapter 18, this volume). The erythema was shown, by intravital microscopy

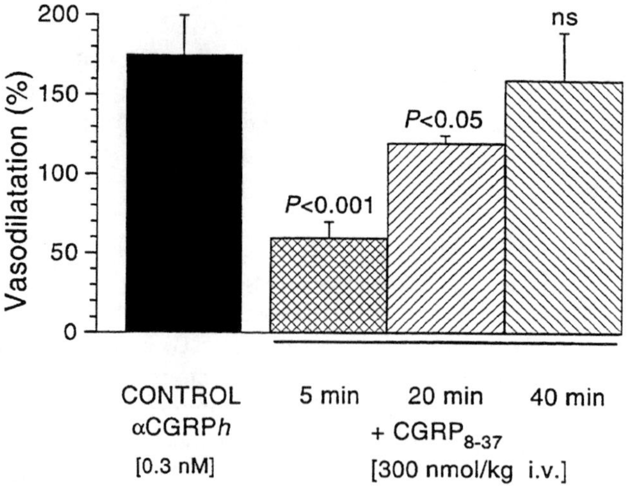

FIGURE 2. The effect of $CGRP_{8-37}$ on CGRP-induced arteriolar dilatation in the cheek pouch microvasculature of the anesthetized hamster *in vivo*. Responses are shown as percent change from control values, measured using an intravital microscopic technique. Arteriolar vasodilator responses following topical application of CGRP (human αCGRP 0.3 n*M*), are significantly ($P < .05$) antagonised 5 and 20 min after intravenous $CGRP_{8-37}$ administration (300 nmol/kg). Results are shown as mean ± SEM, n = 5.

of the hamster cheek pouch microvasculature, to be due to a direct vasodilator effect of CGRP on arterioles[47] and CGRP-evoked microvascular vasodilatation has subsequently been shown to be inhibited by the CGRP receptor antagonist $CGRP_{8-37}$, thereby indicating involvement of $CGRP_1$ receptor activation[48,49] (Figure 2). CGRP also decreases perfusion pressure in isolated perfused organs such as the rat mesentery[50] and kidney[12] via $CGRP_{8-37}$-sensitive receptor activation. More recently, it has been established that CGRP is responsible for the vasodilatation resulting from electrical stimulation of the saphenous nerve in rat skin[49] and capsaicin in rabbit skin;[48] also, perivascular nerve stimulation in the isolated perfused rat mesentery is inhibited by $CGRP_{8-37}$.[50] It is now established that CGRP plays an important role in mediating neurogenic vasodilatation[51] (Figure 3). Of great interest is the proposal that the gastroprotective effect of capsaicin results largely from the hyperemic action of released CGRP (see Reference 52). As well as being a primary mediator of vasodilatation, CGRP contributes to neurogenic inflammation in other ways. Thus, by virtue of its vasodilator activity, CGRP potentiates edema formation, for example in skin and joints, induced by a wide variety of mediators of increased vascular permeability, including substance P[53,54] (Figure 3), and CGRP has been shown to potentiate neutrophil accumulation induced by chemotactic mediators in rabbit skin.[55] CGRP has also been shown to directly stimulate adhesion of neutrophils to human umbilical vein endothelial cells. This latter effect was inhibited by $CGRP_{8-37}$.[56] However, other results on neutrophil accumulation are less clear and the direct effect of CGRP on the cellular component in inflammation is in need of further investigation.

B. OTHER CARDIOVASCULAR EFFECTS

Although the most important effects of CGRP are considered to be on the microvasculature, in isolated tissue studies, CGRP has been demonstrated to act in a potent manner to relax arteries. Examples include intracerebral arteries,[40] coronary arteries,[57] the basilar artery,[58] and the aorta,[44] and high-affinity specific binding sites for CGRP have been demonstrated in vascular tissues including human, bovine, and porcine coronary arteries.[59] In most large blood vessels, the vasodilator effects of CGRP have been shown to be inhibited by the $CGRP_1$ receptor antagonist, $CGRP_{8-37}$.[40,58]

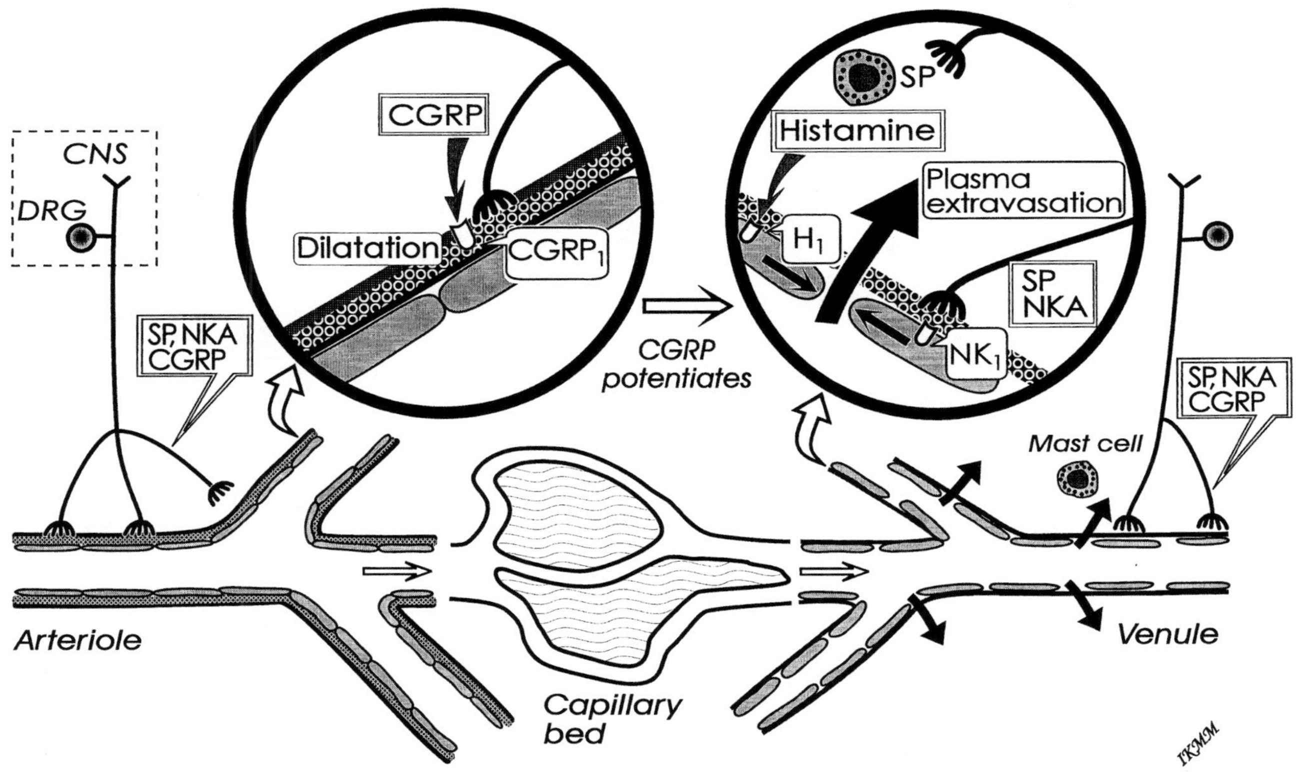

FIGURE 3. CGRP and tachykinin receptors in the skin microvasculature. CGRP mediates prolonged neurogenic vasodilatation in the skin microvasculature via a direct action on $CGRP_1$ receptors on arteriolar smooth muscle. As a consequence of these potent vasodilator properties, CGRP potentiates edema formation produced by mediators that increase microvascular permeability, such as substance P (which is coreleased with CGRP from sensory nerve endings). Substance P and neurokinin A can evoke increased microvascular permeability either directly via NK_1 receptors on endothelial cells or, in the case of substance P, indirectly via histamine released from mast cells acting on endothelial H_1 receptors. DRG, dorsal root ganglion; CNS, central nervous system.

Intravenous injection of CGRP in the conscious rat causes hypotension and an increase in heart rate; the most marked vasodilator effects that contribute to this hypotensive effect are observed in the cerebral circulation and the hindquarters.[60] The hypotensive effect is inhibited by the CGRP receptor antagonist $CGRP_{8\text{-}37}$.[61-62] The tachycardia is thought to result from reflex activation of the sympathetic nervous system.[63]

CGRP-containing fibers have been demonstrated in arteries, the sinoatrial node, and right atrium of the heart,[64] and CGRP binding sites have been demonstrated in the atria.[65] In the atria of some species, including the guinea pig,[16] CGRP increases the rate and force of contraction.

VI. GENITOURINARY TRACT

The pharmacological effects of CGRP in the urinary tract are discussed in detail by Lecci and Maggi (Chapter 16). In the isolated guinea pig urinary bladder preparation, CGRP relaxes electrical or spasmogen-induced tone and interestingly, inhibition of the electrically evoked twitch response in this preparation is not blocked by $CGRP_{8\text{-}37}$.[33] CGRP-LI fibers are found in the ureter and renal pelvis in several species[66-68] and CGRP has been proposed to have a role as an inhibitory neurotransmitter in this structure.[68]

CGRP-LI has been demonstrated in the vas deferens and in the isolated preparation of the rat[29] and mouse,[69] where CGRP has been shown to inhibit the twitch response following sympathetic nerve stimulation. Capsaicin inhibits the twitch response to electrical stimulation in the mouse vas deferens, an effect that is absent in capsaicin-pretreated animals. The inhibitory phase is inhibited by $CGRP_{8\text{-}37}$.[70] CGRP-LI fibers have also been demonstrated in the rat uterus, and in this preparation, CGRP causes relaxation of precontracted tone; this effect is also inhibited by $CGRP_{8\text{-}37}$.[71]

VII. GASTROINTESTINAL TRACT

The effect of CGRP in intestinal smooth muscle has been the subject of recent investigation and is discussed in detail in Chapters 12 and 13. In the gastrointestinal tract, CGRP is present both in extrinsic capsaicin-sensitive sensory nerves and in intrinsic enteric neurons.[72] CGRP has a complex action in the intestinal tract. For example, in the guinea pig small intestine, which has been studied in the most detail, CGRP causes contraction of both the longitudinal and circular muscle layers in the untreated preparation, but relaxes the atropine-treated longitudinal or tetrodotoxin-treated circular muscle.[73] It is thought that the relaxant effect of CGRP is due to a direct effect on the smooth muscle via activation of $CGRP_1$ receptors, whereas the stimulant effects result from release of ACh. Indeed, CGRP has been shown to excite myenteric neurons.[46] The contractile response to CGRP is not inhibited by $CGRP_{8\text{-}37}$.[73] CGRP has also been shown to inhibit ACh release and cholinergic-mediated twitches of the small intestine of the guinea-pig, man, and pig,[74,75] and some of the inhibitory effects of CGRP may be neuronal in origin.[76] In most other smooth muscle preparations, including the rat duodenum[77] and mouse colon,[78] CGRP elicits relaxation.

VIII. AIRWAYS

The effects of CGRP in the airways are discussed in detail by Barnes (Chapter 14). CGRP-LI has been demonstrated in and released from the airways of several species, including man,[79] rat,[80] and guinea pig.[81] In human bronchi, CGRP-evoked constriction has been reported in some studies[79] but not in others,[82] thus there is some confusion. In other species, such as the rat, inhibition of tracheal tone predominates.[83]

IX. CENTRAL NERVOUS SYSTEM

CGRP-immunoreactive fibers and cell bodies have an extensive distribution in sensory and motor nerves in the brain of many species.[3-5] Several autoradiographic studies have identified high-affinity specific binding sites for CGRP in various brain regions.[84-87] In the rat, particularly dense binding was observed in the nucleus accumbens, amygdaloid complex, and mammillary body. The distribution of human αCGRP binding sites exhibited marked phylogenetic differences.[85] Studies involving the central administration of CGRP provide evidence suggesting a role for CGRP in thermoregulation, feeding, control of sympathetic outflow, growth hormone secretion and movement (reviewed by Poyner[5]). In general, the autoradiographic distribution of CGRP binding sites correlates well with the distribution of CGRP-immunoreactive fibers and the effects of centrally administered CGRP. There are, however, few reports of the effects of CGRP receptor antagonists on endogenously released CGRP in the CNS.

At the level of the spinal cord, CGRP-LI predominates in those areas receiving sensory inputs in the dorsal horn, although limited CGRP-LI has been shown in areas associated with autonomic inputs and some ventral horn motor neurons,[4,5] and CGRP binding sites have been reported in the spinal cord of the rat and monkey.[88] At the level of the spinal cord, potentially important tachykinin–CGRP interactions have been demonstrated, especially relating to modulation of nociceptive threshold. For example, CGRP attenuated the tail flick threshold produced by substance P or noxious cutaneous stimulation[89] and CGRP has been shown to facilitate excitation induced by substance P and noxious stimuli in dorsal horn neurons.[90]

X. SKELETAL MUSCLE

CGRP is colocalized with ACh at motor nerve terminals.[91] CGRP increases ACh release at these sites,[38] increases phosphorylation of the nicotinic receptor and functionally enhances the rate of nicotinic receptor desensitization[92] and stimulates nicotinic receptor synthesis and expression.[93] CGRP also has a variety of metabolic effects on skeletal muscle, including inhibition of insulin-stimulated glycogen synthesis[94] and/or insulin-stimulated glucose transport.[95] The metabolic effects of CGRP are shared with amylin.

XI. MISCELLANEOUS EFFECTS

CGRP has been reported to have many additional pharmacological effects, too numerous to be cited in this short account. The mitogenic and trophic actions of CGRP have been discussed by Ziche (Chapter 20). Other effects of CGRP of special interest include increasing amylase secretion from guinea pig pancreatic acini,[96] and inhibition of gastric acid output and secretion of gastrin and pepsin.[97] In addition, CGRP binding sites have been demonstrated on macrophages,[98] mouse bone marrow cells,[99] and rat lymphocytes.[100] The role of CGRP with respect to these cells remains to be established.

XII. CGRP AND DISEASE

The majority of therapeutic uses proposed for CGRP or its antagonists target its potent vasodilator properties. Selective targeting of blood flow in restricted areas, using antagonists in conditions of excess dilatory activity or CGRP or other agonist ligand where vasodilator activity is restricted, has led to proposals of therapeutic usefulness in Raynaud's disease, subarachnoid hemorrhage, congestive heart failure, hemodialysis-induced hypotension, and myocardial infarction (reviewed by Shawket and Brown[101] and Shulks[102]). CGRP receptor antagonists have a potential therapeutic role in conditions resulting from inappropriate neurogenic inflammation, for example in the eye following laser or ocular surgery trauma,[103] in migraine (see Moskowitz et al., Chapter 15) and in the joints in arthritis (see Ferrell and Lam, Chapter 17). In view of the cross-talk between amylin and CGRP, there may be potential use for CGRP/amylin receptor antagonists in maturity-onset diabetes.[8,104] However, until the CGRP receptor gene(s) is cloned, the importance of other endogenous activators (such as amylin and adrenomedullin) of "CGRP" receptors established, and high-affinity receptor antagonists developed, therapeutic possibilities and the roles of CGRP in the body have to remain speculative.

ACKNOWLEDGMENTS

We thank the Wellcome Trust for support, Dr. I.K.M. Morton for preparing the figures, and Dr. I.K.M. Morton and Mr. P. Wilsoncroft for help with the reference list.

REFERENCES

1. Amara, S.G., Jones, V., Rosenfeld, M.G., Ong, E.S., and Evans, R.M., Alternative RNA processing in calcitonin gene expression generates mRNA encoding different polypeptide products, *Nature*, 298, 240, 1982.
2. Amara, S.G., Arriza, J.L., Leff, S.E., Swanson, L.W., Evans, R.M., and Rosenfeld, M.G., Expression in brain of a messenger RNA encoding a novel neuropeptide homologous to calcitonin gene-related peptide, *Science*, 229, 1094, 1985.
3. Goodman, E.C. and Iversen, L.L., Calcitonin gene-related peptide: novel neuropeptide, *Life Sci.*, 38, 2169, 1986.
4. Ishida-Yamamoto, A. and Tohyama, M., Calcitonin gene-related peptide in the nervous tissue, *Prog. Neurobiol.*, 33, 335, 1989.
5. Poyner, D.R., Calcitonin gene-related peptide: multiple actions, multiple receptors, *Pharmacol. Ther.*, 56, 23, 1992.
6. Mulderry, P.K., Ghatei, M.A., Spokes, R.A., Jones, P.M., Pierson, A.M., Hamid, Q.A., Kanse, S., Amara, S.G., Burrin, J.M., Legon, S., Polak, J.M., and Bloom, S.R., Differential expression of α-CGRP and β-CGRP by primary sensory neurones and enteric autonomic neurons of the rat, *Neuroscience*, 25, 195, 1988.
7. Sabate, M.I., Stolarsky, L.S., Polak, J.M., Bloom, S.R., Varndell, I.M., Ghatei, M.A., Evans, R.M., and Rosenfeld, M.G., Regulation of neuroendocrine gene expression by alternative RNA processing. Colocalization of calcitonin and calcitonin gene-related peptide in thyroid C-cells, *J. Biol. Chem.*, 260, 2589, 1985.
8. Rink, T.J., Beaumont, K., Koda, J., and Young, A., Structure and biology of amylin, *Trends Pharmacol. Sci.*, 14, 113, 1993.
9. Kitamura, K., Kangawa, K., Kawamoto, M., Ichiki, Y., Nakamura, S., Matsuo, H., and Eto, T., Adrenomedullin: a novel hypotensive peptide isolated from human pheochromocytoma, *Biochem. Biophys. Res. Commun.*, 192, 553, 1993.
10. Kitamura, K., Kangawa, K., Kojima, M., Ichiki, Y., Matsuo, H., and Eto, T., Complete amino acid sequence of porcine adrenomedullin and cloning of cDNA encoding its precursor, *FEBS Lett.*, 338, 306, 1994.
11. Brain, S.D., Wimalawansa, S., MacIntyre, I., and Williams, T.J., The demonstration of vasodilator activity of pancreatic amylin amide in the rabbit, *Am. J. Pathol.*, 136, 487, 1990.
12. Chin, S.Y., Hall, J.M., Brain, S.D., and Morton, I.K.M., Vasodilator responses to calcitonin gene-related peptide (CGRP) and amylin in the rat isolated perfused kidney are mediated via $CGRP_1$ receptors, *J. Pharmacol. Exp. Ther.*, 269, 989, 1994.
13. Hall, J.M., Siney, L., Lippton, H., Hyman, A., Kang-Chang, J., and Brain, S.D., Interaction of $adrenomedullin_{13-52}$ with calcitonin gene-related receptors in the microvasculature of the rat and hamster, *Br. J. Pharmacol.*, 1994 (in press).
14. Lippton, H., Chang, J.-K., Hao, Q., Summer, W., and Hyman, A.L., Adrenomedullin dilates the pulmonary vascular bed *in vivo*, *J. Appl. Physiol.*, 76, 2154, 1994.
15. Saria, A., Gamse, R., Petermann, J., Fischer, J.A., Theodorsson-Norheim, E., and Lundberg, J.M., Simultaneous release of several tachykinins and calcitonin gene-related peptide from rat spinal cord slices, *Neurosci. Lett.*, 63, 310, 1986.

16. Franco-Cereceda, A., Lundberg, J.M., Saria, A., Schreibmayer, W., and Tritthart, H.A., Calcitonin gene-related peptide: release by capsaicin and prolongation of the action potential in the guinea-pig heart, *Acta Physiol. Scand.*, 132, 181, 1988.
17. Geppetti, P., Sensory neuropeptide release by bradykinin: mechanisms and pathophysiological implications, *Regul. Pept.*, 47, 1, 1993.
18. Lundberg, J.M., Franco-Cereceda, A., Alving, K., Delay-Goyet, P., and Lou, Y.-P., Release of calcitonin gene-related peptide from sensory neurons, *Ann. N.Y. Acad. Sci.*, 657, 187, 1992.
19. Hall, J.M., Bradykinin receptors: pharmacological properties and biological roles, *Pharmacol. Ther.*, 56, 131, 1992.
20. Holzer, P., Local effector functions of capsaicin-sensitive sensory nerve endings: involvement of tachykinins, calcitonin gene-related peptide and other neuropeptides, *Neuroscience*, 24, 739, 1988.
21. Maggi, C.A., The pharmacology of the efferent function of sensory nerves, *J. Auton. Pharmacol.*, 11, 173, 1991.
22. Wimalawansa, S.J., The effects of neonatal capsaicin on plasma levels and tissue contents of CGRP, *Peptides*, 14, 247, 1993.
23. Le Grevès, P., Nyberg, F., Terenius, L., and Hökfelt, T., Calcitonin gene-related peptide is a potent inhibitor of substance P degradation, *Eur. J. Pharmacol.*, 115, 309, 1985.
24. Brain, S.D. and Williams, T.J., Substance P regulates the vasodilator activity of calcitonin gene-related peptide, *Nature*, 335, 73, 1988.
25. Walls, A.F., Brain, S.D., Desai, A., Jose, P.J., Hawkings, E., Church, M.K., and Williams, T.J., Human mast cell tryptase attenuates the vasodilator activity of calcitonin gene-related peptide, *Biochem. Pharmacol.*, 43, 1243, 1992.
26. Katayama, M., Nadel, J.A., Bunnett, N.W., Di Maria, G.U., Haxhiu, M., and Borson, D.B., Catabolism of calcitonin gene-related peptide and substance P by neutral endopeptidase, *Peptides*, 12, 563, 1991.
27. Chatterjee, T.K., Moy, J.A., Cai, J.J., Lee, H.-C., and Fisher, R.A., Solubilization and characterization of a guanine nucleotide-sensitive form of the calcitonin gene-related peptide receptor, *Mol. Pharmacol.*, 43, 167, 1993.
28. Dennis, T., Fournier, A., St.Pierre, S., and Quirion, R., Structure-activity profile of calcitonin gene-related peptide in peripheral and brain tissues. Evidence for receptor multiplicity, *J. Pharmacol. Exp. Ther.*, 251, 718, 1989.
29. Dennis, T., Fournier, A., Cadieu, A., Pomerleau, F., Jolicoeur, F.B., St.Pierre, S., and Quirion, R., $hCGRP_{8-37}$, a calcitonin gene-related peptide antagonist revealing calcitonin gene-related peptide receptor heterogeneity in brain and periphery, *J. Pharmacol. Exp. Ther.*, 254, 123, 1990.
30. Chiba, T., Yamaguchi, A., Yamatani, T., Nakamura, A., Morishita, T., Inui, T., Fukase, M., Noda, T., and Fujita, T., Calcitonin gene-related peptide receptor antagonist human CGRP-(8-37), *Am. J. Physiol.*, 256, E331, 1989.
31. Chakder, S. and Rattan, S., [Tyr^0]-calcitonin gene-related peptide 28-37 (rat) as a putative antagonist of calcitonin gene-related peptide responses on opossum internal anal sphincter smooth muscle, *J. Pharmacol. Exp. Ther.*, 253, 200, 1990.
32. Hall, J.M., Caulfield, M.P., Watson, S.P., and Guard, S., Receptor subtypes or species homologues: relevance to drug discovery, *Trends Pharmacol. Sci.*, 14, 376, 1993.
33. Giuliani, S., Wimalawansa, S.J., and Maggi, C.A., Involvement of multiple receptors in the biological effects of calcitonin gene-related peptide and amylin in rat and guinea-pig preparations, *Br. J. Pharmacol.*, 107, 510, 1992.
34. Tippins, J.R., Di Marzo, V.M., Panico, M., Morris, H.R., and MacIntyre, I., Investigation of the structure activity relationship of human calcitonin gene-related peptide (CGRP), *Biochem. Biophys. Res. Commun.*, 134, 1306, 1986.
35. Maggi, C.A., Rovero, P., Giuliani, S., Evangelista, S., Regoli, D., and Meli, A., Biological activity of N-terminal fragments of calcitonin gene-related peptide, *Eur. J. Pharmacol.*, 179, 217, 1990.
36. Kim, D., Calcitonin gene-related peptide activates the muscarinic-gated K^+ current in atrial cells, *Pfluegers Arch.*, 418, 338, 1991.
37. Crossman, D.C., McEwan, J., MacDermot, J., MacIntyre, I., and Dollery, C.T., Human calcitonin gene-related peptide activates adenylate cyclase and releases prostacyclin from human umbilical vein endothelium, *Br. J. Pharmacol.*, 92, 695, 1987.
38. Kobayashi, H., Hashimoto, K., Uchida, S., Sakuma, J., Takami, K., Tohyama, M., Izumi, F., and Yoshida, H., Calcitonin gene-related peptide stimulates adenylate cyclase activity in rat striated muscle, *Experientia*, 43, 314, 1987.
39. Gray, D.W. and Marshall, I., Nitric oxide synthesis inhibitors attenuate calcitonin gene-related peptide endothelium-dependent vasorelaxation in rat aorta, *Eur. J. Pharmacol.*, 212, 37, 1992.
40. Edwards, R.M., Stack, E.J., and Trizna, W., Calcitonin gene-related peptide stimulates adenylate cyclase and relaxes intracerebral arterioles, *J. Pharmacol. Exp. Ther.*, 257, 1020, 1991.

41. Laufer, R. and Changeux, J.-P., Calcitonin gene-related peptide elevates cAMP levels in chick skeletal muscle: possible neurotropic role for a coexisting neuronal messenger, *EMBO J.*, 6, 901, 1987.
42. Nelson, M.T., Huang, Y., Brayden, J.E., Hescheler, J., and Standen, N.B., Arterial dilations in response to calcitonin gene-related peptide involve activation of K^+ channels, *Nature*, 344, 770, 1990.
43. Hall, J.M. and Brain, S.D., Inhibition by SR140333 of NK_1 tachykinin receptor-evoked, nitric oxide-dependent vasodilatation in the hamster cheek pouch microvasculature *in vivo*, *Br. J. Pharmacol.*, 113, 522, 1994.
44. Hao, H., Fiscus, R.R., Wang, X., and Diana, J.N., N^{omega}-nitro-L-arginine inhibits vasodilations and elevations of both cyclic AMP and cyclic GMP levels in rat aorta induced by calcitonin gene-related peptide (CGRP), *Neuropeptides*, 26, 123, 1994.
45. Gray, D.W. and Marshall, I., Human α-calcitonin gene-related peptide stimulates adenylate cyclase and guanylate cyclase and relaxes rat thoracic aorta by releasing nitric oxide, *Br. J. Pharmacol.*, 107, 691, 1992.
46. Palmer, J.M., Schemann, M., Tamura, K., and Wood, J.D., Calcitonin gene-related peptide excites myenteric neurones, *Eur. J. Pharmacol.*, 132, 163, 1986.
47. Brain, S.D., Williams, T.J., Tippins, J.R., Morris, H.R., and MacIntyre, I., Calcitonin gene-related peptide is a potent vasodilator, *Nature*, 313, 54, 1985.
48. Hughes, S.R. and Brain, S.D., A calcitonin gene-related peptide (CGRP) antagonist ($CGRP_{8-37}$) inhibits microvascular responses induced by CGRP and capsaicin in skin, *Br. J. Pharmacol.*, 104, 738, 1991.
49. Escott, K.J. and Brain, S.D., Effect of a calcitonin gene-related peptide antagonist ($CGRP_{8-37}$) on skin vasodilatation and oedema induced by stimulation of the rat saphenous nerve, *Br. J. Pharmacol.*, 110, 772, 1993.
50. Claing, A., Télémaque, S., Cadieux, A., Fournier, A., Regoli, D., and D'Orléans-Juste, P., Nonadrenergic and noncholinergic arterial dilatation and venoconstriction are mediated by calcitonin gene-related $peptide_1$ and neurokinin-1 receptors, respectively, in the mesenteric vasculature of the rat after perivascular nerve stimulation, *J. Pharmacol. Exp. Ther.*, 263, 1226, 1992.
51. Holzer, P., Peptidergic sensory neurons in the control of vascular functions: mechanisms and significance in the cutaneous and splanchnic vascular beds, *Rev. Physiol. Biochem. Pharmacol.*, 121, 49, 1992.
52. Peskar, B.M., Wong, H.C., Walsh, J.H., and Holzer, P., A monoclonal antibody to calcitonin gene-related peptide abolishes capsaicin-induced gastroprotection, *Eur. J. Pharmacol.*, 250, 201, 1993.
53. Brain, S.D. and Williams, T.J., Inflammatory oedema induced by synergism between calcitonin gene-related peptide (CGRP) and mediators of increased vascular permeability, *Br. J. Pharmacol.*, 86, 855, 1985.
54. Cambridge, H. and Brain, S.D., Calcitonin gene-related peptide increases blood flow and potentiates plasma protein extravasation in the rat knee joint, *Br. J. Pharmacol.*, 106, 746, 1992.
55. Buckley, T.L., Brain, S.D., Rampart, M., and Williams, T.J., Time dependent synergistic interactions between the vasodilator neuropeptide, calcitonin gene-related peptide and mediators of inflammation, *Br. J. Pharmacol.*, 103, 1515, 1991.
56. Sung, C.P., Arleth, A.J., Aiyar, N., Bhatnagar, P.K., Lysko, P.G., and Feuerstein, G., CGRP stimulates the adhesion of leukocytes to vascular endothelial cells, *Peptides*, 13, 429, 1992.
57. Foulkes, R., Shaw, N., Bose, C., and Hughes, B., Differential vasodilator profile of CGRP in porcine large and small diameter coronary artery rings, *Eur. J. Pharmacol.*, 201, 143, 1991.
58. Jansen, I., Characterization of calcitonin gene-related peptide (CGRP) receptors in guinea pig basilar artery, *Neuropeptides*, 21, 73, 1992.
59. Knock, G.A., Wharton, J., Gaer, J.A.R., Yacoub, M.H., Taylor, K.M., and Polak, J.M., Regional distribution and regulation of [^{125}I]calcitonin gene-related peptide binding sites in coronary arteries, *Eur. J. Pharmacol.*, 219, 415, 1992.
60. Gardiner, S.M., Compton, A.M., and Benenett, T., Regional haemodynamic effects of human α- and β-calcitonin gene-related peptide in conscious Wistar rats, *Br. J. Pharmacol.*, 98, 1225, 1989.
61. Donoso, M.V., Fournier, A., St.-Pierre, S., and Huidobro-Toro, J.P., Pharmacological characterization of $CGRP_1$ receptor subtype in the vascular system of the rat: studies with hCGRP fragments and analogs, *Peptides*, 11, 885, 1990.
62. Gardiner, S.M., Compton, A.M., Kemp, P.A., Bennett, T., Bose, C., Foulkes, R., and Hughes, B., Human α-calcitonin gene-related peptide (CGRP)-(8-37), but not -(28-37), inhibits carotid vasodilator effects of human α-CGRP in vivo, *Eur. J. Pharmacol.*, 199, 375, 1991.
63. Sirén, A.-L. and Feuerstein, G., Cardiovascular effects of rat calcitonin gene-related peptide in the conscious rat, *J. Pharmacol. Exp. Ther.*, 247, 69, 1988.
64. Saito, A., Kimura, S., and Goto, K., Calcitonin gene-related peptide as potential neurotransmitter in guinea pig right atrium, *Am. J. Physiol.*, 250, H693, 1986.
65. Sigrist, S., Franco-Cereceda, A., Muff, R., Henke, H., Lundberg, J.M., and Fischer, J.A., Specific receptor and cardiovascular effects of calcitonin gene-related peptide, *Endocrinology*, 119, 381, 1986.
66. Tamaki, M., Iwanaga, T., Sato, S., and Fujita, T., Calcitonin gene-related peptide (CGRP)-immunoreactive nerve plexuses in the renal pelvis and ureter of rats, *Cell Tissue Res.*, 267, 29, 1992.

67. Amann, R., Neural regulation of ureteric motility, in *Autonomic Control of the Urogenital System*, Maggi, C. A., Ed., Harwood Academic Publishers, Reading, U.K., 1993, 209.
68. Maggi, C.A. and Giuliani, S., The neurotransmitter role of calcitonin gene-related peptide in the rat and guinea-pig ureter: effect of a calcitonin gene-related peptide antagonist and species-related differences in the action of omega conotoxin on calcitonin gene-related peptide release from primary afferents, *Neuroscience*, 43, 261, 1991.
69. Al-Kazwini, S.J., Craig, R.K., and Marshall, I., Postjunctional inhibition of contractor responses in the mouse vas deferens by rat and human calcitonin gene-related peptide, *Br. J. Pharmacol.*, 88, 173, 1986.
70. Manzini, S. and Parlani, M., Opposite prejunctional modulation by NK-1 receptor and CGRP of adrenergic control of mouse vas deferens motility, *Regul. Pept.*, 41 (Suppl. 1), S104, 1992.
71. Shew, R.L., Papka, R.E., and McNeill, D.L., Substance P and calcitonin gene-related peptide immunoreactivity in nerves of the rat uterus: localization, colocalization and effects on uterine contractility, *Peptides*, 12, 593, 1991.
72. Sternini, C., Reeve, J.R., and Brecha, N., Distribution and characterization of calcitonin gene-related peptide immunoreactivity in the digestive system of normal and capsaicin-treated rats, *Gastroenterology*, 93, 852, 1987.
73. Barthó, L., Koczan, G., and Maggi, C.A., Studies on the mechanism of the contractile action of rat calcitonin gene-related peptide and of capsaicin on the guinea- pig ileum: effect of hCGRP (8-37) and CGRP tachyphylaxis, *Neuropeptides*, 25, 325, 1993.
74. Barthó, L., Lembeck, F., and Holzer, P., Calcitonin gene-related peptide is a potent relaxant of intestinal muscle, *Eur. J. Pharmacol.*, 135, 449, 1987.
75. Schwörer, H., Schmidt, W.E., Katsoulis, S., and Creutzfeldt, W., Calcitonin gene-related peptide (CGRP) modulates cholinergic neurotransmission in the small intestine of man, pig and guinea-pig via presynaptic CGRP receptors, *Regul. Pept.*, 36, 345, 1991.
76. Sun, Y.-D. and Benishin, C.G., Calcitonin gene-related peptide on longitudinal muscle and myenteric plexus of guinea-pig ileum, *J. Pharmacol. Exp. Ther.*, 259, 947, 1991.
77. Maggi, C.A., Manzini, S., Giuliani, S., Santicioli, P., and Meli, A., Calcitonin gene-related peptide activates non-adrenergic, non-cholinergic relaxations of the rat isolated duodenum, *J. Pharm. Pharmacol.*, 39, 327, 1987.
78. Cadieux, A., Pomerleau, F., St.Pierre, S., and Fournier, A., Inhibitory action of calcitonin gene-related peptide (CGRP) in the mouse colon, *J. Pharm. Pharmacol.*, 42, 520, 1990.
79. Palmer, J.B.D., Cuss, F.M.C., Mulderry, P.K., Ghatei, M.A., Springall, D.R., Cadieux, A., Bloom, S.R., Polak, J.M., and Barnes, P.J., Calcitonin gene-related peptide is localized to human airway nerves and potently constricts human airway smooth muscle, *Br. J. Pharmacol.*, 91, 95, 1987.
80. Hua, X.-Y. and Yaksh, T.L., Release of calcitonin gene-related peptide and tachykinins from the rat trachea, *Peptides*, 13, 113, 1992.
81. Luts, A., Widmark, E., Ekman, R., Waldeck, B., and Sundler, F., Neuropeptides in guinea pig trachea: distribution and evidence for the release of CGRP into tracheal lumen, *Peptides*, 11, 1211, 1990.
82. Martling, C.-R., Saria, A., Fischer, J.A., Hökfelt, T., and Lundberg, J.M., Calcitonin gene-related peptide and the lung: neuronal coexistence with substance P, release by capsaicin and vasodilator effect, *Regul. Pept.*, 20, 125, 1988.
83. Cadieux, A., Lanoue, C., Sirois, P., and Barabé, J., Carbamylcholine- and 5-hydroxytryptamine-induced contraction in rat isolated airways: inhibition by calcitonin gene-related peptide, *Br. J. Pharmacol.*, 101, 193, 1990.
84. Chatterjee, T.K. and Fisher, R.A., Multiple affinity forms of the calcitonin gene-related peptide receptor in rat cerebellum, *Mol. Pharmacol.*, 39, 798, 1991.
85. Dennis, T., Fournier, A., Guard, S., St.Pierre, S., and Quirion, R., Calcitonin gene-related peptide (hCGRPα) binding sites in the nucleus accumbens. Atypical structural requirements and marked phylogenic differences, *Brain Res.*, 539, 59, 1991.
86. Mimeault, M., Fournier, A., Dumont, Y., St.-Pierre, S., and Quirion, R., Comparative affinities and antagonistic potencies of various human calcitonin gene-related peptide fragments on calcitonin gene-related peptide receptors in brain and periphery, *J. Pharmacol. Exp. Ther.*, 258, 1084, 1991.
87. Sexton, P.M., McKenzie, J.S., Mason, R.T., Moseley, J.M., Martin, T.J., and Mendelsohn, F.A., Localization of binding sites for calcitonin gene-related peptide in rat brain by *in vitro* autoradiography, *Neuroscience*, 19, 1235, 1986.
88. Yashpal, K., Kar, S., Dennis, T., and Quirion, R., Quantitative autoradiographic distribution of calcitonin gene-related peptide (hCGRPα) binding sites in the rat and monkey spinal cord, *J. Comp. Neurol.*, 322, 224, 1992.
89. Cridland, R.A. and Henry, J.L, Intrathecal administration of CGRP in the rat attenuates a facilitation of the tail flick reflex induced by either substance P or noxious cutaneous stimulation, *Neurosci. Lett.*, 102, 241, 1989.

90. Biella, G., Panara, C., Pecile, A., and Sotgiu, M.L., Facilitatory role of calcitonin gene-related peptide (CGRP) on excitation induced by substance P (SP) and noxious stimuli in rat spinal dorsal horn neurons. An iontophoretic study *in vivo*, *Brain Res.*, 559, 352, 1991.
91. Mora, M., Marchi, M., Polak, J.M., Gibson, S.J., and Cornelio, F., Calcitonin gene-related peptide immunoreactivity of the human neuromuscular junction, *Brain Res.*, 492, 404, 1989.
92. Mulle, C., Benoit, P., Pinse, C., Roa, M., and Changeux, J.-P., Calcitonin gene-related peptide enhances the rate of desensitisation of the nicotinic acetylcholine receptor in cultured mouse muscle cells, *Proc. Natl. Acad. Sci. U.S.A.*, 85, 5728, 1988.
93. New, H.V. and Mudge, A.W., Calcitonin gene-related peptide regulates muscle acetylcholine receptor synthesis, *Nature*, 323, 809, 1986.
94. Leighton, B. and Cooper, G.J., Pancreatic amylin and calcitonin gene-related peptide cause resistance to insulin in skeletal muscle *in vitro*, *Nature*, 335, 632, 1988.
95. Hothersall, J.S., Muirhead, R.P., and Wimalawansa, S., The effect of amylin and calcitonin gene-related peptide on insulin-stimulated glucose transport in the diaphragm, *Biochem. Biophys. Res. Commun.*, 169, 451, 1990.
96. Maton, P.N., Pradhan, T., Zhou, Z.-C., Gardner, J.D., and Jensen, R.T., Activities of calcitonin gene-related peptide (CGRP) and related peptides at the CGRP receptor, *Peptides*, 11, 485, 1990.
97. Kraenzlin, M.E., Ch'ng, J.L., Mulderry, P.K., Ghatei, M.A., and Bloom, S.R., Infusion of a novel peptide, calcitonin gene-related peptide (CGRP) in man. Pharmacokinetics and effects on gastric acid secretion and on gastrointestinal hormones, *Regul. Pept.*, 10, 189, 1985.
98. Nong, Y.-H., Titus, T.G., Ribeiro, J.M., and Remold, H.G., Peptides encoded by the calcitonin gene inhibit macrophage function, *J. Immunol.*, 143, 45, 1989.
99. Mullins, M.W., Ciallella, J., Rangnekar, V., and McGillis, J.P., Characterization of a calcitonin gene-related peptide (CGRP) receptor on mouse bone marrow cells, *Regul. Pept.*, 49, 65, 1993.
100. McGillis, J.P., Humphreys, S., and Reid, S., Characterization of functional calcitonin gene-related peptide receptors on rat lymphocytes, *J. Immunol.*, 147, 3482, 1991.
101. Shawket, S. and Brown, M.W., Pathogenetic and therapeutic implications of the calcitonin gene-related peptide in the cardiovascular system, *Trends Cardiovasc. Med.*, 1, 211, 1991.
102. Shulkes, A., Calcitonin gene-related peptide. Potential role in vascular disorders, *Drugs Aging*, 3, 189, 1993.
103. Krootila, K., Oksala, O., Zschauer, A., Palkama, A., and Uusitalo, H., Inhibitory effect of methysergide on calcitonin gene-related peptide-induced vasodilatation and ocular irritative changes in the rabbit, *Br. J. Pharmacol.*, 106, 404, 1992.
104. Cooper, G.J.S., Amylin compared with calcitonin gene-related peptide: structure, biology, and relevance to metabolic disease, *Endocr. Rev.*, 15, 163, 1994.

Chapter 10

PEPTIDASE MODULATION OF NEUROGENIC INFLAMMATION

Jay A. Nadel

CONTENTS

I. NEUROGENIC INFLAMMATION

Neuropeptides are released from a selective population of sensory neurons when they are stimulated with capsaicin or with electrical stimulation. When these nerves are stimulated, release of neuropeptides occurs, including substance P (SP), neurokinin A (NKA), as well as calcitonin gene-related peptide (CGRP). Local release of these neuropeptides results in a series of inflammatory tissue responses, termed "neurogenic inflammation." These responses occur in multiple organs, including airways, eye, genitourinary and gastrointestinal tracts, and heart.[1] In airways, these responses include vasodilation, vascular extravasation, leukocyte adhesion to the endothelium, gland secretion, bronchial smooth muscle contraction, cough, and effects on lymphocytes and mast cells. A variety of stimuli, including irritants such as tobacco smoke and toluene diisocyanate, hypertonic saline, and mediators such as bradykinin, serotonin, histamine, prostanoids, and high potassium cause the neural release of sensory neuropeptides.

0-8493-7646-7/96/$0.00+$.50

II. MODULATION OF NEUROGENIC INFLAMMATION BY PEPTIDASES

There is no evidence that the clearance of neuropeptides occurs due to a reuptake process. Rather, the biological actions of neuropeptides are terminated by degradative cleavage by enzymes. Tachykinins can be degraded by serine proteases, mast cell chymase, and calpains (reviewed in Reference 2). However, present evidence suggests that two peptidases play important roles in limiting the actions of tachykinins after their release from sensory nerves: neutral endopeptidase (NEP; also called enkephalinase, EC 3.4.24.11) and angiotensin converting enzyme (ACE; also called kininase II, dipeptidyl carboxypeptidase 1, peptidyl dipeptidase A, EC 3.4.15.1).

A. CHARACTERISTICS OF THE PEPTIDASES

Both peptidases are zinc metalloproteases with variable glycosylation, and they are anchored to the cell by a single hydrophobic membrane-spanning domain, with their catalytic domain facing the extracellular milieu, where it cleaves peptides as they diffuse through tissue. The zinc ion, an essential component of the catalytic site, is a common target of inhibitors that utilize phosphate (phosphoramidon), thiol (thiorphan, captopril), or carboxyalkyl (enalapril, lisinopril) zinc-coordinating groups.

NEP has been cloned in various species, including man,[3] and does not appear to exist as isozyme variants. In the human DNA, the open reading frame encodes a 742-amino acid polypeptide (molecular weight, ~94 kDa). NEP has an N-terminal cytoplasmic domain and a large C-terminal extracellular domain containing the catalytic site. The enzyme is highly conserved among species (homology >90%). The human NEP gene is located on chromosome 3.[4]

In contrast to NEP, two ACE isozymes are reported: the larger form (150 to 180 kDa) is widely distributed in vascular endothelium; the lungs and kidneys contain the highest concentrations.[5] The smaller form (90 to 100 kDa) occurs in the testis.[6] Unlike NEP, ACE is probably anchored to the cell surface by the hydrophobic region located near its C-terminal domain.[6] ACE contains two highly symmetric extracellular domains,[7] each of which contains putative zinc-binding sequences. NEP and testicular ACE contain a single extracellular domain and a high degree of homology among species. Endothelial ACE shows a lower degree of homology; most of the differences are found in the region linking the two highly conserved extracellular domains.[8]

B. MECHANISMS OF ACTION OF THE PEPTIDASES

1. Specificity

Generally, the activity of both enzymes is limited to peptides less than 3 kDa, but some larger molecules are hydrolyzed slowly by NEP. NEP cleaves internal peptide bonds on the amino side of hydrophobic amino acids, and hence the name "endopeptidase"; ACE cleaves C-terminal di- or tripeptides.[9] NEP cleaves substance P at the Gln^6–Phe^7, Phe^7–Phe^8, and Gly^9–Leu^{10} bonds[9]. In the latter case, the principal catabolic fragment is $SP_{1\text{-}9}$, which is biologically inactive. Two other tachykinins, NKA and NKB, are hydrolyzed by NEP but not by ACE. Calcitonin gene-related peptide (CGRP), another neuropeptide released from sensory nerves, is cleaved only slowly by NEP.[10] Many other peptides are cleaved effectively by both enzymes, including bradykinin, angiotensins, gastrin, and the bacterial chemotactic peptide *N*-formyl-Met-Leu-Phe (FMLP). NEP also cleaves atrial natriuretic peptide (ANP) and endothelins.

3. Localization

Like other receptor–ligand interactions, the rank order of potency of the effects of neuropeptides on cells depends in part on the receptor affinity and density. However, because neuropeptides such as SP are cleaved potently by both NEP and ACE, their effects are also modulated by peptidases. These peptidases are anchored to cells on which they are expressed and can therefore cleave neuropeptides in close association with the surfaces of these cells. Therefore, the exact location and degree of expression of NEP and ACE are major determinants of neuropeptide actions.

ACE is concentrated on the luminal surface of the vascular endothelium,[11] so its modulation would be predicted to be predominantly on vascular responses (i.e., vascular extravasation, vasomotor tone, and leukocyte adhesion to the endothelium). On the other hand, because NEP is widely distributed on a variety of cells, including airway epithelium, smooth muscle, submucosal glands, and vascular endothelium, NEP modulates neurogenic inflammatory responses in a wide variety of cell types.

NEP is located on the surfaces of cells that contain tachykinin receptors, and studies suggest that NEP regulates the SP binding to its receptor by decreasing the amount of SP available to the receptor without significantly changing the affinity or the number of receptors.[66] When neuropeptides are released from nerves, they diffuse through tissues toward the neuropeptide receptor on the surface of a target cell, where the enzyme and the receptor compete for the peptide. Some molecules of the neuropeptide are cleaved and inactivated, thus decreasing the cellular response. In fact, with high expression of NEP, no cellular responses occur. Further evidence for competition between neuropeptide receptors and NEP on cells is demonstrated by the observation that responses to SP are smaller in isolated cells engineered to express both neurokinin receptors and NEP than in cells expressing only neurokinin receptors incubated together with adjacent cells that express only NEP.[12]

In addition, the sensory nerve terminals that release neuropeptides in airways are located predominantly in the airway epithelium in close contact with the basal cells (which contain NEP). Thus, NEP can also modulate neurogenic inflammatory responses at sites of neuropeptide release (Figure 1A and B). Evidence for this is as follows: when the epithelium is removed from airways *in vitro*, smooth muscle responses to SP and to capsaicin are increased because the epithelium is normally in close contact with capsaicin-sensitive nerve terminals. Figure 2 illustrates the concept of peptidase modulation at sites of neuropeptide release and on target cells containing neuropeptide receptors. The distribution of sites of neuropeptide release may vary in airways of different sizes, among organs and among species, and these can profoundly affect cellular responses. Furthermore, in chronic inflammatory states, proliferation of sensory nerves containing neuropeptides may alter and exaggerate neurogenic inflammatory responses.

C. PHYSIOLOGIC ROLES OF THE PEPTIDASES

Although NEP was discovered in the brush border epithelium of the kidney[13] and subsequently in the brain,[14] it was a decade later before the physiologic modulation of neurogenic inflammation by NEP was recognized. Selective inhibitors of NEP and ACE have been developed, and they have been useful in assessing the role of peptidases in modulating neurogenic inflammation. It was suggested that NEP might modulate neuropeptide actions in airways, reasoning that, if NEP modulates neurogenic inflammation, then selective inhibition of NEP should potentiate the effects of tachykinins.[15] This first study showed that NEP inhibition caused marked potentiation of tachykinin effects on airway submucosal gland secretion. It was hypothesized that peptidase modulation could critically modulate all neurogenic inflammatory responses,[16] and subsequent studies have confirmed this hypothesis in a variety of species and tissues. Some of these responses in airways will be reviewed.

1. Airway Smooth Muscle

Ferret tracheal smooth muscle is contracted by exogenous tachykinins and by electrical field stimulation. These responses are potentiated by inhibition of NEP,[17] and NEP-like

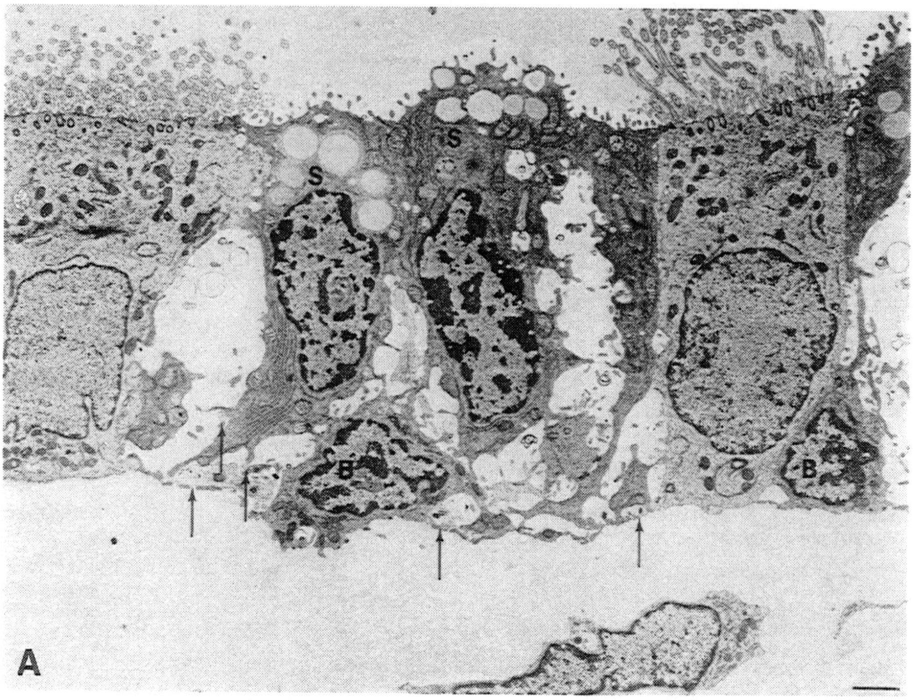

FIGURE 1. (A) Electron micrograph of tracheal epithelium showing at least five intraepithelial neurites (arrows) all of which are located at the base of the epithelium. Also visible are several prominent changes resulting from 2 min of vagal stimulation: the shrunken appearance of the secretory cells (S), the clustering of secretory granules at the apex of secretory cells, and the conspicuous widening of the spaces next to secretory cells and basal cells (B) and beneath the epithelium. The ciliated cells appear to be unaffected by the vagal stimulation. Bar = 2 μm. (Reproduced with permission from McDonald, D. M. et al., *J. Neurocytol.*, 17, 605, 1988.) (B) A region of Figure 1A enlarged to show a neuronal varicosity (V), which contains both small clear vesicles and large dense-cored vesicles, and two tiny neurites without vesicles (arrows) at the base of the tracheal epithelium. Arrowheads mark the basal lamina of the epithelium. The extracellular space between epithelial cells and the space beneath the epithelial basal lamina are conspicuously enlarged as a result of vagal stimulation for 2 min. B, epithelial basal cell; S, epithelial secretory cell. Bar = 0.5 μm. (Reproduced with permission from McDonald, D. M. et al., *J. Neurocytol.*, 17, 605, 1988.)

activity exists in airway smooth muscle.[18] In guinea pigs *in vivo*, increasing concentrations of aerosolized SP causes no airway narrowing in the control state, but after preincubation with an NEP inhibitor, SP causes a dose-dependent airway narrowing. This illustrates the potency of NEP in modulating inflammatory responses.

NK_2 receptors are believed to play a major role in neurogenic bronchomotor responses,[19] but there is some evidence that NK_1 receptors also play a role.[20,21] Inhibitors of ACE have no effect on smooth muscle contraction in response to aerosolized SP,[22] and this is compatible with the selective localization of ACE in the vascular bed. Thus, the major peptidase modulation of airway smooth muscle is via NEP, but some effects of ACE are demonstrable.[23-25]

2. Plasma Extravasation and Leukocyte Adhesion

Postcapillary venules are the target cells for plasma extravasation and are also the sites of leukocyte adhesion to the endothelium. Substance P and sensory nerve stimulation cause vascular extravasation.[26] These responses are markedly potentiated by NEP inhibitors.[27,28] Both NEP and ACE contribute to the modulation of vascular extravasation induced by SP.[29]

Substance P and electrical nerve stimulation also cause adhesion of leukocytes to postcapillary venules in airways. This effect is potentiated by both phosphoramidon[28,30] and captopril.[28,30] Simultaneous inhibition of both peptidases potentiates SP-induced responses more than a

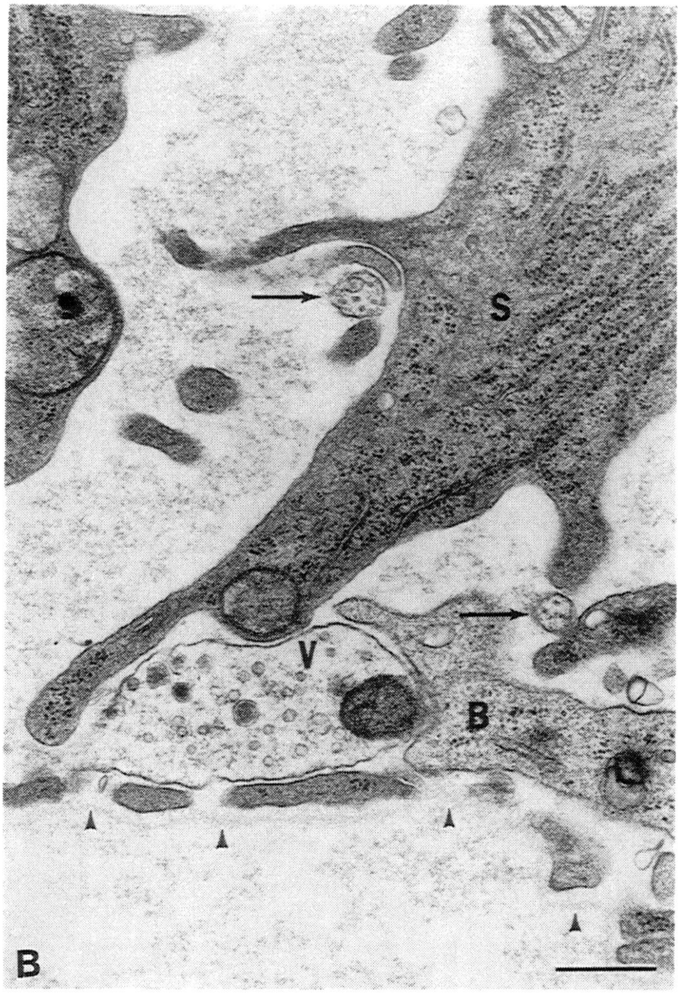

FIGURE 1. (Continued).

maximally effective dose of either inhibitor alone.[28,30] These results implicate both NEP and ACE in the modulation of leukocyte adhesion. Neurogenic inflammation causes rapid adherence of leukocytes to the postcapillary endothelium, an effect which is mediated by NK_1 receptors.[31] Adhesion is followed by a gradual reentry of the neutrophils into the circulation and comparatively little neutrophil migration into airway tissue. However, when NEP is inhibited, neurogenic inflammatory stimuli cause leukocyte adhesion followed by migration.[32]

Inhalation of hypertonic saline (the putative mechanism for exercise-induced asthma) results in plasma extravasation in rats, an effect that is abolished by sensory denervation with capsaicin[33] or by pretreatment with a selective NK_1 receptor antagonist[34] and is potentiated by an NEP inhibitor,[33] implicating neurogenic inflammation in the response. Inhalation of cold air also causes vascular extravasation in rats (presumably by evaporative water loss in airways), and this response is also increased by an NEP inhibitor and abolished by an NK_1 receptor antagonist.[35]

3. Airway Blood Flow

Although changes in blood flow must play an important role in inflammation, limited attention has been given to this area. Blood flow affects the recruitment of cells and mediators

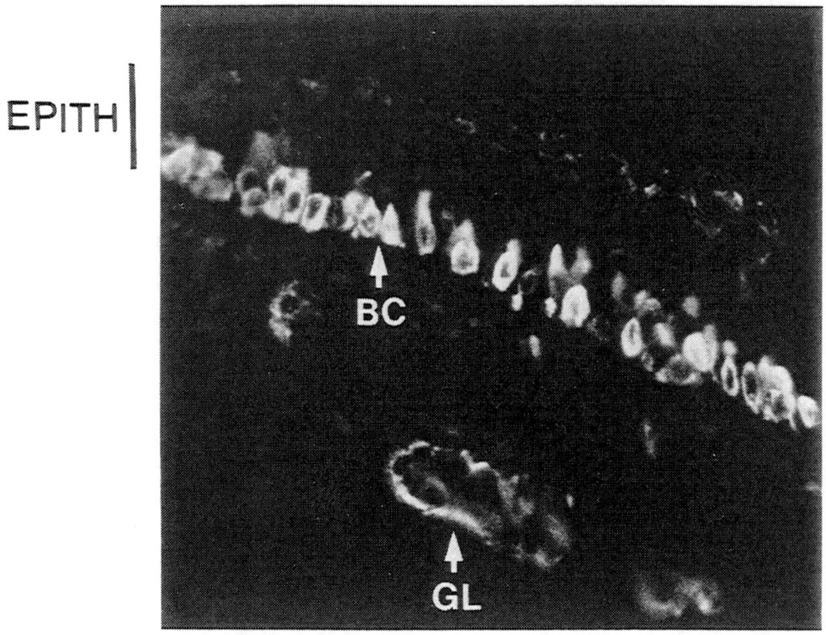

FIGURE 2. Immunocytochemical localization of neutral endopeptidase in rat trachea, showing staining with fluorescein-labeled neutral endopeptidase antibody in basal cells (BC) of the epithelium (EPITH) and glands (GL). (Reproduced with permission from Nadel, J. A., *Eur. Respir. J.*, 4, 745, 1991.)

and the production of tissue edema, and it affects the clearance of inflammatory mediators and drugs. Stimulation of capsaicin-sensitive nerves profoundly increases airway extrapulmonary blood flow in rats; low concentrations of capsaicin increase airway blood flow without circulatory effects in many other organs.[36] It is interesting that neurogenic inflammatory vasodilation occurs at concentrations that have no effect on vascular extravasation. Thus, mild inflammatory stimuli may be eliminated by neurogenic vasodilation, whereas stronger stimuli induce additional defense mechanisms (vascular extravasation, leukocyte recruitment). Substance P injected intravenously reproduces the effect of nerve stimulation, but CGRP is without effect.[36] Neuropeptides have no effect on intrapulmonary blood flow.[36]

Like other vascular effects of airway sensory nerves, neurogenic vasodilation in airways and nose is due to stimulation of NK_1 receptors.[36,37] Both phosphoramidon[36-38] and captopril[38] potentiate the vasodilator action of SP in extrapulmonary airways, indicating that vasomotor tone in these airways is modulated by both NEP and ACE.

Bradykinin also causes airway vasodilation mediated via B_2 receptors, and this effect is potentiated by inhibition of ACE and to a lesser extent by inhibition of NEP.[39] This is compatible with the anatomic distribution of the two enzymes. The effects of peptidases on bradykinin-induced responses may be due in part to direct stimulation of bradykinin, but some of the effects may be indirect, due to the release by bradykinin of neuropeptides from sensory nerves.

4. Cough

Cough is a prominent symptom in various airway diseases, but the underlying mechanisms are unknown, so effective therapy remains elusive. In unanesthetized guinea pigs, inhalation of SP aerosols causes cough. This effect is potentiated by NEP inhibitors[40] and is abolished by pretreatment with recombinant NEP.[41] If cough is the result of irritant-induced release of SP or other neuropeptides from sensory nerves, then peptidases such as NEP may modulate the responses. Downregulation or inhibition of peptidases could exaggerate cough responses, and inhalation of recombinant enzyme could abolish cough. The cough that is reported to occur during therapy with ACE inhibitors may be due to the blockade of breakdown of cough-inducing peptides such as SP or bradykinin.

5. Mucus Secretion and Mucociliary Clearance

Immunocytochemical staining shows NEP localization in airway submucosal glands, and tachykinins stimulate mucus secretion[42] via NK_1 receptor activation.[43] The first studies of peptidase modulation of neurogenic inflammation showed that inhibitors of NEP potentiate the effect of SP and other tachykinins on mucus secretion in the ferret trachea,[15] while inhibitors of ACE and various proteases had no effect. The fragment SP_{1-9} obtained after cleavage by NEP was inactive, providing further evidence that NEP cleavage generates inactive SP metabolites. Tachykinins also increase ciliary beat frequency via an NK_1 receptor effect.[44] The increased mucus secretion and acceleration of ciliary activity, as well as cough, presumably assist in the clearance of causative irritants.

6. Immune Responses

Receptors for neuropeptides are present in a wide variety of cells important for immunologic responses, but only limited knowledge exists about their functions and the roles of peptidases. Best studied are allergic responses in rodents, and this review is limited to that subject. A review of peptidases and immune cells can be found elsewhere.[45] Inhalation of aerosolized antigen in sensitized guinea pigs causes bronchoconstriction[46,47] and plasma extravasation.[48] Pretreatment with tachykinin receptor antagonists markedly inhibits the "late phase" of both responses,[46–48] indicating that tachykinin release plays an important role in these antigen-induced airway responses. Bronchoconstriction and vascular extravasation in response to small doses of antigen are increased dramatically by an inhibitor of NEP, confirming that peptides are involved in the response.[45,47] Present evidence suggests that antigen–antibody interactions in airways cause kinin release, which causes the subsequent release of tachykinins.[46] When bradykinin is applied topically to the airways, most of its effects are mediated by the stimulation of sensory nerves and the subsequent release of tachykinins.[49] The effects of bradykinin are also modulated by ACE and NEP.[39] Thus, modulation of antigen-induced responses in airways could be due to effects on both tachykinins and kinins.

D. MODULATION OF EXPRESSION OF THE PEPTIDASES

Sensory nerves, by releasing neuropeptides, produce a series of responses to irritants that are protective: stimulation of mucus secretion, ciliary activity, and cough provide mechanisms for clearance of inhaled irritants. Vascular effects (vasodilation, vascular extravasation, and leukocyte recruitment) provide protective inflammatory responses. Peptidases (NEP and ACE) provide mechanisms for modulating and thus limiting these responses. Peptidase downregulation, by limiting peptidase cleavage and inactivation of neuropeptides, exaggerates neurogenic inflammatory responses. Upregulation, by increasing neuropeptide cleavage, decreases neurogenic responses. Thus, the level of peptidase activity profoundly affects neuropeptide actions, and thus may have important pathophysiologic and therapeutic implications.

1. Downregulation

a. Respiratory Viral Infections

Infection of ferret airway tissues with influenza virus increases the contractile response of airway smooth muscle to SP, and this is associated with decreased NEP activity in the infected tissues.[50] The effect is not due to the production of an endogenous NEP inhibitor. Similarly, airway inoculation with Sendai virus in pathogen-free guinea pigs increases the bronchomotor responses to SP and to stimulation of capsaicin-sensitive nerves, but responses to acetylcholine are unaffected. Like influenza infections, Sendai virus-infected guinea pig airways show a marked decrease in NEP activity in the tracheal epithelial layer.[51]

SP responses in airway blood vessels are also exaggerated by respiratory infections. Sendai virus, rat coronavirus, and *Mycoplasma pulmonis* infections cause exaggerated increases in plasma extravasation in response to SP injected intravenously,[52] and this is due to decreased NEP activity in the airways.[27] Similar effects on extravasation occur following capsaicin challenge in rats inoculated with Sendai virus.[53] *Mycoplasma pulmonis* infection causes long-lasting potentiation of neurogenic extravasation, and this could be due, at least in part, to the proliferation of new blood vessels that occurs with this infection.[54] Sendai virus infection potentiates the increase in airway blood flow induced by SP but not by histamine. The exaggerated neurogenic blood flow responses appear to be due to decreases in both NEP and ACE activities.[38]

These findings implicate downregulation of peptidases in the potentiation of various neuropeptide responses in airway tissues during respiratory viral infections. Viral infections trigger asthma attacks and increase bronchomotor responsiveness. Exaggerated neurogenic inflammatory responses may play a role in these effects. Studies with selective drugs will be required in humans to determine the role of neuropeptides in disease.

b. *Inhaled Irritants*

Not surprisingly, a variety of inhaled irritants induce inflammatory responses. Neurogenic inflammation has been implicated in these responses. Furthermore, some of these irritants cause a loss of peptidase activity and exaggerated neurogenic inflammatory responses. Two examples are given: toluene diisocyanate (TDI), a widely used industrial chemical, causes activation of capsaicin-sensitive nerves and subsequent neurogenic inflammatory bronchoconstriction.[55,56] TDI also exaggerates this neurogenic response by decreasing NEP activity in airways.[57] Similarly, in guinea pigs inhalation of cigarette smoke increases neurogenic bronchoconstriction (Figure 3) due to a decrease in NEP activity.[58] Present evidence suggests that the decrease in NEP activity is due to the generation of free radicals by the cigarette smoke.[58]

2. Upregulation

a. *Aerosolized Recombinant NEP*

If inflammatory peptides play important roles in airway disease, then degradative cleavage of these peptides might have important therapeutic effects. Two examples are given.

Cough can be provoked at low concentrations of inhaled capsaicin or SP in guinea pigs,[40] and inhalation of aerosolized recombinant human NEP inhibits neuropeptide-induced cough dose-dependently.[41] Similarly, recombinant human NEP prevents SP-induced plasma extravasation in guinea pig skin dose-dependently. Thus, recombinant NEP may be useful therapeutically in preventing the inflammatory effects of neuropeptides, kinins, and other proinflammatory peptides.

b. *Glucocorticoids*

ACE is stimulated by glucocorticoids in endothelial cells and rat lungs,[59] in alveolar macrophages[60] and monocytes;[61] NEP gene expression is increased by glucocorticoids in human airway epithelial cells.[62] Dexamethasone inhibits SP-induced vascular extravasation in rats,[63] but has no effect on PAF-induced extravasation, suggesting that this steroid effect is selective. The SP-induced extravasation is reversed completely by simultaneously inhibiting both NEP and ACE activities with peptidase inhibitors, providing strong evidence that the suppressive effect of dexamethasone is due to increased peptidase activity. The time course of dexamethasone-induced suppression of neurogenic extravasation is slow and similar to the

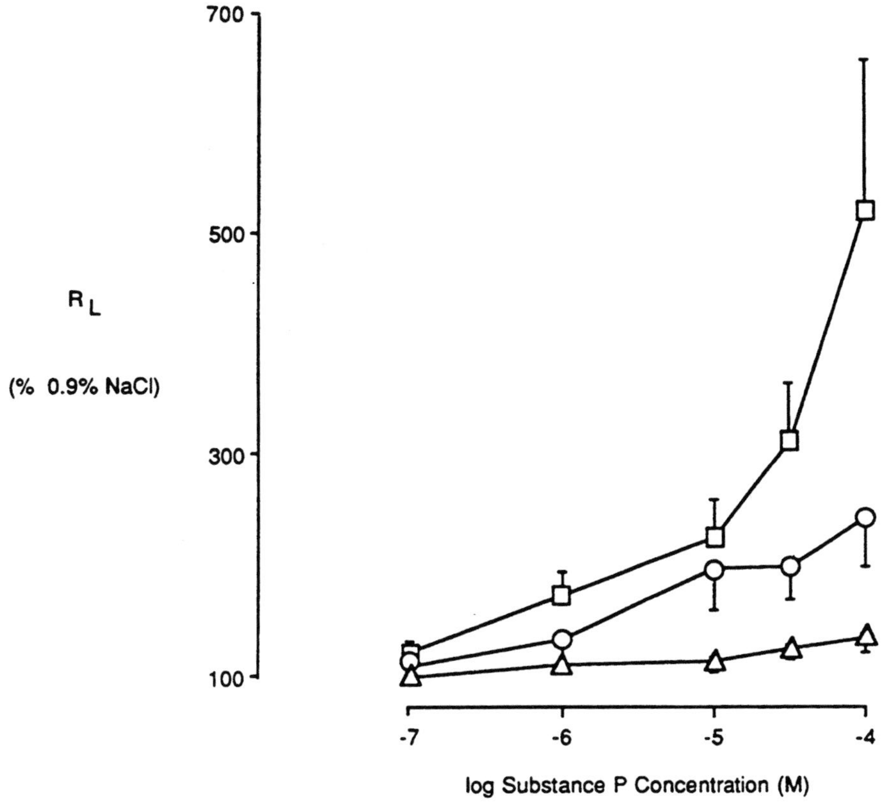

FIGURE 3. Concentration-response curves to aerosolized substance P (seven breaths) after exposure to air (triangles) or after acute exposure to the smoke from one (circles) or two (squares) cigarettes in anesthetized guinea pigs. Each curve represents data collected from five different animals. Total pulmonary resistance (R_L) is expressed (mean $\pm$ 1 SE) as the percent of the response to aerosolized 0.9% NaCI solution (seven breaths) given prior to the administration of substance P. After exposure to cigarette smoke, substance P-induced responses were greater than in air-exposed animals ($p < .001$). Cigarette smoke-induced responses to substance P after exposure to two cigarettes was greater than after one cigarette ($p < .001$). (Reproduced with permission from Dusser, D. et al., *J. Clin. Invest.*, 84, 900, 1989.)

time course of increased peptidase expression in cultured cells treated with corticosteroids. Glucocorticoid pretreatment also inhibits neurogenic inflammatory extravasation[63] (Figure 4). The effect has a considerable latency, is dose dependent, has significant effects in doses considered to be in the therapeutic range, and can totally abolish neurogenic extravasation.[63]

Effects of increased vascular permeability may play important roles in the pathophysiology of various inflammatory states. Increased peptidase activity, by suppression of neurogenic inflammation, may therefore play a significant role in the "anti-inflammatory" effects of steroids.

III. SUMMARY

NEP and ACE are membrane-bound peptidases that are present in high concentrations in the respiratory tract. The peptidases are anchored to the surfaces of cells that have peptide receptors. The active site faces the extracellular milieu. By cleaving and thus inactivating peptides close to the surfaces of cells, the peptidase competes for peptides whose receptors exist on the cell surface. ACE is concentrated on the luminal surface of vascular endothelium, and its effects are mainly limited to the vascular bed. NEP is more widely distributed among

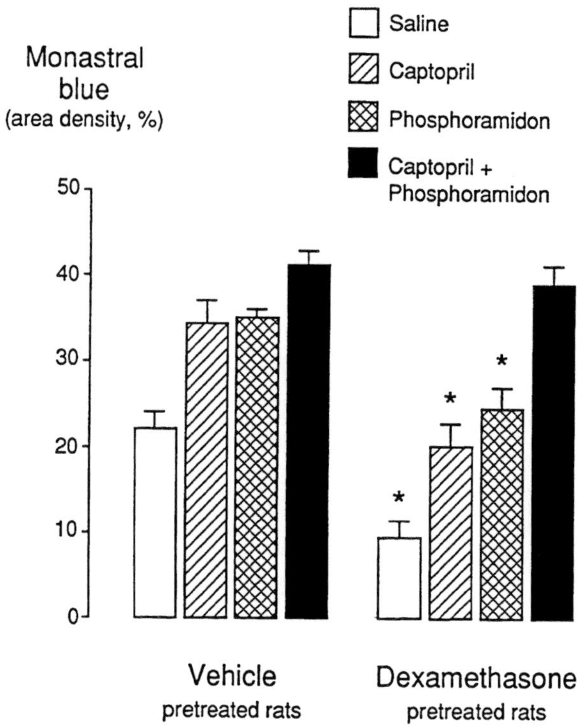

FIGURE 4. Effect of captopril (inhibitor of kininase II; 2.5 mg/kg i.v.), phosphoramidon (inhibitor of neutral endopeptidase; 2.5 mg/kg i.v.), or both on the amount of Monastral blue extravasation from tracheal blood vessels induced by substance P (5 µg/kg i.v.). Rats were pretreated with the vehicle of dexamethasone (left columns) or with dexamethasone (0.5 mg/kg d i.p. for 2 d; right columns). Values are mean ± SEM; n = 5–6 rats per group. Asterisks mark significant differences between dexamethasone-treated rats and the corresponding vehicle-treated controls (*$P < .01$). In vehicle-pretreated rats, substance P-induced extravasation was potentiated by captopril alone or by phosphoramidon alone ($P < .01$) and was further potentiated by the combination of both inhibitors ($P < .05$). Dexamethasone reduced substance P-induced extravasation, and this effect was only partially reversed by captopril alone or by phosphoramidon alone. However, the combination of the two inhibitors completely abolished the inhibitory effect of dexamethasone. Saline (open columns); captopril (hatched columns); phosphoramidon (cross-hatched columns); captopril + phosphoramidon (solid columns). (Reproduced with permission from Piedimonte, G. et al., *J. Clin. Invest.*, 86, 1409, 1990.)

airway cells, and its targets include epithelium, glands, cough receptors, and smooth muscle, as well as blood vessels.

The effects of neurogenic inflammation include cough, hypersecretion, bronchoconstriction, vascular extravasation, leukocyte recruitment, and vasodilation. These responses mimic many aspects of chronic inflammatory airway diseases in humans. Neurogenic inflammation is prominent in rodents, but even in these species tissue responses may be small in healthy animals. However, inhibition of NEP and ACE markedly increases neurogenic inflammatory responses. In healthy humans, neurogenic inflammatory responses appear to be small. However, with chronic inflammation, vascular proliferation and decreased peptidase activity may lead to exaggerated responses, which may play significant pathophysiologic roles. To evaluate the roles of neuropeptides and other peptides (e.g., kinins) in human diseases, the "proof of concept" will depend on the use in patients of drugs that (1) block specific receptors or (2) increase degradation of the inflammatory peptides.

REFERENCES

1. Holzer, P., Capsaicin: cellular targets, mechanisms of action, and selectivity for thin sensory neurons, *Pharmacol. Rev.*, 43, 143, 1991.
2. Nadel, J. A., Modulation of neurogenic inflammation by peptidases, in *Neuropeptides in Respiratory Medicine*, Kaliner, M. A., Barnes, P. J., Kunkel, G. H. H., and Baraniuk, J. N., Eds., Marcel Dekker, New York, 1994, 351.
3. Malfroy, B., Kuang, W.-J., Seeburg, P. H., Mason, A. J., and Schofield, P. R., Molecular cloning and amino acid sequence of human enkephalinase (neutral endopeptidase), *FEBS Lett.*, 229, 206, 1988.
4. Tran, P. R., Willard, H. F., and Letarte, M., The common acute lymphoblastic leukemia antigen (neutral endopeptidase-3.4.24.11) gene is located on human chromosome 3, *Cancer Genet. Cytogenet.*, 42, 129, 1989.
5. Patchett, A. A. and Cordes, E. H., The design and properties of N-carboxyalkyldipeptide inhibitors of angiotensin converting enzyme, *Adv. Enzymol.*, 57, 1, 1985.
6. Velletri, P. A., Testicular angiotensin I-converting enzyme, *Life Sci.*, 36, 1597, 1985.
7. Soubrier, F., Alhenc-Gelas, F., Hubert, C., Allegrini, J., John, M., Tregear, G., and Corvol, P., Two putative active centers in human angiotensin I-converting enzyme revealed by molecular cloning, *Proc. Natl. Acad. Sci. U.S.A.*, 85, 9386, 1988.
8. Bernstein, K. E., Martin, B. M., Edwards, A. S., and Bernstein, E. A., Mouse angiotensin-converting enzyme is a protein composed of two homologous domains, *J. Biol. Chem.*, 264, 11945, 1989.
9. Skidgel, R. A., Engelbrecht, S., Johnson, A. R., and Erdös, E. G., Hydrolysis of substance P and neurotensin by converting enzyme and neutral endopeptidase, *Peptides*, 5, 769, 1984.
10. Katayama, M., Nadel, J. A., Bunnett, N. W., Di Maria, G. U., Haxhiu, M., and Borson, D. B., Catabolism of calcitonin gene-related peptide and substance P by neutral endopeptidase, *Peptides*, 12, 563, 1991.
11. Caldwell, P. R. B., Seegal, B. C., Hsu, K. C., Das, M., and Soffer, R. L., Angiotensin-converting enzyme: vascular endothelial localization, *Science*, 191, 1050, 1976.
12. Okamoto, A., Payan, D. G., and Bunnett, N. W., Interactions between neutral endopeptidase (NEP, EC 3.4.24.11) and the substance P receptor (SPR, NK-1) expressed in mammalian cells, *Gastroenterology*, 104, A560, 1993.
13. Kerr, M. A. and Kenny, A. J., The purification and specificity of a neutral endopeptidase from rabbit kidney brush border, *Biochem. J.*, 137, 477, 1974.
14. Malfroy, B., Swerts, J. P., Guyon, A., Roques, B. Q., and Schwartz, J. C., High-affinity enkephalin-degrading peptidase in brain is increased after morphine, *Nature*, 276, 523, 1978.
15. Borson, D., Corrales, R., Varsano, S., Gold, M., Viro, N., Caughey, G., Ramachandran, J., and Nadel, J., Enkephalinase inhibitors potentiate substance P-induced secretion of $^{35}SO_4$-macromolecules from ferret trachea, *Exp. Lung Res.*, 12, 21, 1987.
16. Nadel, J. A. and Borson, D. B., Modulation of neurogenic inflammation by neutral endopeptidase, *Am. Rev. Respir. Dis.*, 143, S33, 1991.
17. Sekizawa, K., Tamaoki, J., Nadel, J., and Borson, D., Enkephalinase inhibitor potentiates substance P- and electrically induced contraction in ferret trachea, *J. Appl. Physiol.*, 63, 1401, 1987.
18. Dusser, D., Umeno, E., Graf, P., Djokic, T., Borson, D., and Nadel, J., Airway neutral endopeptidase-like enzyme modulates tachykinin-induced bronchoconstriction in vivo, *J. Appl. Physiol.*, 65, 2585, 1988.
19. Regoli, D., Drapeau, G., Dion, S., and Couture, R., New selective agonists for neurokinin receptors: pharmacological tools for receptor characterization, *Trends Pharmacol. Sci.*, 9, 290, 1988.
20. Bertrand, C., Nadel, J. A., Graf, P. D., and Geppetti, P., Capsaicin increases airflow resistance in guinea pigs in vivo by activating both NK2 and NK1 tachykinin receptors, *Am. Rev. Respir. Dis.*, 148, 909, 1993.
21. Maggi, C. A., Patacchini, R., Quartara, L., Rovero, P., and Santicioli, P., Tachykinin receptors in the guinea-pig isolated bronchi, *Eur. J. Pharmacol.*, 197, 167, 1991.
22. Djokic, T., Nadel, J., Dusser, D., Sekizawa, K., Graf, P., and Borson, D., Inhibitors of neutral endopeptidase potentiate electrically and capsaicin-induced noncholinergic contraction in guinea pig bronchi, *J. Pharmacol. Exp. Ther.*, 248, 7, 1989.
23. Shore, S. A., Stimler-Gerard, N. P., Coats, S. R., and Drazen, J. M., Substance P-induced bronchoconstriction in the guinea pig: enhancement by inhibitors of neutral metalloendopeptidase and angiotensin-converting enzyme, *Am. Rev. Respir. Dis.*, 137, 331, 1988.
24. Lötvall, J. O., Tokuyama, K., Löfdahl, C. G., Ullman, A., Barnes, P. J., and Chung, K. F., Peptidase modulation of noncholinergic vagal bronchoconstriction and airway microvascular leakage, *J. Appl. Physiol.*, 70, 2730, 1991.
25. Subissi, A., Guelfi, M., and Criscuoli, M., Angiotensin converting enzyme inhibitors potentiate the bronchoconstriction induced by substance P in the guinea pig, *Br. J. Pharmacol.*, 100, 502, 1990.
26. Lundberg, J. M., Saria, A., Brodin, E., Rosell, S., and Folkers, K., A substance P antagonist inhibits vagally induced increase in vascular permeability and bronchial smooth muscle contraction in the guinea pig, *Proc. Natl. Acad. Sci. U.S.A.*, 80, 1120, 1983.

27. Borson, D., Brokaw, J., Sekizawa, K., McDonald, D., and Nadel, J., Neutral endopeptidase and neurogenic inflammation in rats with respiratory infections, *J. Appl. Physiol.*, 66, 2653, 1989.
28. Umeno, E., Nadel, J., Huang, H.-T., and McDonald, D., Inhibition of neutral endopeptidase potentiates neurogenic inflammation in the rat trachea, *J. Appl. Physiol.*, 66, 2647, 1989.
29. Piedimonte, G., McDonald, D. M., and Nadel, J. A., Neutral endopeptidase and kininase II mediate glucocorticoid inhibition of neurogenic inflammation in the rat trachea, *J. Clin. Invest.*, 88, 40, 1991.
30. Katayama, M., Nadel, J. A., Piedimonte, G., and McDonald, D. M., Peptidase inhibitors reverse steroid-induced suppression of neutrophil adhesion in rat tracheal blood vessels, *Am. J. Physiol.*, 264, L316, 1993.
31. Baluk, P., Bertand, C., Geppetti, P., McDonald, D. M., and Nadel, J. A., NK1 receptors mediate leukocyte adhesion in neurogenic inflammation in the rat trachea, *Am. Physiol. Soc.*, 95, L263, 1995.
32. Umeno, E., Nadel, J. A., and McDonald, D. M., Neurogenic inflammation of the rat trachea: fate of neutrophils that adhere to venules, *J. Appl. Physiol.*, 69, 2131, 1990.
33. Umeno, E., McDonald, D. M., and Nadel, J. A., Hypertonic saline increases vascular permeability in the rat trachea by producing neurogenic inflammation, *J. Clin. Invest.*, 85, 1905, 1990.
34. Piedimonte, G., Bertand, C., Geppetti, P., Snider, R. M., Desai, M. C., and Nadel, J. A., A new NK_1 receptor antagonist (CP-99,994) prevents the increase in tracheal vascular permeability produced by hypertonic saline, *J. Pharmacol. Exp. Ther.*, 266, 270, 1993.
35. Yoshihara, S., Chan, B., Yamawaki, I., Geppetti, P., Ricciardolo, F. L. M., Massion, P. P., and Nadel, J. A., Plasma extravasation in the rat trachea induced by cold air is mediated by tachykinin release from sensory nerves, *Am. J. Respir. Crit. Care Med.*, 151, 1011, 1995.
36. Piedimonte, G., Hoffman, J. I. E., Husseini, W. K., Snider, R. M., Desai, M. C., and Nadel, J. A., NK_1 receptors mediate neurogenic inflammatory increase in blood flow in rat airways, *J. Appl. Physiol.*, 74, 2462, 1993.
37. Piedimonte, G., Hoffman, J. I. E., Husseini, W. K., Bertrand, C., Snider, R. M., Desai, M. C., Petersson, G., and Nadel, J. A., Neurogenic vasodilation in the rat nasal mucosa involves NK_1 tachykinin receptors, *J. Pharmacol. Exp. Ther.*, 265, 36, 1993.
38. Yamawaki, I., Geppetti, P., Bertrand, C., Chan, B., Massion, P., Piedimonte, G., and Nadel, J. A., Sendai virus infection potentiates the increase in airway blood flow induced by substance P in rat, *Am. Rev. Respir. Dis.*, 147, A476, 1993.
39. Yamawaki, I., Geppetti, P., Bertrand, C., Chan, B., and Nadel, J. A., Airway vasodilation by bradykinin is mediated via B2 receptors and modulated by peptidase inhibitors, *Am. J. Physiol.*, 266, L156, 1994.
40. Kohrogi, H., Graf, P., Sekizawa, K., Borson, D., and Nadel, J., Neutral endopeptidase inhibitors potentiate substance P- and capsaicin-induced cough in awake guinea pigs, *J. Clin. Invest.*, 82, 2063, 1988.
41. Kohrogi, H., Nadel, J., Malfroy, B., Gorman, C., Bridenbaugh, R., Patton, J., and Borson, D., Recombinant human enkephalinase (neutral endopeptidase) prevents cough induced by tachykinins in awake guinea pigs, *J. Clin. Invest.*, 84, 781, 1989.
42. Gashi, A. A., Borson, D. B., Finkbeiner, W. E., and Nadel, J. A., Neuropeptides degranulate serous cells of ferret trachea glands, *Am. J. Physiol.*, 251, C223, 1986.
43. Meini, S., Mak, J. C. W., Rohde, J. A. L., and Rogers, D. F., Tachykinin control of ferret airways — mucus secretion, bronchoconstriction and receptor mapping, *Neuropeptides*, 24, 81, 1993.
44. Lindberg, S. and Dolata, J., NK1 receptors mediate the increase in mucociliary activity produced by tachykinins, *Eur. J. Pharmacol.*, 231, 375, 1993.
45. Piedimonte, G. and Nadel, J. A., Peptidases in airway defense mechanisms, in *Airways and Environment: From Injury to Repair*, Chretien, J. and Dusser, D., Eds., Marcel Dekker, New York, in press, 1995.
46. Bertrand, C., Geppetti, P., Graf, P. D., Foresi, A., and Nadel, J. A., Involvement of neurogenic inflammation in antigen-induced bronchoconstriction in guinea pigs, *Am. Physiol. Soc.*, 93, L507, 1993.
47. Ricciardolo, F. L. M., Nadel, J. A., Graf, P. D., Bertrand, C., Yoshihara, S., and Geppetti, P., Role of kinins in anaphylactic-induced bronchoconstriction mediated by tachykinins in guinea pigs, *Br. J. Pharmacol.*, 113, 508, 1994.
48. Bertrand, C., Geppetti, P., Baker, J., Yamwaki, I., and Nadel, J. A., Role of neurogenic inflammation in antigen-induced vascular extravasation in guinea pig trachea, *J. Immunol.*, 150, 1479, 1993.
49. Geppetti, P., Sensory neuropeptide release by bradykinin: mechanism and pathophysiological implications, *Regul. Pept.*, 47, 1, 1993.
50. Jacoby, D. B., Tamaoki, J., Borson, D. B., and Nadel, J. A., Influenza infection causes airway hyperresponsiveness by decreasing enkephalinase, *J. Appl. Physiol.*, 64, 2653, 1988.
51. Dusser, D., Jacoby, D., Djokic, T., Rubinstein, I., Borson, D., and Nadel, J., Virus induces airway hyperresponsiveness to tachykinins: role of neutral endopeptidase, *J. Appl. Physiol.*, 67, 1504, 1989.
52. McDonald, D., Respiratory tract infections increase susceptibility to neurogenic inflammation in the rat trachea, *Am. Rev. Respir. Dis.*, 137, 1432, 1988.
53. Piedimonte, G., Nadel, J. A., Umeno, E., and McDonald, D. M., Sendai virus infection potentiates neurogenic inflammation in the rat trachea, *J. Appl. Physiol.*, 68, 754, 1990.

54. McDonald, D. M., Schoeb, T. R., and Lindsey, J. R., *Mycoplasma pulmonis* infections cause long-lasting potentiation of neurogenic inflammation in the respiratory tract of the rat, *J. Clin. Invest.*, 87, 787, 1991.
55. Mapp, C. E., Graf, P. D., Boniotti, A., and Nadel, J. A., Toluene diisocyanate contracts guinea pig bronchial smooth muscle by activating capsaicin-sensitive sensory nerves, *J. Pharmacol. Exp. Ther.*, 256, 1082, 1991.
56. Thompson, J. E., Scypinski, L., Gordon, T., and Sheppard, D., Tachykinins mediate toluene diisocyanate induced airway hyperresponsiveness in guinea pigs, *Am. Rev. Respir. Dis.*, 136, 43, 1987.
57. Sheppard, D., Thompson, J. E., Scypinski, L., Dusser, D., Nadel, J. A., and Borson, B. D., Toluene diisocyanate increases airway responsiveness to substance P and decreases airway neutral endopeptidase, *J. Clin. Invest.*, 81, 1111, 1988.
58. Dusser, D., Djokic, T., Borson, D., and Nadel, J., Cigarette smoke induces bronchoconstrictor hyperresponsiveness to substance P and inactivates airway neutral endopeptidase in the guinea pig. Possible role of free radicals, *J. Clin. Invest.*, 84, 900, 1989.
59. Mendelsohn, F. A. O., Lloyd, C. J., Kachel, C., and Funder, J. W., Induction by glucocorticoids of angiotensin converting enzyme production from bovine endothelial cells in culture and rat lung in vivo, *J. Clin. Invest.*, 70, 684, 1982.
60. Friedland, J., Setton, C., and Silverstein, E., Angiotensin converting enzyme: induction by steroids in rabbit alveolar macrophages in culture, *Science*, 197, 64, 1977.
61. Vuk-Pavlovic, Z., Kreofsky, T. J., and Rohrbach, M. S., Characteristics of monocyte angiotensin-converting enzyme (ACE) induction by dexamethasone, *J. Leukocyte Biol.*, 45, 503, 1989.
62. Borson, D. B. and Gruenert, D. C., Glucocorticoids induce neutral endopeptidase in transformed human tracheal epithelial cells, *Am. J. Physiol.*, 260, L83, 1991.
63. Piedimonte, G., McDonald, D. M., and Nadel, J. A., Glucocorticoids inhibit neurogenic plasma extravasation and prevent virus-potentiated extravasation in the rat trachea, *J. Clin. Invest.*, 86, 1409, 1990.
64. McDonald, D. M., Mitchell, R. A., Gabella, G., and Haskell, A., Neurogenic inflammation in the rat trachea. II. Identity and distribution of nerves mediating the increase in vascular permeability, *J. Neurocytol*, 17, 605, 1988.
65. Nadel, J. A., Neutral endopeptidase modulates neurogenic inflammation, *Eur. Respir. J.*, 4, 745, 1991.
66. Iwamoto, I. and Nadel, J. A., Tachykinin receptor subtype that mediates the increase in vascular permeability in guinea-pig skin, *Life Sci.*, 44, 1089, 1989.

Part III. Pathophysiology

Chapter 11

SENSORY NEUROPEPTIDES IN THE EYE

Rolf Håkanson and Zun-Yi Wang

CONTENTS

I. STRUCTURE OF THE EYE

The mammalian eye is made up of an outer collagenous coat (cornea and sclera); an intermediate vascularized coat, the uvea; and an inner, highly differentiated, epithelial coat. The densely pigmented uveal coat contains numerous specialized structures, the iris, ciliary body, and choroid. The iris consists mainly of the sphincter and dilator muscles. The ciliary body, situated behind the iris, contains the ciliary muscle and ciliary processes. Acting through zonules connected to the equatorial region of the lens, the ciliary muscles control accommodation. The innermost part of the ciliary body forms heavily vascularized, epithelium-lined radial projections, termed ciliary processes, that secrete the aqueous humor. The secretion of aqueous humor maintains the intraocular pressure (IOP) and the shape of the eye. The aqueous humor also provides nutrition to the lens and cornea.

The integrity of the internal environment of the eye is critically dependent on the continuous formation and drainage of aqueous humor, as well as on its effective separation from the interstitial plasma compartment. A functional permeability barrier, the blood-aqueous barrier (BAB) is formed in the anterior uvea by the ciliary epithelium and the endothelium of the blood vessels, such that the level of aqueous protein does not exceed 1% of that in blood plasma.[1-4] As there is no barrier separating the iris and the ciliary body from the anterior chamber, any agent that is released locally will diffuse into the anterior chamber, making the eye an excellent model for studies of transmitter release.

II. OCULAR RESPONSES TO NOXIOUS STIMULI

Exposing the mammalian eye to various kinds of injury provokes an immediate response characterized by miosis, vasodilation, rise in IOP, and breakdown of the BAB.[1,5-10] These events are quite prominent in the rabbit.[5-10] The first response to injury is dilation of blood vessels in the ciliary body and increased vascular permeability. This causes plasma extravasation, edema, and an influx of plasmoid fluid into the anterior and posterior chambers. The extent of the rise in IOP and breakdown of the BAB depends on the animal species and on the

0-8493-7646-7/96/$0.00+$.50

intensity, duration, and type of the noxious stimulus.[5-10] The acute ocular response to minor injury usually does not involve the invasion of leukocytes into the aqueous humor; neither circulating blood cells nor humoral antibodies participate, and inflammatory cells are not recruited.[10] There is now considerable evidence suggesting that an intact sensory innervation is required for the inflammatory response in the uvea to occur; and the transmitters released from sensory nerve fibers are thought to play a key role in the ocular responses to injury. The term neurogenic inflammation is often used to describe this phenomenon.

III. SENSORY INNERVATION OF THE EYE

Sensory nerves to the eye derive from the trigeminal ganglion.[1,10] The importance of an intact sensory innervation for the ocular response to injury was first recognized by Bruce,[11] who observed that the conjunctival hyperemia induced by nitrogen mustard was greatly reduced after retrobulbar anesthesia and abolished after section and subsequent degeneration of the ophthalmic nerve. The results suggested a local sensory reflex in line with the axon-reflex-evoked vasodilation in the skin.[12,13] Perkins showed that the ocular response to stimulation of the trigeminal nerve is similar to that produced by local injury and resistant to atropine.[14] The local response to noxious stimuli of the rabbit eye can be elicited by antidromic stimulation of the trigeminal nerve.[14,15] Capsaicin is known to stimulate and excite sensory nerves, preferentially C-fibers.[16,17] Intracameral perfusion with capsaicin mimics the entire irritative response pattern.[18,19] After the excitatory phase, capsaicin causes functional impairment of C-fibers, due to depletion of transmitters.[15-17] The response to various chemical irritants can be abolished by coagulation of the trigeminal nerve,[20] supporting the view that sensory nerves play a prominent role in ocular inflammation.

A variety of neuropeptides occurs in the ocular sensory nerve fibers that originate in the trigeminal ganglion. Among these neuropeptides, substance P (SP) and calcitonin gene-related peptide (CGRP) are well documented. SP is an undecapeptide, belonging to the tachykinin family of peptides. CGRP is a 37-amino acid peptide encoded by the calcitonin gene. CGRP occurs in two forms, designated α and β, which differ from each other by three (human) and one (rat) amino acid residues. Based on the immunocytochemical analysis of the trigeminal ganglion, SP-like immunoreactivity (LI) occurs primarily in small neurons;[21-23] CGRP-LI is present in small, but also in larger, neurons.[24] About 20% of the neurons in the trigeminal ganglion are immunoreactive to SP[21,22] and about 40% are immunoreactive to CGRP.[23,24] Those neurons that contain SP seem to contain CGRP as well.[21-24] SP-immunoreactive (IR) and CGRP-IR nerve fibers have been described in the eyes of many mammals, including rat,[23,24] rabbit,[21-28] cat,[27] monkey,[27,29] and man.[27,30] Recently, pituitary adenylate cyclase activating peptide (PACAP)-LI was found in small and middle-sized neurons in the trigeminal ganglion of the rabbit and rat;[31,32] all these neurons also contained CGRP.[31,32]

IV. TACHYKININS

A. SUBSTANCE P (SP)

Substance P immunoreactivity has been demonstrated in nerve fibers in the eyes of rat, guinea pig, rabbit, cat, monkey, and man.[25-30] In the anterior segment of the eye, SP-IR nerve fibers are quite numerous in the sphincter muscle and ciliary processes, while being few in the cornea.[25-30] Using radioimmunoassay, SP-LI was detected in the iris and ciliary body, optic nerve, and retina. The concentration of SP-LI was about 5 pmol/g in the cornea of the rat[33] and rabbit,[34] and about 7 to 8 pmol/g in the anterior uvea of rabbit[34] and man.[35] The trigeminal origin of the SP-LI positive fibers in the anterior uvea is supported by many observations. SP was released into the aqueous humor following antidromic stimulation of the trigeminal nerve[36] and intracameral injection of capsaicin.[37] Diathermic destruction of the trigeminal

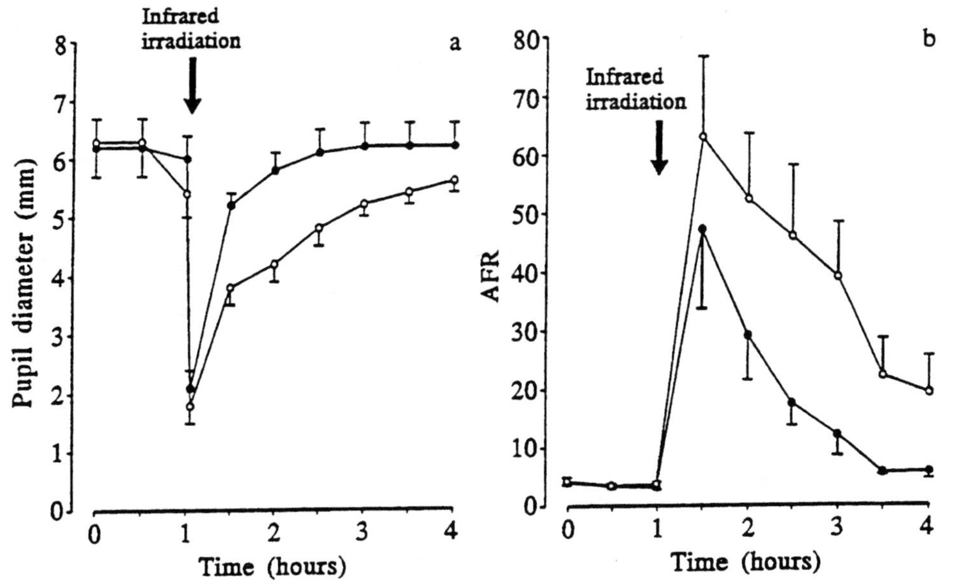

FIGURE 1. The miotic response (a) and aqueous flare response (AFR), (b) to infrared irradiation of the iris in the rabbit eye after prior topical application of 90 nmol (±) CP-96,345 onto the left eye (●) and 0.9% saline onto the right eye (○). (±) CP-96,345 inhibited the inflammatory response with respect to both miosis and AFR.

ganglion reduced the SP-LI concentration by 70% in the anterior uvea.[38] The SP depletion corresponded with the degeneration of sensory fibers and the loss of sensitivity of the eye to irritation.[20] Selective and permanent degeneration of sensory neurons by treatment of neonatal rats with capsaicin prevented the appearance of SP-LI in the iris-ciliary body.[39]

Studies *in vivo* as well as *in vitro* have shown that SP is a potent miotic. In the isolated iris sphincter muscle of rat,[40] rabbit,[41-47] and cow,[47] SP starts to induce contractions at the nanomolar concentration range.[40-47] The contractions are not affected by atropine and the nerve conduction blocker, tetrodotoxin, indicating a direct action on the smooth muscle.[40-47] Intravitreal injection of SP into the rabbit eye induced a dose-dependent miosis (with an ED_{50} value of about 10 pmol) which was quite long lasting (more than 20 h duration).[45] In some reports, the isolated human iris sphincter muscle was found to be insensitive to SP at concentrations up to 1 μM.[47] However, in an eye cup model, SP did induce contraction of the human iris sphincter muscle (EC_{50} value of 0.1 μM).[48] In a study of young healthy volunteers, topical application of SP was found to produce pupillary contraction in a dose-dependent manner.[49]

In the isolated, atropinized rabbit iris sphincter, electrical stimulation, capsaicin, and bradykinin evoke contractions, which can be inhibited by tachykinin receptor antagonists,[50-62] indicating that the responses are mediated by endogenous tachykinins released from sensory nerves. This view is supported by the observation that the three stimuli mentioned above failed to induce contractions of the iris sphincter isolated from the trigeminally denervated eye.[52] The contraction of the iris sphincter induced by either exogenous or endogenous SP can be inhibited by tachykinin receptor antagonists.[42,43,50-64] Pretreatment with a tachykinin receptor antagonist was found to inhibit the miosis induced by infrared irradiation of the iris[65] and by capsaicin[66,67] and bradykinin.[50] Recently, nonpeptide tachykinin receptor antagonists, such as CP-96,345[68] and RP 67580,[69] have been described. Also these compounds were found to block the contractions of the rabbit iris sphincter induced by electrical stimulation or by exogenous tachykinins.[60,64,70] A single topical application of 90 nmol CP-96,345 onto the cornea of rabbits inhibited the miosis induced by infrared irradiation of the iris (Figure 1).[94] By autoradiography and binding experiments, specific and high-affinity SP binding sites have been demonstrated in the eyes of rat,[71,72] rabbit,[72] cow,[73,74] and man.[75] SP binding sites were concentrated to the iris sphincter muscle of the rabbit eye.[72] Based on the study of a series of selective tachykinin receptor agonists as well as selective receptor antagonists, the tachykinin receptor in the iris sphincter is thought to be of the NK_1 (and NK_3) type.[60,61,64,70,76]

Reports concerning the effects of SP on ocular circulation and BAB are inconsistent. When infused intravenously or intracamerally into the rabbit eye, SP has no apparent effect on the anterior uveal blood flow.[41,77,78] In reports from our laboratory, intravitreal injection of SP into the rabbit eye induced a moderate disruption of the BAB.[45] In other reports, intraarterial infusion of large doses of SP was found to induce brief ocular vasodilation.[77] In the rat, SP caused leakage of labeled albumin into the aqueous humor.[79] Furthermore, the involvement of endogenous SP in the disruption of BAB by minor injury received some support from the studies of Holmdahl et al.[65] In those experiments, the aqueous flare response (AFR) (reflecting the leakage of plasma protein into the aqueous humor) to infrared irradiation could be inhibited by pretreating the eyes with a high dose of a tachykinin receptor antagonist[65-67] (see also Figure 1). Thus, SP is likely to be involved in the disruption of BAB in response to injury, although higher amounts of SP may be required than are needed to induce miosis.[45,92] The discrepancy in results between different laboratories may reflect differences in methods and doses used.

In two early reports, SP appeared to cause a rise of IOP, but the rise was small and inconsistent[41] and was attributed to blockage of the aqueous flow causing the iris to protrude, impeding the outflow through the drainage angle.[77] Stone and co-workers observed SP IR fibers passing close to the epithelium of the ciliary processes in several mammals,[27] offering the possibility that the peptide might in some way modulate aqueous humor formation.

B. NEUROKININ A (NKA) AND NEUROKININ B (NKB)

NKA is present in the sensory nerve fibers in the rabbit eye.[45,80] While Taniguchi et al. described relatively large amounts of NKB in the rabbit iris,[80] Beding-Barnekow et al. failed to detect it.[45] In the isolated rabbit iris sphincter, SP, NKA and NKB were equipotent in evoking contractions.[45,59] Intravitreal injection of the three peptides showed that both NKA and NKB were less potent than SP as a miotic and that, unlike SP, NKA, and NKB were inactive in causing the breakdown of BAB.[45] We have also observed that intravenous injection of NKA (5 nmol/kg) in conscious rabbits induced a short-lasting (5 to 10 min) miosis while failing to evoke any noticeable breakdown of the BAB.[94]

V. CALCITONIN GENE-RELATED PEPTIDE (CGRP)

Sensory nerve fibers that contain CGRP-like immunoreactivity can be found throughout the anterior uvea of several mammals (rat, rabbit, guinea pig, and man).[24,29,31,32,81,84] The fibers are numerous and mostly associated with blood vessels.[24,29,31,32,81,84] A network of CGRP-IR fibers in the iris innervates both the sphincter and the dilator muscles.[24,29,32,81,84] In the ciliary body, CGRP-IR nerves appear to form a continuous plexus that sends fibers to the vessels of the ciliary processes. In the human eye, nerve fibers that are immunoreactive for CGRP have also been described in the ciliary muscle and ciliary process and seem to be prominently associated with uveal blood vessels of the ciliary body.[29,84] By radioimmunoassay, the concentrations of CGRP (pmoles per gram) in the rabbit eye were 9, 536, 53, and 35 in the cornea, iris sphincter muscle, ciliary body, and iris dilator muscle, respectively.[32]

CGRP can be released into the aqueous humor of the rabbit following mechanical or electrical stimulation of the trigeminal nerve.[85] Damage to the trigeminal ganglion or treatment of neonatal rats and guinea pig with capsaicin results in substantial reduction of

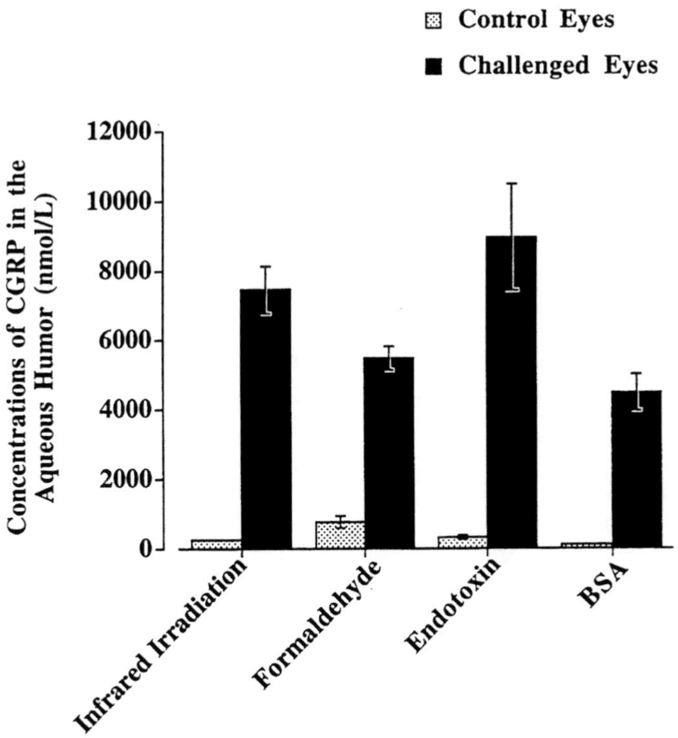

FIGURE 2. Four inflammatory models in the rabbit eye. Inflammatory responses were induced by: (1) infrared irradiation of the iris, (2) topical application of 1% neutral formaldehyde onto the cornea, (3) intravitreal injection of *Shigella* endotoxin, or (4) intravitreal injection of bovine serum albumin into eyes previously sensitized with albumin. Aqueous humor was collected at the time of peak aqueous flare response. The concentrations of CGRP in the aqueous humor were elevated 40- to 60-fold in all four inflammatory models.

CGRP-LI in the iris-ciliary bodies.[24,31,81,82,85] Capsaicin was found to release CGRP-LI from the iris and ciliary body of human[35] and rabbit.[32] These observations affirm the sensory origin of CGRP in the eye.

When administered into the rabbit eye directly or when given by intravenous injection, CGRP induces conjunctival hyperemia, an increase of the blood flow in the iris and ciliary body, a rise in IOP, and breakdown of BAB.[81,84,87-91] CGRP has no apparent effect on the pupil, although it has been reported to enhance the miotic effect of SP.[85] Fluorescein, injected intravenously together with CGRP, occurs promptly in the aqueous humor of the anterior chamber, indicating that CGRP evokes an abrupt disruption of the BAB.[81] These responses can also be observed in the cat after intracameral injection of CGRP.[89,90] In the monkey eye, intracameral administration of CGRP increased the blood flow in the conjunctiva, ciliary body, and sclera.[92] Suprisingly, CGRP had no effect in the rat eye, even though quite high doses were used.[79] The concentration of CGRP in the aqueous humor of the rabbit was greatly increased following infrared irradiation of the iris, topical application of formaldehyde onto the cornea, or intravitreal injection of endotoxin or bovine serum albumin (Figure 2),[32] suggesting its release from ocular C-fibers in response to noxious stimuli. Based on the results of studies of the rabbit eye, it is likely that CGRP is not only released from sensory nerve fibers but that it evokes vascular responses (vasodilation and extravasation of plasma proteins), while SP is responsible mainly for inducing miosis. However, it cannot be excluded that SP and CGRP may cooperate in causing both miosis and vascular responses.

VI. PITUITARY ADENYLATE CYCLASE ACTIVATING PEPTIDE (PACAP)

PACAP is a novel neuropeptide which occurs in two bioactive forms, PACAP-38 and PACAP-27. Immunocytochemistry revealed PACAP-IR nerve fibers in the eye of rat[31] and rabbit.[32] In the rabbit eye, PACAP-IR nerve fibers were found in the iris, ciliary body, cornea, choroid, and sclera.[32] These nerve fibers invariably displayed CGRP-IR.[32] The regional distribution of PACAP-27 in the eye was revealed by radioimmunoassay; the highest concentrations were found in the iris sphincter and ciliary body.[32] The distribution pattern resembled that of CGRP.[32] Treatment of rats with capsaicin was found to reduce the intensity of staining of the PACAP-LI in the rat eye.[31] *In vitro* studies also showed that capsaicin released PACAP from the iris-ciliary body complex of the rabbit into the incubation medium, supporting the sensory origin of PACAP in the eye.[32]

PACAP is potent in causing ocular vasodilation in the rabbit eye.[93] Intravitreal injection of PACAP-27 or PACAP-38 induced conjunctival hyperemia, swelling of the anterior segment of the eye, miosis, and breakdown of the blood-aqueous barrier, which mimicked inflammatory responses.[32] Both PACAP-27 and -38 induced leakage of plasma protein into the aqueous humor in a dose-dependent manner, having similar potency and efficiacy with an ED_{50} value of 0.3 nmol.[32] Tetrodotoxin pretreatment inhibited the conjunctival hyperemia, the swelling of the anterior segment of the eye, and the miosis, but the PACAP-evoked leakage of plasma protein into the aqueous humor was not affected.[32] PACAP was also found to evoke tachykinin-mediated contractions of the isolated iris sphincter muscle, suggesting that PACAP can activate sensory nerve fibers to release transmitters.[32] It should be noted that CGRP has no effect on the contractile activity of the isolated rabbit iris[32] and, thus, this result may indicate that PACAP plays a role distinct from that of CGRP, although both peptides seem to be involved in the vascular responses in the eye. In inflamed eyes following infrared irradiation of the iris, topical application of formaldehyde onto the cornea, or intravitreal injection of endotoxin or bovine serum albumin, the concentration of PACAP in the aqueous humour was increased greatly.[32] These results suggest that PACAP is a sensory neuropeptide in ocular tissues of the rabbit, that it is released from C-fibers in response to noxious stimuli, and is involved in inflammatory responses in the rabbit eye.

VII. CONCLUSIONS

During the past decades, the development of immunohistochemical and immunochemical techniques has allowed the detailed investigation of the innervation of the eye. The neuropeptides of the sensory nerve fibers have been identified, and the pattern of coexisting neuropeptides has been described. Among the neuropeptides of the uvea, tachykinins and CGRP are well documented in sensory fibers, and recently, PACAP has been added to this class of peptides. It is reasonable to assume that the neuropeptides of the sensory fibers in the eye may play different roles in the ocular responses to noxious stimuli. However, although there is increasing evidence that coreleased neuropeptides may cooperate, their precise roles remain unknown. Briefly, the tachykinins are likely to be responsible for injury-evoked miosis. There is some evidence that SP plays a role in the breakdown of the BAB as well, perhaps in conjunction with CGRP and PACAP, both of which are very powerful in this respect.

ACKNOWLEDGMENTS

This study was supported by grants from the Swedish Medical Research Council (04X-1007) and from the Medical Faculty of Lund, Sweden. The authors wish to thank Ms. Bozena Wlosinska and Ms. Britt Carlsson for excellent technical assistance.

REFERENCES

1. Stone, R. A., Kuwayama, Y., and Laties, A. M., Regulatory peptides in the eye, *Experientia*, 43, 791, 1987.
2. Bill, A., Blood circulation and fluid dynamics in the eye, *Physiol. Rev.*, 55, 383, 1975.
3. Raviola, G., The structural basis of the blood-aqueous barriers, *Exp. Eye Res.*, 25 (Suppl.), 27, 1977.
4. Bill, A., Some aspects of the ocular circulation. Friedenwald lecture, *Invest. Ophthalmol. Vis. Sci.*, 26, 410, 1985.
5. Unger, W. G., Perkins, E. S., and Bass, M. S., The response of the rabbit eye to laser irradiation of the iris, *Exp. Eye Res.*, 19, 367, 1974.
6. Unger, W. G., Hammond, B. R., and Cole, D. F., Disruption of the blood-aqueous barrier following paracentesis in the rabbit, *Exp. Eye Res.*, 20, 255, 1975.
7. Jampol, L. M., Neufeld, A. H., and Sears, M. L., Pathways for the response of the eye to injury, *Invest. Ophthalmol.*, 14, 184, 1975.
8. Jampol, L. M., Axelrod, A., and Tessler, H., Pathways of the eye's response to topical nitrogen mustard, *Invest. Ophthalmol.*, 15, 486, 1976.
9. Butler, J. M., Unger, W. G., and Hammond, B. R., Sensory mediation of the ocular response to neutral formaldehyde, *Exp. Eye Res.*, 28, 577, 1979.
10. Unger, W. G., Mediation of the ocular response to injury, *J. Ocul. Pharmacol.*, 6, 337, 1990.
11. Bruce, A. N., Vasodilator axon reflexes, *Q. J. Exp. Physiol.*, 6, 339, 1913.
12. Bayliss, W. M., On the origin from the spinal cord of the vasodilator fibers of the hind limb, and the nature of these fibers, *J. Physiol.*, 26, 173, 1901.
13. Langley, J. N., Antidromic action. Stimulation of the peripheral nerves of the cat's hind foot, *J. Physiol.*, 58, 49, 1923.
14. Perkins, E. S., Influence of the fifth cranial nerve on the intraocular pressure of the rabbit eye, *Br. J. Ophthalmol.*, 41, 257, 1957.
15. Jancsò, N., Jancsò-Gabor, A., and Szolcsányi, J., Direct evidence for neurogenic inflammation and its prevention by denervation and by pretreatment with capsaicin, *Br. J. Pharmacol.*, 31, 138, 1967.
16. Buck, S. H. and Burks, T. F., The neuropharmacology of capsaicin: review of some recent observations, *Pharmacol. Rev.*, 38, 179, 1986.
17. Holzer, P., Capsaicin: cellular targets, mechanisms of action, and selectivity for thin sensory neurons, *Pharmacol. Rev.*, 43, 143, 1991.
18. Camras, C. B. and Bito, L. Z., The pathophysiological effects of nitrogen mustard on the rabbit eye. II. The inhibition of the initial hypertensive phase by capsaicin and the apparent role of substance P, *Invest. Ophthalmol. Vis. Sci.*, 19, 423, 1980.
19. Mandahl, A. and Bill, A., Ocular responses to antidromic trigeminal stimulation, intracameral prostaglandin E_1 and E_2, capsaicin and substance P, *Acta Physiol. Scand.*, 112, 331, 1981.
20. Butler, J. M. and Hammond, B. R., The effects of sensory denervation of the responses of the rabbit eye to prostaglandin E_1, bradykinin and substance P, *Br. J. Pharmacol.*, 69, 495, 1980.
21. Hökfelt, T., Kellerth, J. O., Nilsson, G., and Pernow, B., Experimental immunohistochemical studies on the localisation and distribution of substance P in cat primary sensory neurones, *Brain Res.*, 100, 235, 1975.
22. Tervo, K., Tervo, G., Eränkö, O., and Cuello, A. G., Immunoreactivity for substance P in the Gasserian ganglion, ophthalmic nerve and anterior segment of the rabbit eye, *Histochem. J.*, 13, 435, 1981.
23. Lee, Y., Kawai, Y., Shiosaka, S., Takami, K., Kiyama, H., Hillyard, C. J., Girgis, S., MacIntyre, I., Emson, P. C., and Tohyama, M., Coexistence of calcitonin gene-related peptide and substance P-like peptide in single cells of the trigeminal ganglion of the rat: immunohistochemical analysis, *Brain Res.*, 330, 194, 1985.
24. Terenghi, G., Polak, J. M., Ghatei, M. A., Mulderry, P. K., Butler, J. M., Unger, W. G., and Bloom, S. R., Distribution and origin of calcitonin gene-related peptide (CGRP) immunoreactivity in the sensory innervation of the mammalian eye, *J. Comp. Neurol.*, 233, 506, 1985.
25. Miller, A., Costa, M., Furness, J. B., and Chubb, I. W., Substance P-immunoreactive sensory nerves supply the rat iris and cornea, *Neurosci. Lett.*, 23, 243, 1981.
26. Tornqvist, K., Mandahl, A., Leander, S., Lorén, I., Håkanson, R., and Sundler, F., Substance P-immunoreactive nerve fibers in the anterior segment of the rabbit eye: distribution and possible physiological significance, *Cell Tissue Res.*, 222, 467, 1982.
27. Stone, R. A., Laties, A. M., and Brecha, N., Substance P-like immunoreactivity in the anterior segment of the rabbit, cat and monkey eye, *Neuroscience*, 7, 2459, 1982.
28. Ehinger, B., Sundler, F., Tervo, K., Tervo, T., and Tornquist, K., Substance P fibers in the anterior segment of the rabbit eye, *Acta Physiol. Scand.*, 118, 215, 1983.
29. Stone, R. A. and McGlinn, A., Calcitonin gene-related peptide immunoreactive nerves in human and Rhesus monkey eyes, *Invest. Ophthalmol. Vis. Sci.*, 29, 305, 1988.
30. Tervo, T., Tarkkanen, A., and Tervo, K., Innervation of the human pupillary sphincter muscle by nerve fibers immunoreactive to substance P, *Ophthalmic Res.*, 17, 111, 1985.

31. Moller, K., Zhang, Y.-Z., Håkanson, R., Luts, A., Sjölund, B., Uddman, R., and Sundler, F., Pituitary adenylate cyclase activating peptide is a sensory neuropeptide: immunocytochemical and immunochemical evidence, *Neuroscience,* 57, 725, 1993.
32. Wang, Z.-Y., Alm, P., and Håkanson, R., Distribution and effects of pituitary adenylate cyclase-activating peptide (PACAP) in the rabbit eye, *Neuroscience,* 69, 297, 1995.
33. Gamse, R., Leeman, S. E., Holzer, P., and Lembeck, F., Differential effects of capsaicin on the content of somatostatin, substance P and neurotensin in the nervous system of the rat, *Naunyn-Schmiedeberg's Arch. Pharmacol.,* 317, 140, 1981.
34. Stjernschantz, J. and Sears, M., Identification of substance P in the anterior uvea and retina of the rabbit, *Exp. Eye Res.,* 35, 401, 1982.
35. Geppetti, P., Bianco, E. D., Cecconi, R., Tramontana, M., Romani, A., and Theodorsson, E., Capsaicin releases calcitonin gene-related peptide from the human iris and ciliary body *in vitro, Regul. Pept.,* 41, 83, 1992.
36. Bill, A., Stjernschantz, J., Mandahl, A., Brodin, E., and Nilsson, G., Substance P. Release on trigeminal nerve stimulation, effects in the eye, *Acta Physiol. Scand.,* 106, 371, 1979.
37. Mandahl, A., Brodin, E., and Bill, A., Hypertonic KCl, NaCl and capsaicin intracamerally causes release of substance P-like immunoreactive material into the aqueous humor in rabbits, *Acta Physiol. Scand.,* 120, 579, 1984.
38. Butler, J. M., Powell, D., and Unger, W. G., Substance P levels in normal and sensorily denervated rabbit eyes, *Exp. Eye Res.,* 30, 311, 1980.
39. Terenghi, G., McGregor, G. P., Morrison, J. F. B., Unger, W. G., Bloom, S. R., and Polak, J. M., The effects of neonatal administration of capsaicin on substance P-containing sensory fibers in the rat iris, *Irish J. Med.,* 152 (Suppl.), 60, 1983.
40. Banno, H., Imaizumi, Y., and Watanabe, M., Pharmaco-mechanical coupling in the response to acetylcholine and substance P in the smooth muscle of the rat iris sphincter, *Br. J. Pharmacol.,* 85, 905, 1985.
41. Butler, J. M. and Hammond, B. R., The effects of sensory denervation on the responses of the rabbit eye to prostaglandin E_2, bradykinin and substance P, *Br. J. Pharmacol.,* 69, 495, 1980.
42. Ueda, N., Muramatsu, I., Taniguchi, T., Nakanishi, S., and Fujiwara, M., Effects of neurokinin A, substance P and electrical stimulation on the rabbit iris sphincter muscle, *J. Pharmacol. Exp. Ther.,* 239, 494, 1986.
43. Hosoki, R., Hisayama, T., and Takayanagi, I., Pharmacological evidence for the possible coexistence of multiple receptor sites for mammalian tachykinins in rabbit iris sphincter, *Naunyn-Schmiedeberg's Arch. Pharmacol.,* 335, 290, 1987.
44. Muramatsu, I., Nakanishi, S., and Fujiwara, M., Comparison of the responses to the sensory neuropeptides, substance P, neurokinin A, neurokinin B and calcitonin gene-related peptide, and to trigeminal nerve stimulation in the iris sphincter muscle of the rabbit, *Jpn. J. Pharmacol.,* 44, 85, 1987.
45. Beding-Barnekow, B., Brodin, E., and Håkanson, R., Substance P, neurokinin A and neurokinin B in the ocular response to injury in the rabbit, *Br. J. Pharmacol.,* 95, 259, 1988.
46. Too, H. P., Unger, W. G., and Hanley, M. R., Evidence for multiple tachykinin receptor subtypes on the rabbit iris sphincter muscle, *Mol. Pharmacol.,* 33, 64, 1988.
47. Unger, W. G. and Tighe, J., The response of the isolated iris sphincter muscle of various mammalian species to substance P, *Exp. Eye Res.,* 39, 677, 1984.
48. Anderson, J. A., Malfroy, B., Richard, N. R., Kullerstrand, L., Lucas, C., and Binder, P. S., Substance P contracts the human iris sphincter: possible modulation by endogenous enkephalinase, *Regul. Pept.,* 29, 49, 1990.
49. Alessandri, M., Fusco, B. M., Maggi, C. A., and Fanciullacci, M., *In vivo* pupillary constrictor effects of substance P in man, *Life Sci.,* 48, 2301, 1991.
50. Leander, S., Håkanson, R., Rosell, S., Folkers, K., Sundler, F., and Tornqvist, K., A specific substance P antagonist blocks smooth muscle contractions induced by non-cholinergic, non-adrenergic nerve stimulation, *Nature,* 294, 467, 1981.
51. Bynke, G., Håkanson, R., Hörig, J., and Leander, S., Bradykinin contracts the pupillary sphincter and evokes ocular inflammation through release of neuronal substance P, *Eur. J. Pharmacol.,* 91, 469, 1983.
52. Ueda, N., Muramatsu, I., and Fujiwara, M., Capsaicin and bradykinin-induced substance P-ergic responses in the iris sphincter muscle of the rabbit, *J. Pharmacol. Exp. Ther.,* 230, 469, 1984.
53. Wahlestedt, C., Bynke, G., and Håkanson, R., Pupillary constriction by bradykinin and capsaicin: mode of action, *Eur. J. Pharmacol.,* 106, 577, 1985.
54. Wahlestedt, C., Bynke, G., Beding, B., von Leithner, P., and Håkanson, R., Neurogenic mechanisms in control of the rabbit iris sphincter muscle, *Eur. J. Pharmacol.,* 117, 303, 1985.
55. Håkanson, R., Beding, B., Ljungqvist, A., Chu, J.-Y., Leander, S., Trojnar, J., and Folkers, K., Blockade of sensory nerve mediated contraction of the rabbit iris sphincter by a series of novel tachykinin antagonists, *Regul. Pept.,* 20, 99, 1988.
56. Håkanson, R., Leander, S., Asano, N., Feng, D.-M., and Folkers, K., Spantide II, a novel tachykinin antagonist having high potency and low histamine-releasing effect, *Regul. Pept.,* 31, 75, 1990.

57. Håkanson, R., Wang, Z.-Y., and Folkers, K., Comparison of Spantide II and CP-96,345 for blockade of tachykinin-evoked contractions of smooth muscle, *Biochem. Biophys. Res. Commun.*, 178, 297, 1991.
58. Wang, Z.-Y. and Håkanson, R., The electrically evoked, tachykinin-mediated contractile response of the isolated rabbit iris sphincter muscle involves NK_1 receptors only, *Eur. J. Pharmacol.*, 216, 327, 1992.
59. Hall, J. M., Mitchell, D., and Morton, I. K. M., Tachykinin receptors mediating responses to sensory nerve stimulation and exogenous tachykinins and analogues in the rabbit isolated iris sphincter, *Br. J. Pharmacol.*, 109, 1008, 1993.
60. Wang, Z.-Y., Tung, S. R., Strichartz, G. R., and Håkanson, R., Non-specific actions of the nonpeptide tachykinin receptor antagonists, CP-96,345, RP 67580 and SR 48968, on neurotransmission, *Br. J. Pharmacol.*, 111, 174, 1994.
61. Wang, Z.-Y., Tung, S. R., Strichartz, G. R., and Håkanson, R., Investigation of the specificity of FK 888 as a tachykinin NK_1 receptor antagonist, *Br. J. Pharmacol.*, 111, 1342, 1994.
62. Wang, Z.-Y., Feng, D.-M., Wang, Y.-L., Tung, S. R., Wong, K., Strichartz, G. R., Folkers, K., and Håkanson, R., Pharmacological assessment of spantide II analogues, *Eur. J. Pharmacol.*, 260, 121, 1994.
63. Wang, Z.-Y. and Håkanson, R., (±)-CP-96,345, a selective tachykinin NK_1 receptor antagonist, has non-specific actions on neurotransmission, *Br. J. Pharmacol.*, 107, 762, 1992.
64. Wang, Z.-Y. and Håkanson, R., The rabbit iris sphincter contains NK_1 and NK_3 but not NK_2 receptors: a study with selective agonists and antagonists, *Regul. Pept.*, 44, 269, 1993.
65. Holmdahl, G., Håkanson, R., Leander, S., Rosell, S., Folkers, K., and Sundler, F., A substance P antagonist, D-Pro,[2] D-Trp[7,9]-SP, inhibits inflammatory responses in the rabbit eye, *Science*, 214, 1029, 1981.
66. Bynke, G., Håkanson, R., and Hörig, J., Ocular responses evoked by capsaicin and prostaglandin E_2 are inhibited by a substance P antagonist, *Experientia*, 39, 996, 1983.
67. Mandahl, A. and Bill, A., Effects of the substance P antagonist, (D-Arg,[1] D-Pro,[2] D-Trp,[7,9] Leu[11])-SP on the miotic response to substance P, antidromic trigeminal nerve stimulation, capsaicin, prostaglandin E_1, compound 48/80 and histamine, *Acta Physiol. Scand.*, 120, 27, 1984.
68. Snider, R. M., Constantine, J. W., and Lowe, J. A. III, Longo, K. P., Lebel, W. S., Woody, H. A., Drozda, S. E., Desai, M. C., Vinick, F. J., Spencer, R. W., and Hess, H.-J., A potent nonpeptide antagonist of substance P (NK_1) receptor, *Science*, 251, 435, 1991.
69. Garret, C., Carruette, A., Fardin, V., Moussaoui, S., Peyronel, J.-F., Blanchard, J.-C., and Laduron, P. M., Pharmacological properties of a potent and selective nonpeptide substance P antagonist, *Proc. Natl. Acad. Sci. U.S.A.*, 88, 10208, 1991.
70. Hall, J. M., Mitchell, D., and Morton, I. K. M., Typical and atypical NK_1 tachykinin receptor characteristics in the rabbit isolated iris sphincter, *Br. J. Pharmacol.*, 112, 985, 1994.
71. Lee, C. M. and Cheung, W. T., Effects of neonatal monosodium glutamate treatment on substance P binding sites in the rat retina, *Neurosci. Lett.*, 92, 310, 1988.
72. Denis, P., Fordin, V., Nordmann, J.-P., Elena, P.-P., Laroche, L., Saraux, H., and Rostene, W., Localization and characterization of substance P binding sites in rat and rabbit eyes, *Invest. Ophthalmol. Vis. Sci.*, 32, 1894, 1991.
73. Hirata, H., Baba, S., Mishima, H., and Choschi, K., Specific binding of ^{125}I-substance P in the bovine iris-ciliary body, *Nippon Ganka Gakkai Zasshi*, 10, 395, 1983.
74. Osborne, N. N., Substance P in the bovine retina: localization, identification, release, uptake and receptor analysis, *J. Physiol. (London)*, 349, 83, 1984.
75. Gerhard, F., Kieselbach, G. F., Ragaut, R., Knaus, H. G., König, P., and Wiedermann, C. J., Autoradiographic analysis of binding sites for ^{125}I-Bolton-Hunter-substance P in the human eye, *Peptides*, 11, 655, 1990.
76. Hall, J. M., Mitchell, D., and Morton, I. K. M., Neurokinin receptors in the rabbit iris sphincter characterised by novel agonist ligands, *Eur. J. Pharmacol.*, 199, 9, 1991.
77. Stjernschantz, J., Sears, M., and Stjernschantz, L., Intraocular effects of substance P in the rabbit, *Invest. Ophthalmol. Vis. Sci.*, 20, 53, 1981.
78. Stjernschantz, J., Sears, M., and Mishima, H., Role of substance P in antidromic vasodilation, neurogenic plasma extravasation and disruption of the blood-aqueous barrier in the rabbit eye, *Naunyn-Schmiedeberg's Arch. Pharmacol.*, 32, 329, 1982.
79. Andersson, S. E., Responses to antidromic trigeminal nerve stimulation, substance P, NKA, CGRP and capsaicin in the rat eye, *Acta Physiol. Scand.*, 131, 371, 1987.
80. Taniguchi, T., Fujiwara, M., Masuo, Y., and Kanazawa, I., Levels of neurokinin A, neurokinin B and substance P in rabbit iris sphincter muscle, *Jpn. J. Pharmacol.*, 42, 590, 1986.
81. Unger, W. G., Terenghi, G., Ghatei, M. A., Ennis, K. W., Butler, J. M., Zhang, S. Q., Too, H. P., Polak, J. M., and Bloom, S. R., Calcitonin gene-related peptide as a mediator of the neurogenic ocular injury response, *J. Ocular Pharmacol.*, 1, 189, 1985.
82. Terenghi, G., Zhang, S.-Q., Unger, W. G., and Polak, J. M., Morphological changes of sensory CGRP-immunoreactive and sympathetic nerves in peripheral tissues following chronic denervation, *Histochemistry*, 86, 89, 1986.

83. Geppetti, P., Frilli, S., Renzi, D., Santicioli, P., Maggi, C. A., Theodorsson, E., and Fanciullacci, M., Distribution of calcitonin gene-related peptide-like immunoreactivity in various rat tissues: correlation with substance P and other tachykinins and sensitivity to capsaicin, *Regul. Pept.*, 23, 289, 1988.
84. Uusitalo, H., Krootila, K., and Palkama, A., Calcitonin gene-related peptide (CGRP) immunoreactive sensory nerves in the human and guinea pig uvea and cornea, *Exp. Eye Res.*, 48, 467, 1989.
85. Wahlestedt, C., Beding, B., Ekman, R., Oksala, O., Stjernschantz, J., and Håkanson, R., Calcitonin gene-related peptide in the eye: release by sensory nerve stimulation and effects associated with neurogenic inflammation, *Regul. Pept.*, 16, 107, 1986.
86. Unger, W. G., Terenghi, G., Zhang, S. Q., and Polak, J. M., Alteration in the histochemical presence of tyrosine hydroxylase and CGRP-immunoreactivities in the eye following chronic sympathetic or sensory denervation, *Curr. Eye Res.*, 7, 761, 1988.
87. Krootila, K., Uusitalo, H., and Palkama, A., Effect of neurogenic irritation and calcitonin gene-related peptide (CGRP) on ocular blood flow in the rabbit, *Curr. Eye Res.*, 7, 695, 1988.
88. Krootila, K., CGRP in relation to neurogenic inflammation and cAMP in the rabbit eye, *Exp. Eye Res.*, 47, 307, 1988.
89. Oksala, O., Effects of calcitonin gene-related peptide and substance P on regional blood flow in the cat eye, *Exp. Eye Res.*, 47, 283, 1988.
90. Oksala, O. and Stjernschantz, J., Effects of calcitonin gene-related peptide in the eye. A study in rabbits and cats, *Invest. Ophthalmol. Vis. Sci.*, 29, 1006, 1988.
91. Andersson, S. E. and Bill, A., Effects of intravenous calcitonin gene-related peptide (CGRP) and substance P on the blood-aqueous barrier in the rabbit, *Acta Physiol. Scand.*, 135, 349, 1989.
92. Almegard, B. and Andersson, S. E., Vascular effects of calcitonin gene-related peptide (CGRP) and cholecystokinin (CCK) in the monkey eye, *J. Ocul. Pharmacol.*, 9, 77, 1993.
93. Nilsson, S. F. E., PACAP-27 and PACAP-38: vascular effects in the eye and some other tissues in the rabbit, *Eur. J. Pharmacol.*, 253, 17, 1994.
94. Wang, Z.-Y. and Håkanson, R., unpublished observations.

Chapter 12

SENSORY NEURONS IN THE STOMACH

Peter Holzer

CONTENTS

I. INTRODUCTION

The gastric mucosa is capable of resisting the continuous onslaught from its aggressive secretory products, acid and pepsin, by virtue of a multitude of factors that collectively form the gastric mucosal barrier.[1] Although autonomic neurons have long been recognized to influence gastric defence mechanisms, it was only recently that peptidergic afferent neurons were discovered to constitute an important neural emergency system. These neurons participate in the neural control of blood flow, secretory processes, and motor activity in the stomach and regulate a variety of homeostatic functions that can be viewed as increasing the resistance of the tissue to injury and/or facilitating the repair of damaged tissue. Some of the actions exerted by afferent neurons in the stomach, notably those on blood flow, resemble the neurogenic inflammatory reactions seen in other tissues. As in other tissues, the pathophysiological implications of sensory neurons in the stomach have, to a large extent, been discovered by the use of capsaicin, a pharmacological tool with which the activity of primary afferent neurons with unmyelinated or thinly myelinated nerve fibers can selectively be manipulated.[2] The aim of this chapter is to give a brief account of the roles that afferent neurons play locally in the stomach. The reader is also referred to other articles in which certain functional aspects of gastric afferent neurons have been reviewed.[3,4]

II. PEPTIDERGIC AFFERENT NERVE FIBERS IN THE STOMACH

The primary afferent neurons supplying the stomach arise from two different sources (Figure 1). The spinal sensory neurons originate from cell bodies in the dorsal root ganglia and reach the stomach via the splanchnic and mesenteric nerves, while the sensory nerve fibers running in the vagus nerves have their cell bodies in the nodose ganglia. The subgroup of capsaicin-sensitive afferent neurons contains a number of bioactive peptides, including calcitonin gene-related peptide (CGRP) and the tachykinins substance P (SP) and neurokinin A (NKA).[3] In the rat stomach it is particularly the arterial system that receives a dense supply

0-8493-7646-7/96/$0.00+$.50

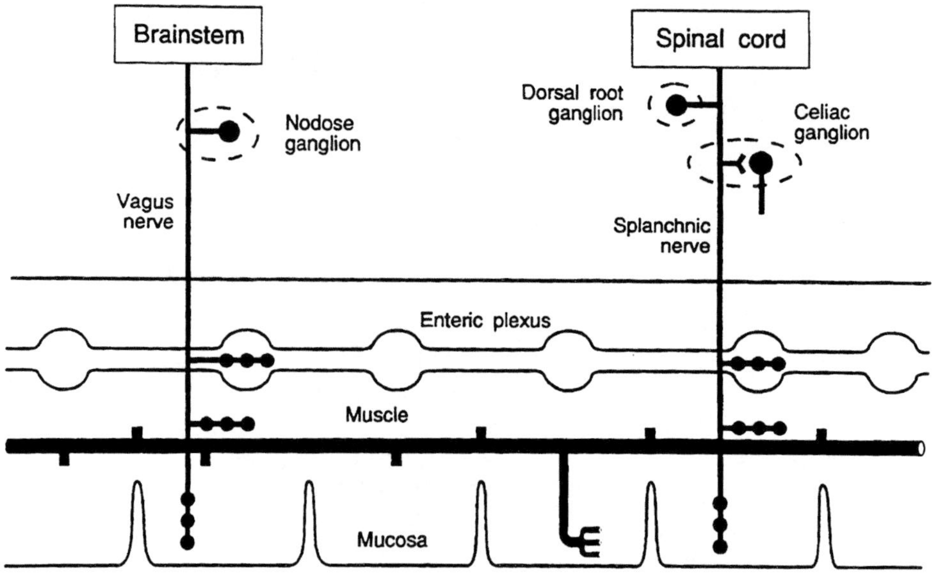

FIGURE 1. Diagram showing the innervation of the stomach by extrinsic primary afferent neurons.

by spinal afferent nerve fibers expressing CGRP, SP, and NKA, whereas the venous system is rather sparsely innervated.[5-7] The peptide-containing axons run primarily in the connective tissue surrounding the vessels and in close proximity to the vascular smooth muscle; in addition, they supply the myenteric plexus, the circular muscle layer, and the gastric mucosa.[5-8] Most, if not all, CGRP-containing nerve fibers in the rat stomach arise from spinal afferent neurons,[6,7,9] whereas vagal afferent nerve fibers seem to contribute little to the CGRP and SP content,[6,9] although this contention has been challenged.[8] On their way to the gastric wall, the fibers of spinal afferent neurons pass through the prevertebral ganglia, where they give off collaterals to form synapses with the sympathetic ganglion cells.[6,9,10] Although the extrinsic autonomic and intrinsic enteric neurons of the gut have many peptide transmitters in common with extrinsic primary afferents, they differ with regard to their chemical coding,[11] the chemical identity of their peptide transmitters,[7,12] and their sensitivity to capsaicin.[2,6,13] Most of the CGRP present in afferent neurons is CGRP-α, whereas the only form of CGRP in enteric neurons is CGRP-β,[7,12] and only the primary afferent neurons are sensitive to capsaicin, whereas the autonomic and enteric neurons are not.[2,6,13]

III. AFFERENT NERVE FIBERS AND GASTRIC BLOOD FLOW

The presence of a dense plexus of peptidergic afferent nerve fibers around gastric arteries and arterioles[6,7] suggests that these fibers are involved in the control of vascular functions in the stomach. Stimulation of these axons by acute administration of capsaicin leads to a marked increase in gastric mucosal blood flow (GMBF),[14,15] which is brought about by dilatation of submucosal arterioles.[16] The vasodilator effect of capsaicin depends on the integrity of the extrinsic afferent innervation, because defunctionalization of afferent neurons by a neurotoxic dose of capsaicin abolishes the hyperemic response to sensory nerve stimulation.[15,17-23] Nerve-selective ablation of afferent nerve fibers has shown that only spinal afferent neurons passing through the celiac ganglion participate in the capsaicin-evoked increment of blood flow through the rat gastric mucosa.[17] The vagus nerves of the rat also contain afferent nerve fibers that give rise to mucosal vasodilatation in the stomach following electrical stimulation,[24] but these fibers do not seem to contribute to the hyperemia caused by intragastric capsaicin.[17]

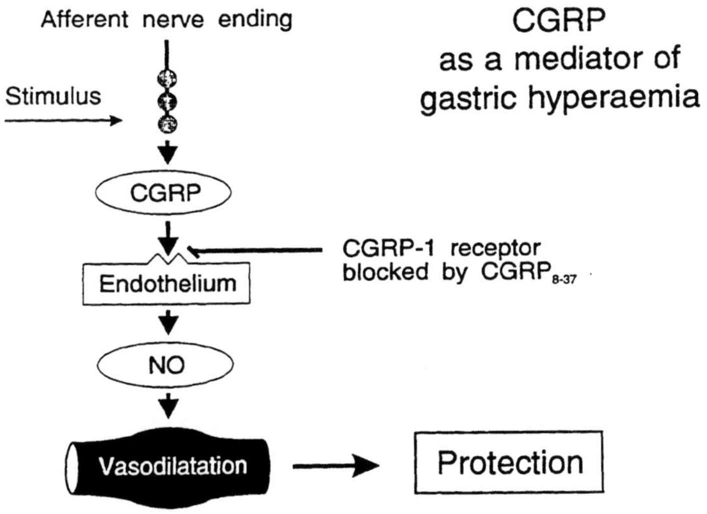

FIGURE 2. Diagram showing the role of calcitonin gene-related peptide (CGRP) in mediating gastric hyperemia and mucosal protection. NO, nitric oxide.

The neural pathways responsible for the capsaicin-evoked rise of GMBF in the rat stomach, which involve ganglionic transmission via nicotinic acetylcholine receptors,[25] are still ill defined. Since noradrenergic neurons and cholinergic vasodilator neurons have been ruled out,[26,27] it has been proposed that the vasodilator response to capsaicin is brought about by a local release of vasodilator peptides from perivascular afferent nerve fibers in the stomach.[14,15] There is conclusive evidence that CGRP plays an important mediator role, because the CGRP receptor antagonist $CGRP_{8\text{-}37}$ prevents the hyperemia caused by intragastric capsaicin.[16,17] This implication of the peptide is consistent with its release[28-32] from extrinsic nerve fibers in the stomach and its high activity in enhancing gastric blood flow[17,21,33-37] through an action on CGRP-1 receptors[16,17,36] that are abundantly present on the muscle and endothelium of gastric arteries and arterioles.[7,38] The vasodilator action of CGRP released by capsaicin in the rat stomach involves the formation of nitric oxide (NO), since blockers of the NO synthase suppress the gastric hyperemic reaction to both capsaicin[21,22] and CGRP.[21,35] As NO is unlikely to be released from the CGRP-containing axons[39] it seems as if CGRP acts via the formation of NO in endothelial cells (Figure 2). In addition to being the final vasodilator messenger of CGRP, NO may also facilitate the release of CGRP from afferent nerve fibers.[40]

SP and NKA, which are also released from sensory nerve fibers in the stomach,[41,42] do not participate in the afferent nerve-mediated increase in GMBF, since SP and NKA fail to augment blood flow in the rat stomach[23,34,36] and a SP antagonist fails to alter the hyperemic reaction to capsaicin.[27] To the contrary, SP can inhibit the capsaicin-evoked vasodilatation in the rat gastric mucosa, a reaction that may come about by the release of CGRP-degrading proteases from mast cells.[23] Vasoactive mediators derived from mast cells such as histamine, however, do not contribute substantially to the capsaicin-evoked rise of GMBF under normal circumstances.[20,23,27] The role of prostaglandins in the capsaicin-evoked rise of GMBF is also unclear. Although capsaicin fails to alter the *ex vivo* formation of prostaglandin E_2, 6-oxo-prostaglandin $F_{1\alpha}$, and leukotriene C_4 in the rat gastric mucosa,[14] indomethacin has been reported to diminish the capsaicin-induced increase in GMBF in the rat,[18,22] but the significance of these findings in terms of an implication of vasodilator prostaglandins has not been ascertained.

There is little evidence that peptidergic afferent nerve fibers play a role in regulating vascular conductance in the stomach under physiological conditions, since defunctionalization of capsaicin-sensitive afferent neurons fails to alter basal GMBF,[15-19,26,27] the gastric vasodilator responses to CGRP, acetylcholine, adenosine,[16] histamine,[19] and pentagastrin[43] and the gastric vasoconstrictor response to adrenaline.[19] However, the autoregulatory escape reaction from adrenaline-induced vasoconstriction in the rat gastric mucosa depends on the integrity of capsaicin-sensitive afferent neurons,[19] which suggests that vasoconstriction is counteracted by afferent nerve-mediated vasodilatation. An analogous explanation may apply to the finding that gastric vasoconstriction caused by platelet-activating factor[44] or blockade of NO synthesis[45] is amplified after capsaicin-induced defunctionalization of afferent neurons. The interaction between the tonically active NO system and the stimulus-driven afferent nerve fibers suggests that these two systems act in concert to provide an active dilator drive on the gastric microcirculation.[4]

IV. AFFERENT NERVE FIBERS AND VASCULAR PERMEABILITY IN THE STOMACH

Neurogenic inflammation in nonvisceral tissues involves a marked increase in venular permeability leading to extravasation of macromolecules, leukocytes, and fluid,[3] whereas in the stomach of the rat stimulation of afferent nerve fibers does not enhance vascular permeability.[46-48]

V. AFFERENT NERVE FIBERS AND GASTRIC SECRETION

A significant aspect of the digestive activity of the stomach is to secrete acid, pepsinogen, mucus, and bicarbonate. Although peptidergic afferent nerve fibers project into the mucosa of the rat stomach,[6-8] capsaicin-evoked stimulation of afferent nerve fibers does not significantly influence the basal secretion of gastric acid, whereas the elimination of acid from the gastric lumen is increased[26,27,49] as a result of enhanced bicarbonate secretion.[50] These effects on the gastric acid/bicarbonate balance are related to the ability of capsaicin to stimulate the release of somatostatin and to inhibit the release of acetylcholine and gastrin in the rat isolated antrum, actions which are, at least in part, mediated by CGRP.[31] The effect of sensory nerve stimulation on gastric mucus secretion has not yet been examined.

Defunctionalization of capsaicin-sensitive afferent neurons fails to change the basal output of gastric acid,[24,26,27,43,51-57] pepsin,[51] and bicarbonate.[50] The basal secretion of mucus in the rat gastric corpus is also unaltered while that in the antrum is reduced.[58] The failure of capsaicin pretreatment to change gastric secretory processes is further reflected by the lack of effect on the permeability of the gastric mucosa to acid and ions.[27,50,59] Fine afferent neurons thus do not contribute to the regulation of basal gastric secretion, but they do participate in physiological alterations of gastric acid and bicarbonate secretion, although this issue is somewhat obscured by discrepant data in the literature. To give an example, acid secretion induced by histamine may be reduced[52,55] or left unchanged[56,60] after pretreatment of rats with the neurotoxin capsaicin, whereas the acid output caused by pentagastrin has been found unaltered in most studies.[43,52,55,56] The acid secretory response to electrical stimulation of the vagus nerves[24,57] also remains unchanged after systemic pretreatment of rats with capsaicin but is attenuated after selective ablation of capsaicin-sensitive vagal afferents.[61] In contrast, capsaicin-sensitive afferent neurons do not play any role in the acid output stimulated by muscarinic acetylcholine receptor agonists.[52,54,56] With regard to the gastric alkaline secretion it has been found that the acid-evoked output of bicarbonate is suppressed in capsaicin-pretreated rats.[50] The transmitters which play a role in the afferent nerve-mediated changes of gastric secretion have not yet been elucidated. CGRP is very potent in inhibiting gastric acid secretion,[62] but the implication of this peptide in afferent nerve-mediated perturbations of gastric secretion remains to be tested.

There is little controversy that extrinsic afferent neurons are involved in the reflex regulation of gastric acid secretion in response to distension of the gastric wall[55,56] or intragastric administration of nutrients,[63] but it is beyond the scope of this chapter to discuss these implications in more detail.

VI. AFFERENT NERVE FIBERS AND GASTRIC MOTOR ACTIVITY

Sensory nerve stimulation with capsaicin has been found to exert both stimulant and inhibitory effects on the motor activity of the stomach.[15,18,64,65] The relaxant action of capsaicin, which may be responsible for the capsaicin-induced inhibition of gastric emptying,[66] depends on the integrity of extrinsic afferent neurons but does not involve SP, CGRP, or vasoactive intestinal polypeptide,[18,64,65] whereas the capsaicin-induced increase of gastric motor activity relies on tachykinins (SP and NKA) as contractile mediators.[65] As is the case with gastric secretion, capsaicin-sensitive afferents do not play a role in the regulation of basal motor activity in the stomach.[18,56,67-71] The inhibition of gastric motility elicited by electrical stimulation of the celiac ganglion[56] is likewise independent of afferent neurons as are the gastric motor stimulant effects of vagus nerve stimulation[72] and administration of SP.[73] In contrast, capsaicin-sensitive afferent neurons contribute to a number of inhibitory gastric motor reflexes, including those evoked by cholecystokinin,[72] intragastric or intraperitoneal acid,[68,72] and intraperitoneal bradykinin.[69] The inhibition of gastric motor activity and/or gastric emptying elicited by a variety of nutrients relies also on afferent neurons.[68,71] Of pathophysiological importance is the finding that the gastroparesis caused by abdominal surgery or irritation involves, at least in part, capsaicin-sensitive afferent pathways[67,70,75] and CGRP as a transmitter substance.[75] It is not yet known, however, whether CGRP (which is a potent relaxant of gastric smooth muscle[62]) and other peptide mediators released from afferent nerve fibers locally in the stomach play a role in the physiological or pathophysiological regulation of gastric motility.

VII. AFFERENT NERVE FIBERS AND GASTRIC DEFENSE MECHANISMS

The first evidence that sensory neurons may strengthen the resistance of the gastric mucosa to damage was obtained by studying the effects of capsaicin-induced ablation of afferent neurons on experimentally imposed injury of the rat stomach. Sensory denervation does not cause damage by itself[27,44,48,54,58] and fails to alter the gastric injury induced by cold and restraint stress[53] but leads to exacerbation of mucosal lesion formation in response to a variety of injurious chemicals. Thus, gastric damage provoked by hydrochloric acid,[51,59,76] aspirin,[77] indomethacin,[32,54,77-79] platelet-activating factor,[44] taurocholate,[80] or ethanol[54,58,77,81,82] is aggravated in capsaicin-pretreated rats. Histological examination of the mucosa shows that not only the extent of macroscopically visible lesions but also the depth of damage is augmented by sensory nerve ablation.[44,58,59,78,81,82] In addition, afferent nerve ablation compromises the activity of cholecystokinin octapeptide,[83] laparotomy,[84] and prostaglandin E_2[85] to protect the gastric mucosa from injury, whereas the gastroprotective effects of atropine,[54] CGRP,[32] and cimetidine[54] are left unaltered in capsaicin-pretreated rats. In contrast, the ability of morphine to exacerbate gastric mucosal damage is halted by sensory nerve ablation which aggravates experimental damage by itself,[81] which is consistent with the inhibitory action of opiates on transmitter release from afferent neurons.[3] Another important interaction relates to the NO system, as blockade of NO synthesis, which by itself does not cause damage, leads to the induction of extensive acid injury in capsaicin-pretreated rats.[78] This finding implies that NO acts in concert with afferent neurons in the control of gastric defence mechanisms, a conjecture that will be further considered below.

A gastroprotective role of sensory neurons is strongly supported by the ability of afferent nerve stimulation to strengthen the resistance of the gastric mucosa to injury. Szolcsányi and Barthó[51] were the first to show that capsaicin-evoked stimulation of gastric afferents reduced the injury caused by acid accumulation in the stomach. In subsequent studies it was found that acute intragastric administration of capsaicin prevented gastric damage caused by a variety of chemicals, including hydrochloric acid,[49,51] aspirin,[86] indomethacin,[32] ethanol,[14,15,18,22,87] and taurocholate.[49] Numerous studies have revealed that the protective effect of capsaicin results in a reduction of the depth of injury while surface damage is not prevented.[14,15,86,87] It is important to stress that, in contradiction of traditional views, intragastric capsaicin does not irritate or damage the gastric mucosa and does not in any way weaken the gastric mucosal barrier,[14,15,18,26,27,48,49,86] as has been assumed for some red pepper spices which contain many substances other than capsaicin.

The gastroprotective activity of capsaicin is due to its stimulant action on afferent nerve fibers, since it is abolished after defunctionalization of extrinsic sensory neurons[15,18,32,48,88] or blockade of nerve conduction with tetrodotoxin.[15] However, the pathways underlying sensory nerve-mediated gastric mucosal protection in the stomach have not yet been fully delineated. As acute vagotomy, acute extirpation of the celiac ganglion, and acute elimination of the adrenal glands are devoid of any influence on the gastroprotective effect of capsaicin,[48] it has been proposed that the ability of capsaicin to strengthen gastric mucosal resistance to injury depends on the local release of transmitter substances from sensory nerve fibers within the gastric wall. In keeping with this contention, there is considerable evidence that afferent nerve-mediated gastroprotection is mediated by CGRP which in the rat stomach occurs exclusively in extrinsic afferent nerve fibers.[6,7] Close arterial administration of this peptide to the rat stomach prevents gastric damage induced by ethanol, acidified aspirin, or endothelin-1.[33,37,87,89] The effect of CGRP to reduce ethanol-induced injury seems to be mediated by CGRP-1 receptors as it is prevented by $CGRP_{8-37}$,[87] whereas the ability of CGRP to inhibit aspirin-induced damage is not antagonized by $CGRP_{8-37}$.[89] Most importantly, the gastroprotective effects of both capsaicin and CGRP are blocked by the CGRP receptor antagonist $CGRP_{8-37}$[87] and by immunoblockade of the peptide,[87,93] while active immunization of rats against CGRP may exacerbate ethanol-induced damage.[94]

The gastroprotective action of capsaicin is also suppressed by blockade of the formation of NO.[22,87] Since the ability of CGRP to strengthen gastric mucosal defense against injurious factors is likewise suppressed by blockers of the NO synthase,[87] it would appear that stimulation of sensory nerve fibers by capsaicin results in the release of CGRP which, via formation of NO, augments the resistance of the gastric mucosa against experimental injury. Such a role of NO is consistent with its ability to prevent gastric mucosal injury[95] and with the effect of NO synthase inhibitors to make the gastric mucosa more vulnerable by acid.[78] Together with the findings on blood flow, it is evident, therefore, that CGRP and NO play an important mediator role in both the protective and hyperemic responses to afferent nerve stimulation in the stomach (Figure 2). The identity of the mediators involved in the two processes suggests a close relationship between the increase in GMBF and the increase in mucosal resistance to injury, a conclusion that applies to many other vasodilator principles.[1] The rise of GMBF is indeed also seen when capsaicin is administered together with an injurious concentration of ethanol, and there is a remarkable correlation between the hyperemic response to capsaicin and the concomitant reduction of gastric mucosal erosions.[15,22] This parallelism appears to signify that the gastric hyperemia may be the chief mechanism that is responsible for afferent nerve-mediated defense against injury. It should not be disregarded, however, that mechanisms other than a rise of GMBF, such as increases in the mucus and bicarbonate[50] secretion, may also play a significant role.[37,49]

Substance P is devoid of a protective action[90] and has, in fact, been found to exaggerate mucosal damage by degranulation of mast cells.[91] Neurokinin A, though, and related analogs acting preferentially on tachykinin NK_2 receptors do enhance the resistance of the gastric

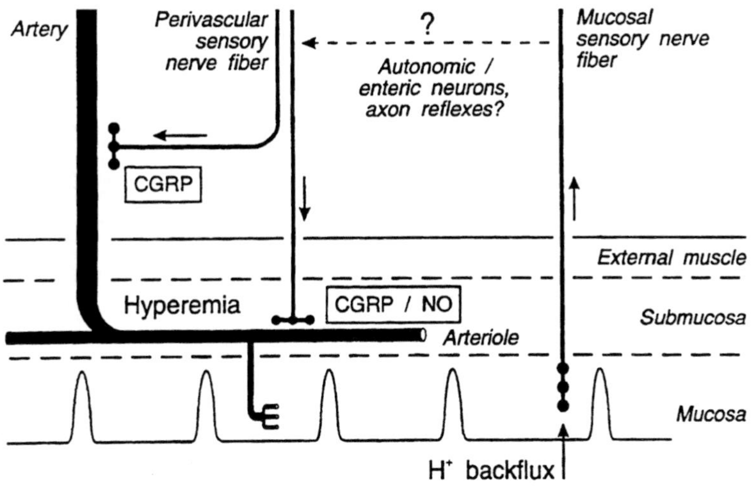

FIGURE 3. Diagram showing the neural mechanisms that may underly the gastric hyperemic response to acid backflux through a disrupted gastric mucosal barrier. CGRP, calcitonin gene-related peptide; NO, nitric oxide.

mucosa to injury,[90,92] but their possible implication in the gastroprotective action of capsaicin has not yet been tested. Prostaglandins have been ruled out as mediators, since capsaicin given intragastrically does not affect the *ex vivo* formation of prostaglandin E_2, 6-oxo-prostaglandin $F_{1\alpha}$, and leukotriene C_4.[87] Indomethacin, though, was found to reduce the gastroprotective action of capsaicin in some,[18,22,88] but not all,[14] studies, but the precise relevance of these findings has not yet been verified.

In addition to regulating the defense against acutely pending injury, capsaicin-sensitive afferent neurons may also be of relevance for the repair of the damaged gastric mucosa. Although sensory neurons play little role in the process of rapid restitution of the superficially injured gastric mucosa,[82] they promote the healing of the injured mucosa, as the rate of repair of gastric ulcers induced by hydrochloric acid,[96] acetic acid,[97] and ethanol[98] is delayed after sensory nerve ablation.

VIII. PATHOPHYSIOLOGICAL IMPLICATIONS

The pathophysiological implications of peptidergic afferent neurons in the regulation of GMBF and gastric mucosal resistance to injury are put into proper perspective if their role in the gastric responses to acid challenge is considered. Acid back-diffusion through a disrupted gastric mucosal barrier causes a prompt increase in GMBF,[1,36,59] which is accomplished by neural pathways. Thus, the acid-evoked rise of gastric blood flow is depressed by both tetrodotoxin[59] and capsaicin-induced ablation of spinal afferent nerve fibers[59,76,80] that pass through the celiac ganglion.[99] As suppression of the acid-evoked rise of GMBF is associated with an aggravation of gastric damage,[59,99] it is evident that the hyperemia is an important mechanism of protection which, by facilitating the disposal of acid, prevents the build-up of an injurious concentration of hydrogen ions in the tissue. The inhibitory effects of tetrodotoxin[59] and the autonomic ganglion blocking drug hexamethonium[92] suggest that the increment of GMBF in the face of pending acid injury results from a reflex-like mechanism. However, the individual reflex arcs which rely on intact nerve conduction through afferent/efferent pathways in the splanchnic nerves and through the celiac ganglion[100] are not yet understood.[3] It is conceivable that the acid-evoked rise of GMBF is relayed by a peripheral neural circuitry which depends on excitatory or inhibitory inputs from splanchnic and/or enteric nerve fibers (Figure 3).[3] This contention receives major support from the evidence that CGRP mediates the gastric vasodilator response to acid challenge of the mucosa.[36,101] As is the case with the vasodilator response to exogenous CGRP, the acid-evoked rise of blood flow through the gastric mucosa depends on the formation of NO,[36,102] whereas the increase of blood flow through the left gastric artery does not (Figure 3).[36] It follows that the effector mechanisms of neurogenic vasodilatation differ in the various levels of the gastric arterial tree. Substance P acting via NK_1 receptors,[36] histamine acting via histamine H_1 receptors,[59] prostaglandins,[102] cholinergic and noradrenergic neurons[59,100] have been ruled out to take part in the gastric hyperemic reaction to acid challenge. Bradykinin, to the contrary, may participate in the protective rise of GMBF caused by severe acid challenge of the gastric mucosa, which is in keeping with the formation of this kinin in response to injury or exposure of the tissue to low pH.[103]

While the physiological and pathophysiological roles of afferent nerve fibers in the control of vessel diameter and blood flow through the stomach have been fairly well studied, there is a lack of information regarding the regulation of other vascular functions by peptides released locally from afferent nerve fibers. Capsaicin-sensitive afferent neurons have been implicated in the control of hemostatic mechanisms[104] and in the regulation of the adherence, migration, and activity of leukocytes,[3] but it remains to be determined whether in the stomach these processes are also under the control of afferent neurons. Another pathophysiologically relevant implication of afferent neurons concerns their ability to stimulate the secretion of bicarbonate from the rat gastric mucosa[50] and the possible role of CGRP to inhibit gastric acid secretion.[62]

IX. SUMMARY

It is obvious from these findings that peptidergic afferent nerve fibers constitute a neural emergency system in the stomach, which is called into operation, e.g., in the face of pending injury to the mucosa. As a result, blood flow to the stomach is greatly augmented, an effect that will add to the removal of injurious factors from the mucosa and promote a wide range of processes that either reduce the vulnerability, or aid the repair, of the gastric mucosa. Apart from facilitation of gastric blood flow, augmentation of bicarbonate secretion as well as inhibition of acid output and motor activity are also likely to contribute to the overall protective role of peptide-containing afferent neurons. The ability of afferent neurons to maintain mucosal homeostasis seems to be due largely to the local release of peptide transmitters in the gastric wall, although their participation in autonomic and neuroendocrine reflexes that regulate the secretory and motor activity of the stomach need also to be considered.

With the gastroprotective role of peptidergic afferent neurons in mind it is logical to assume that improper functioning of this neural emergency system is liable to weaken the resistance of the tissue to injurious stimuli and thus to be an etiologic factor in gastroduodenal ulcer disease. Evidence in favor of this conjecture is indeed accumulating. For instance, chronic intake of nicotine by rats has been found to suppress the increase in mucosal blood flow due to gastric acid back-diffusion,[105] and the gastropathy associated with experimental cirrhosis,[106] kidney failure,[107] and aging[108,109] is associated with an inadequate rise of GMBF in response to acid back-diffusion through a leaky gastric mucosal barrier. It will be important, therefore, to determine how the neural emergency system in the stomach operates under a variety of conditions and pathological circumstances and, if malfunction of the system occurs, which mechanisms fail at the neural, vascular, or mucosal level.

ACKNOWLEDGMENTS

Work performed in the author's laboratory was supported by the Austrian Science Foundation (Grants 9473 and 9823), the Austrian National Bank (Grants 4207 and 4905), and the Franz Lanyar Foundation at the Medical Faculty of the University of Graz. The author is grateful to Dr. Ulrike Holzer-Petsche for drawing the graphs and to Irmgard Russa for secretarial help.

REFERENCES

1. Allen, A., Flemström, G., Garner, A., and Kivilaakso, E., Gastroduodenal mucosal protection, *Physiol. Rev.*, 73, 823, 1993.
2. Holzer, P., Capsaicin: cellular targets, mechanisms of action, and selectivity for thin sensory neurons, *Pharmacol. Rev.*, 43, 143, 1991.
3. Holzer, P., Peptidergic sensory neurons in the control of vascular functions: mechanisms and significance in the cutaneous and splanchnic vascular beds, *Rev. Physiol. Biochem. Pharmacol.*, 121, 49, 1992.
4. Whittle, B.J.R., Neuronal and endothelium-derived mediators in the modulation of the gastric microcirculation: integrity in the balance, *Br. J. Pharmacol.*, 110, 3, 1993.
5. Furness, J.B., Papka, R.E., Della, N.G., Costa, M., and Eskay, R.L., Substance P-like immunoreactivity in nerves associated with the vascular system of guinea-pigs, *Neuroscience*, 7, 447, 1982.
6. Green, T. and Dockray, G.J., Characterization of the peptidergic afferent innervation of the stomach in the rat, mouse, and guinea-pig, *Neuroscience*, 25, 181, 1988.
7. Sternini, C., Enteric and visceral afferent CGRP neurons. Targets of innervation and differential expression patterns, *Ann. N.Y. Acad. Sci.*, 657, 170, 1992.
8. Böck, N., Ahonen, M., Häppölä, O., Kivilaakso, E., and Kiviluoto, T., Effect of vagotomy on expression of neuropeptides and histamine in rat oxyntic mucosa, *Dig. Dis. Sci.*, 39, 353, 1994.
9. Lee, Y., Shiotani, Y., Hayashi, N., Kamada, T., Hillyard, C.J., Girgis, S.I., MacIntyre, I., and Tohyama, M., Distribution and origin of calcitonin gene-related peptide in the rat stomach and duodenum: an immunohistochemical study, *J. Neur. Transmiss.*, 68, 1, 1987.
10. Lindh, B., Hökfelt, T., and Elfvin, L.-G., Distribution and origin of peptide-containing nerve fibers in the celiac superior mesenteric ganglion of the guinea-pig, *Neuroscience*, 26, 1037, 1988.
11. Costa, M., Furness, J.B., and Gibbins, I.L., Chemical coding of enteric neurons, *Prog. Brain Res.*, 68, 217, 1986.
12. Mulderry, P.K., Ghatei, M.A., Spokes, R.A., Jones, P.M., Pierson, A.M., Hamid, Q.A., Kanse, S., Amara, S.G., Burrin, J.M., Legon, S., Polak, J.M., and Bloom, S.R., Differential expression of α-CGRP and β-CGRP by primary sensory neurons and enteric autonomic neurons of the rat, *Neuroscience*, 25, 195, 1988.
13. Barthó, L. and Holzer, P., Search for a physiological role of substance P in gastrointestinal motility, *Neuroscience*, 16, 1, 1985.
14. Holzer, P., Pabst, M.A., Lippe, I.Th., Peskar, B.M., Peskar, B.A., Livingston, E.H., and Guth, P.H., Afferent nerve-mediated protection against deep mucosal damage in the rat stomach, *Gastroenterology*, 98, 838, 1990.
15. Holzer, P., Livingston, E.H., Saria, A., and Guth, P.H., Sensory neurons mediate protective vasodilatation in rat gastric mucosa, *Am. J. Physiol.*, 260, G363, 1991.
16. Chen, R.Y.Z., Li, D.-S., and Guth, P.H., Role of calcitonin gene-related peptide in capsaicin-induced gastric submucosal arteriolar dilation, *Am. J. Physiol.*, 262, H1350, 1992.
17. Li, D.-S., Raybould, H.E., Quintero, E., and Guth, P.H., Role of calcitonin gene-related peptide in gastric hyperemic response to intragastric capsaicin, *Am. J. Physiol.*, 261, G657, 1991.
18. Takeuchi, K., Niida, H., Matsumoto, J., Ueshima, K., and Okabe, S., Gastric motility changes in capsaicin-induced cytoprotection in the rat stomach, *Jpn. J. Pharmacol.*, 55, 147, 1991.
19. Leung, F.W., Modulation of autoregulatory escape by capsaicin-sensitive afferent nerves in the rat stomach, *Am. J. Physiol.*, 262, H562, 1992.
20. Wallace, J.L., McKnight, G.W., and Befus, A.D., Capsaicin-induced hyperemia in the stomach: possible contribution of mast cells, *Am. J. Physiol.*, 263, G209, 1992.
21. Whittle, B.J.R., Lopez-Belmonte, J., and Moncada, S., Nitric oxide mediates rat mucosal vasodilatation induced by intragastric capsaicin, *Eur. J. Pharmacol.*, 218, 339, 1992.
22. Brzozowski, T., Drozdowicz, D., Szlachcic, A., Pytko-Polonczyk, J., Majka, J., and Konturek, S., Role of nitric oxide and prostaglandins in gastroprotection induced by capsaicin and papaverine, *Digestion*, 54, 24, 1993.
23. Grönbech, J.E. and Lacy, E.R., Substance P attenuates gastric mucosal hyperemia after stimulation of sensory neurons in the rat stomach, *Gastroenterology*, 106, 440, 1994.
24. Thiefin, G., Raybould, H.E., Leung, F.W., Taché, Y., and Guth, P.H., Capsaicin-sensitive afferent fibers contribute to gastric mucosal blood flow response to electrical vagal stimulation, *Am. J. Physiol.*, 259, G1037, 1990.

25. Leung, F.W., Inhibition of spinal afferent nerve-mediated gastric hyperemia by nicotine: role of ganglionic blockade, *Am. J. Physiol.*, 264, H1087, 1993.
26. Lippe, I.Th., Pabst, M.A., and Holzer, P., Intragastric capsaicin enhances rat gastric acid elimination and mucosal blood flow by afferent nerve stimulation, *Br. J. Pharmacol.*, 96, 91, 1989.
27. Matsumoto, J., Takeuchi, K., and Okabe, S., Characterization of gastric mucosal blood flow response induced by intragastric capsaicin in rats, *Jpn. J. Pharmacol.*, 57, 205, 1991.
28. Holzer, P., Peskar, B.M., Peskar, B.A., and Amann, R., Release of calcitonin gene-related peptide induced by capsaicin in the vascularly perfused rat stomach, *Neurosci. Lett.*, 108, 195, 1990.
29. Geppetti, P., Tramontana, M., Evangelista, S., Renzi, D., Maggi, C.A., Fusco, B.M., and Del Bianco, E., Differential effect on neuropeptide release of different concentrations of hydrogen ions on afferent and intrinsic neurons of the rat stomach, *Gastroenterology*, 101, 1505, 1991.
30. Inui, T., Kinoshita, Y., Yamaguchi, A., Yamatani, T., and Chiba, T., Linkage between capsaicin-stimulated calcitonin gene-related peptide and somatostatin release in rat stomach, *Am. J. Physiol.*, 261, G770, 1991.
31. Ren, J.Y., Young, R.L., Lassiter, D.C., and Harty, R.F., Calcitonin gene-related peptide mediates capsaicin-induced neuroendocrine responses in rat antrum, *Gastroenterology*, 104, 485, 1993.
32. Gray, J.L., Bunnett, N.W., Orloff, S.L., Mulvihill, S.J., and Debas, H.T., A role for calcitonin gene-related peptide in protection against gastric ulceration, *Ann. Surg.*, 219, 58, 1994.
33. Lippe, I.Th., Lorbach, M., and Holzer, P., Close arterial infusion of calcitonin gene-related peptide into the rat stomach inhibits aspirin- and ethanol-induced hemorrhagic damage, *Regul. Pept.*, 26, 35, 1989.
34. Holzer, P. and Guth, P.H., Neuropeptide control of rat gastric mucosal blood flow. Increase by calcitonin gene-related peptide and vasoactive intestinal polypeptide, but not substance P and neurokinin A, *Circ. Res.*, 68, 100, 1991.
35. Holzer, P., Lippe, I.Th., Jocic, M., Wachter, Ch., Erb, R., and Heinemann, A., Nitric oxide-dependent and -independent hyperemia due to calcitonin gene-related peptide in the rat stomach, *Br. J. Pharmacol.*, 110, 404, 1993.
36. Holzer, P., Wachter, Ch., Jocic, M., and Heinemann, A., Vascular bed-dependent roles of the peptide CGRP and nitric oxide in acid-evoked hyperemia of the rat stomach, *J. Physiol. (London)*, 480, 575, 1994.
37. Lopez-Belmonte, J. and Whittle, B.J.R., The paradoxical vascular interactions between endothelin-1 and calcitonin gene-related peptide in the rat gastric mucosal microcirculation, *Br. J. Pharmacol.*, 110, 496, 1993.
38. Gates, T.S., Zimmerman, R.P., Mantyh, C.R., Vigna, S.R., and Mantyh, P.W., Calcitonin gene-related peptide-α receptor binding sites in the gastrointestinal tract, *Neuroscience*, 31, 757, 1989.
39. Forster, E.R. and Southam, E., The intrinsic and vagal extrinsic innervation of the rat stomach contains nitric oxide synthase, *NeuroReport*, 4, 275, 1993.
40. Holzer, P. and Jocic, M., Cutaneous vasodilatation induced by nitric oxide-evoked stimulation of afferent nerves in the rat, *Br. J. Pharmacol.*, 112, 1181, 1994.
41. Renzi, D., Evangelista, S., Mantellini, P., Santicioli, P., Maggi, C.A., Geppetti, P., and Surrenti, C., Capsaicin-induced release of neurokinin A from muscle and mucosa of gastric corpus: correlation with capsaicin-evoked release of calcitonin gene-related peptide, *Neuropeptides*, 19, 137, 1991.
42. Kwok, Y.N. and McIntosh, C.H.S., Release of substance P-like immunoreactivity from the vascularly perfused rat stomach, *Eur. J. Pharmacol.*, 180, 201, 1990.
43. Livingston, E.H. and Holzer, P., Gastric hyperemia accompanying acid secretion is not mediated by sensory nerves, *Dig. Dis. Sci.*, 38, 1190, 1993.
44. Piqué, J.M., Esplugues, J.V., and Whittle, B.J.R., Influence of morphine or capsaicin pretreatment on rat gastric microcirculatory response to PAF, *Am. J. Physiol.*, 258, G352, 1990.
45. Tepperman, B.L. and Whittle, B.J.R., Endogenous nitric oxide and sensory neuropeptides interact in the modulation of the rat gastric microcirculation, *Br. J. Pharmacol.*, 105, 171, 1992.
46. Lundberg, J.M., Brodin, E., Hua, X.-Y., and Saria, A., Vascular permeability changes and smooth muscle contraction in relation to capsaicin-sensitive substance P afferents in the guinea-pig, *Acta Physiol. Scand.*, 120, 217, 1984.
47. Saria, A., Lundberg, J.M., Skofitsch, G., and Lembeck, F., Vascular protein leakage in various tissues induced by substance P, capsaicin, bradykinin, serotonin, histamine and by antigen challenge, *Naunyn-Schmiedeberg's Arch. Pharmacol.*, 324, 212, 1983.
48. Holzer, P. and Lippe, I.Th., Stimulation of afferent nerve endings by intragastric capsaicin protects against ethanol-induced damage of gastric mucosa, *Neuroscience*, 27, 981, 1988.
49. Sullivan, T.R., Milner, R., Dempsey, D.T., and Ritchie, W.P., Effect of capsaicin on gastric mucosal injury and blood flow following bile acid exposure, *J. Surg. Res.*, 52, 596, 1992.
50. Takeuchi, K., Ueshima, K., Matsumoto, J., and Okabe, S., Role of capsaicin-sensitive sensory nerves in acid-induced bicarbonate secretion in rat stomach, *Dig. Dis. Sci.*, 37, 737, 1992.
51. Szolcsányi, J. and Barthó, L., Impaired defense mechanism to peptic ulcer in the capsaicin-desensitized rat, in *Gastrointestinal Defense Mechanisms*, Mózsik, G., Hänninen, O., and Jávor, T., Eds., Pergamon Press, Oxford and Akadémiai Kiadó, Budapest, 1981, 39.

52. Alföldi, P., Obal, F., Toth, E., and Hideg, J., Capsaicin pretreatment reduces the gastric acid secretion elicited by histamine but does not affect the responses to carbachol and pentagastrin, *Eur. J. Pharmacol.*, 123, 321, 1986.
53. Dugani, A.M. and Glavin, G.B., Capsaicin effects on stress pathology and gastric acid secretion in rats, *Life Sci.*, 39, 1531, 1986.
54. Holzer, P. and Sametz, W., Gastric mucosal protection against ulcerogenic factors in the rat mediated by capsaicin-sensitive afferent neurons, *Gastroenterology*, 91, 975, 1986.
55. Raybould, H.E. and Taché, Y., Capsaicin-sensitive vagal afferent fibers and stimulation of gastric acid secretion in anesthetized rats, *Eur. J. Pharmacol.*, 167, 237, 1989.
56. Esplugues, J.V., Ramos, E.G., Gil, L., and Esplugues, J., Influence of capsaicin-sensitive afferent neurones on the acid secretory responses of the rat stomach in vivo, *Br. J. Pharmacol.*, 100, 491, 1990.
57. Evangelista, S., Santicioli, P., Maggi, C.A., and Meli, A., Increase in gastric secretion induced by 2-deoxy-D-glucose is impaired in capsaicin pretreated rats, *Br. J. Pharmacol.*, 98, 35, 1989.
58. Uchida, M., Yano, S., and Watanabe, K., Involvement of CGRP, substance P and blood circulation in aggravating mechanism of absolute ethanol-induced antral lesions by capsaicin treatment in rats, *Jpn. J. Pharmacol.*, 62, 123, 1993.
59. Holzer, P., Livingston, E.H., and Guth, P.H., Sensory neurons signal for an increase in rat gastric mucosal blood flow in the face of pending acid injury, *Gastroenterology*, 101, 416, 1991.
60. Takeuchi, K., Matsumoto, J., Ueshima, K., Ohuchi, T., and Okabe, S., Induction of duodenal ulcers in sensory deafferented rats following histamine infusion, *Digestion*, 51, 203, 1992.
61. Sharkey, K.A., Oland, L.D., Kirk, D.R., and Davison, J.S., Capsaicin-sensitive vagal stimulation-induced gastric acid secretion in the rat: evidence for cholinergic vagal afferents, *Br. J. Pharmacol.*, 103, 1997, 1991.
62. Holzer, P., Calcitonin gene-related peptide, in *Gut Peptides: Biochemistry and Physiology*, Walsh, J.H. and Dockray, G.J., Eds., Raven Press, New York, 1994, 493.
63. Ramos, E.G., Esplugues, J., and Esplugues, J.V., Gastric acid secretory responses induced by peptone are mediated by capsaicin-sensitive sensory afferent neurons, *Am. J. Physiol.*, 262, G835, 1992.
64. Lefebvre, R.A., De Beurme, F.A., and Sas, S., Relaxant effect of capsaicin in the rat gastric fundus, *Eur. J. Pharmacol.*, 195, 131, 1991.
65. Holzer-Petsche, U., Seitz, H., and Lembeck, F., Effect of capsaicin on gastric corpus smooth muscle of the rat in vitro, *Eur. J. Pharmacol.*, 162, 29, 1989.
66. Horowitz, M., Wishart, J., Maddox, A., and Russo, A., The effect of chilli on gastrointestinal transit, *J. Gastroenterol. Hepatol.*, 7, 52, 1992.
67. Holzer, P., Lippe, I.Th., and Holzer-Petsche, U., Inhibition of gastrointestinal transit due to surgical trauma or peritoneal irritation is reduced in capsaicin-treated rats, *Gastroenterology*, 91, 360, 1986.
68. Forster, E.R., Green, T., Elliot, M., Bremner, A., and Dockray, G.J., Gastric emptying in rats: role of afferent neurons and cholecystokinin, *Am. J. Physiol.*, 258, G552, 1990.
69. Holzer-Petsche, U., Blood pressure and gastric motor responses to bradykinin and hydrochloric acid injected into somatic or visceral tissues, *Naunyn-Schmiedeberg's Arch. Pharmacol.*, 346, 219, 1992.
70. Barquist, E., Zinner, M., Rivier, J., and Taché, Y., Abdominal surgery-induced delayed gastric emptying in rats: role of CRF and sensory neurons, *Am. J. Physiol.*, 262, G616, 1992.
71. Raybould, H.E., Capsaicin-sensitive vagal afferents and CCK in inhibition of gastric motor function induced by intestinal nutrients, *Peptides*, 12, 1279, 1991.
72. Raybould, H.E. and Taché, Y., Cholecystokinin inhibits gastric motility and emptying via a capsaicin-sensitive vagal pathway in rats, *Am. J. Physiol.*, 255, G242, 1988.
73. Holzer-Petsche, U., Modulation of gastric contractions in response to tachykinins and bethanechol by extrinsic nerves, *Br. J. Pharmacol.*, 103, 1958, 1991.
74. Holzer, P., Lippe, I.Th., and Amann, R., Participation of capsaicin-sensitive afferent neurons in gastric motor inhibition caused by laparotomy and intraperitoneal acid, *Neuroscience*, 48, 715, 1992.
75. Plourde, V., Wong, H.C., Walsh, J.H., Raybould, H.E., and Taché, Y., CGRP antagonists and capsaicin on celiac ganglia partly prevent postoperative gastric ileus, *Peptides*, 14, 1225, 1993.
76. Matsumoto, J., Ueshima, K., Ohuchi, T., Takeuchi, K., and Okabe, S., Induction of gastric lesions by 2-deoxy-D-glucose in rats following chemical ablation of capsaicin-sensitive sensory neurons, *Jpn. J. Pharmacol.*, 60, 43, 1992.
77. Evangelista, S., Maggi, C.A., Giuliani, S., and Meli, A., Further studies on the role of the adrenals in the capsaicin-sensitive "gastric defense mechanism", *Int. J. Tissue React.*, 10, 253, 1988.
78. Whittle, B.J.R., Lopez-Belmonte, J., and Moncada, S., Regulation of gastric mucosal integrity by endogenous nitric oxide: interactions with prostanoids and sensory neuropeptides in the rat, *Br. J. Pharmacol.*, 99, 607, 1990.
79. Whittle, B.J.R. and Lopez-Belmonte, J., Interactions between the vascular peptide endothelin-1 and sensory neuropeptides in gastric mucosal injury, *Br. J. Pharmacol.*, 102, 950, 1991.
80. Takeuchi, K., Ohuchi, T., Narita, M., and Okabe, S., Capsaicin-sensitive sensory nerves in recovery of gastric mucosal integrity after damage by sodium taurocholate in rats, *Jpn. J. Pharmacol.*, 63, 479, 1993.

81. Esplugues, J.V. and Whittle, B.J.R., Morphine potentiation of ethanol-induced gastric mucosal damage in the rat. Role of local sensory afferent neurons, *Gastroenterology*, 98, 82, 1990.
82. Pabst, M.A., Schöninkle, E., and Holzer, P., Ablation of capsaicin-sensitive afferent nerves impairs defense but not rapid repair of rat gastric mucosa, *Gut*, 34, 897, 1993.
83. Evangelista, S. and Maggi, C.A., Protection induced by cholecystokinin-8 (CCK-8) in ethanol-induced gastric lesions is mediated via vagal capsaicin-sensitive fibers and CCK-A receptors, *Br. J. Pharmacol.*, 102, 119, 1991.
84. Yonei, Y., Holzer, P., and Guth, P.H., Laparotomy-induced gastric protection against ethanol injury is mediated by capsaicin-sensitive sensory neurons, *Gastroenterology*, 99, 3, 1990.
85. Esplugues, J.V., Whittle, B.J.R., and Moncada, S., Modulation by opioids and by afferent sensory neurones of prostanoid protection of the rat gastric mucosa, *Br. J. Pharmacol.*, 106, 846, 1992.
86. Holzer, P., Pabst, M.A., and Lippe, I.Th., Intragastric capsaicin protects against aspirin-induced lesion formation and bleeding in the rat gastric mucosa, *Gastroenterology*, 96, 1425, 1989.
87. Lambrecht, N., Burchert, M., Respondek, M., Müller, K.M., and Peskar, B.M., Role of calcitonin gene-related peptide and nitric oxide in the gastroprotective effect of capsaicin in the rat, *Gastroenterology*, 104, 1371, 1993.
88. Uchida, M., Yano, S., and Watanabe, K., The role of capsaicin-sensitive afferent nerves in protective effect of capsaicin against absolute ethanol-induced gastric lesions in rats, *Jpn. J. Pharmacol.*, 55, 279, 1991.
89. Evangelista, S., Tramontana, M., and Maggi, C.A., Pharmacological evidence for the involvement of multiple calcitonin gene-related peptide (CGRP) receptors in the antisecretory and antiulcer effect of CGRP in rat stomach, *Life Sci.*, 50, PL13, 1991.
90. Evangelista, S., Lippe, I.Th., Rovero, P., Maggi, C.A., and Meli, A., Tachykinins protect against ethanol-induced gastric lesions in rats, *Peptides*, 10, 79, 1989.
91. Karmeli, F., Eliakim, R., Okon, E., and Rachmilewitz, D., Gastric mucosal damage by ethanol is mediated by substance P and prevented by ketotifen, a mast cell stabilizer, *Gastroenterology*, 100, 1206, 1991.
92. Evangelista, S., Maggi, C.A., Rovero, P., Patacchini, R., Giuliani, S., and Giachetti, A., Analogs of neurokinin A(4-10) afford protection against gastroduodenal ulcers in rats, *Peptides*, 11, 293, 1990.
93. Peskar, B.M., Wong, H.C., Walsh, J.H., and Holzer, P., A monoclonal antibody to calcitonin gene-related peptide abolishes capsaicin-induced gastroprotection, *Eur. J. Pharmacol.*, 250, 201, 1993.
94. Forster, E.R. and Dockray, G.J., The role of calcitonin gene-related peptide in gastric mucosal protection in the rat, *Exp. Physiol.*, 76, 623, 1991.
95. Lopez-Belmonte, J., Whittle, B.J.R., and Moncada, S., The actions of nitric oxide donors in the prevention or induction of injury to the rat gastric mucosa, *Br. J. Pharmacol.*, 108, 73, 1993.
96. Takeuchi, K., Ueshima, K., Ohuchi, T., and Okabe, S., The role of capsaicin-sensitive sensory neurons in healing of HCl-induced gastric mucosal lesions in rats, *Gastroenterology*, 106, 1524, 1994.
97. Tramontana, M., Renzi, D., Calabro, A., Panerai, C., Milani, S., Surrenti, C., and Evangelista, S., Influence of capsaicin-sensitive afferent fibers on acetic acid-induced chronic gastric ulcers in rats, *Scand. J. Gastroenterol.*, 29, 406, 1994.
98. Peskar, B.M., Lambrecht, N., Stroff, T., Respondek, M., and Müller, K.-M., Functional ablation of sensory neurons impairs the healing of acute gastric mucosal damage in rats, *Dig. Dis. Sci.*, in press, 1995.
99. Raybould, H.E., Sternini, C., Eysselein, V.E., Yoneda, M., and Holzer, P., Selective ablation of spinal afferent neurons containing CGRP attenuates gastric hyperemic response to acid, *Peptides*, 13, 249, 1992.
100. Holzer, P. and Lippe, I.Th., Gastric mucosal hyperemia due to acid back-diffusion depends on splanchnic nerve activity, *Am. J. Physiol.*, 262, G505, 1992.
101. Li, D.-S., Raybould, H.E., Quintero, E., and Guth, P.H., Calcitonin gene-related peptide mediates the gastric hyperemic response to acid back-diffusion, *Gastroenterology*, 102, 1124, 1992.
102. Lippe, I.Th. and Holzer, P., Participation of endothelium-derived nitric oxide but not prostacyclin in the gastric mucosal hyperemia due to acid back-diffusion, *Br. J. Pharmacol.*, 105, 708, 1992.
103. Pethö, G., Jocic, M., and Holzer, P., Role of bradykinin in the hyperemia following acid challenge of the rat gastric mucosa, *Br. J. Pharmacol.*, 113, 1036, 1994.
104. Lippe, I.Th., Sametz, W., Sabin, K., and Holzer, P., Inhibitory role of capsaicin-sensitive afferent neurons and nitric oxide in hemostasis, *Am. J. Physiol.*, 265, H1864, 1993.
105. Battistel, M., Plebani, M., Di Mario, F., Jocic, M., Lippe, I.Th., and Holzer, P., Chronic nicotine intake causes vascular dysregulation in the rat gastric mucosa, *Gut*, 34, 1688, 1993.
106. Nishizaki, Y., Kaunitz, J.D., Oda, M., and Guth, P.H., Impairment of gastric mucosal defenses measured in vivo in cirrhotic rats, *Hepatology*, 20, 445, 1994.
107. Quintero, E., Kaunitz, J., Nishizaki, Y., De Giorgio, R., Sternini, C., and Guth, P.H., Uremia increases gastric mucosal permeability and acid back-diffusion injury in the rat, *Gastroenterology*, 103, 1762, 1992.
108. Grönbech, J.E. and Lacy, E.R., Impaired gastric defense mechanisms in aged rats: role of sensory neurons, blood flow, restitution, and prostaglandins, *Gastroenterology*, 106, A84, 1994.
109. Miyake, H., Takeuchi, K., and Okabe, S., Derangement of gastric mucosal blood flow responses in aged rats: relation to capsaicin-sensitive sensory neurons, *Gastroenterology*, 106, A141, 1994.

Chapter 13

SENSORY NEURONS IN THE INTESTINE

Peter Holzer and Lorand Barthó

CONTENTS

I. INTRODUCTION

The regulation of intestinal functions by the sympathetic and parasympathetic divisions of the extrinsic autonomic nervous system and the intrinsic neurons of the enteric system has long been established. The functional roles of the extrinsic primary afferent neurons in the gut have been comparatively less studied, and it was only fairly recently that the relative abundance of afferent fibers in the extrinsic nerves supplying the gut (70 to 90% afferent fibers in the vagus nerves, 7 to 20% afferent fibers in the splanchnic nerves, 30 to 34% afferent fibers in the pelvic nerves) was fully recognized.[1-3] Apart from their implication in visceral sensation and in the autonomic reflex regulation of digestive activity, it has also increasingly been appreciated that extrinsic afferent neurons exert a direct control over a number of intestinal functions, a role that is embodied in the term "neurogenic inflammation". Much as is the case in the stomach (see Chapter 12), these neurons may regulate blood flow, secretory processes, and motor activity of the intestine in a homeostatic fashion. This chapter briefly reviews the roles that afferent neurons play locally in the gut, describes the elucidation of these roles by the use of capsaicin, addresses the mechanisms by which afferent neurons participate in intestinal physiology, and points out their pathophysiological implications.

II. INNERVATION OF THE INTESTINE BY PEPTIDERGIC AFFERENT NERVES

The gut is supplied by vagal and spinal afferent neurons. The spinal afferents originate in the dorsal root ganglia and reach the gut via sympathetic (splanchnic, colonic, and hypogastric) and sacral parasympathetic (pelvic) nerves while passing through prevertebral ganglia and forming collateral synapses with sympathetic ganglion cells.[3-5] The vagal afferents have their cell bodies in the nodose ganglia and supply the digestive tract down to the transverse colon.[6-9] Most of the spinal afferents from the gut contain bioactive peptides, including calcitonin gene-related peptide (CGRP), the tachykinins substance P (SP) and neurokinin A (NKA), and many others,[4,5,10-19] and are sensitive to the neurotoxic action of capsaicin.[17-19] The colocalization, density, and presence of these peptides vary between species and regions of the digestive tract. The identity of the transmitters in the vagal afferents from the gut has not yet been fully explored (see Chapter 12).

0-8493-7646-7/96/$0.00+$.50

As in the stomach, afferent nerve fibers expressing CGRP and SP project primarily to arteries and arterioles in the intestine, but supply myenteric and submucosal plexus, circular muscle, and mucosa as well.[10,12,14-22] In addition, afferent nerve fibers immunoreactive for CGRP and SP are present in all layers of the gallbladder and biliary system and are particularly numerous around blood vessels.[22,23] Peptidergic afferent axons are also found in the pancreas,[15,18,21] in which they supply the vasculature, the exocrine and endocrine compartments, and the intrapancreatic ganglia in a species-related manner.[24,25]

The fact that peptide-containing nerve fibers in the digestive system arise from three different groups of neurons (extrinsic afferent, extrinsic autonomic, and intrinsic enteric neurons[13,26]) is a complicating factor in the study of any of these neuron populations. However, as outlined in Chapter 12, extrinsic afferent neurons differ not only in terms of their projections but also with regard to their exclusive sensitivity to the excitatory and neurotoxic actions of capsaicin[27,28] and their chemical coding. Thus, SP and CGRP are coexpressed in afferent nerves only, but do not coexist in enteric neurons,[13,23] and the chemical identity of peptide transmitters such as CGRP is different, because most CGRP present in afferent neurons is CGRP-α, whereas the only form of CGRP in enteric neurons is CGRP-β.[29,30]

III. REGULATION OF VASCULAR FUNCTIONS

Splanchnic resistance vessels receive a dense supply by perivascular afferent nerve fibers immunoreactive for CGRP and SP, which run primarily in the connective tissue surrounding the vessels and in close proximity to the vascular smooth muscle.[10,11,20,22] There is ample evidence to conclude that the nonadrenergic noncholinergic (NANC) dilatation of mesenteric arteries in response to electrical or chemical stimulation of perivascular nerves is mediated by capsaicin-sensitive afferent nerve fibers.[31] Capsaicin-evoked activation of these fibers dilates the superior mesenteric artery of the rat[32] and dog[33] and thus mimicks both the hyperpolarizing[34] and dilator response to electrical NANC nerve stimulation in the mesenteric,[34-36] hepatic, and splenic[37] arteries of the rat and guinea pig. Conversely, capsaicin-induced defunctionalization of afferent nerve fibers inhibits the mesenteric vasodilator response to NANC nerve stimulation,[34,37-40] the autoregulatory escape from sympathetic vasoconstriction,[41] the hyperemia following sympathetic nerve stimulation,[42] the reactive hyperemia following mesenteric arterial occlusion,[43] the hyperemia caused by intrajejunal bile-oleate,[44] and the hyperemia caused by intestinal warming.[45] The afferent nerve-mediated rise of mesenteric blood flow is aided by inhibition of haemostasis.[46]

There is good reason to assume that CGRP is the major transmitter of neurogenic dilatation of the mesenteric arteries.[31] CGRP is released by stimulation of mesenteric perivascular nerves with electrical impulses or capsaicin application[36,47-49] and is most potent in dilating precontracted mesenteric, splenic, or hepatic arteries isolated from rats,[20,35,38-40,50,51] guinea pigs,[20] rabbits,[52,53] and humans.[54] Like the dilator response to perivascular nerve stimulation, the dilatation due to CGRP is independent of the endothelium in all these vessels.[37,51,53,55] The mediator role of CGRP has been corroborated by the findings that the neurogenic dilatation of rat mesenteric arteries is blocked by desensitization to CGRP,[56] the CGRP antagonist $CGRP_{8-37}$,[48,56,57] and a polyclonal antibody to CGRP.[38]

Although SP can be released from perivascular nerves,[47,49] tachykinins acting via NK_1 and NK_3 receptors have been ruled out as mediators of neurogenic dilatation in the rat mesenteric arteries,[57] and SP and NKA are inactive in dilating the rat isolated mesenteric vascular bed.[11,22,39,55,57,58] SP does play a role, however, in the NANC constriction of rat mesenteric veins.[57,58] Unlike mesenteric arteries from the rat, mesenteric arteries isolated from guinea pigs,[59] rabbits,[60] and humans[54] are relaxed by SP in an endothelium-dependent manner, and *in vivo* hyperemic effects of SP have been seen in the dog[33] and pig[61] mesenteric artery. The proposed role of SP, vasoactive intestinal polypeptide, and cholecystokinin in the mesenteric vasodilator effect of capsaicin in the dog[33] remains to be confirmed.

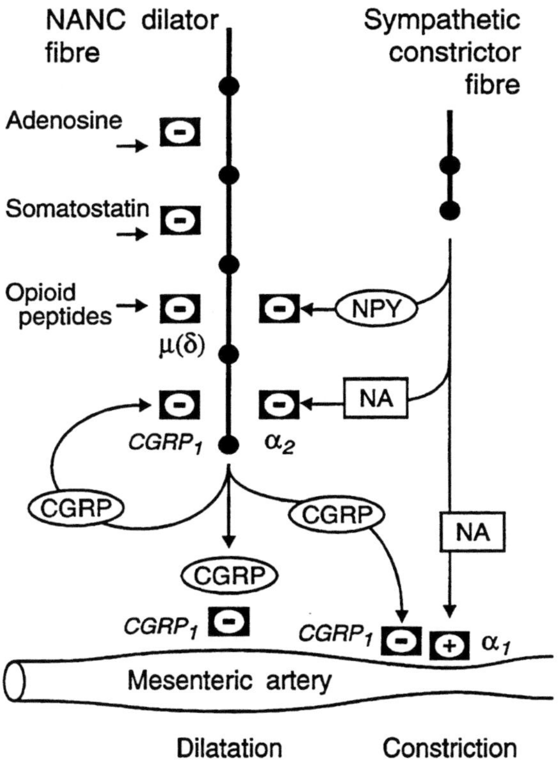

FIGURE 1. Diagram illustrating the neural control of mesenteric arterial tone, in which primary afferent NANC vasodilator nerve fibers releasing CGRP play a central role. Sympathetic neurons can inhibit the release of the vasodilator peptide CGRP by way of prejunctional α_2 adrenoceptors and receptors for neuropeptide Y (NPY), while CGRP inhibits the vasoconstrictor effect of noradrenaline (NA) by a postjunctional site of action. In addition, the release of CGRP is under the inhibitory control of CGRP autoreceptors, prejunctional μ/δ opioid receptors, prejunctional somatostatin receptors, and prejunctional adenosine A_1 receptors.

The CGRP-containing afferent nerve fibers in the rat mesenteric arteries are part of a complex vascular control system (Figure 1). Mesenteric arterial tone is regulated not only by afferent vasodilator fibers but also by sympathetic vasoconstrictor fibers containing noradrenaline and neuropeptide Y and by other (possibly enteric) nerves releasing opioid peptides, somatostatin, vasoactive intestinal polypeptide, and adenosine, these components interacting with each other in a mutual fashion.[31] On the one hand, the afferent CGRP-containing fibers increase mesenteric blood flow both by causing vasodilatation themselves and by inhibiting the activity of sympathetic vasoconstrictor nerves at a postjunctional level.[34,37-39,41,55,62] On the other hand, noradrenaline released from sympathetic nerve fibers reduces NANC vasodilatation by an action on presynaptic α_2-adrenoceptors, thereby inhibiting the release of CGRP from afferent nerve fibers.[48,63] This presynaptic action of noradrenaline is shared by the sympathetic cotransmitter neuropeptide Y.[40,48] Other factors that inhibit the release of CGRP from NANC vasodilator fibers at a prejunctional level include endogenous opioid peptides acting via prejunctional μ and possibly δ opioid receptors,[40,64,65] somatostatin,[33] and adenosine acting via A_1 purinoceptors.[66] In addition, the CGRP-containing vasodilator fibers in the rat mesenteric arteries bear CGRP autoreceptors which regulate transmitter release via a negative feedback mechanism.[67]

The role of afferent nerve fibers in the intestinal microcirculation is still little studied. Local exposure of the intestine to capsaicin increases blood flow in the mesenteric arteries of the dog[68] and rat,[32,44] and in the mucosa of the rat duodenum[69] and colon,[70] and dilates submucosal arterioles in the guinea pig ileum.[71] Capsaicin-sensitive afferent neurons are involved in the hyperemic response of the rat duodenal mucosa to acid challenge[69] and the rat colonic mucosa to irritation with acetic acid.[70] Capsaicin-evoked afferent nerve stimulation releases CGRP from the guinea pig gallbladder[72] and rabbit colon,[73] and SP from the guinea pig small intestine,[74] but the vasodilator mediators released from afferent nerve fibers in the intestinal mucosa have not yet been identified. Mucosal blood flow in the rabbit duodenum is augmented by human CGRP-α, but not CGRP-β,[75] whereas submucosal arterioles in the guinea pig small intestine are dilated by human CGRP-β,[71] but not CGRP-α.[76] Mucosal blood flow in the rat small and large intestine is not altered by rat CGRP-α,[51] although CGRP-α receptors are present throughout the rat digestive tract.[77] Conversely, SP is able to increase blood flow in the feline small intestine[78] and colon[79] and to dilate submucosal arterioles in the guinea pig isolated ileum.[71,76]

Neurogenic inflammation in nonvisceral tissues involves a marked increase in venular permeability leading to extravasation of macromolecules, leukocytes, and fluid.[31] To the contrary, electrical or capsaicin-induced stimulation of intestinal afferent nerve fibers in the vagus, splanchnic, and pelvic nerves fails to increase vascular permeability in the rat small and proximal large intestine to any significant extent, whereas protein extravasation is clearly seen in the rectum, biliary system, and mesentery.[80-83] A mediator role of tachykinins in afferent nerve-mediated protein leakage can be deduced from the finding that the regional activity of SP in increasing vascular permeability is similar to that of electrical or chemical stimulation of afferent nerves.[80-82] A role of sensory neuropeptides in the adhesion of leukocytes to the endothelium[84] has not yet been studied in the intestinal circulation.

IV. REGULATION OF SECRETORY PROCESSES

The digestive activity of the intestine and pancreas includes the secretion of fluid, ions, and enzymes, and there is increasing evidence that capsaicin-sensitive afferent neurons participate in the regulation of these secretory processes. In the rat duodenum, sensory nerve activation by capsaicin increases the output of bicarbonate[85,86] and the secretion of mucus from goblet cells, an action that is mimicked by SP.[87] Capsaicin-induced defunctionalization of afferent neurons does not affect basal alkaline secretion of the rat duodenum,[69,85,86] but inhibits the stimulation of bicarbonate output caused by intraluminal acidification.[85,86]

Exposure of the rat colon to capsaicin acting from the serosal side first enhances and then lowers ion secretion as determined by the short circuit current method.[88,89] The secretory response to capsaicin is blocked by desensitization to SP,[89] which is consistent with the effect of tachykinins to stimulate intestinal electrolyte secretion both by an indirect action involving enteric neurons and a direct action on intestinal epithelial cells.[90] Whether CGRP plays a role as well has not yet been possible to show because the antisecretory effect of low concentrations of CGRP and the secretory effect of higher concentrations of the peptide[91] are not antagonized by $CGRP_{8-37}$ or CGRP desensitization.[89] Secretory effects of CGRP are also seen in the rat colon *in vivo*[92] and in the guinea pig colon *in vitro*.[93] Pathophysiologically relevant are the findings that capsaicin-sensitive afferent neurons participate in the delayed secretory response to bradykinin in the rat colon[88] and in the secretory and inflammatory effects of *Clostridium difficile* toxin A in the rat ileum[94] through a mechanism that involves tachykinins acting via NK_1 receptors.[95] In contrast, the secretory response to cholera toxin depends on intrinsic enteric,[96] but not extrinsic afferent,[94] neurons.

There is evidence that pancreatic secretion is also under the control of capsaicin-sensitive afferents.[97,98] Intraduodenal application of capsaicin enhances the exocrine secretion from the pancreas via a cholecystokinin-dependent mechanism.[97] The secretory effect of endogenous cholecystokinin is likewise brought about by a vagovagal reflex involving capsaicin-sensitive afferent neurons,[98] whereas the pancreatic secretion caused by 2-deoxy-D-glucose[97,98] and insulin[24] is not altered by sensory nerve ablation. Capsaicin-sensitive afferent nerve fibers, however, participate in the secretion of glucagon and in the hyperglycemia caused by 2-deoxy-D-glucose in the mouse.[24] The physiological relevance of the inhibitory action of CGRP on pancreatic secretion[99,100] is not yet understood.

V. REGULATION OF MOTOR ACTIVITY

An important aspect of intestinal function is the digestion-related regulation of motor activity. Capsaicin-sensitive afferent nerve fibers do not seem to be overtly implicated in the physiological control of propulsive motility in the rat gut *in vivo* but can modulate intestinal motor activity under certain conditions.[101-109] On the one hand, these afferents participate in the reflex sympathetic inhibition of intestinal propulsion after laparotomy or peritoneal irritation.[102,104,105,107] On the other hand, local release of sensory neuropeptides in the intestinal wall influences motor activity in a complex manner. Sensory nerve stimulation with capsaicin exerts both excitatory and inhibitory effects on the motility of the longitudinal and circular muscle as studied in isolated segments of the guinea pig,[110-115] rabbit, [73,116] rat,[117] and human[118-120] small and large intestine. Similar effects are elicited by electrical stimulation of intestinal afferents when sympathetic neurotransmission has been blocked. The ability of mesenteric nerve stimulation to cause contraction[111,116,121-125] and/or relaxation of the precontracted or spontaneously active muscle[126] in the guinea pig, rat, and rabbit intestine is prevented by capsaicin-induced defunctionalization of afferent neurons.

The mechanisms that underlie the motor effects of sensory nerve stimulation in the intestine are multifactorial. If all available information is taken together it can be concluded that capsaicin and mesenteric nerve stimulation first release transmitter substances from afferent nerve endings, which in turn activate excitatory enteric motor neurons, releasing acetylcholine and tachykinins, and inhibitory enteric motor neurons, releasing NANC transmitters, and which, in addition, themselves influence the activity of intestinal muscle (Figure 2). Both the electrical and capsaicin-evoked contraction of the intestine involves activation of enteric neurons, as shown by depolarization of myenteric neurons,[127,128] by release of acetylcholine from the myenteric plexus,[73,129] and by inhibition of the motor response by atropine/hyoscine.[73,110,111,113,114,116,121-124,130-132] This circumstance obscures the source of SP[72-74] and CGRP,[72,73,133] which are released by capsaicin from the intestinal wall, because in the gut most of the SP[10,13,17,134] and CGRP[15,19,21] is contained in enteric, but not necessarily afferent, neurons. The contribution of tachykinins to the contractile effect of electrical and capsaicin-evoked afferent nerve stimulation in the guinea pig, rat, and rabbit intestine is well documented, as both the cholinergic[73,114,122,124,130,131,136] and particularly the noncholinergic components[123,130-132,135] of the contractile response to sensory nerve activation are inhibited by SP desensitization and tachykinin antagonists, the tachykinin receptors involved being of the NK_1 and NK_2 types.[135] The mediator role of tachykinins is consistent with the established action of these peptides to stimulate intestinal motor activity both by activation of enteric neurons and a direct excitatory effect on the muscle.[27,90]

A synopsis of the described effects of atropine/hyoscine, tachykinin antagonists, and tetrodotoxin[130,135] has led to the proposal that the tachykinins involved in the motor response to capsaicin are released primarily from enteric neurons and activate NK_1/NK_2 receptors on the intestinal smooth muscle.[135] As enteric tachykinin-releasing neurons are themselves not sensitive to capsaicin,[137] it is held that excitation of enteric neurons is the consequence of transmitter release from afferent nerve fibers.[110,116,130,135] The identity of these transmitters is uncertain, although tachykinins have long been suspected to be involved, given that activation of NK_3 receptors is known to release acetylcholine[138] and tachykinins[139] from enteric neurons. However, a novel nonpeptide antagonist of NK_3 receptors, SR-142,801,[138] fails to block the contractile motor response of the guinea pig ileum to capsaicin.[140] Hence it has to be inferred, at the present stage, that the afferent nerve-derived transmitters which stimulate enteric

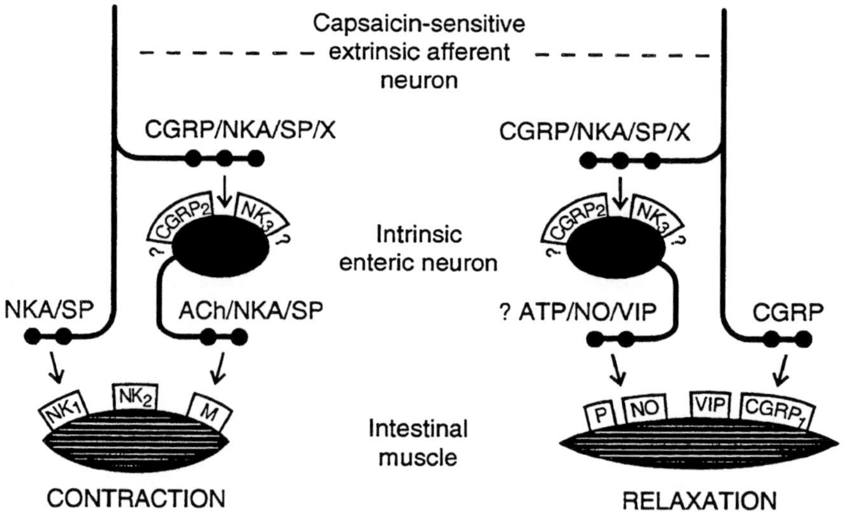

FIGURE 2. Diagram illustrating the potential transmitter mechanisms by which capsaicin-sensitive afferent neurons control motor activity of the guinea-pig,[113,130,135,140,142] rat,[117] and rabbit[73] intestine. Stimulation of extrinsic afferent nerve fibers releases transmitter substances (CGRP, NKA, SP, and X = unknown substances) which stimulate excitatory and/or inhibitory enteric motor neurons. At present there is no positive evidence that CGRP ($CGRP_2$)[142] or NK_3 receptors[140] on enteric neurons are stimulated by afferent nerve-derived peptides. The excitatory enteric neurons cause contraction of the intestinal muscle via release of acetylcholine (ACh) and tachykinins[130,135] and stimulation of muscarinic (M) acetylcholine[130] and tachykinin (NK_1 and NK_2)[135] receptors. The inhibitory enteric neurons may relax the muscle via release of adenosine triphosphate (ATP), nitric oxide (NO), and/or vasoactive intestinal polypeptide (VIP) and stimulation of the respective receptors on the muscle (P, purine receptors).[113,117] In addition, afferent nerve-derived SP and NKA may directly contract the muscle via stimulation of muscular NK_1 and NK_2 receptors,[135] while CGRP can relax the active muscle via muscular $CGRP_1$ receptors.[142]

neurons are non-tachykinin in nature, and remain to be identified. Although cholecystokinin-like peptides contribute to the contraction of the guinea pig ileum elicited by mesenteric nerve stimulation, they do not originate from capsaicin-sensitive neurons.[141] An involvement of CGRP in the contractile effect of sensory nerve stimulation has been disproved.[142]

There is good evidence, though, that the relaxant effect of sensory nerve stimulation in the gut involves release of CGRP, presumably from the afferent nerve fibers themselves. Capsaicin-induced inhibition of motor activity in the guinea pig, rat, and rabbit intestine is antagonized by tachyphylaxis to CGRP,[73,103,112,117,143] immunoblockade of CGRP,[136] and the CGRP antagonist, $CGRP_{8\text{-}37}$.[144] The relaxation of the precontracted guinea pig ileum caused by mesenteric nerve stimulation is also diminished by CGRP desensitization.[126] This mediator role of CGRP is consistent with its ability to relax the precontracted or spontaneously active muscle of the guinea pig, rat, and rabbit gut, although in general the motor actions of CGRP in the digestive tract are complex and involve a variety of pathways and mechanisms.[100] The relaxant effect of sensory nerve-derived CGRP may be due to a direct inhibitory action on the muscle, which is brought about by $CGRP_1$ receptors sensitive to the antagonistic action of $CGRP_{8\text{-}37}$[112,135,142,144] (Figure 2).

The dual excitatory/inhibitory motor effects of capsaicin in the guinea pig gallbladder appear likewise to involve tachykinins and CGRP as excitatory and inhibitory mediators, respectively,[72] while neither of these peptides seems to play a role in the capsaicin-evoked changes of motor activity in isolated segments from the human small and large intestine.[118-120] In particular, capsaicin fails to release CGRP, and pharmacological antagonism of CGRP is unable to prevent the inhibitory motor actions of capsaicin.[118-120,145] There is good evidence, however, that vasoactive intestinal polypeptide, which is released by capsaicin, is responsible for the relaxation of the human ileum and colon in response to sensory nerve stimulation.[118-120]

Although it is obvious that capsaicin-evoked and electrical stimulation of afferent nerve fibers causes profound alterations of intestinal motor activity it is still uncertain as to whether these effects play any role in the physiological regulation of intestinal motility. It has been reported that capsaicin-sensitive afferent nerve fibers participate in the peristaltic movements of the rat isolated colon in response to muscle stretch but not mechanical stimulation of the mucosa[146] and that CGRP released from afferent fibers functions as transmitter of the sensory pathway in the peristaltic reflex.[133] There is little evidence, though, that capsaicin-sensitive afferent neurons are essential to the principal mechanisms of intestinal propulsive activity which is coordinated by the enteric nervous system.[13,27] Thus, motor responses mediated by enteric neurons in the guinea pig and rabbit small and large intestine,[111,116,121,137] enteric motor reflexes involved in the peristaltic reflex of the rat and guinea pig small intestine,[143,147] and peristalsis in the guinea pig small intestine[143a] are not inhibited by capsaicin-induced defunctionalization of afferent neurons. In keeping with these *in vitro* observations are the *in vivo* findings that propulsive motor activity in the gut is not altered in rats pretreated with a neurotoxic dose of capsaicin.[101-104,108,109]

Increasing evidence indicates that capsaicin-sensitive afferent nerve fibers and the transmitters released from them have a bearing on pathophysiological alterations of intestinal motor activity. For instance, local release of CGRP in the gut may be responsible for postoperative ileus[107] and SP acting via NK_1 receptors may contribute to the motor disturbances caused by intestinal anaphylaxis.[108] It is conceivable, therefore, that sensory neuropeptides are also factors in the motor disturbances seen in the irritable bowel syndrome or in inflammatory bowel disease, when irritative or inflammatory changes in the tissue activate sensory nerve fibers or alter their excitability by other stimuli.

VI. PATHOPHYSIOLOGICAL IMPLICATIONS

There is mounting evidence to indicate that capsaicin-sensitive afferent neurons subserve a homeostatic role in the intestine, promoting normalization of tissue functions after their perturbation, but may also contribute to pathological changes in gut physiology. The ability of afferent neurons to facilitate intestinal secretion and tissue blood flow may be viewed as an instance of their homeostatic role (Figure 3). The intestinal diarrhea and inflammation induced by *Clostridium difficile* toxin A, which involves afferent nerve fibers[94] and SP acting via NK_1 receptors,[95] may be a mechanism to expel the pathogen and to initiate the repair of injury. The acid-evoked secretion of bicarbonate in the duodenum is essential for the protection of the mucosa from the deleterious action of gastric acid, since ablation of capsaicin-sensitive afferent neurons results in the aggravation of duodenal acid injury.[69,82,148] Sensory denervation, which does not cause damage by itself, also exacerbates experimental damage in the rat small intestine[149] and in the rat[70,150,151] and rabbit[152,153] colon.

The conjecture that sensory neurons strengthen the resistance of the intestinal mucosa to various insults is strongly supported by the ability of capsaicin-evoked stimulation of afferent nerves to reduce experimentally imposed damage in the rat duodenum[69] and colon.[151,154,155] The protective effects of capsaicin in the rat duodenum are mimicked by analogs of NKA,[156] whereas CGRP has not yet been tested for such an activity. The surmise that CGRP and SP are protective factors in the gut is indirectly related to the findings that experimental damage in the rat duodenum is associated with a selective depletion of CGRP[157] and SP,[158] whereas the tissue content of NKA, vasoactive intestinal polypeptide, galanin, and neuropeptide Y does not change.[157] Experimental injury of the rat and rabbit colon likewise leads to a depletion of CGRP and SP while the level of vasoactive intestinal polypeptide may be left unaltered.[151-153,159-161] Changes in peptide contents have also been found in humans suffering from Crohn's disease,[152,162,163] ulcerative colitis,[152,162-165] and severe constipation,[165] but it is not known whether the alterations reflect changes in afferent and/or enteric neurons.

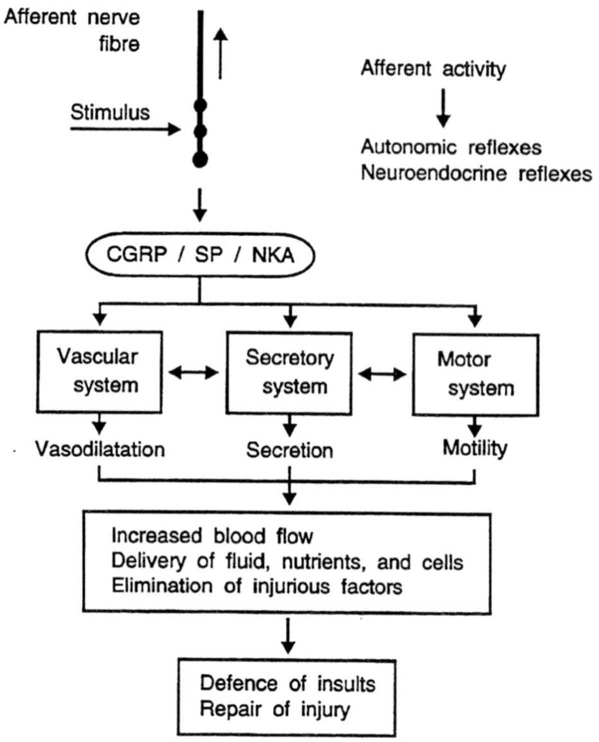

FIGURE 3. Diagram illustrating the pathophysiological roles of extrinsic afferent nerve fibers in gastrointestinal homeostasis.

NK_1 receptors associated with arterioles, venules, and lymphoid tissue are markedly upregulated in ulcerative colitis and Crohn's disease, whereas receptors for NKA, CGRP, and other afferent nerve-derived transmitters are not altered.[166,167] It has been hypothesized that these alterations contribute to the functional disturbances in inflammatory bowel disease and other gut disorders. Further relevant to intestinal disease are the findings that the secretory[168] and motor[108] changes associated with intestinal anaphylaxis are ameliorated in capsaicin-pretreated rats. It is possible that mast cells participate in these nerve-mediated reactions, as nerve fibers containing SP and/or CGRP can occur in close vicinity to mucosal mast cells.[161] Another interaction between afferent nerve fibers and the immune system may be reflected by the upregulation of SP levels in the inflamed jejunum of rats infected with *Trichinella spiralis*, a change that is prevented in capsaicin-pretreated rats.[169] It is also worth noting that capsaicin-sensitive afferent nerve fibers may play a role in certain forms of liver disease[170] and in the hyperkinetic splanchnic circulation of rats with experimental portal hypertension or cirrhosis,[171] although another study was unable to confirm the latter finding.[172]

In conclusion, peptidergic afferent nerve fibers in the gut can subserve a physiologically homeostatic role in circumstances of an insult to the tissue, whereas in the diseased intestine a deranged afferent nerve-peptide effector system may be a factor in the dysfunction of the gut. The homeostatic function is accomplished by increasing blood flow and delivery of nutrients and immune cells to the site of injury, by enhancing secretion to eliminate the pathogen from the intestinal lumen, and by appropriately changing intestinal motility. Both local release of peptide transmitters in the gut wall as well as initiation of autonomic and neuroendocrine reflexes are likely to contribute to the physiological and pathophysiological reactions involving afferent neurons (Figure 3).

ACKNOWLEDGMENTS

Work performed in the authors' laboratory was supported by the Austrian-Hungarian Foundation, the Austrian Science Foundation, the Hungarian Research Foundations ETT and OTKA, and the Franz Lanyar Foundation at the Medical Faculty of the University of Graz. The authors are grateful to Dr. Ulrike Holzer-Petsche for drawing the graphs and to Irmgard Russa for secretarial help.

REFERENCES

1. Cervero, F., Sensory innervation of the viscera: peripheral basis of visceral pain, *Physiol. Rev.*, 74, 95, 1994.
2. Mayer, E.A. and Gebhart, G.F., Basic and clinical aspects of visceral hyperalgesia, *Gastroenterology*, 107, 271, 1994.
3. Sengupta, J.N. and Gebhart, G.F., Gastrointestinal afferent fibers and sensation, in *Physiology of the Gastrointestinal Tract*, 3rd ed., Johnson, L.R., Ed., Raven Press, New York, 1994, 483.
4. Lee, Y., Hayashi, N., Hillyard, C.J., Girgis, S.I., MacIntyre, I., Emson, P.C., and Tohyama, M., Calcitonin gene-related peptide-like immunoreactive sensory fibers form synaptic contact with sympathetic neurons in the rat celiac ganglion, *Brain Res.*, 407, 149, 1987.
5. Lindh, B., Hökfelt, T., and Elfvin, L.-G., Distribution and origin of peptide-containing nerve fibers in the celiac superior mesenteric ganglion of the guinea-pig, *Neuroscience*, 26, 1037, 1988.
6. Kirchgessner, A.L. and Gershon, M.D., Identification of vagal efferent fibers and putative target neurons in the enteric nervous system of the rat, *J. Comp. Neurol.*, 285, 38, 1989.
7. Berthoud, H.-R., Jedrzejewska, A., and Powley, T.L., Simultaneous labeling of vagal innervation of gut and afferent projections from the visceral forebrain with DiI injected into the dorsal vagal complex in the rat, *J. Comp. Neurol.*, 301, 65, 1990.
8. Berthoud, H.-R., Carlson, N.R., and Powley, T.L., Topography of efferent vagal innervation of rat gastrointestinal tract, *Am. J. Physiol.*, 260, R200, 1991.
9. Zhang, X., Fogel, R., Simpson, P., and Renehan, W., The target specificity of the extrinsic innervation of the rat small intestine, *J. Auton. Nerv. Syst.*, 32, 53, 1991.
10. Furness, J.B., Papka, R.E., Della, N.G., Costa, M., and Eskay, R.L., Substance P-like immunoreactivity in nerves associated with the vascular system of guinea-pigs, *Neuroscience*, 7, 447, 1982.
11. Barja, F., Mathison, R., and Huggel, H., Substance P-containing nerve fibers in large peripheral blood vessels of the rat, *Cell Tissue Res.*, 229, 411, 1983.
12. Gibbins, I.L., Furness, J.B., Costa, M., MacIntyre, I., Hillyard, C.J., and Girgis, S., Co-localization of calcitonin gene-related peptide-like immunoreactivity with substance P in cutaneous, vascular and visceral sensory neurons of guinea pigs, *Neurosci. Lett.*, 57, 125, 1985.
13. Costa, M., Furness, J.B., and Gibbins, I.L., Chemical coding of enteric neurons, *Prog. Brain Res.*, 68, 217, 1986.
14. Gibbins, I.L., Furness, J.B., and Costa, M., Pathway-specific patterns of the co-existence of substance P, calcitonin gene-related peptide, cholecystokinin and dynorphin in neurons of the dorsal root ganglia of the guinea-pig, *Cell Tissue Res.*, 248, 417, 1987.
15. Su, H.C., Bishop, A.E., Power, R.F., Hamada, Y., and Polak, J.M., Dual intrinsic and extrinsic origins of CGRP- and NPY-immunoreactive nerves of rat gut and pancreas, *J. Neurosci.*, 7, 2674, 1987.
16. Chéry-Croze, S., Bosshard, A., Martin, H., Cuber, J.C., Charnay, Y., and Chayvialle, J.A., Peptide immunocytochemistry in afferent neurons from lower gut in rats, *Peptides*, 9, 873, 1988.
17. Papka, R.E., Furness, J.B., Della, N.G., Murphy, R., and Costa, M., Time course of effect of capsaicin on ultrastructure and histochemistry of substance P-immunoreactive nerves associated with the cardiovascular system of the guinea-pig, *Neuroscience*, 12, 1277, 1984.
18. Sharkey, K.A., Williams, R.G., and Dockray, G.J., Sensory substance P innervation of the stomach and pancreas. Demonstration of capsaicin-sensitive sensory neurons in the rat by combined immunohistochemistry and retrograde tracing, *Gastroenterology*, 87, 914, 1984.
19. Sternini, C., Reeve, J.R., and Brecha, N., Distribution and characterization of calcitonin gene-related peptide immunoreactivity in the digestive system of normal and capsaicin-treated rats, *Gastroenterology*, 93, 852, 1987.
20. Uddman, R., Edvinsson, L., Ekblad, E., Håkanson, R., and Sundler, F., Calcitonin gene-related peptide (CGRP): perivascular distribution and vasodilatory effects, *Regul. Pept.*, 15, 1, 1986.

21. Sternini, C., De Giorgio, R., and Furness, J.B., Calcitonin gene-related peptide neurons innervating the canine digestive system, *Regul. Pept.*, 42, 15, 1992.
22. Kawasaki, H., Takasaki, K., Saito, A., and Goto, K., Calcitonin gene-related peptide acts as a novel vasodilator neurotransmitter in mesenteric resistance vessels of the rat, *Nature*, 335, 164, 1988.
23. Goehler, L.E., Sternini, C., and Brecha, N.C., Calcitonin gene-related peptide immunoreactivity in the biliary pathway and liver of the guinea-pig: distribution and colocalization with substance P, *Cell Tissue Res.*, 253, 145, 1988.
24. Karlsson, S., Sundler, F., and Ahrén, B., Neonatal capsaicin-treatment in mice: effects on pancreatic peptidergic nerves and 2-deoxy-D-glucose-induced insulin and glucagon secretion, *J. Auton. Nerv. Syst.*, 39, 51, 1992.
25. Sternini, C., De Giorgio, R., Anderson, K., Watt, P.C., Brunicardi, F.C., Widdison, A.L., Wong, H., Reber, H.A., Walsh, J.H., and Go, V.L.W., Species differences in the immunoreactive patterns of calcitonin gene-related peptide in the pancreas, *Cell Tissue Res.*, 269, 447, 1992.
26. Kirchgessner, A.L., Dodd, J., and Gershon, M.D., Markers shared between dorsal root and enteric ganglia, *J. Comp. Neurol.*, 276, 607, 1988.
27. Barthó, L. and Holzer, P., Search for a physiological role of substance P in gastrointestinal motility, *Neuroscience*, 16, 1, 1985.
28. Holzer, P., Capsaicin: cellular targets, mechanisms of action, and selectivity for thin sensory neurons, *Pharmacol. Rev.*, 43, 143, 1991.
29. Mulderry, P.K., Ghatei, M.A., Spokes, R.A., Jones, P.M., Pierson, A.M., Hamid, Q.A., Kanse, S., Amara, S.G., Burrin, J.M., Legon, S., Polak, J.M., and Bloom, S.R., Differential expression of α-CGRP and β-CGRP by primary sensory neurons and enteric autonomic neurons of the rat, *Neuroscience*, 25, 195, 1988.
30. Sternini, C. and Anderson, K., Calcitonin gene-related peptide-containing neurons supplying the rat digestive system: differential distribution and expression patterns, *Somatosensory Motor Res.*, 9, 45, 1992.
31. Holzer, P., Peptidergic sensory neurons in the control of vascular functions: mechanisms and significance in the cutaneous and splanchnic vascular beds, *Rev. Physiol. Biochem. Pharmacol.*, 121, 49, 1992.
32. Hottenstein, O.D., Pawlik, W.W., Remak, G., and Jacobson, E.D., Capsaicin-sensitive nerves modulate resting blood flow and vascular tone in rat gut, *Naunyn-Schmiedeberg's Arch. Pharmacol.*, 343, 179, 1991.
33. Rózsa, Z., Varró, V., and Jancsó, G., Use of immunoblockade to study the involvement of peptidergic afferent nerves in the intestinal vasodilatory response to capsaicin in the dog, *Eur. J. Pharmacol.*, 115, 59, 1985.
34. Meehan, A.G., Hottenstein, O.D., and Kreulen, D.L., Capsaicin-sensitive nerves mediate inhibitory junction potentials and dilatation in guinea-pig mesenteric artery, *J. Physiol. (London)*, 443, 161, 1991.
35. Manzini, S. and Perretti, F., Vascular effects of capsaicin in isolated perfused rat mesenteric bed, *Eur. J. Pharmacol.*, 148, 153, 1988.
36. Fujimori, A., Saito, A., Kimura, S., and Goto, K., Release of calcitonin gene-related peptide (CGRP) from capsaicin-sensitive vasodilator nerves in the rat mesenteric artery, *Neurosci. Lett.*, 112, 173, 1990.
37. Bråtveit, M. and Helle, K.B., Vasodilation by calcitonin gene-related peptide (CGRP) and by transmural stimulation of the methoxamine-contracted rat hepatic artery after pretreatment with guanethidine, *Scand. J. Clin. Lab. Invest.*, 51, 395, 1991.
38. Han, S.-P., Naes, L., and Westfall, T.C., Calcitonin gene-related peptide is the endogenous mediator of nonadrenergic noncholinergic vasodilation in rat mesentery, *J. Pharmacol. Exp. Ther.*, 255, 423, 1990.
39. Kawasaki, H., Nuki, C., Saito, A., and Takasaki, K., Role of calcitonin gene-related peptide-containing nerves in the vascular adrenergic neurotransmission, *J. Pharmacol. Exp. Ther.*, 252, 403, 1990.
40. Li, Y.J. and Duckles, S.P., Differential effects of neuropeptide Y and opioids on neurogenic responses of the perfused rat mesentery, *Eur. J. Pharmacol.*, 195, 365, 1991.
41. Remak, G., Hottenstein, O.D., and Jacobson, E.D., Sensory nerves mediate neurogenic escape in rat gut, *Am. J. Physiol.*, 258, H778, 1990.
42. Hottenstein, O.D., Remak, G., and Jacobson, E.D., Peptidergic nerves mediate post-nerve stimulation hyperemia in rat gut, *Am. J. Physiol.*, 263, G29, 1992.
43. Hottenstein, O.D., Pawlik, W.W., Remak, G., and Jacobson, E.D., Capsaicin-sensitive nerves modulate reactive hyperemia in rat gut, *Proc. Soc. Exp. Biol. Med.*, 199, 311, 1992.
44. Rózsa, Z. and Jacobson, E.D., Capsaicin-sensitive nerves are involved in bile-oleate-induced intestinal hyperemia, *Am. J. Physiol.*, 256, G476, 1989.
45. Rózsa, Z., Mattila, J., and Jacobson, E.D., Substance P mediates a gastrointestinal thermoreflex in rats, *Gastroenterology*, 95, 265, 1988.
46. Lippe, I.Th., Sametz, W., Sabin, K., and Holzer, P., Inhibitory role of capsaicin-sensitive afferent neurons and nitric oxide in hemostasis, *Am. J. Physiol.*, 265, H1864, 1993.
47. Del Bianco, E., Perretti, F., Tramontana, M., Manzini, S., and Geppetti, P., Calcitonin gene-related peptide in rat arterial and venous vessels: sensitivity to capsaicin, bradykinin and FMLP, *Agents Actions*, 34, 376, 1991.
48. Kawasaki, H., Nuki, C., Saito, A., and Takasaki, K., NPY modulates neurotransmission of CGRP-containing vasodilator nerves in rat mesenteric arteries, *Am. J. Physiol.*, 261, H683, 1991.

49. Manzini, S., Perretti, F., Tramontana, M., Del Bianco, E., Santicioli, P., Maggi, C.A., and Geppetti, P., Neurochemical evidence of calcitonin gene-related peptide-like immunoreactivity (CGRP-LI) release from capsaicin-sensitive nerves in rat mesenteric arteries and veins, *Gen. Pharmacol.*, 22, 275, 1991.
50. Marshall, I., Al-Kazwini, S.J., Holman, J.J., and Craig, R.K., Human and rat alpha-CGRP but not calcitonin cause mesenteric vasodilatation in rats, *Eur. J. Pharmacol.*, 123, 217, 1986.
51. Bråtveit, M., Haugan, A., and Helle, K.B., Effects of calcitonin gene-related peptide (CGRP) on regional haemodynamics and on selected hepato-splanchnic arteries from the rat: a comparison with VIP and atriopeptin II, *Scand. J. Clin. Lab. Invest.*, 51, 167, 1991.
52. Nelson, M.T., Huang, Y., Brayden, J.E., Hescheler, J., and Standen, N.B., Arterial dilations in response to calcitonin gene-related peptide involve activation of K^+ channels, *Nature*, 344, 770, 1990.
53. Brizzolara, A.L. and Burnstock, G., Endothelium-dependent and endothelium-independent vasodilatation of the hepatic artery of the rabbit, *Br. J. Pharmacol.*, 103, 1206, 1991.
54. Törnebrandt, K., Nobin, A., and Owman, Ch., Contractile and dilatory action of neuropeptides on isolated human mesenteric blood vessels, *Peptides*, 8, 251, 1987.
55. Li, Y.J. and Duckles, S.P., Effect of endothelium on the actions of sympathetic and sensory nerves in the perfused rat mesentery, *Eur. J. Pharmacol.*, 210, 23, 1992.
56. Han, S.-P., Naes, L., and Westfall, T.C., Inhibition of periarterial nerve stimulation-induced vasodilation of the mesenteric arterial bed by CGRP (8-37) and CGRP receptor desensitization, *Biochem. Biophys. Res. Commun.*, 168, 786, 1990.
57. Claing, A., Télémaque, S., Cadieux, A., Fournier, A., Regoli, D., and D'Orléans-Juste, P., Nonadrenergic and noncholinergic arterial dilatation and venoconstriction are mediated by calcitonin gene-related peptide$_1$ and neurokinin-1 receptors, respectively, in the mesenteric vasculature of the rat after perivascular nerve stimulation, *J. Pharmacol. Exp. Ther.*, 263, 1226, 1992.
58. D'Orléans-Juste, P., Claing, A., Télémaque, S., Warner, T.D., and Regoli, D., Neurokinins produce selective venoconstriction via NK-3 receptors in the rat mesenteric vascular bed, *Eur. J. Pharmacol.*, 204, 329, 1991.
59. Bolton, T.B. and Clapp, L.H., Endothelial-dependent relaxant actions of carbachol and substance P in arterial smooth muscle, *Br. J. Pharmacol.*, 87, 713, 1986.
60. Stewart-Lee, A. and Burnstock, G., Actions of tachykinins on the rabbit mesenteric artery: substance P and [Glp^6,L-Pro^9]SP$_{6\text{-}11}$ are potent agonists for endothelial neurokinin-1 receptors, *Br. J. Pharmacol.*, 97, 1218, 1989.
61. Schrauwen, E. and Houvenaghel, A., Substance P: a powerful intestinal vasodilator in the pig, *Pfluegers Arch.*, 386, 281, 1980.
62. Stewart-Lee, A.L., Aberdeen, J., and Burnstock, G., The effect of atherosclerosis on neuromodulation of sympathetic neurotransmission by neuropeptide Y and calcitonin gene-related peptide in the rabbit mesenteric artery, *Eur. J. Pharmacol.*, 216, 167, 1992.
63. Kawasaki, H., Nuki, C., Saito, A., and Takasaki, K., Adrenergic modulation of calcitonin gene-related peptide (CGRP)-containing nerve-mediated vasodilation in the rat mesenteric resistance vessel, *Brain Res.*, 506, 287, 1990b.
64. Li, Y.J. and Duckles, S.P., Effect of opioid receptor antagonists on vasodilator nerve actions in the perfused rat mesentery, *Eur. J. Pharmacol.*, 204, 323, 1991.
65. Ralevic, V., Rubino, A., and Burnstock, G., Prejunctional modulation of sensory-motor nerve-mediated vasodilatation of the rat mesenteric arterial bed by opioid peptides, *J. Pharmacol. Exp. Ther.*, 268, 772, 1994.
66. Rubino, A., Ralevic, V., and Burnstock, G., The P$_1$-purinoceptors that mediate the prejunctional inhibitory effect of adenosine on capsaicin-sensitive nonadrenergic noncholinergic neurotransmission in the rat mesenteric arterial bed are of the A$_1$-subtype, *J. Pharmacol. Exp. Ther.*, 267, 1100, 1993.
67. Nuki, C., Kawasaki, H., Takasaki, K., and Wada, A., Pharmacological characterization of presynaptic calcitonin gene-related peptide (CGRP) receptors on CGRP-containing vasodilator nerves in rat mesenteric resistance vessels, *J. Pharmacol. Exp. Ther.*, 268, 59, 1994.
68. Rózsa, Z., Sharkey, K.A., Jancsó, G., and Varró, V., Evidence for a role of capsaicin-sensitive mucosal afferent nerves in the regulation of mesenteric blood flow in the dog, *Gastroenterology*, 90, 906, 1986.
69. Leung, F.W., Primary sensory neurons mediate in part the protective mesenteric hyperemia after intraduodenal acidification in rats, *Gastroenterology*, 105, 1737, 1993.
70. Leung, F.W., Role of capsaicin-sensitive afferent nerves in mucosal injury and injury-induced hyperemia in rat colon, *Am. J. Physiol.*, 262, G332, 1992.
71. Vanner, S., Mechanism of action of capsaicin on submucosal arterioles in the guinea pig ileum, *Am. J. Physiol.*, 265, G51, 1993.
72. Maggi, C.A., Santicioli, P., Renzi, D., Patacchini, R., Surrenti, C., and Meli, A., Release of substance P- and calcitonin gene-related peptide-like immunoreactivity and motor response of the isolated guinea pig gallbladder to capsaicin, *Gastroenterology*, 96, 1093, 1989.
73. Mayer, E.A., Koelbel, C.B.M., Snape, W.J., Eysselein, V., Ennes, H., and Kodner, A., Substance P and CGRP mediate motor response of rabbit colon to capsaicin, *Am. J. Physiol.*, 259, G889, 1990.

74. Donnerer, J., Barthó, L., Holzer, P., and Lembeck, F., Intestinal peristalsis associated with release of immunoreactive substance P, *Neuroscience*, 11, 913, 1984.
75. Bauerfeind, P., Hof, R., Cucala, M., Siegrist, S., von Ritter, Ch., Fischer, J.A., and Blum, A.L., Effects of hCGRP I and II on gastric blood flow and acid secretion in anesthetized rabbits, *Am. J. Physiol.*, 256, G145, 1989.
76. Galligan, J.J., Jiang, M.-M., Shen, K.-Z., and Surprenant, A., Substance P mediates neurogenic vasodilatation in extrinsically denervated guinea-pig submucosal arterioles, *J. Physiol. (London)*, 420, 267, 1990.
77. Sternini, C., Enteric and visceral afferent CGRP neurons. Targets of innervation and differential expression patterns, *Ann. N.Y. Acad. Sci.*, 657, 170, 1992.
78. Grönstad, K.O., Dahlström, A., Jaffe, B.M., Zinner, M.J., and Ahlman, H., Studies on the mucosal hyperemia of the feline small intestine observed at endoluminal perfusion with substance P, *Acta Physiol. Scand.*, 128, 97, 1986.
79. Hellström, P.M., Söder, O., and Theodorsson, E., Occurrence, release, and effects of multiple tachykinins in cat colonic tissues and nerves, *Gastroenterology*, 100, 431, 1991.
80. Saria, A., Lundberg, J.M., Skofitsch, G., and Lembeck, F., Vascular protein leakage in various tissues induced by substance P, capsaicin, bradykinin, serotonin, histamine and by antigen challenge, *Naunyn-Schmiedeberg's Arch. Pharmacol.*, 324, 212, 1983.
81. Lundberg, J.M., Brodin, E., Hua, X.-Y., and Saria, A., Vascular permeability changes and smooth muscle contraction in relation to capsaicin-sensitive substance P afferents in the guinea-pig, *Acta Physiol. Scand.*, 120, 217, 1984.
82. Maggi, C.A., Evangelista, S., Abelli, L., Somma, V., and Meli, A., Capsaicin-sensitive mechanisms and experimentally induced duodenal ulcers in rats, *J. Pharm. Pharmacol.*, 39, 559, 1987.
83. Szolcsányi, J., Antidromic vasodilatation and neurogenic inflammation, *Agents Actions*, 23, 4, 1988.
84. Zimmerman, B.J., Anderson, D.C., and Granger, D.N., Neuropeptides promote neutrophil adherence to endothelial cell monolayers, *Am. J. Physiol.*, 263, G678, 1992.
85. Takeuchi, K., Matsumoto, J., Ueshima, K., and Okabe, S., Role of capsaicin-sensitive afferent neurons in alkaline secretory response to luminal acid in the rat duodenum, *Gastroenterology*, 101, 954, 1991.
86. Hamlet, A., Jönson, C., and Fändriks, L., The mediation of increased duodenal alkaline secretion in response to 10 mM HCl in the anaesthetized rat. Support for the involvement of capsaicin-sensitive nerve elements, *Acta Physiol. Scand.*, 146, 519, 1992.
87. Laporte, J.-L., Dauge-Geffroy, M.-C., Chariot, J., Rozé, C., and Potet, F., Capsaicin-sensitive sensory neurons can modulate duodenal mucus secretion. A morphometric study in the rat, *Gastroenterol. Clin. Biol.*, 17, 535, 1993.
88. Perkins, M.N., Forster, P.L., and Dray, A., The involvement of afferent nerve terminals in the stimulation of ion transport by bradykinin in rat isolated colon, *Br. J. Pharmacol.*, 94, 47, 1988.
89. Yarrow, S., Ferrar, J.A., and Cox, H.M., The effects of capsaicin upon electrogenic ion transport in rat descending colon, *Naunyn-Schmiedeberg's Arch. Pharmacol.*, 344, 557, 1991.
90. Maggi, C.A., Patacchini, R., Rovero, P., and Giachetti, A., Tachykinin receptors and tachykinin receptor antagonists, *J. Auton. Pharmacol.*, 13, 23, 1993.
91. Cox, H.M., Ferrar, J.A., and Cuthbert, A.W., Effects of α- and β-calcitonin gene-related peptides upon ion transport in rat descending colon, *Br. J. Pharmacol.*, 97, 996, 1989.
92. Rolston, R.K., Ghatei, M.A., Mulderry, P.K., and Bloom, S.R., Intravenous calcitonin gene-related peptide stimulates net water secretion in rat colon *in vivo*, *Dig. Dis. Sci.*, 34, 612, 1989.
93. McCulloch, C.R. and Cooke, H.J., Human α-calcitonin gene-related peptide influences colonic secretion by acting on myenteric neurons, *Regul. Pept.*, 24, 87, 1989.
94. Castagliuolo, I., LaMont, J.T., Letourneau, R., Kelly, C., O'Keane, J.C., Jaffer, A., Theoharides, T.C., and Pothoulakis, C., Neuronal involvement in the intestinal effects of *Clostridium difficile* toxin A and *Vibrio cholerae* enterotoxin in rat ileum, *Gastroenterology*, 107, 657, 1994.
95. Pothoulakis, C., Castagliuolo, I., LaMont, J.T., Jaffer, A., O'Keane, C.J., Snider, R.M., and Leeman, S.E., CP-96,345, a substance P antagonist, inhibits rat intestinal responses to *Clostridium difficile* toxin A but not cholera toxin, *Proc. Natl. Acad. Sci. U.S.A.*, 91, 947, 1994.
96. Cassuto, J., Siewert, A., Jodal, M., and Lundgren, O., The involvement of intramural nerves in cholera toxin-induced intestinal secretion, *Acta Physiol. Scand.*, 117, 195, 1983.
97. Gicquel, N., Nagain, C., Chariot, J., Tsocas, A., Levenez, F., Corring, T., and Rozé, C., Modulation of pancreatic secretion by capsaicin-sensitive sensory neurons in the rat, *Pancreas*, 9, 203, 1994.
98. Li, Y. and Owyang, C., Endogenous cholecystokinin stimulates pancreatic enzyme secretion via vagal afferent pathway in rats, *Gastroenterology*, 107, 525, 1994.
99. Li, Y., Kolligs, F., and Owyang, C., Mechanism of action of calcitonin gene-related peptide in inhibiting pancreatic enzyme secretion in rats, *Gastroenterology*, 105, 194, 1993.
100. Holzer, P., Calcitonin gene-related peptide, in *Gut Peptides: Biochemistry and Physiology*, Walsh, J.H. and Dockray, G.J., Eds., Raven Press, New York, 1994, 493.

101. Holzer, P., Capsaicin-sensitive afferent neurones and gastrointestinal propulsion in the rat, *Naunyn-Schmiedeberg's Arch. Pharmacol.*, 332, 62, 1986.
102. Holzer, P., Lippe, I.Th., and Holzer-Petsche, U., Inhibition of gastrointestinal transit due to surgical trauma or peritoneal irritation is reduced in capsaicin-treated rats, *Gastroenterology*, 91, 360, 1986.
103. Maggi, C.A., Giuliani, S., Santicioli, P., Patacchini, R., and Meli, A., Neural pathways and pharmacological modulation of defecation reflex in rats, *Gen. Pharmacol.*, 19, 517, 1988.
104. Maggi, C.A., Giuliani, S., Santicioli, P., and Meli, A., Propagated motor activity in the small intestine of urethane-anaesthetized rats: inhibitory action of sympathetic and capsaicin-sensitive nerves, *Gen. Pharmacol.*, 19, 525, 1988.
105. Mizutani, M., Neya, T., and Nakayama, S., Capsaicin-sensitive afferents activate a sympathetic intestinointestinal reflex in dogs, *J. Physiol. (London)*, 425, 133, 1990.
106. Horowitz, M., Wishart, J., Maddox, A., and Russo, A., The effect of chilli on gastrointestinal transit, *J. Gastroenterol. Hepatol.*, 7, 52, 1992.
107. Plourde, V., Wong, H.C., Walsh, J.H., Raybould, H.E., and Taché, Y., CGRP antagonists and capsaicin on celiac ganglia prevent postoperative gastric ileus, *Peptides*, 14, 1225, 1993.
108. Fargeas, M.J., Fioramonti, J., and Buéno, L., Involvement of capsaicin-sensitive afferent nerves in the intestinal motor alterations induced by intestinal anaphylaxis in rats, *Int. Arch. Allergy Immunol.*, 101, 190, 1993.
109. Bonnafous, C., Scatton, B., and Buéno, L., Benzodiazepine-induced intestinal motor disturbances in rats: mediation by ω_2 (BZ_2) sites on capsaicin-sensitive afferent neurones, *Br. J. Pharmacol.*, 113, 268, 1994.
110. Barthó, L. and Szolcsányi, J., The site of action of capsaicin on the guinea-pig isolated ileum, *Naunyn-Schmiedeberg's Arch. Pharmacol.*, 305, 75, 1978.
111. Szolcsányi, J. and Barthó, L., Capsaicin-sensitive innervation of the guinea-pig taenia caeci, *Naunyn-Schmiedeberg's Arch. Pharmacol.*, 309, 77, 1979.
112. Barthó, L., Pethö, G., Antal, A., Holzer, P., and Szolcsányi, J., Two types of relaxation due to capsaicin in the guinea pig isolated ileum, *Neurosci. Lett.*, 81, 146, 1987.
113. Maggi, C.A., Meli, A., and Santicioli, P., Four motor effects of capsaicin on guinea-pig distal colon, *Br. J. Pharmacol.*, 90, 651, 1987.
114. Takaki, M., Jin, J.-G., and Nakayama, S., Effects of capsaicin on the circular muscle motility of the isolated guinea-pig ileum, *Acta Med. Okayama*, 43, 353, 1989.
115. Barthó, L., Calcitonin gene-related peptide and capsaicin inhibit the circular muscle of the guinea-pig ileum, *Regul. Pept.*, 35, 43, 1991.
116. Barthó, L. and Szolcsányi, J., The mechanism of the motor response to periarterial nerve stimulation in the small intestine of the rabbit, *Br. J. Pharmacol.*, 70, 193, 1980.
117. Maggi, C.A., Manzini, S., Giuliani, S., Santicioli, P., and Meli, A., Extrinsic origin of the capsaicin-sensitive innervation of rat duodenum: possible involvement of calcitonin gene-related peptide (CGRP) in the capsaicin-induced activation of intramural non-adrenergic non-cholinergic neurons, *Naunyn-Schmiedeberg's Arch. Pharmacol.*, 334, 172, 1986.
118. Maggi, C.A., Patacchini, R., Santicioli, P., Giuliani, S., Turini, D., Barbanti, G., Beneforti, P., Misuri, D., and Meli, A., Specific motor effects of capsaicin on human jejunum, *Eur. J. Pharmacol.*, 149, 393, 1988.
119. Maggi, C.A., Giuliani, S., Santicioli, P., Patacchini, R., Said, S.I., Theodorsson, E., Turini, D., Barbanti, G., Giachetti, A., and Meli, A., Direct evidence for the involvement of vasoactive intestinal polypeptide in the motor response of the human isolated ileum to capsaicin, *Eur. J. Pharmacol.*, 185, 169, 1990.
120. Maggi, C.A., Theodorsson, E., Santicioli, P., Patacchini, R., Barbanti, G., Turini, D., Renzi, D., and Giachetti, A., Motor response of the human isolated colon to capsaicin and its relationship to release of vasoactive intestinal polypeptide, *Neuroscience*, 39, 833, 1990.
121. Szolcsányi, J. and Barthó, L., New type of nerve-mediated cholinergic contractions of the guinea-pig small intestine and its selective blockade by capsaicin, *Naunyn-Schmiedeberg's Arch. Pharmacol.*, 305, 83, 1978.
122. Grbović, L. and Radmanović, B.Z., The influence of substance P on the response of guinea-pig isolated ileum to periarterial nerve stimulation, *Br. J. Pharmacol.*, 78, 681, 1983.
123. Wali, F.A., Possible involvement of substance P in the contraction produced by periarterial nerve stimulation in the rat ileum, *J. Auton. Pharmacol.*, 5, 143, 1985.
124. Takaki, M., Jin, J.-G., and Nakayama, S., Involvement of hexamethonium-sensitive and capsaicin-sensitive components in cholinergic contractions induced by mesenteric nerve stimulation in guinea-pig ileum, *Biomed. Res.*, 8, 195, 1987.
125. Jin, J.-G., Takaki, M., and Nakayama, S., Ruthenium red prevents capsaicin-induced neurotoxic action on sensory fibers of the guinea pig ileum, *Neurosci. Lett.*, 106, 152, 1989.
126. Takaki, M., Jin, J.-G., and Nakayama, S., Possible involvement of calcitonin gene-related peptide (CGPR) in non-cholinergic non-adrenergic relaxation induced by mesenteric nerve stimulation in guinea pig ileum, *Brain Res.*, 478, 199, 1989.
127. Takaki, M. and Nakayama, S., Effects of mesenteric nerve stimulation on the electrical activity of myenteric neurons in the guinea pig ileum, *Brain Res.*, 442, 351, 1988.

128. Takaki, M. and Nakayama, S., Effects of capsaicin on myenteric neurons of the guinea pig ileum, *Neurosci. Lett.*, 105, 125, 1989.
129. Barthó, L. and Vizi, E.S., Neurochemical evidence for the release of acetylcholine from the guinea-pig ileum myenteric plexus by capsaicin, *Eur. J. Pharmacol.*, 110, 125, 1985.
130. Barthó, L., Holzer, P., Lembeck, F., and Szolcsányi, J., Evidence that the contractile response of the guinea-pig ileum to capsaicin is due to substance P, *J. Physiol. (London)*, 332, 157, 1982.
131. Chahl, L.A., Evidence that the contractile response of the guinea-pig ileum to capsaicin is due to substance P release, *Naunyn-Schmiedeberg's Arch. Pharmacol.*, 319, 212, 1982.
132. Björkroth, U., Inhibition of smooth muscle contractions induced by capsaicin and electrical transmural stimulation by a substance P antagonist, *Acta Physiol. Scand. Suppl.*, 515, 11, 1983.
133. Grider, J.R., CGRP as a transmitter in the sensory pathway mediating peristaltic reflex, *Am. J. Physiol.*, 266, G1139, 1994.
134. Holzer, P., Gamse, R., and Lembeck, F., Distribution of substance P in the rat gastrointestinal tract — lack of effect of capsaicin pretreatment, *Eur. J. Pharmacol.*, 61, 303, 1980.
135. Barthó, L., Maggi, C.A., Wilhelm, M., and Patacchini, R., Tachykinin NK_1 and NK_2 receptors mediate atropine-resistant ileal circular muscle contractions evoked by capsaicin, *Eur. J. Pharmacol.*, 259, 187, 1994.
136. Maggi, C.A., Patacchini, R., Santicioli, P., Theodorsson, E., and Meli, A., Several neuropeptides determine the visceromotor response to capsaicin in the guinea-pig isolated ileal longitudinal muscle, *Eur. J. Pharmacol.*, 148, 43, 1988.
137. Barthó, L., Sebök, B., and Szolcsányi, J., Indirect evidence for the inhibition of enteric substance P neurones by opiate agonists but not capsaicin, *Eur. J. Pharmacol.*, 77, 273, 1982.
138. Emonds-Alt, X., Bichon, D., Ducoux, J.P., Heaulme, M., Miloux, B., Poncelet, M., Proietto, V., Van Broeck, D., Vilain, P., Neliat, G., Soubrie, P., Le Fur, G., and Brelière, J.C., SR 142801, the first potent non-peptide antagonist of the tachykinin NK_3 receptor, *Life Sci.*, 56, PL27, 1994.
139. Guard, S. and Watson, S.P., Evidence for neurokinin-3 receptor-mediated tachykinin release in the guinea-pig ileum, *Eur. J. Pharmacol.*, 144, 409, 1987.
140. Patacchini, R., Barthó, L., Holzer, P., and Maggi, C.A., Activity of SR 142801 at peripheral tachykinin receptors, *Eur. J. Pharmacol.*, 278, 17, 1995.
141. Barthó, L., Pharmacological evidence for the presence of cholecystokinin-containing neurones in the mesenteric nerves supplying the guinea-pig ileum, *Neuropharmacology*, 28, 643, 1989.
142. Barthó, L., Kóczán, G., and Maggi, C.A., Studies on the mechanism of the contractile action of rat calcitonin gene-related peptide and of capsaicin on the guinea-pig ileum: effect of hCGRP(8-37) and CGRP tachyphylaxis, *Neuropeptides*, 25, 325, 1993.
143. Jin, J.-G., Takaki, M., and Nakayama, S., Inhibitory effect of capsaicin on the ascending pathway of the guinea-pig ileum and antagonism of this effect by ruthenium red, *Eur. J. Pharmacol.*, 180, 13, 1990.
143a. Barthó, L., and Holzer, P., unpublished data.
144. Barthó, L., Kóczán, G., Holzer, P., Maggi, C.A., and Szolscányi, J., Antagonism of the effects of calcitonin gene-related peptide and of capsaicin on the guinea-pig isolated ileum by human α-calcitonin gene-related peptide(8-37), *Neurosci. Lett.*, 129, 156, 1991.
145. Maggi, C.A., Santicioli, P., Del Bianco, E., Geppetti, P., Barbanti, G., Turini, D., and Meli, A., Release of VIP- but not CGRP-like immunoreactivity by capsaicin from the human isolated small intestine, *Neurosci. Lett.*, 98, 317, 1989.
146. Grider, J.R. and Jin, J.-G., Distinct populations of sensory neurons mediate the peristaltic reflex elicited by muscle stretch and mucosal stimulation, *J. Neuroscience*, 14, 2854, 1994.
147. Allescher, H.D., Sattler, D., Piller, C., Schusdziarra, V., and Classen, M., Ascending neural pathways in the rat ileum *in vitro*: effect of capsaicin and involvement of nitric oxide, *Eur. J. Pharmacol.*, 217, 153, 1992.
148. Takeuchi, K., Matsumoto, J., Ueshima, K., Ohuchi, T., and Okabe, S., Induction of duodenal ulcers in sensory deafferented rats following histamine infusion, *Digestion*, 51, 203, 1992.
149. Evangelista, S., Maggi, C.A., and Meli, A., Involvement of capsaicin-sensitive mechanism(s) in the antiulcer defence of intestinal mucosa in rats, *Proc. Soc. Exp. Biol. Med.*, 184, 264, 1987.
150. Evangelista, S. and Meli, A., Influence of capsaicin-sensitive fibers on experimentally-induced colitis in rats, *J. Pharm. Pharmacol.*, 41, 574, 1989.
151. Evangelista, S. and Tramontana, M., Involvement of calcitonin gene-related peptide in rat experimental colitis, *J. Physiol. (Paris)*, 87, 277, 1993.
152. Eysselein, V.E., Reinshagen, M., Patel, A., Davis, W., Nast, C., and Sternini, C., Calcitonin gene-related peptide in inflammatory bowel disease and experimentally induced colitis, *Ann. N.Y. Acad. Sci.*, 657, 319,1992.
153. Reinshagen, M., Patel, A., Sottili, M., Nast, C., Davis, W., Mueller, K., and Eysselein, V.E., Protective function of extrinsic sensory neurons in acute rabbit experimental colitis, *Gastroenterology*, 106, 1208, 1994.
154. Endoh, K. and Leung, F.W., Topical capsaicin protects the distal but not the proximal colon against acetic acid injury, *Gastroenterology*, 98, A446, 1990.

155. Goso, C., Evangelista, S., Tramontana, M., Manzini, S., Blumberg, P.M., and Szallasi, A., Topical capsaicin administration protects against trinitrobenzene sulfonic acid-induced colitis in the rat, *Eur. J. Pharmacol.*, 249, 185, 1993.
156. Evangelista, S., Maggi, C.A., Rovero, P., Patacchini, R., Giuliani, S., and Giachetti, A., Analogs of neurokinin A(4-10) afford protection against gastroduodenal ulcers in rats, *Peptides*, 11, 293, 1990.
157. Evangelista, S., Renzi, D., Tramontana, M., Surrenti, C., Theodorsson, E., and Maggi, C.A., Cysteamine-induced duodenal ulcers are associated with a selective depletion in gastric and duodenal calcitonin gene-related peptide-like immunoreactivity in rats, *Regul. Pept.*, 39, 19, 1992.
158. Evangelista, S., Renzi, D., Mantellini, P., Surrenti, C., and Meli, A., Duodenal SP-like immunoreactivity is decreased in experimentally-induced duodenal ulcers, *Neurosci. Lett.*, 112, 352, 1992.
159. Renzi, D., Tramontana, M., Panerai, C., Surrenti, C., and Evangelista, S., Decrease of calcitonin gene-related peptide, but not vasoactive intestinal polypeptide and substance P, in the TNB-induced experimental colitis in rats, *Neuropeptides*, 22, 56, 1992.
160. Miampamba, M., Chéry-Croze, S., and Chayvialle, J.A., Spinal and intestinal levels of substance P, calcitonin gene-related peptide and vasoactive intestinal polypeptide following perendoscopic injection of formalin in rat colonic wall, *Neuropeptides*, 22, 73, 1992.
161. Sharkey, K.A., Substance P and calcitonin gene-related peptide (CGRP) in gastrointestinal inflammation, *Ann. N.Y. Acad. Sci.*, 664, 425, 1992.
162. Bernstein, C.N., Robert, M.E., and Eysselein, V.E., Rectal substance P concentrations are increased in ulcerative colitis but not in Crohn's disease, *Am. J. Gastroenterol.*, 88, 864, 1992.
163. Kimura, M., Masuda, T., Hiwatashi, N., Toyota, T., and Nagura, H., Changes in neuropeptide-containing nerves in human colonic mucosa with inflammatory bowel disease, *Pathol. Int.*, 44, 624, 1994.
164. Koch, T.R., Carney, J.A., and Go, V.L.W., Distribution and quantitation of gut neuropeptides in normal intestine and inflammatory bowel diseases, *Dig. Dis. Sci.*, 32, 369, 1987.
165. Goldin, E., Karmeli, F., Selinger, Z., and Rachmilewitz, D., Colonic substance P levels are increased in ulcerative colitis and decreased in chronic severe constipation, *Dig. Dis. Sci.*, 34, 754, 1989.
166. Mantyh, C.R., Gates, T.S., Zimmerman, R.P., Welton, M.L., Passaro, E.P., Vigna, S.R., Maggio, J.E., Kruger, L., and Mantyh, P.W., Receptor binding sites for substance P, but not substance K or neuromedin K, are expressed in high concentrations by arterioles, venules, and lymph nodules in surgical specimens obtained from patients with ulcerative colitis and Crohn disease, *Proc. Natl. Acad. Sci. U.S.A.*, 85, 3235, 1988.
167. Mantyh, C.R., Vigna, S.R., Maggio, J.E., Mantyh, P.W., Bollinger, R.R., and Pappas, T.N., Substance P binding sites on intestinal lymphoid aggregates and blood vessels in inflammatory bowel disease correspond to authentic NK-1 receptors, *Neurosci. Lett.*, 178, 255, 1994.
168. Crowe, S.E., Sestini, P., and Perdue, M.H., Allergic reactions of rat jejunal mucosa. Ion transport responses to luminal antigen and inflammatory mediators, *Gastroenterology*, 99, 74, 1990.
169. Swain, M.G., Agro, A., Blennerhassett, P., Stanisz, A., and Collins, S.M., Increased levels of substance P in the myenteric plexus of trichinella-infected rats, *Gastroenterology*, 102, 1913, 1992.
170. Casini, A., Lippe, I.Th., Evangelista, S., Geppetti, P., Santicioli, P., Urso, C., Paglierani, M., Maggi, C.A., and Surrenti, C., Effect of sensory denervation with capsaicin on liver fibrosis induced by common bile duct ligation in rat, *J. Hepatol.*, 11, 302, 1990.
171. Lee, S.S. and Sharkey, K.A., Capsaicin treatment blocks development of hyperkinetic circulation in portal hypertensive and cirrhotic rats, *Am. J. Physiol.*, 264, G868, 1993.
172. Fernández, M., Casadevall, M., Schuligoi, R., Pizcueta, P., Panés, J., Barrachina, M.D., Donnerer, J., Piqué, J.M., Esplugues, J.V., Bosch, J., Rodés, J., and Holzer, P., Neonatal capsaicin pretreatment does not prevent the splanchnic vasodilatation in portal hypertensive rats, *Hepatology*, 20, 1609, 1994.

Chapter 14

SENSORY NEUROPEPTIDES AND AIRWAY DISEASES

Peter J. Barnes

CONTENTS

I. INTRODUCTION

The peptides substance P (SP), neurokinin A (NKA), and calcitonin gene-related peptide (CGRP) are localized to a population of sensory neurons in the respiratory tract.[1-3] These peptides have potent effects on bronchomotor tone, airway secretions, the bronchial circulation, and on inflammatory and immune cells. Although some clues to the physiological and pathophysiological role of these peptides are provided by their localization and functional effects, the most useful information is provided by depletion studies using capsaicin, and increasingly by the use of potent and specific receptor antagonists. Many of the inflammatory and functional effects of sensory neuropeptides are relevant to asthma and there is compelling evidence for the involvement of neuropeptides in the pathophysiology and symptomatology of asthma.[4] The purpose of this chapter is to discuss effects of sensory neuropeptides that are relevant to the pathophysiology of airway diseases.

II. EXPERIMENTAL APPROACHES

Several approaches have been used to investigate the role of sensory neuropeptides in airways. The effects of exogenous sensory neuropeptides on various target cells relevant to asthma *in vitro* and their effects on airway function *in vivo* have been widely studied in

0-8493-7646-7/96/$0.00+$.50

animals and humans.[2] This approach is valuable in revealing the potential effects of a particular peptide, but it is not possible to know exactly what the local concentration of a particular peptide might be. Furthermore, there are striking differences between species. Even data in normal human airways may not be relevant to the situation in the diseased airway, where there might be alterations in neuropeptide receptor expression and metabolic breakdown.

A more informative approach is to investigate the action of specific blockers or enhancers, or to study depletion of the relevant peptide, since this can reveal the role of the endogenous neuropeptide. Again, it is possible that the disease state may alter the synthesis, release, or metabolism of a particular peptide or its receptors and therefore produce changes in the effects of blocking drugs. It is only recently that potent specific tachykinin receptor blockers have become available and these will prove to be increasingly important tools in the investigation of the role of neuropeptides in disease.

Several animal models of asthma have been investigated, but none of these closely mimic the chronic eosinophilic inflammation characteristic of asthma and they have been poorly predictive of drugs that will have clinical efficacy.[5] The only certain way to evaluate the role of neuropeptides in asthma is to study the effect of specific antagonists or inhibitors in patients with the disease, and this is discussed in more detail in Chapter 22. Specific neuropeptide antagonists suitable for clinical use are now under development and studies are already underway in asthma. Again there may be pitfalls in this approach, as it is the usual practice to select patients with mild asthma for such studies. It is possible that neuropeptides are relevant only in certain types of asthma or in more severe and intractable disease. Furthermore, it may be difficult to evaluate the effects of neuropeptides on airway function in clinical studies if their main action is on mucosal inflammation, mucus secretion, or on airway blood flow, since techniques to evaluate these responses in patients are difficult.

III. TACHYKININS

SP and NKA, but not neurokinin B, are localized to sensory nerves in the airways of several species. SP-immunoreactive nerves are abundant in rodent airways, but are very sparse in human airways.[6-8] Rapid enzymatic degradation of SP in airways, and the fact that SP concentrations may decrease with age and possibly after cigarette smoking, could explain the difficulty in demonstrating this peptide in some studies. SP-immunoreactive nerves in the airway are found beneath and within the airway epithelium, around blood vessels, and, to a lesser extent, within airway smooth muscle. SP-immunoreactive nerve fibers also innervate parasympathetic ganglia, suggesting a sensory input which may modulate ganglionic transmission and so result in ganglionic reflexes.

SP in the airways is localized predominantly to capsaicin-sensitive unmyelinated nerves in the airways, but chronic administration of capsaicin only partially depletes the lung of tachykinins, indicating the presence of a population of capsaicin-resistant SP-immunoreactive nerves, as in the gastrointestinal tract.[9,10] Similar capsaicin denervation studies are not possible in human airways, but after extrinsic denervation by heart-lung transplantation there appears to be a loss of SP-immunoreactive nerves in the submucosa.[11]

A. EFFECTS ON AIRWAYS

Tachykinins have many different effects on the airways which may be relevant to asthma and these effects are mediated via NK_1 and NK_2 receptors (activated by NKA), whereas there is no evidence of NK_3 receptors. Tachykinins constrict smooth muscle of human airways *in vitro* via NK_2 receptors.[12,13] The contractile response to NKA is significantly greater in smaller human bronchi than in more proximal airways, indicating that tachykinins may have a more important constrictor effect in peripheral airways,[14] whereas cholinergic constriction tends to be more pronounced in proximal airways. This is consistent with the autoradiographic distribution of tachykinin receptors, which are distributed to small and large airways. *In vivo* SP does not cause bronchoconstriction or cough, either by intravenous infusion,[15,16] or by inhalation,[15,17] whereas NKA causes bronchoconstriction after both intravenous administration[16] and inhalation in asthmatic subjects.[17] Mechanical removal of airway epithelium potentiates the bronchoconstrictor response to tachykinins,[18,19] largely because the ectoenzyme neutral endopeptidase (NEP), which is a key enzyme in the degradation of tachykinins in airways, is strongly expressed on epithelial cells.

SP stimulates mucus secretion from submucosal glands in ferret and human airways *in vitro*[20,21] and is a potent stimulant to goblet cell secretion in guinea pig airways.[22] Indeed SP is likely to mediate the increase in goblet cell discharge after vagus nerve stimulation and exposure to cigarette smoke.[23,24]

Stimulation of the vagus nerve in rodents causes microvascular leakage, which is prevented by prior treatment with capsaicin or by a tachykinin antagonist, indicating that release of tachykinins from sensory nerves mediates this effect. Among the tachykinins, SP is most potent at causing leakage in guinea pig airways[25] and NK_1 receptors have been localized to postcapillary venules in the airway submucosa.[26] Inhaled SP also causes microvascular leakage in guinea pigs and its effect on the microvasculature is more marked than its effect on airway smooth muscle.[27] It is difficult to measure airway microvascular leakage in human airways, but SP causes a wheal in human skin when injected intradermally, indicating the capacity to cause microvascular leak in human postcapillary venules; NKA is less potent, indicating that an NK_1 receptor mediates this effect.[28]

Tachykinins have potent effects on airway blood flow. Indeed the effect of tachykinins on airway blood flow may be the most important physiological and pathophysiological role of tachykinins in airways. In canine and porcine trachea both SP and NKA cause a marked increase in blood flow.[29,30] Tachykinins also dilate canine bronchial vessels *in vitro,* probably via an endothelium-dependent mechanism.[31] Tachykinins also regulate bronchial blood flow in pig; stimulation of the vagus nerve causes a vasodilatation mediated by the release of sensory neuropeptides, and it is likely that CGRP as well as tachykinins are involved.[30]

Tachykinins may also interact with inflammatory and immune cells,[32,33] although whether this is of pathophysiological significance remains to be determined. There is likely to be increasing research in the area of neuroimmune interaction and in some species there is already evidence for neuropeptide innervation of bronchus-associated lymphoid tissue.[34] SP degranulates certain types of mast cell, such as those in human skin, although this is not mediated via a tachykinin receptor.[35] There is no evidence that tachykinins degranulate lung mast cells.[36] SP has a degranulating effect on eosinophils;[37] again, the degranulation is related to high concentrations of peptide and, as for mast cells, is not mediated via a tachykinin receptor. At lower concentrations tachykinins have been reported to enhance eosinophil chemotaxis. Tachykinins may activate alveolar macrophages[38] and monocytes to release inflammatory cytokines, such as interleukin (IL)-6.[39] Tachykinins and vagus nerve stimulation also cause transient vascular adhesion of neutrophils in the airway circulation.[40]

In guinea pig trachea tachykinins also potentiate cholinergic neurotransmission at postganglionic nerve terminals, and an NK_2 receptor appears to be involved.[41] There is also potentiation at the ganglionic level,[42,43] which appears to be mediated via a NK_1 receptor.[43] Endogenous tachykinins may also facilitate cholinergic neurotransmission, since capsaicin pretreatment results in a significant reduction in cholinergic neural responses both *in vitro* and *in vivo.*[44,45] However, in human airways there is no evidence for a facilitatory effect on cholinergic neurotransmission,[46] although such an effect has been reported in the presence of potassium channel blockers.[47]

In conscious guinea pigs very low concentrations of inhaled SP are reported to cause cough and this effect is potentiated by NEP inhibition.[48] Citric acid-induced cough is blocked by a

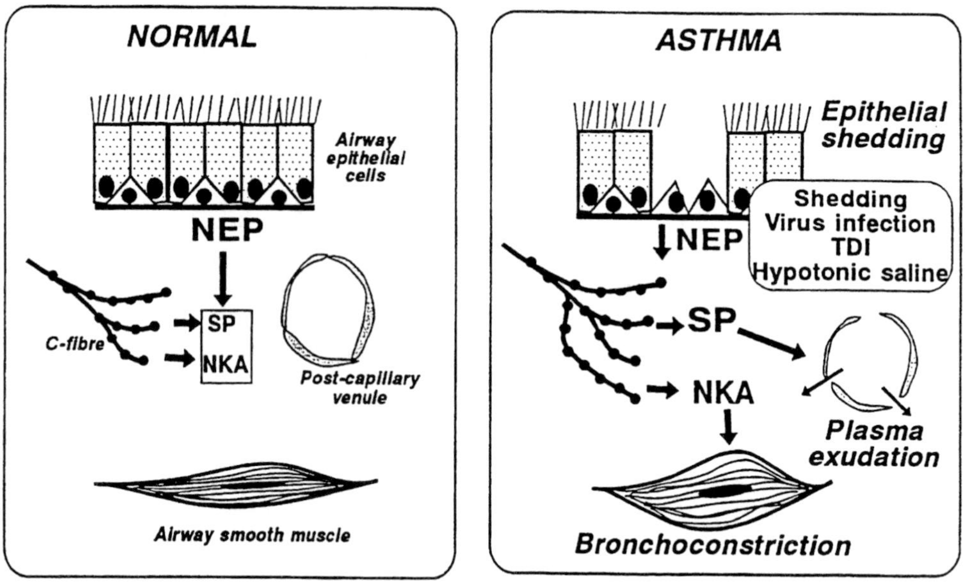

FIGURE 1. Interaction of tachykinins with airway epithelium. When epithelium is intact neutral endopeptidase (NEP) degrades substance P (SP) and neurokinin A (NKA) released from sensory nerves (left panel). In asthmatic airways when epithelium is shed or NEP downregulated any tachykinins released will have an exaggerated effect (right panel).

nonpeptide NK_2 receptor antagonist (SR 48968), suggesting the presence of NK_2 receptors on airway sensory nerves.[49]

B. METABOLISM

Tachykinins are subject to degradation by at least two enzymes, angiotensin converting enzyme (ACE) and NEP.[50] ACE is predominantly localized to vascular endothelial cells and therefore breaks down intravascular peptides. ACE inhibitors, such as captopril, enhance bronchoconstriction due to intravenous SP,[51,52] but not inhaled SP.[53] NKA is not a good substrate for ACE, however. NEP appears to be the most important enzyme for the breakdown of tachykinins in tissues. Inhibition of NEP by phosphoramidon or thiorphan markedly potentiates bronchoconstriction *in vitro* in animal[54] and human airways[55] and after inhalation *in vivo.*[53] NEP inhibition also potentiates mucus secretion in response to tachykinins in human airways.[20] NEP inhibition enhances excitatory nonadrenergic, noncholinergic (e-NANC)- and capsaicin-induced bronchoconstriction, due to the release of tachykinins from airway sensory nerves.[18,56]

The activity of NEP in the airways appears to be an important factor in determining the effects of tachykinins; any factors that inhibit the enzyme or its expression may be associated with increased effects of exogenous or endogenously released tachykinins. Several of the stimuli known to induce bronchoconstrictor responses in asthmatic patients have been found to reduce the activity of airway NEP[50] (Figure 1).

IV. CALCITONIN GENE-RELATED PEPTIDE

CGRP-immunoreactive nerves are abundant in the respiratory tract of several species. CGRP is costored and colocalized with SP in afferent nerves.[57] CGRP has been extracted from and is localized to human airways.[8,58] CGRP-immunoreactive nerve fibers appear to be more abundant than SP fibers, possibly because CGRP has greater stability, and is also present in

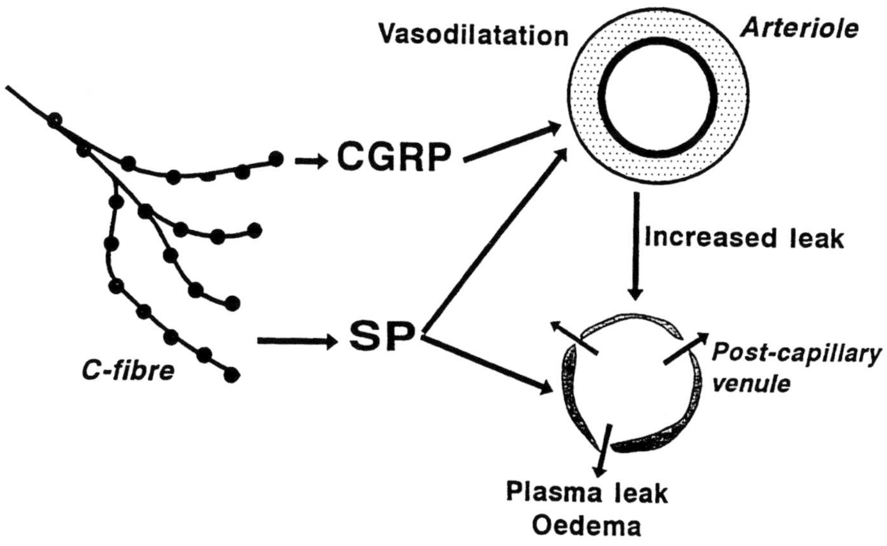

FIGURE 2. Effect of sensory neuropeptides in airway vessels. Substance P causes vasodilatation and plasma exudation, whereas calcitonin gene-related peptide causes vasodilatation of arterioles, which may theoretically increase plasma extravasation by increasing blood delivery to leaky postcapillary venules.

some nerves that do not contain SP. CGRP is found in trigeminal, nodose-jugular, and dorsal root ganglia[59] and has also been detected in neuroendocrine cells of the lower airways.

CGRP is a potent vasodilator, which has long-lasting effects. CGRP is an effective dilator of human pulmonary vessels *in vitro* and acts directly on receptors on vascular smooth muscle.[60] It also potently dilates bronchial vessels *in vitro*[60] and produces a marked and long-lasting increase in airway blood flow in anesthetized dogs[61] and conscious sheep *in vivo*.[62] Receptor mapping studies have demonstrated that CGRP receptors are localized predominantly to bronchial vessels rather than to smooth muscle or epithelium in human airways.[63] It is possible that CGRP may be the predominant mediator of arterial vasodilatation and increased blood flow in response to sensory nerve stimulation in the bronchi.[30] CGRP may be an important mediator of airway hyperemia in asthma.

By contrast, CGRP has no direct effect of airway microvascular leak.[25] In the skin, CGRP potentiates the leakage produced by SP, presumably by increasing the blood delivery to the sites of plasma extravasation in the postcapillary venules.[64] This does not occur in guinea pig airways when CGRP and SP are coadministered, possibly because blood flow in the airways is already high,[25] although an increased leakage response has been reported in rat airways.[65] It is possible that potentiation of leak may occur when the two peptides are released together from sensory nerves (Figure 2).

CGRP causes constriction of human bronchi *in vitro*.[58] This is surprising, since CGRP normally activates adenylyl cyclase, an event that is usually associated with bronchodilatation. Receptor mapping studies suggest few, if any, CGRP receptors over airway smooth muscle in human or guinea pig airways and this suggests that the paradoxical bronchoconstrictor response reported in human airways may be mediated indirectly. In guinea pig airways, CGRP has no consistent effect on tone.[66]

CGRP has a weak inhibitory effect on cholinergically stimulated mucus secretion in ferret trachea[67] and on goblet cell discharge in guinea pig airways.[22] This is probably related to the low density of CGRP receptors on mucus secretory cells, but does not preclude the possibility that CGRP might increase mucus secretion *in vivo* by increasing blood flow to submucosal glands.

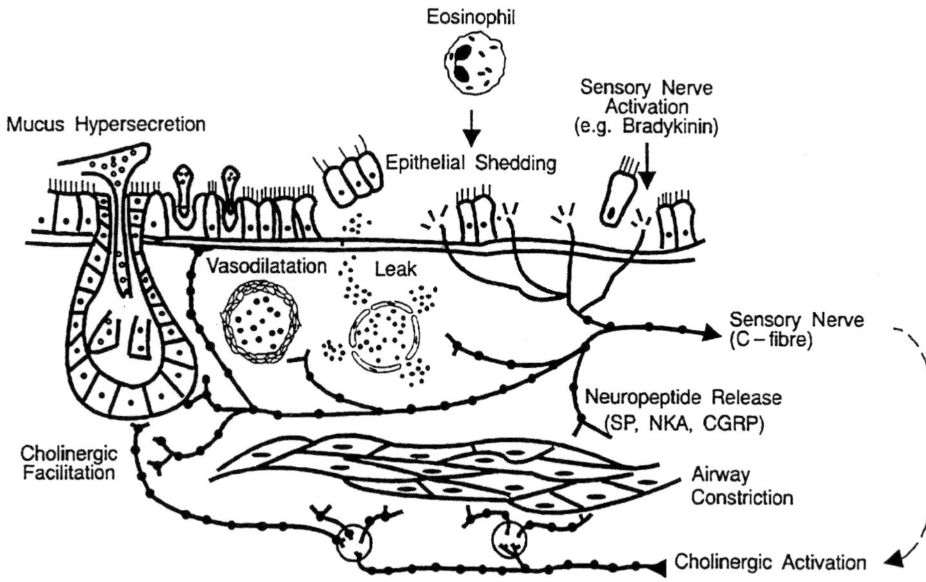

FIGURE 3. Possible neurogenic inflammation in asthmatic airways via retrograde release of peptides from sensory nerves via an axon reflex. Substance P (SP) causes vasodilatation, plasma exudation, and mucus secretion, whereas neurokinin A (NKA) causes bronchoconstriction and enhanced cholinergic reflexes and calcitonin gene-related peptide (CGRP) vasodilatation.

CGRP injection into human skin causes a persistent flare, but biopsies have revealed an infiltration of eosinophils.[68] CGRP itself does not appear to be chemotactic for eosinophils, but proteolytic fragments of the peptide are active,[69] suggesting that CGRP released into the tissues may lead to eosinophilic infiltration.

CGRP inhibits the proliferative response of T-lymphocytes to mitogens and specific receptors have been demonstrated on these cells.[70] CGRP also inhibits macrophage secretion and the capacity of macrophages to activate T-lymphocytes.[71] This suggests that CGRP has potential anti-inflammatory actions in the airways.

V. NEUROGENIC INFLAMMATION IN AIRWAYS

In rodents there is now considerable evidence for neurogenic inflammation in the airways due to the antidormic release of neuropeptides from nociceptive nerves or C-fibers via an axon reflex[72-74] and it is possible that it may contribute to the inflammatory response in asthma and chronic bronchitis[75,76] (Figure 3).

A. NEUROGENIC AIRWAY INFLAMMATION IN ANIMAL MODELS

There are several lines of evidence that neurogenic inflammation may be important in animal models that may have relevance to asthma. These models have usually been in rodents where tachykinin effects are pronounced and may not be predictive of the role of tachykinins in human airways, however. There are four main experimental approaches which have been used to assess the role of sensory neuropeptides in animal models of asthma; these include studies of depletion with capsaicin, enhancement with inhibitors of NEP, tachykinin receptor antagonists, and inhibitors of sensory neuropeptide release.

B. CAPSAICIN DEPLETION STUDIES

Capsaicin pretreatment depletes neuropeptides from C-fibers, either in neonatal animals (which results in degeneration of C-fibers) or acute treatment in adult animals (resulting in depletion of sensory neuropeptides). In rat trachea capsaicin pretreatment inhibits the microvascular leakage induced by irritant gases, such as cigarette smoke,[77] and inhibits goblet cell discharge and microvascular leak induced by cigarette smoke in guinea pigs.[24] Capsaicin-sensitive nerves may also contribute to the bronchoconstriction and microvascular leak induced by isocapnic hyperventilation,[78] hypocapnia,[79] inhaled sodium metabisulfite,[80] and nebulized hypertonic saline[81] and toluene diisocyanate[82] in rodents. In guinea pigs capsaicin pretreatment has little or no effect on the acute bronchoconstrictor or plasma exudation response to allergen inhalation in sensitized animals.[83] Administration of capsaicin increases airway responsiveness in guinea pigs to cholinergic agonists, and this effect is prevented by prior treatment with capsaicin, suggesting that capsaicin-sensitive nerves release products that increase airway responsiveness.[84] In pigs capsaicin pretreatment inhibits the vasodilator response to allergen (which may be mediated by the release of CGRP).[85] In allergic sheep capsaicin pretreatment prevents the airway hyperresponsiveness to both allergen and cholinergic agonists.[86] In a model of chronic allergen exposure in guinea pigs capsaicin pretreatment results in complete inhibition of airway hyperresponsiveness, without any change in the eosinophil inflammatory response.[87] In rabbits neonatal capsaicin treatment inhibits the airway hyperresponsiveness associated with neonatal allergen sensitization, although this does not appear to be associated with any change in content of sensory neuropeptides in lung tissue.[88] This suggests that capsaicin-sensitive nerves may play a role in chronic inflammatory responses to allergen.

There has been speculation that mast cells in the airways might be influenced by capsaicin-sensitive nerves. Histological studies have demonstrated a close proximity between mast cells and sensory nerves in airways.[89] There is also evidence that antidromic stimulation of the vagus nerve leads to mast cell mediator release in canine airways.[90] Furthermore, allergen exposure has effects on ion transport in guinea pig airways which are dependent on capsaicin-sensitive nerves.[91]

C. INHIBITION OF NEUROPEPTIDE METABOLISM

The activity of NEP may be an important determinant of the extent of neurogenic inflammation in airways and inhibition of NEP by thiorphan or phosphoramidon has been shown to enhance neurogenic inflammation in various rodent models. NEP is not specific to tachykinins and is also involved in the metabolism of other bronchoactive peptides, including kinins and endothelins. Certain virus infections enhance e-NANC responses in guinea pigs[92] and mycoplasma infection enhances neurogenic microvascular leakage in rats,[73] an effect that is mediated by inhibition of NEP activity. Influenza virus infection of ferret trachea *in vitro* and of guinea pigs *in vivo* inhibits the activity of epithelial NEP and markedly enhances the bronchoconstrictor responses to tachykinins.[93] Similarly, Sendai virus infection potentiates neurogenic inflammation in rat trachea.[94] This may explain why respiratory tract virus infections are so deleterious to patients with asthma. Hypertonic saline also impairs epithelial NEP function, leading to exaggerated tachykinin responses,[81] and cigarette smoke exposure has a similar effect which can be explained by an oxidizing effect on the enzyme.[95] Toluene diisocyanate, albeit at rather high doses, also reduces NEP activity and this may be a mechanism contributing to the airway hyperresponsiveness that may follow exposure to this chemical.[96] Inhalation of IL-1a is associated with increased responsiveness to bradykinin and this may be due to inhibition of NEP expression.[97] Thus, many of the agents that lead to exacerbations of asthma appear to reduce the activity of NEP at the airway surface, thus leading to exaggerated responses to tachykinins (and other peptides) and so to increased airway inflammation.

D. INHIBITION OF SENSORY NEUROPEPTIDE EFFECTS

Specific antagonists of tachykinin receptors have now been developed and provide a more specific tool to investigate the role of tachykinins in animal models. Several highly potent and stable peptide and nonpeptide tachykinin antagonists have recently been developed which are

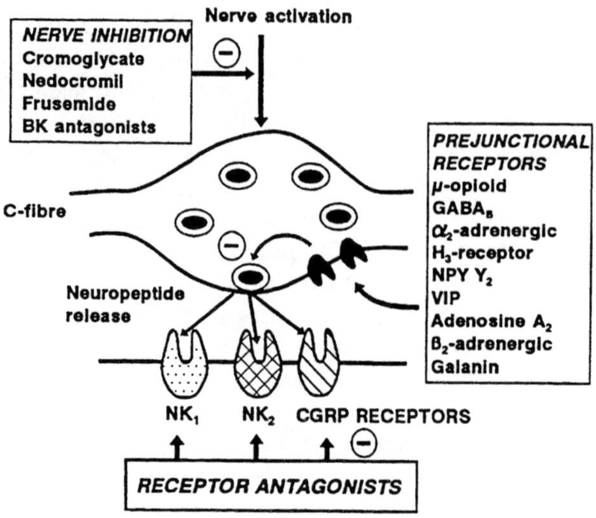

FIGURE 4. Modulation of neurogenic inflammation in airway sensory nerves.

highly selective for either NK_1 or NK_2 receptors.[98] The NK_1 receptor antagonist CP 96,345 is able to block the plasma exudation response to vagus nerve stimulation and to cigarette smoke in guinea pig airways,[99,100] without affecting the bronchoconstrictor response, which is blocked by the NK_2 antagonist SR 48,968.[13] Similar results have been obtained with the very potent NK_1 selective antagonist FK 888.[101] CP 96,345 also blocks hyperpnea- and bradykinin-induced plasma exudation in guinea pigs,[102,103] but has no effect on the acute plasma exudation induced by allergen in sensitized animals.[103] These specific antagonists are very useful new tools in probing the involvement of tachykinins in disease, and will be invaluable in clinical studies in the future.

E. INHIBITION OF SENSORY NEUROPEPTIDE RELEASE

Several agonists act on prejunctional receptors on airway sensory nerves to inhibit the release of neuropeptides and neurogenic inflammation[104] (Figure 4). Opioids are the most effective inhibitory agonists, acting via prejunctional μ-receptors and have been shown to inhibit cigarette smoke-induced discharge from goblet cells in guinea pig airways *in vivo*[105] and to inhibit ozone-induced hyperreactivity in guinea pigs, which appears to be mediated via sensory nerves.[106] Several other agonists are also effective and may act by opening a common calcium-activated large conductance potassium channel in sensory nerves.[107] Openers of other potassium channels, which achieve the same hyperpolarization of the sensory nerve, are also effective in blocking neurogenic inflammation in rodents[108] and have been shown to block cigarette smoke-induced goblet cell secretion in guinea pigs.[109]

VI. NEUROGENIC INFLAMMATION IN AIRWAY DISEASES

A. ASTHMA

Although it was proposed several years ago that neurogenic inflammation and peptides released from sensory nerves might be important as an amplifying mechanism in asthmatic inflammation,[75] there is little evidence to date to support this idea, despite the extensive work in rodent models. This is partly because it has proved difficult to apply the same approaches to human volunteers.

1. Sensory Nerves in Human Airways

In comparison with rodent airways, SP- and CGRP-immunoreactive nerves are very sparse in human airways. Quantitative studies indicate that SP-immunoreactive fibers constitute only 1% of the total number of intraepithelial fibers, whereas in guinea pig they make up 60% of the fibers.[110] This raises the possibility that sensory nerves in humans may contain some unidentified transmitter which may be involved in neurogenic inflammation. Chronic inflammation may lead to changes in the pattern of innervation, through the release of neurotrophic factors from inflammatory cells. Thus, in chronic arthritis and inflammatory bowel disease there is an increase in the density of SP-immunoreactive nerves.[111,112] A striking increase in SP-like immunoreactive nerves has been reported in the airway of patients with fatal asthma.[113] This increased density of nerves is particularly noticeable in the submucosa. Whether this apparent increase is due to proliferation of sensory nerves or is due to increased synthesis of tachykinins has not yet been established. Elevated concentrations of SP in bronchoalveolar lavage of patients with asthma have been reported, with a further rise after allergen challenge,[114] suggesting that there may be an increase in SP in the airways of asthmatic patients. Similarly, SP has been detected in the sputum of asthmatic patients after hypertonic saline inhalation.[115] SP also increases in the bronchoalveolar lavage of normal volunteers exposed to ozone, possible because of a reduction in NEP activity.[116]

Cultured sensory neurons are stimulated by nerve growth factor (NGF), which markedly increases the transcription of preprotachykinin (PPT)-A gene, the major precursor peptide for tachykinins.[117] Similarly, adjuvant-induced inflammation in rat spinal cord increases the gene expression of PPT-A.[118] Preliminary studies suggest that allergen challenge is associated with a doubling in PPT-A mRNA-positive neurons in nodose ganglia of guinea pigs and an increase in SP immunoreactivity in the lungs.[119] However, bronchial biopsies of mildly asthmatic patients have not revealed any evidence of increased SP-immunoreactive nerves.[120] This may indicate that the increased innervation[113] may be a feature of either prolonged or severe asthma and indicate the need for more studies.

2. Sensory Nerve Activation

Sensory nerves may be activated in airway disease. In asthmatic airways the epithelium is often shed, thereby exposing sensory nerve endings. Sensory nerves in asthmatic airways may be "hyperalgesic" as a result of exposure to inflammatory mediators such as prostaglandins and certain cytokines, such as IL-1α and tumor necrosis factor (TNF)-α.[121]

Capsaicin induces bronchoconstriction and plasma exudation in guinea pigs[9] and increases airway blood flow in pigs.[85] In humans capsaicin inhalation causes cough and a *transient* bronchoconstriction, which is inhibited by cholinergic blockade and is probably due to a laryngeal reflex.[122,123] This suggests that neuropeptide release does not occur in human airways, although it is possible that insufficient capsaicin reaches the lower respiratory tract because the dose is limited by coughing. In patients with asthma, there is no evidence that capsaicin induces a greater degree of bronchoconstriction than in normal individuals.[122]

Bradykinin is a potent bronchoconstrictor in asthmatic patients and also induces coughing and a sensation of chest tightness, which closely mimics a naturally occurring asthma attack.[124,125] Yet it is a weak constrictor of human airways *in vitro,* suggesting that its potent constrictor effect is mediated indirectly. Bradykinin is a potent activator of bronchial C-fibers in dogs,[126] and releases sensory neuropeptides from perfused rodent lungs.[127] In guinea pigs bradykinin instilled into the airways causes bronchoconstriction, which is reduced significantly by a cholinergic antagonist (as in asthmatic patients[125]), and also by capsaicin pretreatment.[128] The plasma leakage induced by inhaled bradykinin is inhibited by an NK_1 antagonist and bronchoconstriction by an NK_2 antagonist.[103,129] This indicates that bradykinin activates sensory nerves in the airways and that part of the airway response is mediated by release of constrictor peptides from capsaicin-sensitive nerves. In asthmatic patients an inhaled nonselective tachykinin antagonist FK 224 has been shown to reduce the bronchoconstrictor response to inhaled bradykinin and also to block the cough response in those patients who coughed in response to bradykinin.[130]

3. Studies with NEP Inhibitors

In rodents inhibition of NEP with thiorphan or phosphoramidon results in striking potentiation of tachykinin- and sensory nerve-induced effects, and has been used as an approach to explore the potential for neurogenic inflammation in disease.[50] Intravenous acetorphan, which is hydrolyzed to thiorphan, was administered to asthmatic subjects and while there was potentiation of the wheal and flare response to intradermal SP, there was no effect on baseline airway caliber or on bronchoconstriction induced by a "neurogenic" trigger sodium metabisulfite.[131] The lack of effect could be due to inadequate inhibition of NEP in the airways, and particularly at the level of the epithelium. Nebulized thiorphan has been shown to potentiate the bronchoconstrictor response to inhaled NKA in normal and asthmatic subjects,[132,133] but there was no effect on baseline lung function in asthmatic patients,[133] indicating that there is unlikely to be any basal release of tachykinins. NEP is strongly expressed in the human airway,[134] but there is no evidence, based on immunocytochemical staining or *in situ* hybridization, that it is defective in asthmatic airways,[133a] and the fact that after inhaled thiorphan the bronchoconstrictor response to inhaled NKA is further enhanced in asthmatic subjects provides supportive functional data that NEP function may not be impaired, at least in mild asthma.[133] Of course, it is possible that NEP may become dysfunctional after viral infections or exposure to oxidants and thus contribute to asthma exacerbations.

4. Tachykinin Responsiveness

In inflammatory bowel disease there is evidence for a marked upregulation of tachykinin receptors, particularly in the vasculature, suggesting that chronic inflammation may lead to changes in tachykinin receptor expression.[135,136] In patients with allergic rhinitis an increased vascular response to nasally applied SP is observed.[137] There is evidence that NK_1 receptor gene expression may be increased in the lungs of asthmatic patients.[138] This might be due to increased transcription in response to activation of transcription factors, such as AP-1, which are activated in human lung by cytokines such as TNF-α.[139] A consensus sequence for AP-1 binding has been identified upstream of the NK_1 receptor gene.[140] Corticosteroids, conversely, reduce NK_1 receptor gene expression,[141] presumably via an inhibitory effect on AP-1 activation.

5. Modulation of Neurogenic Inflammation

Apart from tachykinin receptor antagonists neurogenic inflammation may be modulated by either preventing activation of sensory nerves, or preventing the release of neuropeptides. Both approaches may be tried in asthmatic patients, using currently available drugs, although these approaches are not as specific as tachykinin antagonists, as the drugs used have additional effects.

Activation of sensory nerves may be inhibited by local anesthetics, but it has proved to be very difficult to achieve adequate local anesthesia of the respiratory tract. Inhalation of local anesthetics, such as lidocaine, have not been found to have consistent inhibitory effects on various airway challenges, and indeed may even promote bronchoconstriction in some patients with asthma.[142] This paradoxical bronchoconstriction may be due to the greater anesthesia of laryngeal afferents which are linked to a tonic nonadrenergic bronchodilator reflex.[143,144] Other drugs may inhibit the activation of airway sensory nerves. Cromolyn sodium and nedocromil sodium may have direct effects on airway C-fibers[145,146] and this might contribute to their antiasthma effect. Nedocromil sodium is highly effective against bradykinin-induced and sulfur dioxide-induced bronchoconstriction in asthmatic patients,[145,147] which are believed to be mediated by activation of sensory nerves in the airways. In addition, nedocromil sodium and, to a much lesser extent, cromolyn sodium inhibit the e-NANC neural bronchoconstriction due to tachykinin release from sensory nerves in guinea pig bronchi *in vitro*, indicating an effect on release of sensory neuropeptides as well as on activation.[148] The loop diuretic furosemide (frusemide) given by nebulization, behaves in a similar fashion to nedocromil sodium and inhibits metabisulfite-induced bronchoconstriction in asthmatic patients[149] and also e-NANC and cholinergic bronchoconstriction in guinea pig airways *in vitro*.[150] In addition, nebulized furosemide also inhibits certain types of cough,[151] providing further evidence for an effect on sensory nerves.

Many drugs act on prejunctional receptors to inhibit the release of neuropeptides, as discussed above. Opioids are the most effective inhibitors, but an inhaled μ-opioid agonist, the pentapeptide BW 443C, was found to be ineffective in inhibiting metabisulfite-induced bronchoconstriction, which is believed to act via neural mechanisms.[152] One problem with BW 443C is that it may be degraded by NEP in the airway epithelium and may not, therefore, reach a high enough concentration in the vicinity of the airway sensory nerves. Another agent that has a prejunctional modulatory effect in guinea pigs is the H_3 receptor agonist α-methyl histamine.[153] However, inhalation of α-methyl histamine had no effect on either resting tone or metabisulfite-induced bronchoconstriction in asthmatic patients.[154]

B. CHRONIC BRONCHITIS

Chronic bronchitis is usually due to cigarette smoking and is characterized by chronic mucus hypersecretion. It is possible that the irritant effect of cigarette smoke activates neurogenic inflammation, resulting in increased release of SP that acts on submucosal glands and goblet cells to increase mucus secretion. Studies in guinea pigs show that cigarette smoke activates goblet cell secretion via an axon reflex that releases SP.[24] This is blocked by NK_1 receptor antagonists, suggesting that such drugs may be useful in controlling mucus hypersecretion in chronic bronchitis. Indeed, preliminary clinical data indicate that a nonselective tachykinin antagonist, FK-224, taken over 4 weeks has an inhibitory effect on cough and mucus production in patients with chronic bronchitis.[155]

REFERENCES

1. Uddman, R., Hakanson, R., Luts, A., and Sundler, F., Distribution of neuropeptides in airways, in *Autonomic Control of the Respiratory System*, Barnes, P. J., Ed., Horwood Academic, London, 1995, chap. 2.
2. Barnes, P. J., Baraniuk, J., and Belvisi, M. G., Neuropeptides in the respiratory tract, *Am. Rev. Respir. Dis.*, 144, 1187, 1991.
3. Kaliner, M., Barnes, P. J., Kunkel, G. H. H., and Baraniuk, J. N., *Neuropeptides in Respiratory Medicine*, Marcel Dekker, New York, 1994.
4. Barnes, P. J., Neuropeptides and asthma, *Am. Rev. Respir. Dis.*, 143, S28, 1991.
5. Smith, H., Animal models of asthma, *Pulm. Pharmacol.*, 2, 59, 1989.
6. Martling, C. R., Theodorsson-Norheim, E., and Lundberg, J. M., Occurrence and effects of multiple tachykinins: substance P, neurokinin A, and neuropeptide K in human lower airways, *Life Sci.*, 40, 1633, 1987.
7. Laitinen, L. A., Laitinen, A., Panula, P. A., Partenen, M., Tervo, K., and Tervo, T., Immunohistochemical demonstration of substance P in the lower respiratory tract of the rabbit and not of man, *Thorax*, 38, 531, 1983.
8. Komatsu, T., Yamamoto, M., Shimokata, K., and Nagura, H., Distribution of substance-P-immunoreactive and calcitonin gene-related peptide-immunoreactive nerves in normal human lungs, *Int. Arch. Allergy Appl. Immunol.*, 95, 23, 1991.
9. Lundberg, J. M., Saria, A., Lundblad, L., Angaard, A., Marting, C.-R., Theodorsson-Norheim, E., Stjarne, P., and Hokfelt, T., Bioactive peptides in capsaicin-sensitive C-fiber afferents of the airways: functional and pathophysiological implications, in *Neural Control of Airway*, Kaliner, M. A. and Barnes, P. J., Eds., Marcel Decker, New York, 1987, 417-445.

10. Dey, R. D., Altemus, J. B., and Michalkiewicz, M., Distribution of vasoactive intestinal peptide- and substance P-containing nerves originating from neurons of airway ganglia in cat bronchi, *J. Comp. Neurol.*, 304, 330, 1991.
11. Springall, D. R., Polak, J. M., Howard, L., Power, R. F., Krausz, T., Manickan, S., Banner, N. R., Khagani, A., Rose, M., and Yacoub, M. H., Persistence of intrinsic neurones and possible phenotypic changes after extrinsic denervation of human respiratory tract by heart-lung transplantation, *Am. Rev. Respir. Dis.*, 141, 1538, 1990.
12. Naline, E., Devillier, P., Drapeau, G., Totly, L., Bakdach, H., Regoli, D., and Advenier, C., Characterization of neurokinin effects on receptor selectivity in human isolated bronchi, *Am. Rev. Respir. Dis.*, 140, 679, 1989.
13. Advenier, C., Naline, E., Toty, L., Bakdach, H., Emors-Alt, X., Vilain, P., Breliere, J.-C., and Le Fur, G., Effects on the isolated human bronchus of SR 48968, a potent and selective nonpeptide antagonist of the neurokinin A (NK_2) receptors, *Am. Rev. Respir. Dis.*, 146, 1177, 1992.
14. Frossard, N. and Barnes, P. J., Effects of tachykinins on small human airways and the influence of thiorphan, *Am. Rev. Respir. Dis.*, 137, 195A, 1988.
15. Fuller, R. W., Maxwell, D. L., Dixon, C. M. S., McGregor, G. P., Barnes, V. F., Bloom, S. R., and Barnes, P. J., The effects of substance P on cardiovascular and respiratory function in human subjects, *J. Appl. Physiol.*, 62, 1473, 1987.
16. Evans, T. W., Dixon, C. M., Clarke, B., Conradson, T. B., and Barnes, P. J., Comparison of neurokinin A and substance P on cardiovascular and airway function in man, *Br. J. Pharmacol.*, 25, 273, 1988.
17. Joos, G., Pauwels, R., and van der Straeten, M. E., Effect of inhaled substance P and neurokinin A in the airways of normal and asthmatic subjects, *Thorax*, 42, 779, 1987.
18. Frossard, N., Rhoden, K. J., and Barnes, P. J., Influence of epithelium on guinea pig airway responses to tachykinins: role of endopeptidase and cyclooxygenase, *J. Pharmacol. Exp. Ther.*, 248, 292, 1989.
19. Devillier, P., Advenier, C., Drapeau, G., Marsac, J., and Regoli, D., Comparison of the effects of epithelium removal and of an enkephalinase inhibitor on the neurokinin-induced contractions of guinea pig isolated trachea, *Br. J. Pharmacol.*, 94, 675, 1988.
20. Rogers, D. F., Aursudkij, B., and Barnes, P. J., Effects of tachykinins on mucus secretion on human bronchi *in vitro*, *Eur. J. Pharmacol.*, 174, 283, 1989.
21. Ramnarine, S. I., Hirayama, Y., Barnes, P. J., and Rogers, D. F., "Sensory-efferent" neural control of mucus secretion: characterization using tachykinin receptor antagonists in ferret trachea *in vitro*, *Br. J. Pharmacol.*, 113, 1183, 1994.
22. Kuo, H.-P., Rhode, J. A. L., Tokuyama, K., Barnes, P. J., and Rogers, D. F., Capsaicin and sensory neuropeptide stimulation of goblet cell secretion in guinea pig trachea, *J. Physiol.*, 431, 629, 1990.
23. Tokuyama, K., Kuo, H.-P., Rohde, J. A. L., Barnes, P. J., Rogers, D. F., Neural control of goblet cell secretion in guinea pig airways, *Am. J. Physiol.*, 259, L108, 1990.
24. Kuo, H.-P., Barnes, P. J., and Rogers, D. F., Cigarette smoke-induced airway goblet cell secretion: dose dependent differential nerve activation, *Am. J. Physiol.*, 7, L161, 1992.
25. Rogers, D. F., Belvisi, M. G., Aursudkij, B., Evans, T. W., and Barnes, P. J., Effects and interactions of sensory neuropeptides on airway microvascular leakage in guinea pigs, *Br. J. Pharmacol.*, 95, 1109, 1988.
26. Sertl, K., Wiedermann, C. J., Kowalski, M. L., Hurtado, S., Plutchok, J., Linnoila, I., Pert, C. B., and Kaliner, M. A., Substance P: the relationship between receptor distribution in rat lung and the capacity of substance P to stimulate vascular permeability, *Am. Rev. Respir. Dis.*, 138, 151, 1988.
27. Lötvall, J. O., Lemen, R. J., Hui, K. P., Barnes, P. J., and Chung, K. F., Airflow obstruction after substance P aerosol: contribution of airway and pulmonary edema, *J. Appl. Physiol.*, 69, 1473, 1990.
28. Fuller, R. W., Conradson, T.-B., Dixon, C. M. S., Crossman, D. C., and Barnes, P. J., Sensory neuropeptide effects in human skin, *Br. J. Pharmacol.*, 92, 781, 1987.
29. Salonen, R. O., Webber, S. E., and Widdicombe, J. G., Effects of neuropeptides and capsaicin on the canine tracheal vasculature in vivo, *Br. J. Pharmacol.*, 95, 1262, 1988.
30. Matran, R., Alving, K., Martling, C. R., Lacroix, J. S., and Lundberg, J. M., Effects of neuropeptides and capsaicin on tracheobronchial blood flow in the pig, *Acta Physiol. Scand.*, 135, 335, 1989.
31. McCormack, D. G., Salonen, R. O., and Barnes, P. J., Effect of sensory neuropeptides on canine bronchial and pulmonary vessels *in vitro*, *Life Sci.*, 45, 2405, 1989.
32. McGillis, J. P., Organist, M. L., and Payan, D. G., Substance P and immunoregulation, *Fed. Proc.*, 14, 120, 1987.
33. Daniele, R. P., Barnes, P. J., Goetzl, E. J., Nadel, J., O'Dorisio, S., Kiley, J., and Jacobs, T., Neuroimmune interactions in the lung, *Am. Rev. Respir. Dis.*, 145, 1230, 1992.
34. Nohr, D. and Weihe, E., The neuroimmune link in the bronchus-associated lymphoid tissue (BALT) of cat and rat: peptides and neural markers, *Brain Behav. Immunol.*, 5, 84, 1991.
35. Lowman, M. A., Benyon, R. C., and Church, M. K., Characterization of neuropeptide-induced histamine release from human dispersed skin mast cells, *Br. J. Pharmacol.*, 95, 121, 1988.

36. Ali, H., Leung, K. B. I., Pearce, F. L., Hayes, N. A., and Foremean, J. C., Comparison of histamine releasing activity of substance P on mast cells and basophils from different species and tissues, *Int. Arch. Allergy,* 79, 121, 1986.
37. Kroegel, C., Giembycz, M. A., and Barnes, P. J., Characterization of eosinophil activation by peptides. Differential effects of substance P, mellitin, and f-met-leu-phe, *J. Immunol.,* 145, 2581, 1990.
38. Brunelleschi, S., Vanni, L., Ledda, F., Giotti, A., Maggi, C. A., and Fantozzi, R., Tachykinins activate guinea pig alveolar macrophages: involvement of NK_2 and NK_1 receptors, *Br. J. Pharmacol.,* 100, 417, 1990.
39. Lotz, M., Vaughn, J. H., and Carson, D. M., Effect of neuropeptides on production of inflammatory cytokines by human monocytes, *Science,* 241, 1218, 1988.
40. Umeno, E., Nadel, J. A., Huang, H. T., and McDonald, D. M., Inhibition of neutral endopeptidase potentiates neurogenic inflammation in the rat trachea, *J. Appl. Physiol.,* 66, 2647, 1989.
41. Hall, A. K., Barnes, P. J., Meldrum, L. A., and Maclagan, J., Facilitation by tachykinins of neurotransmission in guinea-pig pulmonary parasympathetic nerves, *Br. J. Pharmacol.,* 97, 274, 1989.
42. Undem, B. J., Myers, A. C., Barthlow, H., and Weinreich, D., Vagal innervation of guinea pig bronchial smooth muscle, *J. Appl. Physiol.,* 69, 1336, 1991.
43. Watson, N., Maclagan, J., and Barnes, P. J., Endogenous tachykinins facilitate transmission through parasympathetic ganglia in guinea-pig trachea, *Br. J. Pharmacol.,* 109, 751, 1993.
44. Martling, C., Saria, A., Andersson, P., and Lundberg, J. M., Capsaicin pretreatment inhibits vagal cholinergic and noncholinergic control of pulmonary mechanisms in guinea pig, *Naunyn-Schmiedeberg's Arch. Pharmacol.,* 325, 343, 1984.
45. Stretton, C. D., Belvisi, M. G., and Barnes, P. J., The effect of sensory nerve depletion on cholinergic neurotransmission in guinea pig airways, *Br. J. Pharmacol.,* 98, 782P, 1989.
46. Belvisi, M. G., Patacchini, R., Barnes, P. J., and Maggi, C. A., Facilitatory effects of selective agonists for tachykinin receptors on cholinergic neurotransmission: evidence for species differences, *Br. J. Pharmacol.,* 111, 103, 1994.
47. Black, J. L., Johnson, P. R., Alouan, L., and Armour, C. L., Neurokinin A with K^+ channel blockade potentiates contraction to electrical stimulation in human bronchus, *Eur. J. Pharmacol.,* 180, 311, 1990.
48. Kohrogi, H., Graf, P. P. D., Sekizawa, K., Borson, D. B., and Nadel, J. A., Neutral endopeptidase inhibitors potentiate substance P and capsaicin-induced cough in awake guinea pigs, *J. Clin. Invest.,* 82, 2063, 1988.
49. Advenier, C., Girard, V., Naline, E., Vilain, P., and Emons-Alt, X., Antitussive effect of SR 48968, a non-peptide tachykinin NK_2 receptor antagonist, *Eur. J. Pharmacol.,* 250, 169, 1992.
50. Nadel, J. A., Neutral endopeptidase modulates neurogenic inflammation, *Eur. Respir. J.,* 4, 745, 1991.
51. Shore, S. A., Stimler-Gerard, N. P., Coats, S. R., and Drazen, J. M., Substance P induced bronchoconstriction in guinea pig. Enhancement by inhibitors of neutral metalloendopeptidase and angiotensin converting enzyme, *Am. Rev. Respir. Dis.,* 137, 331, 1988.
52. Martins, M. A., Shore, S. A., Gerard, N. P., Gerald, C., and Drazen, J. M., Peptidase modulation of the pulmonary effects of tachykinins superfused guinea pig lungs, *J. Clin. Invest.,* 85, 170, 1990.
53. Lötvall, J. O., Skoogh, B.-E., Barnes, P. J., and Chung, K. F., Effects of aerosolized substance P on lung resistance in guinea pigs: a comparison between inhibition of neutral endopeptidase and angiotensin-converting enzyme, *Br. J. Pharmacol.,* 100, 69, 1990.
54. Sekizawa, K., Tamaoki, J., Graf, P. D., Basbaum, C. B., Borson, D. B., and Nadel, J. A., Enkephalinase inhibitors potentiate mammalian tachykinin-induced contraction in ferret trachea, *J. Pharmacol. Exp. Ther.,* 243, 1211, 1987.
55. Black, J. L., Johnson, P. R. A., and Armour, C. L., Potentiation of the contractile effects of neuropeptides in human bronchus by an enkephalinase inhibitor, *Pulm. Pharmacol.,* 1, 21, 1988.
56. Djokic, T. D., Nadel, J. A., Dusser, D. J., Sekizawa, K., Graf, P. D., and Borson, D. B., Inhibitors of neutral endopeptidase potentiate electrically and capsaicin-induced non-cholinergic contraction in guinea pig bronchi, *J. Pharmacol. Exp. Ther.,* 248, 7, 1989.
57. Martling, C. R., Sensory nerves containing tachykinins and CGRP in the lower airways: functional implications for bronchoconstriction, vasodilation, and protein extavasation, *Acta Physiol. Scand. Suppl.,* 563, 1, 1987.
58. Palmer, J. B. D., Cuss, F. M. C., Mulderry, P. K., Ghatei, M. A., Springall, D. R., Cadieux, A., Bloom, S. R., Polak, J. M., and Barnes, P. J., Calcitonin gene-related peptide is localized to human airway nerves and potently constricts human airway smooth muscle, *Br. J. Pharmacol.,* 91, 95, 1987.
59. Uddman, R., Luts, A., and Sundler, F., Occurrence and distribution of calcitonin gene related peptide in the mammalian respiratory tract and middle ear, *Cell Tissue Res.,* 214, 551, 1985.
60. McCormack, D. G., Mak, J. C. W., Coupe, M. O., and Barnes, P. J., Calcitonin gene-related peptide vasodilation of human pulmonary vessels: receptor mapping and functional studies, *J. Appl. Physiol.,* 67, 1265, 1989.
61. Salonen, R. O., Webber, S. E., and Widdicombe, J. G., Effects of neuropeptides and capsaicin on the canine tracheal vasculature *in vivo, Br. J. Pharmacol.,* 95, 1262, 1988.

62. Parsons, G. H., Nichol, G. M., Barnes, P. J., and Chung, K. F., Peptide mediator effects on bronchial blood velocity and lung resistance in conscious sheep, *J. Appl. Physiol.*, 72, 1118, 1992.
63. Mak, J. C. W. and Barnes, P. J., Autoradiographic localization of calcitonin gene-related peptide binding sites in human and guinea pig lung, *Peptides*, 9, 957, 1988.
64. Khalil, Z., Andrews, P. V., and Helme, R. D., VIP modulates substance P induced plasma extravasation in vivo, *Eur. J. Pharmacol.*, 151, 281, 1988.
65. Brockaw, J. J. and White, G. W., Calcitonin gene-related peptide potentiates substance P-induced plasma extravasation in the rat trachea, *Lung*, 170, 89, 1992.
66. Martling, C. R., Saria, A., Fischer, J. A., Hokfelt, T., and Lundberg, J. M., Calcitonin gene related peptide and the lung: neuronal coexistence and vasodilatory effect, *Regul. Peptides*, 20, 125, 1988.
67. Webber, S. G., Lim, J. C. S., and Widdicombe, J. G., The effects of calcitonin gene related peptide on submucosal gland secretion and epithelial albumin transport on ferret trachea in vitro, *Br. J. Pharmacol.*, 102, 79, 1991.
68. Pietrowski, W. and Foreman, J. C., Some effects of calcitonin gene related peptide in human skin and on histamine release, *Br. J. Dermatol.*, 114, 37, 1986.
69. Haynes, L. W. and Manley, C., Chemotactic response of guinea pig polymorphonucleocytes in vivo to rat calcitonin gene related peptide and proteolytic fragments, *J. Physiol.*, 43, 79P, 1988.
70. Umeda, Y. and Arisawa, H., Characterization of the calcitonin gene related peptide receptor in mouse T lymphocytes, *Neuropeptides*, 14, 237, 1989.
71. Nong, Y. H., Titus, R. G., Riberio, J. M., and Remold, H. G., Peptides encoded by the calcitonin gene inhibit macrophage function, *J. Immunol.*, 143, 45, 1989.
72. Barnes, P. J., Neurogenic inflammation in airways and its modulation, *Arch. Int. Pharmacodyn.*, 303, 67, 1990.
73. McDonald, D. M., Neurogenic inflammation in the respiratory tract: actions of sensory nerve mediators on blood vessels and epithelium of the airway mucosa, *Am. Rev. Respir. Dis.*, 136, S65, 1987.
74. Solway, J. and Leff, A. R., Sensory neuropeptides and airway function, *J. Appl. Physiol.*, 71, 2077, 1991.
75. Barnes, P. J., Asthma as an axon reflex, *Lancet*, i, 242, 1986.
76. Barnes, P. J., Sensory nerves, neuropeptides and asthma, *Ann. N.Y. Acad. Sci.*, 629, 359, 1991.
77. Lundberg, J. M. and Saria, A., Capsaicin-induced desensitization of the airway mucosa to cigarette smoke, mechanical and chemical irritants, *Nature*, 302, 251, 1983.
78. Ray, D. W., Hernandez, C., Leff, A. R., Drazen, J. M., and Solway, J., Tachykinins mediate bronchoconstriction elicited by isocapnic hyperpnea in guinea pigs, *J. Appl. Physiol.*, 66, 1108, 1989.
79. Reynolds, A. M. and McEvoy, R. D., Tachykinins mediate hypocapnia-induced bronchoconstriction in guinea pigs, *J. Appl. Physiol.*, 67, 2454, 1989.
80. Sakamoto, T., Elwood, W., Barnes, P. J., and Chung, K. F., Pharmacological modulation of inhaled metabisulphite-induced airway microvascular leakage and bronchoconstriction in guinea pig, *Br. J. Pharmacol.*, 107, 481, 1992.
81. Umeno, E., McDonald, D. M., and Nadel, J. A., Hypertonic saline increases vascular permeability in the rat trachea by producing neurogenic inflammation, *J. Clin. Invest.*, 85, 1905, 1990.
82. Thompson, J. E., Scypinski, L. A., Gordon, T., and Sheppard, D., Tachykinins mediate the acute increase in airway responsiveness by toluene diisocyanate in guinea-pigs, *Am. Rev. Respir. Dis.*, 136, 43, 1987.
83. Lötvall, J. O., Hui, K. P., Löfdahl, C.-G., Barnes, P. J., and Chung, K. F., Capsaicin pretreatment does not inhibit allergen-induced airway microvascular leakage in guinea pig, *Allergy*, 46, 105, 1991.
84. Hsiug, T.-R., Garland, A., Ray, D. W., Hershenson, M. B., Leff, A. R., and Solway, J., Endogenous sensory neuropeptide release enhances non specific airway responsiveness in guinea pigs, *Am. Rev. Respir. Dis.*, 146, 148, 1992.
85. Alving, K., Matran, R., Lacroix, J. S., and Lundberg, J. M., Allergen challenge induces vasodilation in pig bronchial circulation via a capsaicin sensitive mechanism, *Acta Physiol. Scand.*, 134, 571, 1988.
86. Abraham, W. M., Ahmed, A., Cortes, A., and Delehunt, J. C., C-fiber desensitization prevents hyperresponsiveness to cholinergic and antigenic stimuli after antigen challenge in allergic sheep, *Am. Rev. Respir. Dis.*, 147, A478, 1993.
87. Matsuse, T., Thomson, R. J., Chen, X.-R., Salari, H., and Schellenberg, R. R., Capsaicin inhibits airway hyperresponsiveness, but not airway lipoxygenase activity nor eosinophilia following repeated aerosolized antigen in guinea pigs, *Am. Rev. Respir. Dis.*, 144, 368, 1991.
88. Riccio, M. M., Manzini, S., and Page, C. P., The effect of neonatal capsaicin in the development of bronchial hyperresponsiveness in allergic rabbits, *Eur. J. Pharmacol.*, 232, 89, 1993.
89. Bienenstock, J., Perdue, M., Blennerghassett, M., Stead, R., Kakuta, N., Sestini, P., Vancheri, C., and Marshall, J., Inflammatory cells and epithelium: mast cell/nerve interactions in lung in vitro and in vivo, *Am. Rev. Respir. Dis.*, 138, S31, 1988.
90. Leff, A. R., Stimler, N. P., Munoz, N. M., Shioya, T., Tallet, J., and Dame, C., Augmentation of respiratory mast cell secretion of histamine caused by vagal nerve stimulation during antigen challenge, *J. Immunol.*, 136, 1066, 1982.

91. Sestini, P., Bienenstock, J., Crowe, S. E., Marshall, J. S., Stead, R. M., Kakuta, Y., and Perdue, M. H., Ion transport in rat tracheal ganglion in vitro. Role of capsaicin-sensitive nerves in allergic reactions, *Am. Rev. Respir. Dis.*, 141, 393, 1990.
92. Saban, R., Dick, E. C., Fishlever, R. I., and Buckner, C. K., Enhancement of parainfluenze 3 infection of contractile responses to substance P and capsaicin in airway smooth muscle from guinea pig, *Am. Rev. Respir. Dis.*, 136, 586, 1987.
93. Jacoby, D. B., Tamaoki, J., Borson, D. B., and Nadel, J. A., Influenza infection increases airway smooth muscle responsiveness to substance P in ferrets by decreasing enkephalinase, *J. Appl. Physiol.*, 64, 2653, 1988.
94. Piedimonte, G., Nadel, J. A., Umeno, E., and McDonald, D. M., Sendai virus infection potentiates neurogenic inflammation in the rat trachea, *J. Appl. Physiol.*, 68, 754, 1990.
95. Dusser, D. J., Djoric, T. D., Borson, D. B., and Nadel, J. A., Cigarette smoke induces bronchoconstrictor hyperresponsiveness to substance P and inactivates airway neutral endopeptidase in the guinea pig, *J. Clin. Invest.*, 84, 900, 1989.
96. Sheppard, D., Thompson, J. E., Scypinski, L., Dusser, D. J., Nadel, J. A., and Borson, D. B., Toluene diisocyanate increases airway responsiveness to substance P and decreases airway neutral endopeptidase, *J. Clin. Invest.*, 81, 1111, 1988.
97. Tsukagoshi, H., Sakamoto, T., Xu, W., Barnes, P. J., and Chung, K. F., Effect of interleukin-1β on airway hyperresponsiveness and inflammation in sensitized and non-sensitized Brown-Norway rats, *J. Allergy Clin. Immunol.*, 93, 103, 1994.
98. Watling, K. J., Nonpeptide antagonists heralded a new era in tachykinin research, *Trends Pharmacol. Sci.*, 13, 266, 1992.
99. Lei, Y.-H., Barnes, P. J., and Rogers, D. F., Inhibition of neurogenic plasma exudation in guinea pig airways by CP-96,345, a new non-peptide NK_1-receptor antagonist, *Br. J. Pharmacol.*, 105, 261, 1992.
100. Delay-Goyet, P. and Lundberg, J. M., Cigarette smoke-induced airway oedema is blocked by the NK_1-antagonist CP-96,345, *Eur. J. Pharmacol.*, 203, 157, 1991.
101. Hirayama, Y., Lei, Y. H., Barnes, P. J., and Rogers, D. F., Effects of two novel tachykinin antagonists FK 224 and FK 888 on neurogenic plasma exudation, bronchoconstriction and systemic hypotension in guinea pigs in vivo, *Br. J. Pharmacol.*, 108, 844, 1993.
102. Solway, J., Kao, B. M., Jordan, J. E., Gitter, B., Rodger, I. W., Howbert, J. J., Alger, L. E., Necheles, J., Leff, A. R., and Garland, A., Tachykinin receptor antagonists inhibit hypernea-induced bronchoconstriction in guinea pigs, *J. Clin. Invest.*, 92, 315, 1993.
103. Sakamoto, T., Barnes, P. J., and Chung, K. F., Effect of CP-96,345, a non-peptide NK_1-receptor antagonist against substance P-, bradykinin-, and allergen-induced airway microvascular leak and bronchoconstriction in the guinea pig, *Eur. J. Pharmacol.*, 231, 31, 1993.
104. Barnes, P. J., Belvisi, M. G., and Rogers, D. F., Modulation of neurogenic inflammation: novel approaches to inflammatory diseases, *Trends Pharmacol. Sci.*, 11, 185, 1990.
105. Kuo, H.-P., Rohde, J., Barnes, P. J., and Rogers, D. F., Differential effects of opioids on cigarette smoke, capsaicin and electrically-induced goblet cell secretion in guinea pig trachea, *Br. J. Pharmacol.*, 105, 361, 1992.
106. Yeadon, M., Wilkinson, D., Darley-Usmar, V., O'Leary, V. J., and Payne, A. N., Mechanisms contributing to ozone-induced bronchial hyperreactivity in guinea pigs, *Pulm. Pharmacol.*, 5, 39, 1992.
107. Stretton, C. D., Miura, M., Belvisi, M. G., and Barnes, P. J., Calcium-activated potassium channels mediate prejunctional inhibition of peripheral sensory nerves, *Proc. Natl. Acad. Sci. U.S.A.*, 89, 1325, 1992.
108. Ichinose, M. and Barnes, P. J., A potassium channel activator modulates both noncholinergic and cholinergic neurotransmission in guinea pig airways, *J. Pharmacol. Exp. Ther.*, 252, 1207, 1990.
109. Kuo, H.-P., Rohde, J. A. L., Barnes, P. J., and Rogers, D. F., K^+ channel activator inhibition of neurogenic goblet cell secretion in guinea pig trachea, *Eur. J. Pharmacol.*, 221, 385, 1992.
110. Bowden, J. and Gibbins, I. L., Relative density of substance P-immunoreactive nerve fibers in the tracheal epithelium of a range of species, *FASEB J.*, 6, A1276, 1992.
111. Levine, J. D., Dardick, S. J., Roizan, M. F., Helms, C., and Basbaum, A. I., Contribution of sensory afferents and sympathetic efferents to joint injury in experimental arthritis, *J. Neurosci.*, 6, 3423, 1986.
112. Holzer, P., Local effector functions of capsaicin-sensitive sensory nerve endings: involvement of tachykinins, calcitonin gene related peptide, and other neuropeptides, *Neuroscience*, 24, 739, 1988.
113. Ollerenshaw, S. L., Jarvis, D., Sullivan, C. E., and Woolcock, A. J., Substance P immunoreactive nerves in airways from asthmatics and non-asthmatics, *Eur. Respir. J.*, 4, 673, 1991.
114. Nieber, K., Baumgarten, C. R., Rathsack, R., Furkert, J., Oehame, P., and Kunkel, G., Substance P and β-endorphin-like immunoreactivity in lavage fluids of subjects with and without asthma, *J. Allergy Clin. Immunol.*, 90, 646, 1992.
115. Tomaki, M., Ichinose, M., Nakajima, N., Miura, M., Yamauchi, H., Inoue, H., and Shirato, K., Elevated substance P concentration in sputum after hypertonic saline inhalation in asthma and chronic bronchitis patients, *Am. Rev. Respir. Dis.*, 147, A478, 1993.

116. Hazbun, M. E., Hamilton, R., Holian, A., and Eschenbacher, W. L., Ozone-induced increases in substance P and 8 epi-prostaglandin F_{2a} in the airways of human subjects, *Am. J. Respir. Cell. Mol. Biol.*, 9, 568, 1993.
117. Lindsay, R. M. and Harmar, A. J., Nerve growth factor regulates expression of neuropeptide genes in sensory neurons, *Nature*, 337, 362, 1989.
118. Minami, M., Kuraishi, Y., Kawamura, M., Yamguchi, T., Masu, Y., and Nakanishi, S., Enhancement of preprotachykinin A gene expression by adjuvant-induced inflammation in the rat spinal cord: possible inducement of substance P-containing spinal neurons in nociceptor, *Neurosci. Lett.*, 98, 105, 1989.
119. Fischer, A., Philippin, B., Saria, A., McGregor, G., and Kummer, W., Neuronal plasticity in sensitized and challenged guinea pigs: neuropeptides and neuropeptide gene expression, *Am. J. Respir. Crit. Care Med.*, 149, A890, 1994.
120. Howarth, P. H., Djukanovic, R., Wilson, J. W., Holgate, S. T., Springall, D. R., and Polak, J. M., Mucosal nerves in endobronchial biopsies in asthma and non-asthma, *Int. Arch. Allergy Appl. Immunol.*, 94, 330, 1991.
121. Cunha, F. Q., Poole, S., Lorenzetti, B. B., and Ferreira, S. H., The pivotal role of tumour necrosis factor α in the development of inflammatory hyperalgesia, *Br. J. Pharmacol.*, 107, 660, 1992.
122. Fuller, R. W., Dixon, C. M. S., and Barnes, P. J., The bronchoconstrictor response to inhaled capsaicin in humans, *J. Appl. Physiol.*, 85, 1080, 1985.
123. Midgren, B., Hansson, L., Karlsson, J. A., Simonsson, B. G., and Persson, C. G. A., Capsaicin-induced cough in humans, *Am. Rev. Respir. Dis.*, 146, 347, 1992.
124. Barnes, P. J., Bradykinin and asthma, *Thorax*, 47, 979, 1992.
125. Fuller, R. W., Dixon, C. M. S., Cuss, F. M. C., and Barnes, P. J., Bradykinin-induced bronchoconstriction in man: mode of action, *Am. Rev. Respir. Dis.*, 135, 176, 1987.
126. Kaufman, M. P., Coleridge, H. M., Coleridge, J. C. G., and Baker, D. G., Bradykinin stimulates afferent vagal C-fibres in intrapulmonary airways of dogs, *J. Appl. Physiol.*, 48, 511, 1980.
127. Saria, A., Martling, C. R., Yan, Z., Theodorsson-Norheim, E., Gamse, R., and Lundberg, J. M., Release of multiple tachykinins from capsaicin-sensitive nerves in the lung by bradykinin, histamine, dimethylphenylpiperainium, and vagal nerve stimulation, *Am. Rev. Respir. Dis.*, 137, 1330, 1988.
128. Ichinose, M., Belvisi, M. G., and Barnes, P. J., Bradykinin-induced bronchoconstriction in guinea-pig *in vivo:* role of neural mechanisms, *J. Pharmacol. Exp. Ther.*, 253, 1207, 1990.
129. Sakamoto, T., Tsukagoshi, H., Barnes, P. J., and Chung, K. F., Role played by NK_2 receptors and cyclooxygenase activation in bradykinin B_2 receptor-mediated airway effects in guinea pigs, *Agents Actions*, 111, 117, 1993.
130. Ichinose, M., Nakajima, N., Takahashi, T., Yamauchi, H., Inoue, H., and Takishima, T., Protection against bradykinin-induced bronchoconstriction in asthmatic patients by a neurokinin receptor antagonist, *Lancet*, 340, 1248, 1992.
131. Nichol, G. M., O'Connor, B. J., Le Compte, J. M., Chung, K. F., and Barnes, P. J., Effect of neutral endopeptidase inhibitor on airway function and bronchial responsiveness in asthmatic subjects, *Eur. J. Clin. Pharmacol.*, 42, 495, 1992.
132. Cheung, D., Bel, E. H., den Hartigh, J., Dijkman, J. H., and Sterk, P. J., An effect of an inhaled neutral endopeptidase inhibitor, thiorphan, on airway responses to neurokinin A in normal humans in vivo, *Am. Rev. Respir. Dis.*, 145, 1275, 1992.
133. Cheung, D., Timmers, M. C., Bel, E. H., den Hartigh, J., Dijuman, J. H., and Sterk, P. J., An isolated neutral endopeptidase inhibitor, thiorphan, enhances airway narrowing to neurokin A in asthmatic subjects in vivo, *Am. Rev. Respir. Dis.*, 195, A682, 1992.
133a. Baraniuk, J. and Barnes, P. J., unpublished data.
134. Baraniuk, J. N., Ohkubo, O., Kwon, O. J., Mak, J. C. W., Davis, R., Twort, C., Kaliner, M., Letarte, M., and Barnes, P. J., Localization of neutral endopeptidase (NEP) mRNA in human bronchi, *Eur. Respir. J.*, 8, 1458, 1995.
135. Mantyh, C. R., Gates, T. S., Zimmerman, R. P., Welton, M. L., Passard, E. P., Vigna, S. R., Maggio, J. E., Druger, L., and Mantyh, P. W., Receptor binding sites for substance P but not substance K or neuromedin K are expressed in high concentrations by arterioles, venules and lymph nodes in surgical specimens obtained from patients with ulcerative colitis and Crohns disease, *Proc. Natl. Acad. Sci. U.S.A.*, 85, 3235, 1988.
136. Mantyh, P. W., Substance P and the inflammatory and immune response, *Ann. N.Y. Acad. Sci.*, 632, 263, 1991.
137. Devillier, P., Dessanges, J. F., Rakotashanaka, F., Ghaem, A., Boushey, H. A., and Lockhart, A., Nasal response to substance P and methacholine with and without allergic rhinitis, *Eur. Respir. J.*, 1, 356, 1988.
138. Adcock, I. M., Peters, M., Gelder, C., Shirasaki, H., Brown, C. R., and Barnes, P. J., Increased tachykinin receptor gene expression in asthmatic lung and its modulation by steroids, *J. Mol. Endocrinol.*, 11, 1, 1993.
139. Adcock, I. M., Gelder, C. M., Shirasaki, H., Yacoub, M., and Barnes, P. J., Effects of steroids on transcription factors in human lung, *Am. Rev. Respir. Dis.*, 145, A834, 1992.
140. Nakanishi, S., Mammalian tachykinin receptors, *Annu. Rev. Neurosci.*, 14, 123, 1991.
141. Ihara, H. and Nakanishi, S., Selective inhibition of expression of the substance P receptor mRNA in pancreatic acinar AR42J cells by glucocorticoids, *J. Biol. Chem.*, 36, 22,441, 1990.
142. McAlpine, L. G. and Thomson, N. C., Lidocaine-induced bronchoconstriction in asthmatic patients. Relation to histamine airway responsiveness and effect of preservative, *Chest*, 96, 1012, 1989.

143. Lammers, J.-W. J., Minette, P., McCusker, M., Chung, K. F., and Barnes, P. J., Nonadrenergic bronchodilator mechanisms in normal human subjects in vivo, *J. Appl. Physiol.*, 64, 1817, 1988.
144. Lammers, J.-W. J., Minette, P., McCusker, M., Chung, K. F., and Barnes, P. J., Capsaicin-induced bronchodilatation in mild asthmatic subjects: possible role of nonadrenergic inhibitory system, *J. Appl. Physiol.*, 67, 856, 1989.
145. Dixon, N., Jackson, D. M., and Richards, I. M., The effect of sodium cromoglycate on lung irritant receptors and left ventricular receptors in anesthetized dogs, *Br. J. Pharmacol.*, 67, 569, 1979.
146. Jackson, D. M., Norris, A. A., and Eady, R. P., Nedocromil sodium and sensory nerves in the dog lung, *Pulm. Pharmacol.*, 2, 179, 1989.
147. Dixon, C. M. S., Fuller, R. W., and Barnes, P. J., The effect of nedocromil sodium on sulphur dioxide induced bronchoconstriction, *Thorax*, 42, 462, 1987.
148. Verleden, G. M., Belvisi, M. G., Stretton, C. D., and Barnes, P. J., Nedocromil sodium modulates non-adrenergic non-cholinergic bronchoconstrictor nerves in guinea-pig airways *in vitro*, *Am. Rev. Respir. Dis.*, 143, 114, 1991.
149. Nichol, G. M., Alton, E. W. F. W., Nix, A., Geddes, D. M., Chung, K. F., and Barnes, P. J., Effect of inhaled furosemide on metabisulfite- and methacholine induced bronchoconstriction and nasal potential difference in asthmatic subjects, *Am. Rev. Respir. Dis.*, 142, 576, 1990.
150. Elwood, W., Lötvall, J. O., Barnes, P. J., and Chung, K. F., Loop diuretics inhibit cholinergic and non-cholinergic nerves in guinea-pig airways *in vitro*, *Am. Rev. Respir. Dis.*, 143, 1340, 1991.
151. Ventresca, G. P., Nichol, G. M., Barnes, P. J., and Chung, K. F., Inhaled furosemide inhibits cough induced by low chloride content solutions but not by capsaicin, *Am. Rev. Respir. Dis.*, 142, 143, 1990.
152. O'Connor, B. J., Chen-Wordsell, M., Barnes, P. J., and Chung, K. F., Effects of an inhaled opioid peptide on airway responses to sodium metabisulphite in asthma, *Thorax*, 46, 294, 1991.
153. Ichinose, M., Belvisi, M. G., and Barnes, P. J., Histamine H_3-receptors inhibit neurogenic microvascular leakage in airways, *J. Appl. Physiol.*, 68, 21, 1990.
154. O'Connor, B. J., Lecomte, J. M., and Barnes, P. J., Effect of an inhaled H_3-receptor agonist on airway responses to sodium metabisulphite in asthma, *Br. J. Clin. Pharmacol.*, 35, 55, 1993.
155. Ichinose, M., Katsumata, U., Kikuchi, R., Fukushima, T., Ishi, M., Inoue, C., Shirato, K., and Takashima, T., Effect of tachykinin receptor antagonist on chronic bronchitis patients, *Am. Rev. Respir. Dis.*, 147, A318, 1993.

Chapter 15

SENSORY NEUROPEPTIDES IN MIGRAINE

Michael A. Moskowitz, Won S. Lee, and F. Michael Cutrer

CONTENTS

I. INTRODUCTION

Since the time of Wolff almost half a century ago, it has been universally accepted that dilated and engorged cranial blood vessels caused migrainous pain, and that vasoconstriction provided headache relief. Over the past several years, however, accumulating evidence has shifted the emphasis away from the vascular smooth muscle hypothesis and towards mechanisms related to activation of meningeal afferents, neuropeptide release, and neurogenic inflammation.

Neurogenic inflammation is known to occur in many organ systems. The headache of migraine may turn out to be only one variation on this ancient evolutionary theme, which for unknown reasons develops periodically in susceptible individuals. This chapter describes the evidence that relates neurogenic inflammation and neuropeptides to the pathophysiology and pharmacology of migraine headaches.

II. ANATOMY OF A HEADACHE

Epilepsy surgery performed early in this century provided important information about the origins of intracranial pain in awake patients. During craniotomies under local anesthesia, it was found that the meninges and particularly large meningeal blood vessels (dura and pia mater) produced severe, migraine-like head pain when electrically or mechanically stimulated. The pain was ipsilateral to the stimulation and described as aching, boring, or throbbing; the brain parenchyma was essentially insensate under the same experimental conditions.[1] These observations, consistent with what is now known about the course and distribution of sensory axons in the meninges, led to the inference (which is probably incorrect), that blood vessels are the only source of pain within the cranium. We now know that neurophysiological events within the cortex, such as spreading depression, may lead to

0-8493-7646-7/96/$0.00+$.50

meningeal afferent activation. Hence, a link has been established between events within the brain and the development of headache.

Pseudounipolar neurons originating primarily within the first division of the trigeminal ganglia send small-caliber C-fibers along blood vessels to innervate meningeal tissues.[2-4] To reach the pia mater, small axons traverse the cavernous sinus to join the pericarotid plexus, then course along the internal carotid artery to the circle of Willis. (The importance of the cavernous sinus anatomy to the pathophysiology of cluster headache has been emphasized.[5]) By contrast, dural innervation derives from all three trigeminal divisions, including trigeminal afferents traversing the foramen spinosum with the middle meningeal artery. For both tissues, the blood vessels provide the route by which traversing sensory axons reach the meninges. Very few large-diameter myelinated afferents innervate the meninges.

Headache pain arises predominantly from intracranial structures. When trigeminovascular fibers become activated, impulses are transmitted centrally toward the first synapse within lamina I, II_0 of the trigeminal nucleus caudalis.[6,7] In addition, depolarization and antidromic conduction cause the release of sensory neuropeptides from widely branching perivascular trigeminal axons. These neuropeptides, which include substance P (SP), calcitonin gene-related peptide (CGRP), and neurokinin A (NKA), mediate neurogenic inflammation. In some trigeminal neurons, CGRP coexists with SP. In others, SP coexists with NKA.[8]

Neurogenic inflammation (NI) develops within the dura mater following chemical or electrical trigeminal stimulation. It results from neuropeptide release from capsaicin-sensitive perivascular afferent axons. NI may represent an adaptive response to real or threatened tissue injury and is characterized by endothelial, platelet, and mast cell activation. Endothelial activation is accompanied by plasma leakage, which can be detected by the presence of systemically administered horseradish peroxidase within the perivascular space. An increase in tissue oncotic pressure promotes water transport into the vessel wall to cause edema. Tissue water facilitates dilution and removal of foreign substances. Increased blood flow and mast cell activation deliver nutrient substrates and enhance local cellular immune defenses, respectively.[9]

The resulting sterile inflammation may lower the threshold for subsequent neuronal activation, leading to the phenomenon of sensitization. In fact, the throbbing quality of migraine pain may develop as a consequence of stimulating sensitized nerve fibers during the course of normal hemodynamic events (e.g., vessel pulsations). It has been postulated that the prolongation of pain in many migraine attacks occurs with continued activation of primary afferents as a consequence of secondary changes after neuropeptide release.

NI has never been searched for or identified in human meninges during a migraine headache, possibly because imaging methods are not sensitive enough to resolve the small volume of meningeal tissue. The available evidence supporting the importance of the meninges is largely inferential and based on two observations: (1) the meninges are the only tissues within the cranium that receive a sensory innervation, (2) agents that block the development of NI in rat and guinea pig dura mater are effective agents for acute treatment of headache in man.

III. TRIGEMINAL ELECTRICAL STIMULATION: A MODEL TO STUDY NEUROGENIC INFLAMMATION

In this model, neuropeptide release is evoked in rodent meninges by either electrical stimulation of the trigeminal ganglion, or capsaicin administration (either intravenously or via injection into the trigeminal ganglion). With stimulation of trigeminal afferents, neuropeptides (i.e., CGRP, SP) become elevated in draining venous effluent (sagittal sinus).[10] These findings are consistent with observed elevations in CGRP levels after electrical stimulation of sagittal sinus in cat.[11] Once released, the neuropeptides initiate a cascade of events within dura mater characterized by:

1. An increase in leakage of plasma from postcapillary venules
2. The formation of endothelial microvilli, endothelial vesicles, and vacuoles specifically within postcapillary venules[12]
3. Activation and degranulation of mast cells[13]
4. Platelet aggregation
5. Vasodilation

NK_1 receptor blockers inhibit neurogenic plasma protein extravasation completely, thereby suggesting the importance of SP and the NK_1 receptor subtype. Vasodilation is most likely CGRP-mediated, although no studies have established this point, to our knowledge. Vasodilation is a prominent component of neurogenic inflammation, and its presence is undeniable in a minority of headache patients.

The NI model quantitates the amount of plasma protein leakage within the meninges induced by either unilateral electrical trigeminal ganglion stimulation or intravenous administration of capsaicin in anesthetized rats or guinea pigs. The assay is highly reproducible in our hands, with more than 12,000 rodents studied to date. Electrical stimulation (5 Hz, 5 ms, 0.6 to 1 mA, 5 min) is preceded or followed by a femoral vein injection of [^{125}I]-albumin. This stimulus is sufficient to activate small unmyelinated C-fibers. After 10 min, the animals are perfused to remove isotope from the intravascular compartment and the dural tissues are harvested from both sides. The radioactivity is then compared between the two sides and expressed as a ratio. In the rat and guinea pig, this ratio is approximately 1.6 to 1.7. The model, developed by Markowitz and colleagues in 1987, emphasizes unique features of trigeminovascular pharmacology.

The trigeminal model is relevant to headache and its acute pharmacotherapy for the following reasons:

1. The tissue is an important source of cephalic pain.
2. The sterile inflammatory process may well sustain pain long after the initial trigger and contributes to hyperalgesia and the sensitization of polymodal nociceptors.[14]
3. Increases in neuropeptides (e.g., CGRP, SP) in draining venous blood[10,63] during electrical trigeminal stimulation also are similar to the CGRP increases reported in migraine patients during an acute attack.[15]
4. The stimulation-evoked increases in CGRP in animals[10] and migraine-evoked increases in humans[15] are blocked by sumatriptan and dihydroergotamine (DHE; rats and guinea pigs) and sumatriptan (humans) (see Figures 1 and 2).
5. The same medications (i.e., sumatriptan and DHE) also block the neurogenic inflammatory response selectively in dura mater but not extracranial tissues (see Figure 3). These drugs do not appear to be analgesic in other pain states. (Please note that one publication reported that sumatriptan blocked NI within peripheral nerve roots.[16])
6. The dosages in most instances approximate the amounts required to treat humans during an acute migraine attack.
7. The onset of action in both cases is rapid (within minutes) and the blockade of NI occurs even when the drugs are administered after trigeminal ganglion stimulation.
8. Ketorolac (a nonsteroidal anti-inflammatory drug) and corticosteroids, both of which relieve migraine headaches, also inhibit the development of NI.

A. SEROTONIN RECEPTORS AND BLOCKADE OF DURAL NI

Sumatriptan, DHE, and ergotamine tartrate possess high-affinity binding to the 5-hydroxytryptamine ($5\text{-}HT_{1D}$) receptor subtype. Prejunctional $5\text{-}HT_{1D}$receptors on trigeminovascular fibers inhibit neuropeptide release and thereby block the development of the inflammatory response (Figure 3). Supportive evidence is as follows:

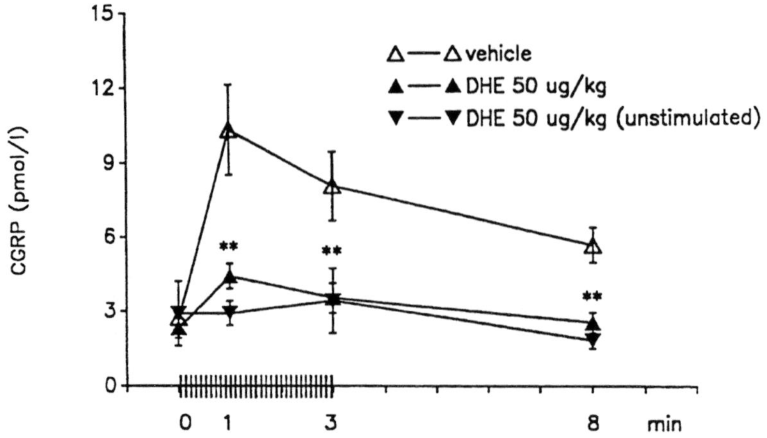

FIGURE 1. The administration of dihydroergotamine (DHE) (50 µg/kg, i.p.; n = 17) attenuated the increases in levels of CGRP in plasma of the superior sagittal sinus, during electrical stimulation of the trigeminal ganglion, at the indicated times, without significantly changing baseline levels when administered 10 min prior to stimulation (3 min, 0.3 mA, 5 ms, 5 Hz). Means ± SEM; $**P < .01$ as compared to vehicle-treated animals (n = 17). (From Buzzi, M. G., *Neuropharmacology,* 30, 1196, 1991. With permission.)

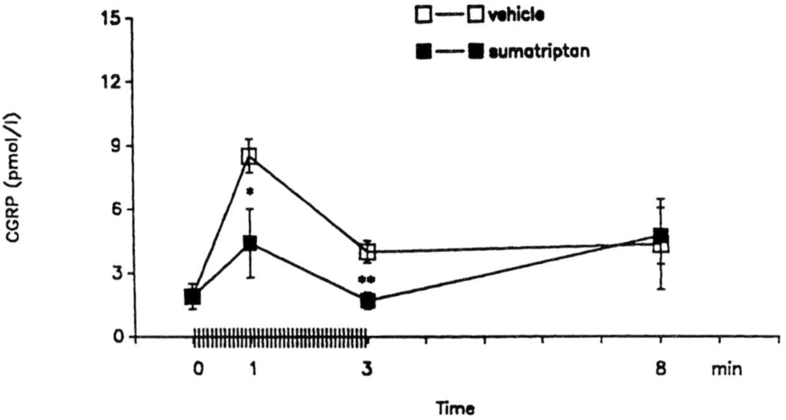

FIGURE 2. Levels of CGRP in plasma from superior sagittal sinus after the administration of sumatriptan (300 µg/kg, i.p.; n = 4), demonstrate significant attenuation at both 1 and 3 min during electrical stimulation of the trigeminal ganglion (0.1 mA), Means ± SEM; $*P < .05$, and $**P < .01$ compared to levels in vehicle-treated animals. (From Buzzi, M. G., *Neuropharmacology,* 30, 1997, 1991. With permission.)

1. As mentioned above, sumatriptan or DHE attenuate elevations in sagittal sinus CGRP levels,[10] plasma protein leakage,[17] and histopathological changes induced by electrical stimulation.
2. SP-induced plasma protein extravasation is not blocked by sumatriptan, DHE, or CP-93,129 (a selective 5-HT_{1B} agonist) but is blocked by NK_1 receptor antagonists, RP-67,580[18] and RPR 100893.[19]
3. Expression of the gene encoding 5-HT_{1D} receptors has been identified in the human and guinea pig using the reverse transcriptase-polymerase chain reaction and by *in situ* hybridization techniques.[20,21]
4. The 5-$HT_{1B/D}$ receptor subtype mediates inhibition of neurotransmitter release in both the peripheral and central nervous systems.[22]

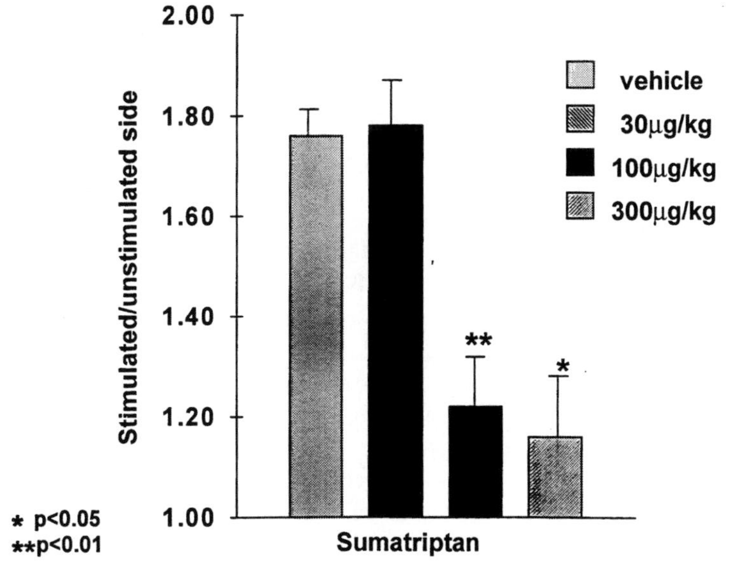

FIGURE 3. Sumatriptan significantly reduced plasma protein extravasation ([^{125}I]-albumin) following electrical stimulation of the right trigeminal ganglion in the rat ipsilateral dura mater. Animals were injected (i.v.) with 30, 100, or 300 μg/kg sumatriptan 15 min before electrical stimulation (1.2 mA, 5 ms, 5 Hz) and 5 min before [^{125}I]-albumin (50 μCi/kg). Means ± SEM. (Adapted from Buzzi, M. G. et al., *Br. J. Pharmacol.*, 99, 206, 1990.)

The receptor subtype mediating the blockade of plasma protein extravasation in the rat is most consistent with the 5-HT_{1B} receptor subtype. The analogous receptor subtype in guinea pig and man is 5-HT_{1D}.[23] CP-93,129, a highly selective and specific 5-HT_{1B} receptor agonist[24] blocks plasma protein extravasation in rats but not guinea pigs. Sumatriptan, which has a 30-fold higher affinity for the 5-HT_{1D} than the 5-HT_{1B} receptor, is 30-fold more potent in guinea pig than in rat.

There is now evidence that a previously unrecognized but related receptor(s) may be important in the blockade of plasma extravasation. This evidence is as follows:

1. 5-Carboxamidotryptamine (5-CT) has a significantly greater potency in the extravasation model than one would expect based on the known affinity of 5-CT for the 5-HT_{1D} receptor subtype.
2. The potency in the plasma extravasation model does not correlate well with the affinities of the same compounds for 5-HT_{1D} receptor binding sites (as determined by ligand binding studies in bovine caudate nucleus). For example, novel conformationally restricted sumatriptan analogs, CP-122,288 and CP-122,638, are 1,800 and 30,000 times more potent than sumatriptan in the extravasation model, yet they are no more potent than sumatriptan at displacing [^{3}H]-5-HT at the 5-HT_{1D} receptor binding site.[25]
3. Methiotepin and metergoline, both 5-HT_1 antagonists, behave as partial agonists and do not block the effects of sumatriptan or 5-CT, respectively.

1. The 5-HT_{1D} Receptor

The 5-HT_{1D} receptor possesses seven transmembrane spanning domains, is coupled to G proteins,[26] and is comprised of two closely related proteins designated 5-$HT_{1D\alpha}$ and 5-$HT_{1D\beta}$. The 5-$HT_{1D\beta}$ receptor exhibits remarkable sequence homology to the rat 5-HT_{1B} receptor, although the pharmacology appears quite distinct. Since 5-HT_{1B} receptor-selective agents such as CP-93,129 blocked dural plasma extravasation in rat, it was anticipated that among the two

5-HT_{1D} receptor subtypes, the $5\text{-HT}_{1D\beta}$ would inhibit plasma extravasation in guinea pig and possibly in man. Using polymerase chain reaction amplification strategies and oligonucleotide probes complementary to the unique coding sequences for the α and β subtypes, it was found that the mRNA encoding the $5\text{-HT}_{1D\alpha}$ receptor subtype was selectively expressed within human and guinea pig trigeminal ganglia.[20] Hamel et al. reported that the $5\text{-HT}_{1D\beta}$ gene is selectively expressed within pial vascular smooth muscle (large arteries but not small arterioles).[27] Hence, there is reason to believe that the pre- and postjunctional binding sites for sumatriptan can be differentiated pharmacologically. Thus, the potential exists for the development of clinically useful compounds which do not constrict pial or coronary arteries.

Drugs that block plasma protein extravasation within dura mater but do not bind to 5-HT_1 receptors have also been identified. Prejunctional receptors are likely to mediate blockade by UK-14,304 (an α_2-adrenoceptor agonist), *R*(–)-α-methyl-histamine (α-MeHA), (a histamine H_3 receptor agonist), and SMS 210-995 (a synthetic octapeptide somatostatin analog).[28] The effects of UK-14,304 and α-MeHA are reversed by specific antagonists (idazoxan and thioperamide, respectively). None of the agents blocked the extravasation caused by the administration of the neuropeptide mediator, substance P.[28]

2. The NK_1 Receptor and Neurogenic Inflammation

A nonpeptide NK_1 receptor antagonist belonging to the perhydroisoindolone series, RPR 100,893, decreases dural plasma extravasation induced both by electrical trigeminal stimulation and intravenous capsaicin.[19] The effect is mediated via blockade at postjunctional SP receptors, which are present on the endothelium and coupled to the release of nitric oxide (NO). In fact, treatment with an NO synthesis inhibitor (L-NAME) decreased neurogenic plasma protein extravasation and the effects of exogenously administered SP (unpublished data). The effects of RPR 100893 were remarkably potent after either oral or intravenous dosing. The ED_{50} was 0.5 μg kg^{-1} and 2.5 ng kg^{-1} by oral and i.v. routes, respectively, as compared to sumatriptan (ED_{50} 915 ng kg^{-1} i.v.).[19] Inhibition of NI continued even when the drug was administered up to 80 min after terminating trigeminal stimulation. Phase II clinical trials are near completion in acute migraine headaches.

IV. C-FOS MODEL

Many of the compounds that block neurogenic inflammation also attenuate c-fos protein-like immunoreactivity in the trigeminal nucleus caudalis following noxious meningeal stimulation. The expression of c-fos, an immediate early gene, is well characterized in the brain and is a marker of neuronal activation.[29,30] In spinal cord, c-fos protein-like immunoreactivity (c-fos-LI) develops within the nuclei of postsynaptic neurons in the superficial laminae of the dorsal horn following noxious heat and chemical stimulation.[31,32] This response can be suppressed by subcutaneously or intraventricularly administered opiates in a dose-dependent, naloxone-reversible manner.[33,34]

We have developed an animal model which measures the number of neurons expressing c-fos-LI within laminae I, II_o induced by noxious chemical stimulation of meninges (e.g., autologous blood, carrageenin, capsaicin). Once injected intracisternally, the irritants activate primary sensory afferents supplying the meninges. Two hours later, the animals are sacrificed, perfused, and the brainstems with attached spinal cord are dissected. Fifty-micron axial sections through the medulla and upper cervical spinal cord are then processed immunohistochemically and the number of nuclei expressing c-fos immunoreactivity are counted in lamina I, II_o of the medullary trigeminal nucleus caudalis (TNC) and the dorsal horn of the upper cervical spinal cord. C-fos expression has been demonstrated previously in the dorsal horn of the spinal cord after subcutaneous injection of formalin or other chemical irritants.[31–33]

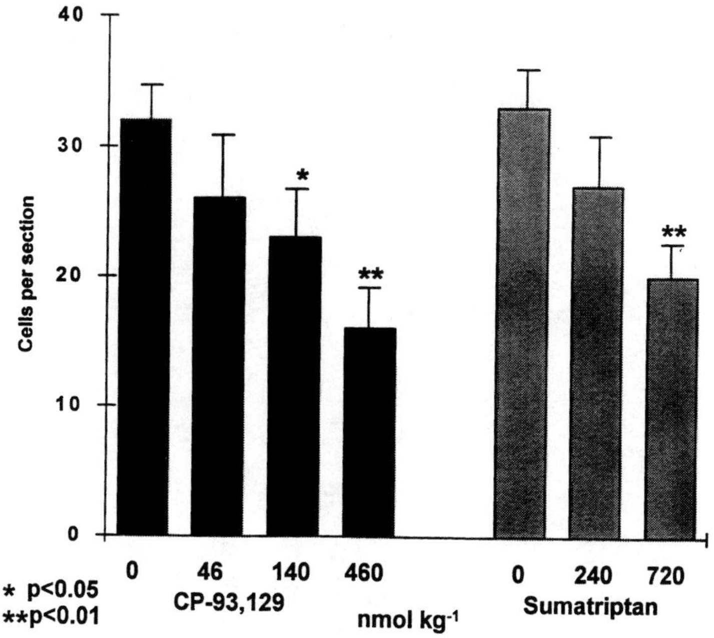

FIGURE 4. CP-93,129 and sumatriptan decrease the numbers of c-fos immunoreactive cells per 50-μm section within lamina I, II_0 of trigeminal nucleus caudalis (TNC) after intracisternal blood injection. CP-93,129 [vehicle (*n* = 11), 46 nmol/kg (*n* = 3), 140 nmol/kg (*n* = 5), or 460 nmol/kg (*n* = 7)] or sumatriptan [vehicle (*n* = 9), 240 nmol/kg (*n* = 3), or 720 nmol/kg (*n* = 7)] were injected (i.v.) at 60 and 10 minutes before blood injection. The number of c-fos positive cells per section observed in sham animals (6 ± 3, *n* = 5) is subtracted from each number. Means ± SEM. (Adapted from Nozaki, K. et al., *Br. J. Pharmacol.*, 106, 409, 1992.)

Nozaki et al. have shown that following the intracisternal instillation of autologous blood or carrageenin in rats, positive cells appear in lamina I, II_0 within 1 h, peak at 2 h, and decline thereafter.[35] The number of cells corresponded to the amount but not volume of blood injected. Destruction of unmyelinated fibers by neonatal capsaicin treatment significantly decreased the number of positive cells within lamina I, II_0, as did surgical transection of trigeminal meningeal afferents. Other labeled nuclei included the nucleus of the solitary tract (NTS), the lateral reticular nucleus (LRN), area postrema (AP), and the parabrachial nucleus (PBN).

A. 5-HT AND ATTENUATION OF C-FOS EXPRESSION

Pretreatment with sumatriptan, dihydroergotamine, and a selective 5-HT_{1B} receptor agonist, CP-93,129, significantly and dose dependently reduced the number of stained cells in TNC but not NTS or AP[36] (see Figures 4 and 5). Sumatriptan did not suppress c-fos expression in TNC following formalin application to the nasal mucosa. Hence, blockade appeared specifically within trigeminovascular projections.

Recent experiments have been carried out in guinea pigs. Guinea pigs possess the 5-HT_1 receptor subtype that is more closely related to man (5-HT_{1D} receptors). Dilute capsaicin solution (0.1 ml; 0.1 m*M*) placed intracisternally evokes c-fos expression. Although the number of labeled cells is greater (218 cells per section after capsaicin vs. 37 cells per section after subarachnoid blood), the temporal and spatial characteristics are similar to that observed after subarachnoid blood injection.[37]

In guinea pigs, the response was significantly reduced by very small doses of CP-122,288 (100 pmol/kg i.v.), a sumatriptan analog. CP-122,288 is significantly more potent than sumatriptan in the plasma protein extravasation model (see above), but binds with similar affinities to 5-HT_{1D} receptors in ligand binding studies.[38] If vasoconstriction and the

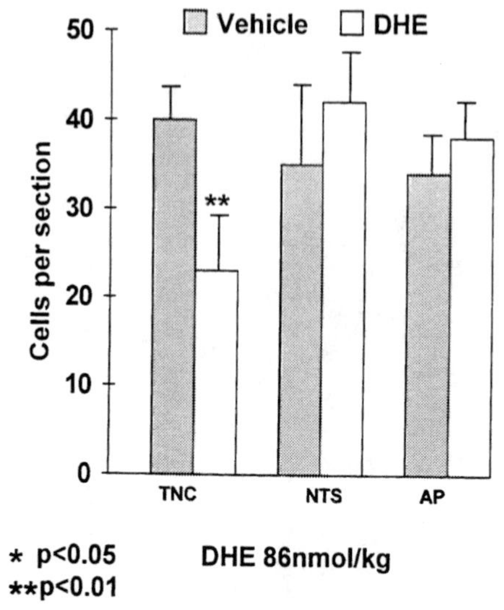

FIGURE 5. C-fos immunoreactive cells were counted per 50-μm section within lamina I, II_o of trigeminal nucleus caudalis (TNC), nucleus of the solitary tract (NTS), and area postrema (AP) in animals that were pretreated with dihydroergotamine (DHE) (86 nmol/kg; $n = 4$) or vehicle ($n = 3$) before intracisternal blood injection. The number of c-fos positive cells per section observed in sham animals (6 ± 3 for TNC, 4 ± 2 for NTS, 6 ± 4 for AP) are subtracted for each number. Means ± SEM. (Adapted from Nozaki, K. et al., *Br. J. Pharmacol.*, 106, 409, 1992.)

attenuation of c-fos expression were mediated by the same receptor, then the marked increase in potency in the c-fos model should be accompanied by an increase in vasoconstrictor potency. However, CP-122,288 and sumatriptan constrict cat pial vessels with similar potency. Therefore, the neurogenic and vascular actions of sumatriptan may be mediated by distinct receptors.[39]

B. NK_1 RECEPTOR AND C-FOS

Nonpeptide antagonists that block NK_1 receptors also attenuate c-fos expression after capsaicin injection. For example, capsaicin-induced c-fos expression is significantly reduced after pretreatment with RPR 100,893 (1 μg/kg i.v.)[37] (see Figure 6). The effects of RPR 100893 were stereoselective and potent. RPR 100893 exhibited micromolar affinity at L-type calcium channels[40] and did not lower mean arterial blood pressure. The effects are most likely mediated by blockade of the NK_1 receptor within trigeminal nucleus caudalis.

The role of NK_1 antagonists as antinociceptive agents has been controversial. The nonpeptide NK_1 receptor antagonist CP-96,345 binds with high-affinity and specificity.[41-44] A racemic mixture of CP-96,345 (8 to 73 μg kg^{-1} s.c.) blocked the formalin response and abolished carrageenin-induced mechanical hyperalgesia in rats.[45] However, CP-96,345, like nifedipine and verapamil, binds to L-type calcium channels in a nonstereoselective manner and lowers blood pressure.[46] The latter compounds also inhibit nociceptive activity after formalin injection.[47,48] CP-96,345 was about 50 times more potent than lidocaine in blocking voltage-dependent Na^+ currents.[49] Finally, Nagy et al. reported that postsynaptic firing induced by capsaicin activation of dorsal root ganglion cells was not significantly decreased by CP-96,345 (100 μg kg^{-1} i.v.) in rats.[50] Together, the evidence raises strong doubts as to whether the antinociceptive effects of CP-96,345 are, in fact, NK_1 receptor mediated.[49]

A recent report by Rupniak et al.[51] raised similar concerns about another NK_1 receptor antagonist, RP 67580. Using the racemate, RP 67580 exhibited antinociceptive behavior in the writhing and formalin paw tests, but reportedly displaced [^{3}H]-diltiazem binding to rat skeletal membranes, inhibited calcium entry into depolarized strips of intestinal smooth muscle, and

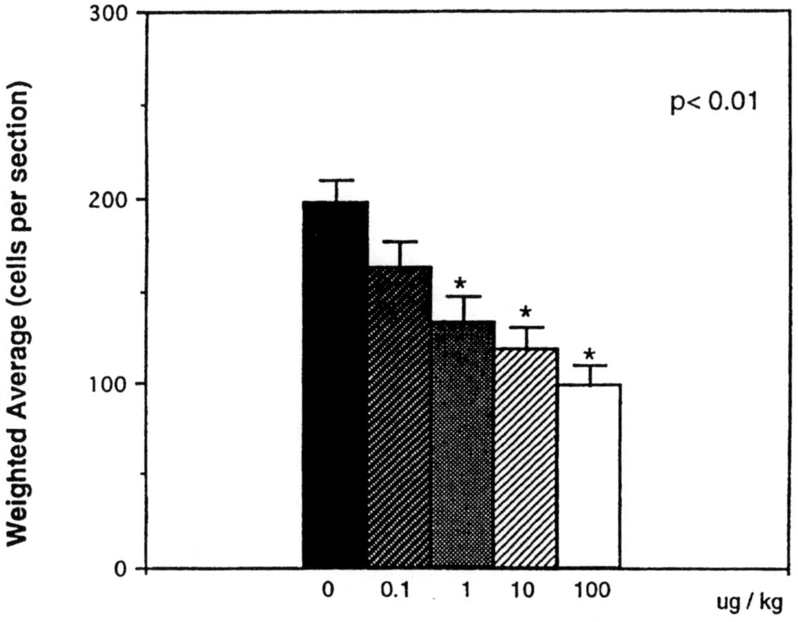

FIGURE 6. Pretreatment with RPR 100893 dose dependently decreases the c-fos immunoreactive cells evoked by intracisternal capsaicin injection. Cell numbers are given per 50-μm section within the trigeminal nucleus caudalis (TNC) (lamina I, II_0) as determined by a weighted average method. Vehicle (n = 12), 0.1 μg kg^{-1} (n = 6), 1.0 μg kg^{-1} (n = 8), 10 μg kg^{-1} (n = 9), or 100 μg kg^{-1} (n = 10) was injected 30 min prior to capsaicin and the animals sacrificed 120 min later. Means ± SEM; *$P < .01$ as compared to vehicle-treated group.

depressed high-threshold calcium currents in cultured rat neurons.[51] Garret and colleagues[52] found that RP 67580, but not its enantiomer, RP 68651, blocked the biphasic nociceptive response that follows the injection of 5% formalin in the paw of rats. Wang et al.[53] showed that large concentrations of both drugs exhibit nonspecific inhibitory effects on neurotransmission (e.g., blockade of frog sciatic action potentials similar to local anesthetics) in addition to blocking tachykinin receptors.

V. TRIGGERING MECHANISMS

How might trigeminovascular fibers be triggered in migraine? Recent experiments addressed the importance of events within the cortical brain surface (i.e., spreading depression [SD]). To address this, we examined c-fos expression in TNC induced by neocortical SD, an electrophysiological event felt by many investigators to underlie migraine aura.[54] SD (evoked by microinjections of KCl into the left parietal cortex at 9-min intervals over 1 h) caused an increase in c-fos expression in TNC ipsilateral to KCl microinjections compared to the noninjected side.[55] The increase in c-fos expression did not occur when meningeal afferents were surgically transected or when the occurrence of SD was blocked by hyperoxia/hypercapnia in potassium-injected animals. Sumatriptan (720 nmol/kg i.v.) 30 min prior to SD blocked the rise in c-fos expression in the ipsilateral TNC but not the ability of KCl to induce spreading depression.

The increase in c-fos protein observed in TNC in response to spreading depression suggests the possibility that the trigeminovascular system may be sensitized and/or activated by nociceptive chemicals (i.e., K^+, arachidonic acid, or H^+) released as a result of neurophysiologically driven ionic and metabolic mechanisms within the cortex. After release, these chemicals may accumulate within the perivascular space or within the lumen of draining venules as they enter and exit the cerebral cortex in the Virchow-Robin spaces.[55] Levels of these and other potentially nociceptive compounds rise after SD.[56-58] Because of the anatomical barriers present between the pia-arachnoid and dural vessels, the extent to which SD or some related cortical phenomenon is involved in the generation of dural NI remains to be determined. However, trigeminovascular axon collaterals and veins which bridge between the pial and dural circulations provide the basis by which the two circulations can communicate.

Exogenously administered chemicals may also provoke neuropeptide release from trigeminovascular fibers. Nitrates, long known to generate headaches, are likely examples. The NO donors nitroglycerine and sodium nitroprusside cause both vasodilation and headache. Vasodilator responses to topical nitroprusside and nitroglycerine are markedly depressed on the denervated side after trigeminal ganglionectomy. Topical application of a selective CGRP antagonist, $CGRP_{8-37}$, reduces the response to nitrodilators in normal pial vessels.[59] Holzer and Jocic recently found that the NO synthase inhibitor L-NAME blocked vasodilation in skin following application of the irritant mustard oil, but failed to block vasodilation due to electrical stimulation (saphenous nerve) or CGRP.[60] Moreover, $CGRP_{8-37}$ and chemical deafferentation block the hyperemia caused by intraplantar infusion of sodium nitroprusside. Hence, NO promotes vasodilation primarily by neuropeptide (CGRP) release. Studies by Hughes and Brain agree with these conclusions, although a response to nitroprusside was not noted in their studies[61] or by those of Kitazono et al.[62]

VI. CONCLUSIONS

Our investigations on neurogenic inflammation in the meninges and its involvement in the generation of headache suggest the following conclusions.

1. Sensory neuropeptides are involved in meningeal inflammation associated with headache.
2. Receptor-mediated blockade of neuropeptide release blocks neurogenic inflammation in dura mater.
3. Compounds that block neuropeptide release and neurogenic inflammation attenuate the expression of c-fos in TNC neurons. Other laboratories have shown that c-fos expression is a marker for neuronal activation and provides a good correlate of behavioral tests in animal models of nociception.
4. It appears possible to distinguish between the vasoconstrictive and anti-inflammatory actions of potential antimigraine drugs acting via $5\text{-HT}_{1D/B}$ receptors.
5. The selective blockade of sensory neuropeptides, both their release (prejunctional) and binding (postjunctional), are potential targets for new generations of antimigraine medications which block both NI and pain transmission.

REFERENCES

1. Ray, B. S. and Wolff, H. G., Experimental studies on headache. Pain-sensitive structures of the head and their significance in headache, *Arch. Surg.*, 41, 813, 1940.
2. Mayberg, M. A., Langer, R. S., Zervas, N. T., and Moskowitz, M. A., Perivascular meningeal projection from cat trigeminal ganglia: possible pathway for vascular headache in man, *Science*, 213, 228, 1981.
3. Mayberg, M. R., Zervas, N. T., and Moskowitz, M. A., Trigeminal projections to supratentorial pial and dural blood vessels in cats demonstrated by horseradish peroxidase histochemistry, *J. Comp. Neurol.*, 223, 46, 1984.

4. Arbab, M. A. R., Wiklund, L., and Svendgaard, N. A., Origin and distribution of cerebral vascular innervation from superior cervical, trigeminal and spinal ganglia investigated with retrograde and anterograde WGA-HRP tracing in the rat, *Neuroscience,* 19, 695, 1986.
5. Moskowitz, M. A., Cluster headache: evidence for a pathophysiologic focus in the superior pericarotid cavernous sinus plexus, *Headache,* 28, 584, 1988.
6. Strassman, A., Mason, P., Moskowitz, M., and Maciewicz, R., Response of brainstem trigeminal neurons to electrical stimulation of the dura, *Brain Res.,* 379, 242, 1986.
7. Davis, K. D. and Dostrovsky, J. O., Activation of trigeminal brain stem nociceptive neurons by dural artery stimulation, *Pain,* 25, 395, 1986.
8. Van der Kooy, D., and O'Connor T. P., Pattern of intracranial and extracranial projections of trigeminal ganglion cells, *J. Neurosci.,* 6, 2200, 1986.
9. Moskowitz, M. A., Buzzi, M. G., Sakas, D. E., and Linnik, M. D., Pain mechanisms underlying vascular headaches: progress report, *Rev. Neurol.,* 145, 181, 1989.
10. Buzzi, M. G., Carter, W. B., Shimizu, T., Heath, H. III, and Moskowitz, M. A., Dihydroergotamine and sumatriptan attenuate levels of CGRP in plasma in rat superior sagittal sinus during electrical stimulation of the trigeminal ganglion, *Neuropharmacology,* 30, 1193, 1991.
11. Zagami, A. S., Goadsby, P. J., and Edvinsson, L., Stimulation of the superior sagittal sinus in the cat causes release of vasoactive peptides, *Neuropeptides,* 16, 69, 1994.
12. Dimitriadou, V., Buzzi, M. G., Theoharides, T. C., and Moskowitz, M. A., Ultrastructural evidence for neurogenically mediated changes in blood vessels of the rat dura mater and tongue following antidromic trigeminal stimulation, *Neuroscience,* 48, 187, 1992.
13. Dimitriadou, V., Buzzi, M. G., Moskowitz, M. A., and Theoharides, T. C., Trigeminal sensory fiber stimulation induces morphological changes reflecting secretion in rat dura mater mast cells, *Neuroscience,* 44, 97, 1991.
14. Fields, H. L., *Pain,* McGraw-Hill, New York, 1987, p. 31.
15. Goadsby, P. J., Edvinsson, L., and Ekman, R., Release of vasoactive peptides in the extracerebral circulation in humans and the cat during activation of the trigeminovascular system, *Ann. Neurol.,* 23, 193, 1993.
16. Zochodne, D. W. and Ho, L. T., Sumatriptan blocks neurogenic inflammation in the peripheral nerve trunk, *Neurology,* 44, 161, 1994.
17. Buzzi, M. G. and Moskowitz, M. A., The antimigraine drug sumatriptan, (GR43175), selectively blocks neurogenic plasma extravasation from blood vessels in dura mater, *Br. J. Pharmacol.,* 99, 202, 1990.
18. Shepheard, S. L., Williamson, D. J., Hill, R. G., and Hargreaves, R. J., The non-peptide neurokinin$_1$ receptor antagonist, RP 67580, blocks neurogenic plasma extravasation in the dura mater of rats, *Br. J. Pharmacol.,* 108, 11, 1993.
19. Lee, W. S., Moussaoui, S. M., and Moskowitz, M. A., Oral or parenteral non-peptide NK1 receptor antagonist RPR 100,893 blocks neurogenic plasma extravasation within guinea-pig dura mater and conjunctiva, *Br. J. Pharmacol.,* 112, 920, 1994.
20. Rebeck, G. W., Maynard, K. I., Hyman, B., and Moskowitz, M. A., Selective 5-HT1Dα receptor gene expression in trigeminal ganglia: implications for anti-migraine drug development, *Proc. Natl. Acad. Sci. U.S.A.,* 91, 3666, 1994.
21. Bruinvels, A., 5-HT1D Receptors Reconsidered, Doctoral Dissertation, University of Basel, Basel, Switzerland, 1993.
22. Hoyer, D. and Middlemiss, D. N., The pharmacology of the terminal 5-HT autoreceptors in mammalian brain: evidence for species differences, *Trends Pharmacol. Sci.,* 10, 130, 1989.
23. Waeber, C., Schoeffter, P., Hoyer, D., and Palacios, J. M., The serotonin 5-HT(1D) receptor: a progress review, *Neurochem. Res.,* 15, 567, 1990.
24. Matsubara, T., Moskowitz, M. A., and Byun, B., CP-93,129, a potent and selective 5-HT1B receptor agonist blocks neurogenic extravasation within rat but not guinea-pig dura mater, *Br. J. Pharmacol.,* 104, 3, 1991.
25. Lee, W. S. and Moskowitz, M. A., Conformationally restricted sumatriptan analogues, CP-122,288 and CP-122,638 exhibit enhanced potency against neurogenic inflammation in dura mater, *Brain Res.,* 626, 303, 1993.
26. Hartig, P. R., Branchek, T. A., and Weinshank, R. L., A subfamily of 5-HT1D receptor genes, *Trends Pharmacol. Sci.,* 12, 152, 1992.
27. Hamel, E., Fan, E., Linville, D., Ting, V., Villemure, J. G., and Chia, L. S., Expression of mRNA for the serotonin 5-hydroxytryptamine 1 D_{beta} receptor subtype in human and bovine cerebral arteries, *Mol. Pharmacol.,* 44, 242, 1993.
28. Matsubara, T., Moskowitz, M. A., and Huang, Z., UK-14,394, R(−)-a-methyl-histamine and SMS 201-995 block plasma protein leakage within dura mater by prejunctional mechanisms, *Eur. J. Pharmacol.,* 224, 145, 1992.
29. Morgan, J. I., Cohen, D. R., Hempstead, J. L., and Curran, T., Mapping patterns of c-fos expression in the central nervous system after seizure, *Science,* 237, 192, 1987.

30. Sagar, S. M., Sharp, F. R., and Curran, T., Expression of c-fos protein in brain: metabolic mapping at the cellular level, *Science*, 240, 1328, 1988.
31. Hunt, S. P., Pini, A., and Evan, G., Induction of c-fos-like protein in spinal cord neurons following sensory stimulation, *Nature*, 328, 632, 1987.
32. Menetrey, D., Gannon, A., Levine, J. D., and Basbaum, A. I., The expression of c-fos protein in presumed-nociceptive interneurons and projection neurons of the rat spinal cord: anatomical mapping of the central effects of noxious somatic, articular and visceral stimulation, *J. Comp. Neurol.*, 258, 177, 1989.
33. Gogas, K. R., Presley, R. W., Levine, J. D., and Basbaum, A. I., The antinociceptive action of supraspinal opioids results from an increase in descending inhibitory control: correlation of nociceptive behavior and c-fos expression, *Neuroscience*, 42, 617, 1991.
34. Presley, R. W., Menetrey, D., Levine, J. D., and Basbaum, A. I., Systemic morphine suppresses noxious stimulus-evoked Fos protein-like immunoreactivity in the rat spinal cord, *J. Neurosci.*, 10, 323, 1990.
35. Nozaki, K., Boccalini, P., and Moskowitz, M. A., Expression of c-fos-like immunoreactivity in brainstem after meningeal irritation by blood in the subarachnoid space, *Neuroscience*, 49(3), 669, 1992.
36. Nozaki, K., Moskowitz, M. A., and Boccalini, P., CP-93,129, sumatriptan, dihydroergotamine block c-fos expression within rat trigeminal nucleus caudalis caused by chemical stimulation of the meninges, *Br. J. Pharmacol.*, 106, 409, 1992.
37. Cutrer, F. M., Moussaoui, S., Garret, C., and Moskowitz, M. A., The nonpeptide NK1 antagonist, RPR 100,893 decreases C-fos expression in trigeminal nucleus caudalis following noxious chemical meningeal stimulation, *Neuroscience*, 64, 740, 1995.
38. Macor, J. E., Blank, D. H., Post, R. J., and Ryan, K., The synthesis of a conformationally restricted analog of the anti-migraine drug sumatriptan, *Tetrahedron Lett.*, 33, 8011, 1992.
39. Cutrer, F. M., Schoenfeld, D., Limmroth, V., Panahian, N., and Moskowitz, M. A., The sumatriptan analog CP-122,288 suppresses c-fos immunoreactivity in trigeminal nucleus caudalis induced by intracisternal capsaicin, *Br. J. Pharmacol.*, 114, 987, 1995.
40. Fardin, V., Carruette, A., Menager, J., Bock, M., Flamand, O., Foucault, F., Heuillet, E., Moussaoui, S. M., Tabart, J. F., Peyronel, J. F., and Garret, C., In vitro pharmacological profile of RPR 100893, a novel non-peptide antagonist of the human NK1 receptor, Abstract for European Neuropeptide Club, April 1994, *Neuropeptides*, 26, 51, 1994.
41. Beresford, I. L. M., Birch, P. J., Hagan, R. M., and Ireland, S. J., Investigation into species variants in tachykinin NK_1 receptors by use of the non-peptide antagonist, CP-96,345, *Br. J. Pharmacol.*, 104, 292, 1991.
42. Lecci, A., Giuliani, S., Patacchini, R., Viti, G., and Maggi, C. A., Role of NK_1 tachykinin receptors in thermonociception: effect of (±)-CP-96,345, a non-peptide substance P antagonist, on the hot-plate test in mice, *Neurosci. Lett.*, 129, 299, 1991.
43. Rouissi, N., Gitter, B. D., Waters, D. C., Howbert, J. J., Nixon, J. A., and Regoli, D., Selectivity and specificity of new, non-peptide, quinuclidine antagonists of substance P, *Biochem. Biophys. Res. Commun.*, 176, 894, 1991.
44. Snider, R. M., Constantine, J. W., Lowe, J. A., Longo, K. P., Lebel, W. S., Woody, H. A., Drozda, S. E., Desai, M. C., Vinick, F. G., Spencer, R. W., and Hess, H. J., A potent non-peptide antagonist of the substance P (NK_1) receptor, *Science*, 251, 435, 1991.
45. Birch, P. J., Harrison, S. M., Hayes, A. G., Rogers, H., and Tyers, M. B., The non-peptide NK_1 antagonist, (±)-CP96.345, produces antinociceptive and anti-oedema effects in the rat, *Br. J. Pharmacol.*, 105, 508, 1992.
46. Nagahisa, A., Asai, R., Kanai, Y., Murase, A., Tsuchiya-Nakagaki, M., Nakagaki, T., Shieh, T.-C., and Taniguchi, K., Non-specific activity of (±)-CP96,345 in models of pain and inflammation, *Br. J. Pharmacol.*, 107, 273, 1992.
47. Guard, S. and Watling, K. J., Interaction of the non-peptide NK_1 tachykinin receptor antagonist (±)-CP96,345 with L-type Ca^{2+} channels in rat cerebral cortex, *Br. J. Pharmacol.*, 106, 385, 1992.
48. Schmidt, A. W., McLean, S., and Heym, J., The substance P receptor antagonist CP-96,345 interacts with Ca^{++} channels, *Eur. J. Pharmacol.*, 215, 351, 1992.
49. Caeser, M., Seabrook, G. R., and Kemp, J. A., Block of voltage-dependent sodium currents by the substance P receptor antagonist (±)-CP-96,345 in neurones cultured from rat cortex, *Br. J. Pharmacol.*, 109, 918, 1993.
50. Nagy, I., Maggi, C. A., Dary, A., Woolf, C. J., and Urban, L., The role of neurokinin and N-methyl-D-aspartate receptors in synaptic transmission from capsaicin sensitive primary afferent in the rat spinal cord *in vitro*, *Neuroscience*, 52, 1029, 1993.
51. Rupniak, N. M. J., Boyce, S., Williams, A. R., Cook, G., Longmore, J., Seabrook, G. R., Caesar, M., Iversen, S. D., and Hill, R. G., Antinociceptive activity of NK_1 receptor antagonists: non-specific effects of racemic RP 67580, *Br. J. Pharmacol.*, 110, 1607, 1993.
52. Garret, C., Carruette, A., Fardin, V., Moussaoui, S., Montier, F., Peyronel, J. F., and Laduron, P. M., Antinociceptive properties and inhibition of neurogenic inflammation with potent SP antagonists belonging to perhydroisoindolones, *Regul. Peptides*, 46, 24, 1993.

53. Wang, Z. Y., Tung, S. R., Strichartz, G. R., and Hakanson, R., Non-specific actions of the non-peptide tachykinin receptor antagonists, CP-96,345, RP 67580 and SR 48968, on neurotransmission, *Br. J. Pharmacol.*, 111, 179, 1994.
54. Lauritzen, M., Cerebral blood flow in migraine and cortical spreading depression, *Acta Neurol. Scand.*, 76, (Suppl. 113), 9, 1987.
55. Moskowitz, M. A., Nozaki, K., and Kraig, R. P., Neocortical spreading depression provokes the expression of c-fos protein-like immunoreactivity within trigeminal nucleus caudalis via trigeminovascular mechanisms, *J. Neurosci.*, 13 (3), 1167, 1993.
56. Nicholson, C. and Kraig, R. P., The behavior of extracellular ions during spreading depression, in *The Application of Ion-Selective Microelectrodes*, Zeuthen, T., Ed., Elsevier/North Holland, Amsterdam, 1981, 217.
57. Lauritzen, M., Hansen, A. K., Kronberg, D., and Wieloch, T., Cortical spreading depression is associated with arachidonic acid accumulation and preservation of energy charge, *J. Cereb. Blood Flow Metab.*, 10, 115, 1990.
58. Krivanek, J., Some metabolic changes accompanying Leao's spreading cortical depression in the rat, *J. Neurochem.*, 6, 183, 1961.
59. Wei, E. P., Moskowitz, M. A., Boccalini, P., and Kontos, H. A., Calcitonin gene-related peptide mediates nitroglycerin and sodium nitroprusside-induced vasodilation in feline cerebral arterioles, *Circ. Res.*, 70, 1313, 1992.
60. Holzer, P. and Jocic, M., Cutaneous vasodilatation induced by nitric oxide-evoked stimulation of afferent nerves in the rat, *Br. J. Pharmacol.*, 112, 1181, 1994.
61. Hughes, S. R. and Brain, S. D., Nitric oxide-dependent release of vasodilator quantities of calcitonin gene-related peptide from capsaicin-sensitive nerves in rabbit skin, *Br. J. Pharmacol.*, 104, 738, 1994.
62. Kitazono, T., Heistad, D. D., and Faraci, F. M., Role of ATP sensitive K^+ channels in CGRP-induced dilatation of basilar artery in vivo, *Am. J. Physiol.*, 265, H581, 1993.
63. Moussaoui, S. M., unpublished data.

Chapter 16

SENSORY NEUROPEPTIDES IN THE LOWER URINARY TRACT

Alessandro Lecci and Carlo A. Maggi

CONTENTS

I. INTRODUCTION

The general principles of the contribution of the peripheral nervous system to inflammatory processes in the lower urinary tract (LUT) are essentially similar to those operating in other regions of the body.[1,2] The peripheral nervous system takes a role in the defense of the LUT by activating motor reflexes, coordinated at the spinal or the supraspinal level, which are aimed at removing noxious stimuli. Nociceptive stimulation of sensory fibers also elicits a direct, "efferent" response through the release of neuropeptides from nerve endings, which in turn produce a number of local effects, including

1. A direct effect on target cells producing smooth muscle contraction/relaxation, vasodilatation, etc.
2. A prejunctional modulation of transmitter release from nerve terminals and/or a postjunctional modulation of the effects of the released transmitter
3. Disruption of the barrier between the circulatory system and neighboring tissues to facilitate the infiltration of cells of the immune system
4. The activation of cells of the immune system; these cells, in turn, release cytokines and other mediators that participate in the sensitization of sensory fibers
5. A trophic influence on target tissues through the control of the microcirculation and mitotic cycle

These events deeply affect the physiology of the urinary tract, which becomes hyperresponsive to external (exteroceptive) or internal (proprioceptive) stimuli: a condition of hyperreflexia that parallels what has been defined as "inflammatory pain". Most of these effects have been highlighted by the use of capsaicin, a neurotoxin that selectively affects the function of a subset of sensory nerves. It must be pointed out that capsaicin-sensitive fibers do not represent a homogeneous population of nerves and that not all nerve fibers participating in neurogenic inflammation are capsaicin-sensitive. A population of capsaicin-sensitive afferent fibers containing tachykinins (TKs) and calcitonin gene-related peptide (CGRP) has been clearly identified as contributing to the sensory innervation of the mammalian urinary tract. Throughout the urinary tract, the functional significance and the pharmacological properties of this innervation appear to afford protection toward irritant or noxious stimuli and to facilitate the removal of the *noxae*. The functional responses induced by the activation of this set of nerves vary in different regions of the urinary tract: regional variations are mainly related to postjunctional variations in neuropeptide receptors present in the target tissue.[3,4]

0-8493-7646-7/96/$0.00+$.50

TABLE 1
Summary of the Role and Distribution of Neuropeptide Receptors in the Lower Urinary Tract

Anatomical District	Receptor	Effect	Target Tissue
Renal pelvis			
	NK_1	*Reflexes*	*Afferent terminals*
		Plasma extravasation	Endothelium
	NK_2	NANC contraction	Smooth muscle
		Pacemaker	Smooth muscle
Ureter			
	NK_1	Plasma extravasation	Endothelium
	NK_2	NANC contraction	Smooth muscle
		Pacemaker	Smooth muscle
	CGRP	Hyperpolarization	Smooth muscle
Bladder			
	NK_1	Plasma extravasation	Endothelium
		(NANC contraction)	Smooth muscle
		Reflexes	*Afferent terminals**
		Twitch enhancement	Efferent terminals
			Pelvic ganglion
	NK_2	NANC contraction	Smooth muscle
		Reflexes	*Afferent terminals**
		Twitch enhancement	*Efferent terminals*
			Pelvic ganglion
	CGRP	Vasodilatation	Vascular smooth muscle
		Unknown	Urothelium
		(NANC contraction)	Smooth muscle
Urethra			
	NK_1	(NANC contraction)	Smooth muscle
		Plasma extravasation	Endothelium
	NK_2	NANC contraction	Smooth muscle
	CGRP	NANC relaxation	Smooth muscle

Note: Effects that are species-dependent are indicated in parentheses. In italics are indicated those effects that are not necessarily directly mediated (i.e., there is no evidence of the presence of the given receptor in the target tissue). The asterisk indicates that there are conflicting results about the presence of NK_2 receptors on afferent terminals in the bladder. NANC, nonadrenergic noncholinergic.

Table 1 summarizes functional effects mediated by neuropeptide receptors in the lower urinary tract.

II. NEUROGENIC INFLAMMATION IN THE RENAL PELVIS AND URETER

The renal pelvis collects the urine formed in the kidney and drives ureteral peristaltis through pacemaker cells which initiate a propagated wave of excitation-contraction spreading through the ureter. Although nerve activity is not required to initiate excitation-contraction coupling in the pyeloureteral system, it is now recognized that neuropeptides released by the stimulation of sensory nerves can modulate the myogenic activity of the renal pelvis and ureteral smooth muscle.[5] The anatomical arrangement of TK/CGRP-containing nerve profiles supports this view, since nerve endings are present in the suburothelial layer, and but also within the smooth muscle of the renal pelvis and ureter.[6] Electrical nerve stimulation of the renal pelvis increases the frequency of spontaneous contractions, an effect that is resistant to cholinoceptor and adrenoceptor blockers, but is prevented by capsaicin desensitization indicating the involvement of sensory nerves.[7]

A pharmacological analysis of this effect revealed that the stimulation of pacemaker activity was mainly attributable to tachykinin NK_2 receptors with only a minor contribution of NK_1 receptors.[8] On the other hand the role of NK_1 receptors in the renal pelvis could be more intimately associated with the afferent function of sensory nerves. In fact, administration of substance P (SP) into the renal pelvis increases afferent renal nerve activity, inducing a reflex, controlateral diuretic and natriuretic response. This reflex, which can also be evoked by the stimulation of renal mechanoceptors by increasing intraureteral pressure, is abolished by pretreatment with neurotoxic doses of capsaicin.[9] Likewise, this reno-renal reflex evoked by mechanoceptor stimulation or by SP infusion is selectively reduced by intraureteral administration of the tachykinin NK_1 receptor antagonist CP 96,345.[10] Therefore, the reno-renal reflex could be activated in pathological conditions which increase intraureteral pressure such as hydropyelonephritis or renal stones, or in conditions where normal constituents of the urine (KCl, hyperosmolarity, low pH, or bradykinin), or products of bacterial metabolism, penetrate through the urothelium to stimulate sensory nerves.

The activation of sensory nerves by electrical stimulation or by irritants inhibits spontaneous activity of the ureter.[11-13] This inhibition is caused by CGRP, which directly decreases contractility[14] but also blocks the genesis and the propagation of electrical impulses within the smooth muscle.[15,16] The second effect appears to be especially important, since it enables CGRP to transiently suppress the activity of latent pacemakers in the ureter, thus preventing antiperistalsis and backflow of urine toward the kidney, an event that is important for the pathogenesis of ascending infections of the urinary tract and pyelonephritis. This effect of CGRP is produced through the activation of glibenclamide-sensitive K_{ATP} channels, leading to transient hyperpolarization of the muscle membrane and indirect prevention of the opening of voltage-sensitive Ca^{2+} channels.[17] An elevation of intracellular cAMP seems causally linked to the production of this effect. The different motor effects induced by sensory nerve activation in the renal pelvis and ureter (excitation and inhibition, respectively) depend mainly on the location of sensory neuropeptide receptors: those for TKs predominate in the renal pelvis and those for CGRP in the ureter.[18]

TK NK_1 receptors are involved in mediating plasma protein extravasation (PPE) induced by capsaicin or nerve stimulation.[19,20] The pertinent experiments have clarified the anatomical arrangement of the sensory innervation in the ureter, in which the distal portion is innervated by the pelvic nerve and the proximal one by sympathetic afferents. This arrangement has been recently confirmed in capsaicin-induced PPE experiments, where pelvic ganglia were pretreated with colchicine: under these conditions PPE was abolished in the vesical portion, but remained unaffected in the renal portion of the ureter.[61]

III. NEUROGENIC INFLAMMATION IN THE URINARY BLADDER AND URETHRA

As in the ureter, inflammatory states of the urinary bladder produce two distinct kinds of neurogenically mediated motor responses: a local response and a reflex response involving a coordinated contraction of the detrusor muscle with the opening of urethral sphincters, which occurs at lower filling volumes than in normal conditions (hyperreflexia). These responses have been studied in several experimental models involving intravesical infusion of irritants in normal animals[21-25] or antigen in sensitized animals.[26] Local motor responses of the urinary bladder can be evoked in normal animals by the intravesical application of irritants (capsaicin, xylene, acrolein, turpentine, croton and mustard oil, low pH, hyperosmolarity)[27-29] or by electrical stimulation of intramural nerves.[30] The local motor responses induced by irritants appear to be largely capsaicin-sensitive, with tachykinins mediating the excitation and CGRP the inhibition of smooth muscle contraction.

The quality of the observed effects and the pharmacology of sensory neuropeptides at the bladder level is largely species-dependent and most of observed differences appear to be

TABLE 2
Effect of Various Treatments on Chemically Induced Plasma Extravasation (Evans Blue Technique) in the Rat Urinary Bladder

			Xylene (Intravesical)		Cyclophosphamide (i.p.)	
Treatment	**SP (i.v.)**	**48/80 (i.v.)**	**Early Phase**	**Late Phase**	**Early Phase**	**Late Phase**
48/80	+[a]/–	+	+			
Capsaicin-N		+	++	–/—		
Capsaicin-A		+	+	+	+	–/—
Cim. + Clorph.	–	+/–	–		–	
Methysergide	–	+/–	–	–	–	
Indomethacin	+[a]	–	–	+	–	
(±)CP or RP	+	+[b]/–	+		+	
Hoe 140	–		+		+	

Legend: Inhibition of PPE: +, about 50%; ++, about 100%; –, not effective; – –, aggravates; +/–, conflicting results. SP, substance P (tachykinin agonist); 48/80 (mast cell degranulator); Capsaicin-N (pretreatment with capsaicin in newborns); Capsaicin-A (pretreatment with capsaicin in adults); Cim. + Clorph: cimetidine + clorpheniramine (histamine H_1 and H_2 receptor antagonists); Methysergide (serotonin 5-HT1/2 receptor antagonist); Indomethacin (cyclooxygenase inhibitor); (±) CP or RP: (±) CP 96,345 or RP 67,580 (tachykinin NK_1 receptor antagonists); Hoe 140 (bradykinin B_2 antagonist).

[a] Effect not mediated by tachykinin receptors; [b] study performed with [D-Arg,[1] D-Trp[7,9] Leu[11]]-substance P. Data were taken from References 20, 21, 24, 27, 41–43 and Eglezos, Giuliani, Santicioli, Lecci, and Maggi, unpublished results.

attributable to a differential distribution of receptors. Thus, CGRP has no effect on rat or human bladder motility, whereas it inhibits the evoked motility in the guinea pig bladder.[11,31,32] TKs induce contraction of bladder smooth muscle of most species through NK_2 receptors, while a role of NK_1 receptors in bladder muscle contractility seems to be restricted to rats and guinea pigs.[33-36] A capsaicin-sensitive component of the bladder contraction induced by electrical field stimulation of intramural nerves has been recently demonstrated in rats.[30] However, the characteristics of this response suggest that capsaicin-sensitive nerves do not play a role in the physiological (distension-induced) contraction of the detrusor muscle. Accordingly, the combined administration of tachykinin NK_1 and NK_2 receptor antagonists, which completely blocks the capsaicin-sensitive component of the contraction induced by electrical stimulation, does not modify the distension-induced micturition reflex in normal animals.[37]

This point has also been assessed by comparing the effect of physiological distension, electrical stimulation of the pelvic nerve, or bladder irritation on plasma PPE in the rat bladder: physiological distension activating the micturition reflex had no effect, whereas electrical stimulation or intravesical application of an irritant caused a large, nerve-mediated PPE.[38] PPE induced by electrical stimulation of lumbosacral dorsal roots innervating the urinary bladder (as well as ureteral, urethral, and genital organs) is reduced by pretreatment with neurotoxic doses of capsaicin, thus showing the involvement of mediator(s) released from the peripheral endings of sensory fibers.[39]

Studies with selective agonists and antagonists have shown that TKs, through the stimulation of NK_1 receptors, are the mediators of PPE in the LUT.[20,40] Table 2 summarizes the results of studies on the effect of antagonists of some mediators of inflammation on irritant-induced PPE in the urinary bladder.[20,21,24,27,41-43] It can be noted that the effect of irritants on PPE has been resolved into two phases and that only the early phase is consistently reduced by neurotoxic capsaicin pretreatment, tachykinin NK_1 receptor antagonists, and bradykinin (BK) B_2 antagonists. It is also worth noting that, while SP-induced PPE is partially mediated by prostanoid production through a nontachykinin receptor (an effect mediated by the N-terminal portion of SP), the irritant-induced PPE is not reduced by indomethacin. These results suggest that SP is not the major endogenous mediator of PPE in the LUT, while other TKs could be more important for producing PPE through NK_1 receptors.

The finding that natural TKs are equipotent to induce PPE[20] is in line with this hypothesis, and in view of the tissue levels of natural TKs in the LUT, it could be speculated that NKA is the main mediator to induce PPE in the LUT by NK_1 receptors. The late component of the irritant-induced PPE appears to be capsaicin-resistant, although still largely nerve-mediated. However, the interpretation of these results is complicated by the fact that capsaicin pretreatment abolishes the chemoceptive micturition reflex, thus hindering the quick removal of the irritant from the bladder with the consequent aggravation of the inflammatory response. In fact, although peripheral tachykinins do not contribute to the micturition reflex in physiological conditions, the peripheral administration of TK NK_2 receptor antagonists was found to affect micturition pattern in a model of chemical cystitis,[44,45] suggesting that the tachykinergic component of noncholinergic, nonadrenergic bladder contraction may become more prominent during inflammatory conditions.

Chemical irritation of the bladder also dramatically changes the receptive properties of bladder afferents. First, some unmyelinated (C)-fibers, which normally do not respond to physiological or noxious stimuli (distension or overdistension, respectively), termed "silent" nociceptors, become mechanosensitive after irritation.[46] Second, thin, myelinated (Aδ)-fibers, which respond to graded distension in normal conditions, become hyperresponsive and acquire an ongoing activity even in resting conditions (i.e., when the bladder is empty and the irritant has been removed already).[47] This may represent the electrophysiological substrate for detrusor hyperreflexia. Chemical irritation also induces a delayed (4 h) increase in nerve growth factor (NGF) and NGF mRNA in the mucosa and muscle layers of the bladder,[48] an effect which could contribute to the sensitization of vesical afferents.

The role of capsaicin-sensitive primary afferents in detrusor hyperreflexia is controversial. On the one hand neurotoxic capsaicin pretreatment in the adult is ineffective in preventing xylene- and hyperosmolarity-induced hyperreflexia,[21,25] although the peripheral administration of TK NK_2 receptor antagonists may have some beneficial effects in the former model.[44,45] On the other hand, the same kind of capsaicin pretreatment effectively prevents acidic solution- and cyclophosphamide-induced hyperreflexia,[24,49] although in the latter model the peripheral administration of TK NK_1 or NK_2 antagonists failed to reduce bladder hyperreflexia.[50] Cyclophosphamide-induced hyperreflexia was, instead, reduced by the peripheral administration of a BK B_2 antagonist[51] or by the intrathecal administration of TK NK_1 or NK_2 receptor antagonists,[50] suggesting that the sensory but not the "efferent" function of capsaicin-sensitive afferents is involved. It is not clear to what extent the variable results obtained in various models of cystitis may derive from different mechanisms of action of the irritants used or from the different degrees of severity of hyperreflexia in various models.

An interesting model of urinary bladder inflammation and associated hyperreflexia has been recently introduced: intravesical instillation of ovalbumin in previously sensitized animals.[26] Ovalbumin does not affect micturition in nonsensitized animals and this model has been proposed as an animal counterpart of interstitial cystitis in humans, a disease characterized by an apparently idiopathic bladder hyperreflexia. In sensitized guinea pigs, the intravesical infusion of ovalbumin induces interstitial hemorrhage, leukocyte accumulation, and PPE. All these effects are reproduced by intravesical infusion of SP, which is also effective in nonsensitized animals. *In vitro* exposure to ovalbumin causes release of SP, histamine, peptidoleukotrienes (LTC_4), and prostaglandins (PGD_2) from the bladders of sensitized animals. In turn, histamine, LTC_4, or PGD_2 induce SP release from the bladder of sensitized guinea pigs, therefore establishing a positive feedback between capsaicin-sensitive afferents and cells of the immune system.[52] It is worth mentioning that prostanoids can also sensitize capsaicin-sensitive bladder afferents, thus setting the threshold for micturition. It has been postulated that prostanoids modulate micturition threshold even in physiological conditions: since they are produced during bladder distension, prostanoids could represent a chemical link between mechanical stimuli and the activation of bladder afferents.[53]

At the urethral level, stimulation of capsaicin-sensitive afferents inhibits electrically evoked contractions, an effect mediated by CGRP release.[14] On the other hand, a weak contraction can be elicited by capsaicin in the unstimulated urethra; this effect is mediated by TKs acting mainly through TK NK_2 receptors in humans[54] and both NK_1 and NK_2 receptors in the rat urethra.[55] Although the urethral innervation derives from three different sources (pelvic, hypogastric, and pudendal nerves), the chronic ablation of pelvic ganglia abolishes all the motor effects induced by capsaicin;[56] the same applies to capsaicin-induced PPE, which is very intense at the urethral level.[32] However, not all urethral capsaicin-sensitive fibers derive from the pelvic ganglia; the behavioral response (perineal licking) induced by intravesical instillation of capsaicin is resistant to pelvic ganglion ablation but is abolished by pudendal denervation,[57] further indicating the complexity of urethral innervation. Since the intraurethral instillation of capsaicin activates a urethrovesical inhibitory reflex, whose sensory branch travels in the pudendal nerves,[58] and concomitantly increases intracavernal pressure (in some cases producing erection),[59] it is tempting to speculate that capsaicin-sensitive fibers are also involved in the regulation of sexual functions. These results would suggest the existence of two functionally distinct kinds of capsaicin-sensitive fibers at the urethral level: one mediating local motor responses and PPE and the other subserving a (pure) sensory function.

A model mimicking urethral inflammation induced by catheter insertion has been developed: also in this case neurotoxic pretreatment with capsaicin was found to reduce the inflammatory response. In this model the participation of sympathetic (efferent) fibers in the inflammatory process has been also hypothesized.[60]

In conclusion, sensory neuropeptides (in particular TKs and CGRP) participate in inflammatory processes of urogenital organs, integrating the response between the nervous system and peripheral cells. Neurogenic inflammation in the urinary tract appears to have a pronounced defensive character, designed for the expulsion of irritants, removal of the *noxae*, but also for the prevention of organ damage. Blockade of sensory nerves through the topical instillation of capsaicin into the urinary bladder or blockade of receptors for sensory neuropeptides both appear to be rational approaches for controlling neurogenic inflammation at this level.

REFERENCES

1. Maggi, C. A. and Meli, A., The sensory-efferent function of capsaicin-sensitive sensory neurons, *Gen. Pharmacol.*, 19, 1, 1988.
2. Holzer, P., Local effector functions of capsaicin-sensitive sensory nerve endings: involvement of tachykinins, CGRP and other neuropeptides, *Neuroscience*, 24, 739, 1988.
3. Maggi, C. A., The role of neuropeptides in the regulation of the micturition reflex, *J. Auton. Pharmacol.*, 6, 133, 1986.
4. Maggi, C. A., The role of peptides in the regulation of the micturition reflex: an update, *Gen. Pharmacol.*, 22, 1, 1991.
5. Amann, R., Neural regulation of ureteric motility, in *Nervous Control of the Urogenital System*, Maggi, C. A., Ed., Harwood Academic Publishers, Chur, 1993, chap. 7.
6. Ferguson, M. and Bell, C., Autonomic innervation of the kidney and ureter, in *Nervous Control of the Urogenital System*, Maggi, C. A., Ed., Harwood Academic Publishers, Chur, 1993, chap. 1.
7. Maggi, C. A. and Giuliani, S., Non-adrenergic non-cholinergic excitatory innervation of the guinea-pig isolated renal pelvis: involvement of capsaicin-sensitive primary afferent neurons, *J. Urol.*, 147, 1394, 1992.

8. Maggi, C. A., Patacchini, R., Eglezos, A., Quartara, L., Giuliani, S., and Giachetti, A., Tachykinin receptors in the guinea-pig renal pelvis: activation by exogenous and endogenous tachykinins, *Br. J. Pharmacol.*, 107, 27, 1992.
9. Kopp, U. C. and DiBona, G. F., The neural control of renal function, in *Nervous Control of the Urogenital System*, Maggi, C. A., Ed., Harwood Academic Publishers, Chur, 1993, chap. 5.
10. Kopp, U. C. and Smith, L. A., Effects of the substance P receptor antagonist CP-96,345 on renal sensory receptor activation, *Am. J. Physiol.*, 264, R647, 1993.
11. Maggi, C. A., Santicioli, P., Giuliani, S., Abelli, L., and Meli, A., The motor effect of capsaicin-sensitive inhibitory innervation of the rat ureter, *Eur. J. Pharmacol.*, 126, 333, 1986.
12. Hua, X.-Y. and Lundberg, J. M., Dual effect of capsaicin on ureteric motility: low dose inhibition mediated by calcitonin gene-related peptide and high dose stimulation by tachykinins?, *Acta Physiol. Scand.*, 128, 453, 1986.
13. Hua, X.-Y., Kinn, A. C., and Lundberg, J. M., Capsaicin-sensitive nerves and ureteric motility: opposing effects of tachykinins and calcitonin gene-related peptide, *Acta Physiol. Scand.*, 128, 317, 1986.
14. Maggi, C. A., Giuliani, S., and Santicioli, P., Multiple mechanisms in the smooth muscle relaxant action of calcitonin gene-related peptide in the guinea-pig ureter, *Naunyn-Schmiedeberg's Arch. Pharmacol.*, 350, 537, 1994.
15. Maggi, C. A. and Giuliani, S., Calcitonin gene-related peptide (CGRP) regulates excitability and refractory period of the guinea-pig ureter, *J. Urol.*,, 152, 520-524, 1994.
16. Meini, S., Santicioli, P., and Maggi, C. A., Calcitonin gene-related peptide (CGRP) affects propagation of impulses in the guinea-pig ureter, *Naunyn-Schmiedeberg's Arch. Pharmacol.*, 351, 79, 1995.
17. Santicioli, P. and Maggi, C. A., Calcitonin gene-related peptide acts as inhibitory transmitter by activating glibenclamide-sensitive potassium channels in the guinea-pig ureter, *Br. J. Pharmacol.*, 113, 588-592, 1994.
18. Sann, H., Rossler, W., Hammer, K., and Pierau, F. K., Substance P and calcitonin gene-related peptide in the ureter of chicken and guinea-pig: distribution, binding sites and possible functions, *Neuroscience*, 49, 699, 1992.
19. Saria, A., Lundberg, J. M., Hua, X.-Y., and Lembeck, F., Sensory control of vascular permeability and capsaicin-induced substance P release and in the guinea-pig ureter, *Neurosci. Lett.*, 41, 167, 1983.
20. Abelli, L., Somma, V., Maggi, C. A., Regoli, D., Astolfi, M., Parlani, M., Rovero, P., Conte, B., and Meli, A., Effects of tachykinins and selective tachykinin receptor agonists on vascular permeability in the rat lower urinary tract: evidence for the involvement of NK_1 receptors, *J. Auton. Pharmacol.*, 9, 253, 1989.
21. Maggi, C. A., Abelli, L., Giuliani, S., Santicioli, P., Geppetti, P., Somma, V., Frilli, S., and Meli, A., The contribution of sensory nerves to xylene-induced cystitis in rats, *Neuroscience*,, 26, 709, 1988.
22. McMahon, S. B. and Abel, C., A model for the study of visceral pain states: chronic inflammation of the chronic decerebrate rat urinary bladder by irritant chemicals, *Pain*, 28, 109, 1987.
23. Morikawa, K., Fukuoka, M., Kakiuchi, M., Kato, H., Ito, Y., and Gomi, Y., Detrusor hyperreflexia induced by intravesical instillation of xylene in conscious rats, *Jpn. J. Pharmacol.*, 52, 587, 1990.
24. Maggi, C. A., Lecci, A., Santicioli, P., Del Bianco, E., and Giuliani, S., Cyclophosphamide cystitis in rats: involvement of capsaicin sensitive primary afferents, *J. Auton. Nerv. Syst.*, 38, 201, 1992.
25. Maggi, C. A., Abelli, L., Giuliani, S., Somma, V., Furio, M., Patacchini, R., and Meli, A., Motor and inflammatory effect of hyperosmolar solutions on the rat urinary bladder in relation to capsaicin-sensitive sensory nerves, *Gen. Pharmacol.*, 21, 97, 1990.
26. Kim, Y. S., Longhurst, P. A., Wein, A. J., and Levin, R. M., Effects of sensitization on female guinea pig urinary bladder function: *in vivo* and *in vitro* studies, *J. Urol.*, 146, 454, 1991.
27. Ahluwalia, A., Maggi, C. A., Santicioli, P., Lecci, A., and Giuliani, S., Characterization of the capsaicin-sensitive component of cyclophosphamide-induced inflammation in the rat urinary bladder, *Br. J. Parmacol.*, 111, 1017, 1994.
28. Geppetti, P., Del Bianco, E., Patacchini, R., Santicioli, P., and Maggi, C. A., Low pH-induced release of calcitonin gene-related peptide from capsaicin-sensitive sensory nerves: mechanism of action and biological response, *Neuroscience*, 41, 295, 1991.
29. Patacchini, R., Maggi, C. A., and Meli, A., Capsaicin-like activity of some natural pungent substances on peripheral endings of visceral primary afferents, *Naunyn-Schmiedeberg's Arch. Pharmacol.*, 342, 72, 1990.
30. Meini, S. and Maggi, C. A., Identification of a capsaicin-sensitive, tachykinin-mediated, component in the NANC contraction of the rat urinary bladder to nerve stimulation, *Br. J. Pharmacol.*, 112, 1123-1131, 1994.
31. Maggi, C. A., Patacchini, R., Santicioli, P., Turini, D., Barbanti, G., Beneforti, P., Rovero, P., and Meli, A., Further studies on the motor response of the human isolated urinary bladder to tachykinins, capsaicin and electrical field stimulation, *Gen. Pharmacol.*, 20, 663, 1989.
32. Maggi, C. A., Santicioli, P., Patacchini, R., Geppetti, P., Giuliani, S., Astolfi, M., Baldi, E., Parlani, M., Theodorsson, E., Fusco, B., and Meli, A., Regional differences in the motor response to capsaicin in the guinea pig urinary bladder: relative role of pre- and postjunctional factors related to neuropeptide-containing sensory nerves, *Neuroscience*, 27, 675, 1988.

33. Maggi, C. A., Santicioli, P., Patacchini, R., Cellerini, M., Turini, D., Barbanti, G., Beneforti, P., Rovero, P., and Meli, A., Contractile response of the human isolated urinary bladder to tachykinins, involvement of NK_2 receptors, *Eur. J. Pharmacol.*, 145, 335, 1988.
34. Maggi, C. A., Patacchini, R., Santicioli, P., and Giuliani, S., Tachykinin antagonists and capsaicin-induced contraction of the rat isolated urinary bladder: evidence for tachykinin-mediated cotransmission, *Br. J. Pharmacol.*, 103, 1535, 1991.
35. Shinkai, M. and Takayanagi, I., Characterization of tachykinin receptors in urinary bladder from guinea-pig, *Jpn. J. Pharmacol.*, 54, 241, 1990.
36. Longmore, J. and Hill, R. G., Characterization of neurokinin receptors in the guinea-pig urinary bladder smooth muscle: use of selective antagonists, *Eur. J. Pharmacol.*, 222, 167, 1992.
37. Lecci, A., Giuliani, S., Patacchini, R., and Maggi, C. A., Evidence against a peripheral role of tachykinins in the initiation of micturition reflex, *J. Pharmacol. Exp. Ther.*, 264, 1327, 1993.
38. Koltzenburg, M. and McMahon, S. B., Plasma extravasation in the rat urinary bladder following mechanical, electrical and chemical stimuli: evidence for a new population of chemosensitive primary sensory afferents, *Neurosci. Lett.*, 72, 352, 1986.
39. Szolcsányi, J., Antidromic vasodilatation and neurogenic inflammation, *Agents Actions*, 23, 4, 1988.
40. Eglezos, A., Giuliani, S., Viti, G., and Maggi, C. A., Direct evidence that capsaicin-induced plasma protein extravasation is mediated through tachykinin NK_1 receptors, *Eur. J. Pharmacol.*, 209, 277, 1991.
41. Saria, A., Hua, X.-Y., Skofitsch, G., and Lundberg, J. M., Inhibition of compound 48/80-induced vascular protein leakage by pretreatment with capsaicin and a substance P antagonist, *Naunyn-Schmiedeberg's Arch. Pharmacol.*, 328, 9, 1984.
42. Abelli, L., Nappi, F., Perretti, F., Maggi, C. A., Manzini, S. and Giachetti, A., Microvascular leakage induced by substance P in rat urinary bladder: involvement of cyclo-oxygenase metabolites of arachidonic acid, *J. Auton. Pharmacol.*, 12, 269, 1992.
43. Giuliani, S., Santicioli, P., Lippe, I. T., Lecci, A., and Maggi, C. A., Effect of bradykinin and tachykinin receptor antagonist on xylene-induced cystitis in rats, *J. Urol.*, 150, 1014, 1993.
44. Maggi, C. A., Giuliani, S., Ballati, L., Lecci, A., Manzini, S., Patacchini, R., Renzetti, R., Rovero, P., Quartara, L., and Giachetti, A., *In vivo* evidence for tachykininergic transmission using a new NK_2 receptor selective antagonist, MEN 10,376, *J. Pharmacol. Exp. Ther.*, 257, 1172, 1991.
45. Pietra, C., Bettellini, R., Hagan, R. M., Ward, P., McElroy, A., and Trist, D. G., Effect of selective antagonists at tachykinin NK1 and NK2 receptors on xylene-induced cystitis in rats, *Neuropeptides*, 22 (Suppl.), 52, 1992.
46. Häbler, H.-J., Jänig, W., and Koltzenburg, M., Activation of unmyelinated afferent fibers by mechanical stimuli and inflammation of the urinary bladder in the cat, *J. Physiol.*, 425, 545, 1990.
47. Häbler, H.-J., Jänig, W., and Koltzenburg, M., Receptive properties of myelinated primary afferents innervating the inflamed urinary bladder of the cat, *J. Neurophysiol.*, 69, 395, 1993.
48. Andreev, N. Y., Bennett, D., Priestley, J., Rattray, M., and McMahon, S. B., Nerve growth factor mRNA is increased by experimental inflammation of adult rat urinary bladder, *Soc. Neurosci. Abstr.*, 19, 248, 1993.
49. Muhlhauser, M. A. and Thor, K., Vesicoanal reflex activity in the rat: a model of urinary bladder irritation, *Soc. Neurosci. Abstr.*, 18, 500, 1992.
50. Lecci, A., Giuliani, S., Santicioli, P., and Maggi, C. A., Involvement of spinal tachykinin NK_1 and NK_2 receptors in detrusor hyperreflexia during chemical cystitis in anaesthetized rats, *Eur. J. Pharmacol.*, 259, 129-135, 1994.
51. Maggi, C. A., Santicioli, P., Del Bianco, E., Lecci, A., and Giuliani, S., Evidence for the involvement of bradykinin in chemically-evoked cystitis in anaesthetized rats, *Naunyn-Schmiedeberg's Arch. Pharmacol.*, 347, 432, 1993.
52. Bjorling, D. E., Saban, M. R., Tengowski, M. W., and Saban, R., Neurogenic inflammation of the bladder, *FASEB J.*, 8, A664, 1994.
53. Maggi, C. A., Prostanoids as local modulators of reflex micturition, *Pharmacol. Res.*, 25, 13, 1992.
54. Parlani, M., Conte, B., Majmone, S., Maggi, C. A., Rovero, P., Regoli, D., and Giachetti, A., The contractile effects of tachykinins on human prostatic urethra, involvement of NK_2 receptors, *J. Urol.*, 144, 1543, 1990.
55. Maggi, C. A., Parlani, M., Astolfi, M., Santicioli, P., Rovero, P., Abelli, L., Somma, V., Giuliani, S., Regoli, D., Patacchini, R., and Meli, A., Neurokinin receptors in the rat lower urinary tract, *J. Pharmacol. Exp. Ther.*, 246, 308, 1988.
56. Maggi, C. A., Santicioli, P., Manzini, S., Conti, S., Giuliani, S., Patacchini, R., and Meli, A., Functional studies on the cholinergic and sympathetic innervation of the rat proximal urethra: effect of pelvic ganglionectomy or experimental diabetes, *J. Auton. Pharmacol.*, 9, 231, 1989.
57. Lecci, A., Giuliani, S., Lazzeri, M., Benaim, G., Turini, D., and Maggi, C. A., The behavioral response induced by intravesical instillation of capsaicin in rats is mediated by pudendal urethral sensory fibers, *Life Sci.*, 55, 429-436, 1994.
58. Conte, B., Maggi, C. A., and Meli, A., Vesico-inhibitory responses and capsaicin-sensitive afferents in rats, *Naunyn-Schmiedeberg's Arch. Pharmacol.*, 339, 178, 1989.

59. Lecci, A., Giuliani, S., Barbanti, G., Lazzeri, M., Turini, D., and Maggi, C. A., Intraurethral infusion of capsaicin induces penile erection: effect in awake and in anesthetized rats, presented at 5th World Meeting on Impotence, Milano, September 13 to 17, 1992.
60. Nordling, L., Liedberg, H., Ekman, P., and Lundeberg, T., Influence of the nervous system on experimentally induced urethral inflammation, *Neurosci. Lett.*, 115, 183, 1990.
61. Lecci, A., Giulani, S., and Maggi, C. A., unpublished data.

Chapter 17

SENSORY NEUROPEPTIDES IN ARTHRITIS

William R. Ferrell and Francis Y. Lam

CONTENTS

I. INTRODUCTION

Arthritis is a disease of uncertain etiology that manifests itself as a chronic inflammatory disorder in synovial joints. A large number of inflammatory mediators have been implicated in the perpetuation of synovitis, including arachidonic acid metabolites, vasoactive amines, cytokines, such as tumor necrosis factor-α (TNF-α) and interleukins 1 and 6 (IL-1 and -6), and neuropeptides.[1,2] In rheumatoid arthritis (RA), another hallmark of the disease is the presence of abnormal cellular and humoral immune responses. Thus, autoantibodies, particularly rheumatoid factors (RFs) and antibodies against collagen type II often prevail in these patients, and T lymphocytes frequently accumulate in their synovia.[3] The disease shows synovial hyperplasia at the later stage, which is characterized by fibroblast proliferation, transformed appearance of synovial cells,[4] and large numbers of infiltrating macrophages in the joint.[5]

Much evidence is available to suggest an important role for the nervous system in determining the pattern and severity of joint destruction. Clinical studies have revealed that the onset and, more commonly, exacerbation of rheumatoid arthritis is often preceded by psychological stress.[6] Joints on the paretic side of hemiplegic patients, who later develop rheumatoid arthritis, are either completely spared or develop only mild synovitis and do not develop erosions or nodules.[7] Furthermore, it has been proposed that the distal and symmetric distribution of joint involvement in rheumatoid arthritis in humans (and adjuvant arthritis in rats), could be attributed to the denser sensory innervation of these joints.[8] This idea stems from the concept of neurogenic inflammation described by Lewis[9] when he described the triple response.

Pain and hyperalgesia commonly associated with arthritis is due to activation of nociceptive sensory neurons in the joint by some of the mediators that are released in the inflammatory process. In addition to this nociceptive role, a subset of the sensory neurons, upon activation, could also release neuropeptides, notably substance P (SP) and calcitonin gene-related peptide (CGRP), which have significant proinflammatory effects. Thus, in arthritic joints an environment exists for complex interaction between the synovial innervation and various cell types, both indigenous as well as those invading the joint, including immune cells, as a result of the

0-8493-7646-7/96/$0.00+$.50

inflammatory process. This review focuses on the possible roles of sensory neuropeptides in arthritis.

II. NEUROPEPTIDERGIC INNERVATION OF THE SYNOVIUM

Synovial joints are innervated by both myelinated and unmyelinated afferent nerve fibers.[10] Large fibers are myelinated, arising from specialized encapsulated structures at their peripheral ends[11] and whose function is primarily proprioceptive.[12] Unmyelinated and finely myelinated axons with unmyelinated terminals, which are not associated with specialized receptive structures, are termed free nerve endings. Afferent fibers with these endings provide the overwhelming majority of joint afferent innervation.[13] It is these fibers that are primarily associated with the release of neuropeptides into the surrounding tissue.

Recent developments in immunohistochemical techniques have provided fresh insight into joint innervation. Substantially improved resolution and characterization of individual fibers has been possible using antisera against specific neuronal markers combined with sensitive staining methods. The use of this technique has revealed vastly increased small-diameter nerve fibers in the joint[14] compared with previous studies using standard histological methods.[15] Small-diameter nerve fibers immunoreactive for protein gene product 9.5 were found in all sections of normal joint tissue and were scattered throughout the fibrous capsule, ligaments, tendons, and synovium.[16]

Many of the small-diameter nerves found in normal synovium are immunoreactive for neuropeptides.[16,17] These include fibers containing immunoreactive SP and CGRP, which are considered to be markers of sensory fibers, as well as nerves containing immunoreactive neuropeptide Y and its C flanking peptide, found in most peripheral noradrenergic neurons. Many SP- and CGRP-immunoreactive nerves were found in perivascular areas and numerous free fibers were also present, with some extending through the synovium almost as far as the synovial surface.[16]

A recent study has shown that even in acute stages of inflammation as depicted by monoarthritis of the cat knee joint induced by intra-articular injection of kaolin and carrageenan, the density of CGRP-containing and -releasing terminals is enhanced.[18] Another study also showed increased CGRP-positive dorsal root ganglion (DRG) neurons during acute and chronic phases of inflammatory lesions in rat ankle joints.[19] This upregulation of CGRP occurs more or less in parallel to the upregulation of SP in DRG and peripheral nerves that was described in acute and chronic phases of inflammatory lesions.[20-22] Collectively, these data indicate that changes in the neuropeptide content and in the proportion of neurons producing particular neuropeptides are characteristic neuronal changes associated with inflammation. Thus, with the abundant nerve supply and the possible upregulation of sensory neurons containing proinflammatory neuropeptides in the joint, it is plausible to hypothesize that sensory neuropeptides may contribute to arthritis.

III. SENSORY NEUROPEPTIDES IN SYNOVIAL FLUID

Consistent with the view that SP plays an important role in acute and perhaps chronic phases of inflammation, in rats an upregulation of the synthesis of SP can be detected within the first few hours after the application of an acute inflammatory stimulus,[22] and in the presence of chronic inflammation, the production of SP in peripheral nerves is enhanced over weeks.[20] Thus, it might be expected that elevated concentrations of sensory neuropeptides would be detected in synovial fluid aspirated from inflamed joints. Although a number of such studies have been performed, a clear picture has not yet emerged. While the presence of various neuropeptides, including SP, has been demonstrated in synovial fluid from patients with rheumatoid arthritis,[23-26] other studies showed the existence of CGRP, vasoactive intestinal polypeptide (VIP), and a small amount of neurokinin A (NKA), but not SP.[27,28] This poses difficulties in interpreting the data, especially when true control samples from normal subjects are not available. The issue is complicated further by the fact that there are considerable differences in the rates of degradation of neuropeptides. SP is metabolized much more rapidly than CGRP and NKA and thus, any increased release of SP might not be as readily detectable as other, more stable, peptides. Furthermore, local interaction of neuropeptides can occur and it has been shown that SP can attenuate the vasodilator effect of CGRP,[29,30] and an excess of CGRP may lead to the depletion of SP locally.[30] All these are causative factors in the conflicting reports, but it is likely that variation in experimental conditions and variation in severity of arthritic lesions in different subjects could also contribute to the discrepancies between the different studies.

Despite such discrepancies, substance P-like immunoreactivity (SP-LI) is generally found to be highest in synovial fluids aspirated from patients of rheumatoid arthritis compared to those found in other arthritic diseases such as osteoarthritis and psoriatic arthritis.[26] A significant correlation also exists between the high level of SP in rheumatoid arthritis patients and the increase in erythrocyte sedimentation rate.[26] Recently, using various experimental models of monoarthritis in the rat knee, it has been shown that significant increases of SP-, NKA-, CGRP-, and neuropeptide Y (NPY)-LI were found in knee joints pretreated with inflammatory agents compared to those found in control knee joints pretreated with saline.[31] Thus, in general, these data are consistent with the proinflammatory actions of neuropeptides contained in sensory fibers, and suggest a peptidergic contribution to arthritis.

IV. NEUROGENIC CONTRIBUTION TO JOINT INFLAMMATION

Involvement of the nervous system in inflammation has long been recognized and Lewis[9,32] first suggested that cutaneous wheal and flare responses occurred by the release of substances from the peripheral terminals of nociceptive afferents. Since then, evidence has accumulated in support of the notion that the "axon reflex" is mediated by unmyelinated afferent fibers. Plasma extravasation and vasodilatation occur during antidromic nerve stimulation at C-fiber strength in both the skin[33] and the joint.[34,35] Direct activation of C polymodal nociceptive afferents in the skin also elicits an inflammatory response.[36]

The sympathetic nervous system may also play a part in acute synovitis as vascular permeability is increased by sympathetic nerve stimulation[37] and baseline plasma extravasation in the knee joints of cats is substantially reduced after lumbosacral sympathectomy.[38] In addition, intra-articular infusion of 6-hydroxydopamine, which stimulates sympathetic postganglionic nerves (SPGN) to release the contents of their peripheral terminals, produces a prolonged increase in synovial plasma extravasation.[39] This was inhibited by pretreatment with indomethacin, suggesting involvement of prostaglandins.

A potential model for the role of the somatic sensory and sympathetic nervous systems in arthritis was described by Fitzgerald,[40] who postulated that C-fiber nociceptors activated by the damaged joint will transmit impulses to the spinal cord and, via the axon reflex, release proinflammatory mediators peripherally, which in turn will exacerbate inflammation and further excite C-fibers. At the same time, SPGN are activated, perhaps by somatosympathetic reflexes or by generalized increase in autonomic activity. Sympathetic axons may themselves also release inflammatory mediators and further excite C-nociceptors, thus constituting a positive feedback mechanism leading to enhanced joint pain and inflammation.

Apart from the sensory and sympathetic neurons, recent evidence also indicates a role for the spinal cord in the development of joint inflammation. It has been shown that there is enhanced excitability of spinal cord neurons following tissue injury and inflammation.[41,42] The increased excitability of neurons in the spinal cord may contribute to hyperalgesia in tissue injury and inflammation and this may involve polysynaptic mechanisms. Thus, non-*N*-methyl-D-aspartate (non-NMDA) receptor activation has been implicated during the first stages of the development of inflammation, while secondary hyperalgesia, once developed, is maintained by the activation of NMDA receptors.[43] This is supported by findings which showed that administration of either a non-NMDA receptor antagonist[43] or gamma-amino-butyric acid type A receptor antagonist[44] significantly decreases the extent of carrageenan-induced joint inflammation in the rat. This inflammatory response is attenuated by dorsal rhizotomy (leaving peripheral afferents intact) but unaffected by sympathectomy, and there is evidence to indicate that dorsal root reflexes are initiated by the inflammatory process.[45] These "reflexes" involve antidromic activation of sensory afferents, leading to peripheral release of sensory neuropeptides, and may be generated via primary afferent depolarization of joint afferents by innocuous stimuli to other sensory afferents.

It is beyond the scope of this chapter to discuss all possible neurotransmitters involved in arthritis. In our view, neuropeptides are likely to constitute the most important group of neurogenic mediators of the inflammatory process. Thus, involvement of sensory neuropeptides in arthritis will be discussed in further detail.

V. SENSORY NEUROPEPTIDES IN PAIN TRANSMISSION

Pain is the dominant symptom of all forms of human arthritic disease. Its occurrence in animals is inferred from characteristic behavioral changes following induction of experimental arthritis.[46] Transection of spinal pathways transmitting nociceptive information to higher centers reduces these behaviors,[47] indicating the importance of these centers in the expression of the disease process. These centers are also important in behavioral modification in response to disease, e.g., arthritic rats develop a preference for solutions containing nonsteroidal anti-inflammatory agents or analgesics.[48] Neurophysiological and neurochemical changes in both the peripheral and central nervous systems are responsible for such behavior and these aspects have recently been reviewed.[49] In this chapter we limit ourselves to examination of the role of sensory neuropeptides in nociception.

As articular afferent fibers are involved in nociception and contain SP and other neuropeptides,[16,17] it is reasonable to hypothesize that these could be neurotransmitters at the first synapse in nociceptive pathways. SP-immunoreactive terminals are found in areas where nociceptive primary afferent neurons terminate (laminae I, II, V, and X of the spinal dorsal horn).[50,51] Consistent with the concept of SP as a neurotransmitter is the finding that electrical stimulation of C-fiber afferents[52] or noxious stimulation of the skin[53] enhances SP release in the spinal cord. However, whether released SP originates solely from the central terminations of afferent neurons is uncertain because interneurons and fibers of descending pathways also contain SP.[54] Furthermore, afferent neurons also contain other neuropeptides such as NKA and CGRP which can coexist with SP and could be co-released[55] and function as co-transmitters.[56] Slow excitatory postsynaptic potentials of relatively long latencies (several seconds) and durations (up to 2 min) can be elicited by iontophoretic application of SP to dorsal horn neurons,[57,58] suggesting that SP may have the role of modulating neurotransmission. SP may exert these effects by binding to tachykinin NK_1 receptors which occur in the highest densities in the dorsal horn, including laminae I–II, lamina V, and lamina X,[59] and to which it has the highest affinity.[60,61]

The contribution of SP to increased excitability of spinal neurons in arthritis is highlighted by the finding that binding of Bolton-Hunter-labeled ^{125}I-substance P in the dorsal horn of the spinal cord is altered following unilateral adjuvant-induced inflammation in the hindpaw of the rat.[62] This is a time-dependent phenomenon: 6 h after the induction of inflammation, widespread decreases in ^{125}I-substance P binding occurred on both sides of the dorsal horn of spinal level L4 in comparison to control group, whereas by 2 d, widespread increases in labeled ^{125}I-substance P binding were observed in the same regions. It was suggested that this time-dependent fluctuation was because during the initial hours of induction of peripheral inflammation (< 6 h), SP synaptic transmission is possibly enhanced via increased release of SP. The SP released would activate spinal NK_1 receptors, thereby enhancing the excitability of spinal neurons. This proposal is supported by two other studies which showed that (1) following carrageenan-induced inflammation in the rat hind paw both spontaneous and capsaicin-evoked SP release from dorsal lumbar spinal cord were increased,[63] and (2) following carrageenan-kaolin-induced inflammation in the cat knee joint SP released in response to joint movement was increased.[64] An additional consequence of the initial increased release of SP would be downregulation of SP binding. Acute pretreatment with SP results in downregulation of SP receptors in membranes from rat brain[65] and spinal cord.[66] Thus, the decrease in Bolton-Hunter-labeled ^{125}I-substance P binding at 6 h may be a response to an acute increase in release of SP in the spinal cord. However, 2 d from the onset of inflammation, SP binding is upregulated, apparently by increased affinity of NK_1 receptors, although the factors underlying this are unknown. As a consequence of increased binding, SP neurotransmission is enhanced from 2 to 8 d following induction of inflammation and this would lead to increased excitability of spinal nociceptive neurons. Thus, normally subthreshold inputs from primary afferent neurons would cross threshold and discharge second-order spinal neurons, thereby contributing to hyperalgesia.

It should be noted that changes in SP binding do not account for all of the hyperalgesic response. Sensitization of primary afferent neurons also contributes to the phenomenon. Electrophysiological recordings show that sensory nerve terminals of unmyelinated fibers normally have high thresholds to mechanical stimulation and often only discharge if noxious stimuli (e.g., bending or twisting to the threshold of tissue damage) are applied to the joint,[67] suggesting that these are nociceptors. These receptors are normally relatively quiescent within the non-noxious range of joint movement,[67] but become spontaneously active and discharge even with innocuous movement during inflammation.[68,69] Thus, the mechanical thresholds of many nociceptors are greatly reduced by joint inflammation, which contributes to the tenderness of the inflamed joint to the application of pressure or movement.[68,70] In addition to sensitization of peripheral receptors and modification of spinal cord neurotransmission, changes in descending inhibition of nociceptive pathways are also important.[49]

VI. SENSORY NEUROPEPTIDES AND THE INFLAMMATORY PROCESS

Many of the peptides that have been identified in small unmyelinated sensory nerves have been implicated in the etiology of "neurogenic inflammation". A substantial body of evidence is accumulating to suggest that the neuropeptide SP is of particular importance in mediating neurogenic inflammation occurring in skin[33,71,72] and in the joint.[35,73] Administration of capsaicin on skin causes the release of neuropeptides including SP and CGRP from unmyelinated sensory nerve fibers.[74] These acute events can be observed functionally by an increase in blood flow of the skin and, at higher doses, increased vascular permeability.[75]

Electrical stimulation of nerves supplying the cat knee joint can evoke the release of SP from articular nerve fibers[73] and produce protein extravasation into the synovial cavity.[34] This neurogenically induced plasma extravasation can be abolished by prior intra-articular administration of the SP antagonist D-Pro^4 D-Trp^{7,9,10} SP_{4-11},[34] which suggests that SP is the mediator of the response. Direct topical application or intra-articular injection of SP into the rat knee has been shown to elicit marked inflammatory reactions, including vasodilatation[29,76] and increased capillary permeability.[76-81] The latter response has been associated with the release of inflammatory mediators from mast cells.[79] SP and related peptides have also been shown in joints to exert other proinflammatory actions such as secretion of PGE_2 and collagenase from synoviocytes,[82] and secretion of IL-1-like activity from macrophages.[83] In addition, neurokinins may importantly contribute to activation of the immune system.[84]

Studies on the time course of actions of SP in producing vasodilatation and protein extravasation in normal rat knee joints have revealed that these are transient phenomena. Thus, the vasodilator action of SP lasted no more than 5 min after a single application to the joint,[29] while SP-induced protein extravasation returned to basal level within 30 min, despite continuous perfusion of the drug into the joint.[80] The transient action of SP in the normal joint may be an effective protective mechanism for the animal. Thus, when an acute injury occurs, SP is released from primary afferent fibers along with other neuropeptides such as CGRP to cause vasodilatation of blood vessels supplying the injured area. The presence of SP also increases blood vessel permeability, facilitating the passage of cells such as macrophages and other substances essential for protective and healing processes at the injured site. At the same time, the short-lasting effects of SP could also avoid prolonged extravasation of inflammatory mediators to the injured area, which could have deleterious consequences.

VII. ACTIONS OF SENSORY NEUROPEPTIDES IN THE INFLAMED JOINT

As previously mentioned, the nervous system has been implicated in the pathology of arthritis as hemiparesis minimizes arthritic manifestations on the affected side.[7] This phenomenon is also observed in animal models of arthritis. Systemic administration of adjuvant in rats results in symmetrical involvement of joints but in animals with unilateral nerve section, arthritis is delayed in onset and diminished in severity compared to the intact side.[85]

We have performed similar studies and showed that denervation of the rat knee joint significantly reduced carrageenan-induced inflammation in the joint.[78] Furthermore, in the same study we were able to demonstrate a similar reduction on the inflammatory response to carrageenan by pretreating the joint with capsaicin.[78] This is confirmed by a recent report which showed that pretreatment of the rat knee joint with capsaicin also diminished plasma protein extravasation of the joint induced by passive synovial anaphylaxis.[86] Capsaicin is a neurotoxic agent which is known to cause selective degeneration of unmyelinated sensory nerve fibers in the joint after a single intra-articular injection.[87] In accordance with these findings, earlier works have shown that systemic administration of capsaicin could attenuate adjuvant arthritis in rats.[88] Moreover, perfusion of a joint with SP could enhance cartilage and bone damage in adjuvant arthritis, whereas injection of a SP antagonist does not.[89] We have also demonstrated a near-complete inhibition of carrageenan-induced inflammation in joints that have been pretreated with a SP antagonist.[78]

In spite of the overwhelming evidence in support of an important role of neuropeptides in the pathogenesis of joint diseases, our understanding of the actions of neuropeptides in inflamed joints is limited. To shed further light on this, we have compared the actions of some naturally occurring neuropeptides in normal and acutely inflamed rat knee joints. Acute inflammation of the rat knee was induced by injection of 0.2 ml of 2% carrageenan into the synovial cavity of the knee. Over a 24-h period the carrageenan-injected knee showed a higher basal level of protein extravasation compared to saline-injected or untreated knees.[81] In addition, carrageenan-induced inflammation enhanced SP-induced plasma protein extravasation. The time course of the SP response was also found to be drastically different between normal and inflamed joints. Normally, the response to SP is transient despite continuous perfusion, lasting no more than 30 min even at the highest SP concentration (10 m*M*), whereas in the inflamed joint, extravasation induced by SP persisted for the whole infusion period (>1 h) with all the effective concentrations.[81]

Recently, we have shown that only the specific NK_1 receptor agonist [Sar,[9] Met $(O_2)^{11}$]-substance P, but not the specific NK_2 receptor agonist [Nle[10]]-neurokinin $A_{(4-10)}$, or the specific NK_3 receptor agonist [MePhe[7]]-neurokinin B, was effective in eliciting protein extravasation in the rat knee joint.[90] This is in agreement with our earlier studies using the Evans blue technique.[91] Furthermore, we have demonstrated that in the presence of the specific NK_1 receptor antagonist FK 888[92] protein extravasation induced by the NK_1 agonist in both normal (Figure 1A) and acutely inflamed (Figure 1B) rat knee joints was abolished, indicating that NK_1 receptors are solely responsible for mediating the protein extravasation response.

As previously discussed, the normally transient effect of SP on synovial permeability could be beneficial to the animal by providing mediators for effective healing at the injured area, while also preventing excessive accumulation of such inflammatory mediators. In the acutely inflamed knee joint, the transient response to SP is replaced by a sensitized and sustained response. The mechanism responsible for this change is unknown at present. It is possible that upregulation of SP receptors occurs in the inflamed joint. Also possible is that the inflammatory process alters the binding and dissociation mechanisms of SP to its receptors and thus enables a sustained effect. Alternately, the second messenger system coupled to the SP receptor may have been altered from its normally self-limiting mode to a continuous one as a result of the inflammatory process. Under these conditions the presence of neuropeptides could inflict further damage to the injured tissues instead of serving as a protective mechanism.

In another series of experiments we have investigated the influence of acute joint inflammation on the responses of articular blood flow to neuropeptides and sympathetic nerve stimulation. Alterations in perfusion of the knee joint as a consequence of vasoconstrictor or vasodilator effects could significantly influence the final manifestations of inflammatory processes. In our studies, we used a laser Doppler perfusion imaging system (LDI) to assess changes in joint blood flow. The LDI system employs the same working principle as other conventional laser Doppler flowmeters (LDF). However, conventional LDF only provides measurements of tissue perfusion at a single location, whereas LDI allows averaging and spatial mapping of blood flow changes for a selected tissue area. Using this technique, it was observed that acute inflammation in the rat knee joint induced by prior carrageenan-injection resulted in enhancement of the vasodilator responses to SP and CGRP as compared to normal joints.[76] The response of the rat knee microvasculature to electrical stimulation of the nerve supply to the joint was also substantially altered, with small vasoconstrictor responses occurring in carrageenan-treated knees, whereas potent vasoconstriction is observed in normal knees.[76]

The reduction of the vasoconstrictor response could represent decreased effectiveness of sympathetic neurotransmission, alteration of postsynaptic α-adrenoceptors, or perhaps liberation of vasodilator substances such as nitric oxide, neuropeptides, or PGE_2, any of which could counteract sympathetic vasoconstrictor effects in the inflamed joint. Although we cannot identify the exact mechanism at present, our data indicate that in pathophysiological conditions, the microvasculature of the joint is less influenced by its sympathetic innervation and this may result in a higher articular blood flow in the inflamed joint compared to normal.

Electrical stimulation of the nerves supplying the joint not only activates sympathetic efferent fibers, but also antidromically stimulates peripheral sensory nerve terminals, which results in the release of vasodilator neuropeptides.[73] Normally their release produces injury hyperemia which facilitates the passage of mediators from systemic circulation for repair processes at the site of injury. As a consequence, the sympathetic vasoconstriction of blood vessels in the joint may be reduced. Thus, we observed that nerve-induced vasoconstriction was reduced in the presence of exogenously applied neuropeptides.[76] Our previous findings using LDF also confirm this observation for the normal joint.[93] Of greater importance is that in acutely inflamed joints, sympathetic nerve stimulation in the presence of either SP or CGRP

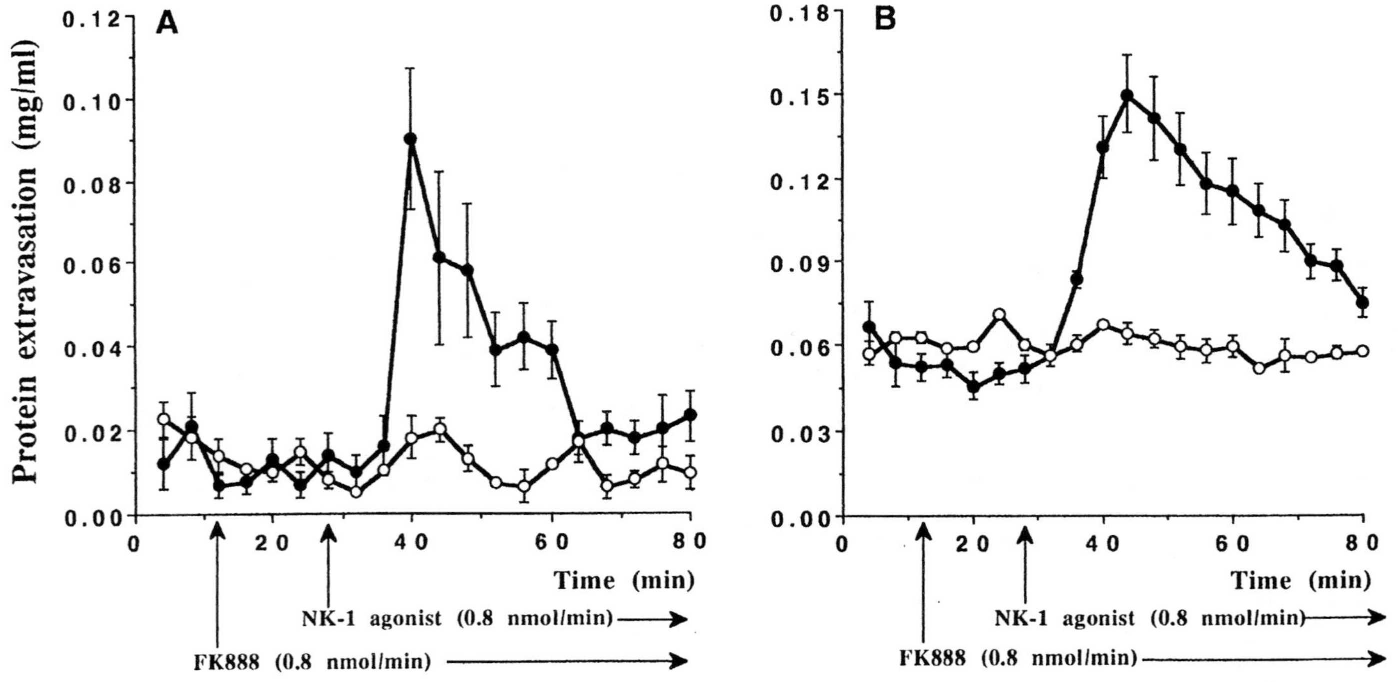

FIGURE 1. Effects of intra-articular perfusion (0.8 nmol/min) of the NK_1 receptor agonist [Sar,^{9} Met$(O_2)^{11}$]-substance P on plasma protein extravasation in the normal rat knee joint (A) and in the acutely inflamed joint (B) in the absence (-●-, n = 5–6) and in the presence of 0.8 nmol/min of the NK_1 antagonist FK 888 (-○-, n = 3). The antagonist was administered 20 min prior to coperfusion with the NK_1 agonist into the joint. Acute inflammation of the rat knee joint was induced by intra-articular injection of 0.2 ml (2%) carrageenan into both knees 24 h prior to the experiment.

produced vasodilatation but not vasoconstriction of the articular blood vessels.[76] These findings are of significance as they imply that the presence of neuropeptides in inflammatory conditions could abolish sympathetic influences on the microvasculature.

It has been shown that mediators such as prostaglandins released from SPGN, as a consequence of their vasodilator effects, could amplify neurogenic inflammation.[94] Thus, it is possible that in our experiments the reduced nerve-mediated vasoconstriction was partly influenced by this mechanism. Alternatively, it is known that carrageenan-induced acute inflammation alters the adrenoceptor profile of synovial blood vessels in rabbit knee joints,[95] with a reduction in the α_1 response and an associated increase in the α_2 response. Whether changes in adrenoceptor profile of synovial blood vessels also occur in carrageenan-injected rat knee joints is uncertain; nevertheless, it may offer another possible explanation for the reduced nerve-mediated vasoconstriction seen in our studies. At present there is little evidence to implicate SPGN in carrageenan-induced joint inflammation, as our studies have shown that this model of inflammation was unaffected by reserpine pretreatment.[96] In addition, nerve-mediated constrictor responses in the acutely inflamed (carrageenan-induced) rabbit knee joint are unaffected by treatment with indomethacin.[97]

Although plasma extravasation in rat skin[98] and in the rat knee joint[91] is mediated solely by the NK_1 receptor subtype, our studies have shown that the specific NK_1 receptor antagonist, CP-96345, selectively inhibits SP-induced vasodilatation while leaving the vasodilator responses to NKA and NKB unaltered.[29] This suggests that, unlike plasma extravasation, blood flow changes in the rat knee joint could be mediated by multiple neurokinin receptor types. We have recently shown that in acutely inflamed rat knee joints (compared to normal joints), the enhancement of the vasodilator response to the NK_2 receptor agonist [Nle^{10}]-neurokinin $A_{(4\text{-}10)}$ surpassed those seen with the NK_1 receptor agonist [$Sar,^9$ Met $(O_2)^{11}$]-substance P,[90] suggesting a greater contribution of both NK_1 and NK_2 receptors in mediating vasodilator responses in the inflamed condition. Furthermore, in both normal and inflamed knee joints, the NK_1 agonist-induced vasodilatation was abolished by FK 888 (Figure 2A), while the NK_2 agonist-induced vasodilator response was inhibited by the specific NK_2 receptor antagonist SR 48968[99] (Figure 2B), confirming their vasodilator responses were mediated by activation of NK_1 and NK_2 receptor subtypes, respectively.

The experiments described above involved acute joint inflammation but do not indicate whether such altered responsiveness would also occur in more chronic models of inflammation. To assess this, we have examined nerve-mediated vasoconstriction and SP-mediated vasodilatation in adjuvant monoarthritic rat knees.[100] Again, vasoconstrictor responses were substantially attenuated (Figure 3A) but even more striking was that SP-mediated vasodilatation was completely abolished (Figure 3B). Although only shown for 1 week post-induction of inflammation, these effects were still present at 3 weeks post-induction. The reduction in vasoconstriction could be explained by nerve depletion, which is known to occur in the synovium of adjuvant arthritic rats.[101] Abolition of the SP response may be related to reduction of endothelial NK_1 receptors in areas of the synovium with intense cellular infiltration.[102] Such differences in response to SP serve to highlight the fact that acute inflammation is not a truncated version of the chronic inflammatory process but that these processes may be generated by different mechanisms. Acute inflammatory reactions may involve release of proinflammatory neuropeptides, whereas in chronic inflammation, loss of innervation and tachykinin receptors could result in defective regulation of the synovial microcirculation, which could contribute to the degenerative changes characteristic of chronic joint disease. This is not to say that both processes are mutually exclusive: both could coexist even within the same joint, with the balance being altered by various factors, including therapeutic intervention. This may account for the "flare-up" which often occurs in rheumatoid arthritis — an acute exacerbation of what is fundamentally a chronic inflammatory process.

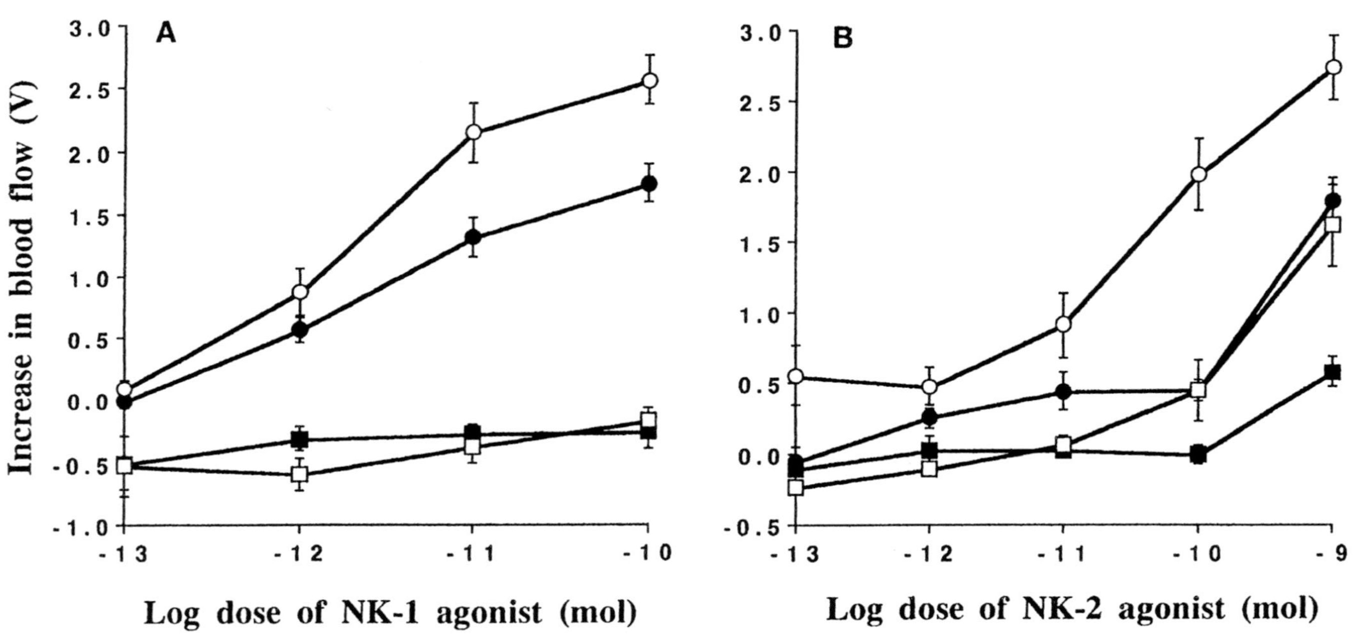

FIGURE 2. Effects of topical bolus administration of (A) the NK_1 receptor agonist [Sar,^{9} Met(O_2)^{11}]-substance P and (B) the NK_2 receptor agonist [Nle^{10}]-neurokinin $A_{(4-10)}$ on blood flow in normal (closed symbols) and inflamed (open symbols) rat knee joint in the absence (circle symbols) and in the presence of (A) 2×10^{-9} mol of the NK_1 receptor antagonist FK 888 (square symbols) and (B) 2×10^{-11} mol of the NK_2 receptor antagonist SR 48969 (square symbols). The antagonists (10^{-9} mol and 10^{-11} mol, respectively) were applied 5 min prior to co-administration with the agonist. $n = 6–13$. Change in blood flow from control (pre-SP administration) is given as change in laser Doppler signal (in volts).

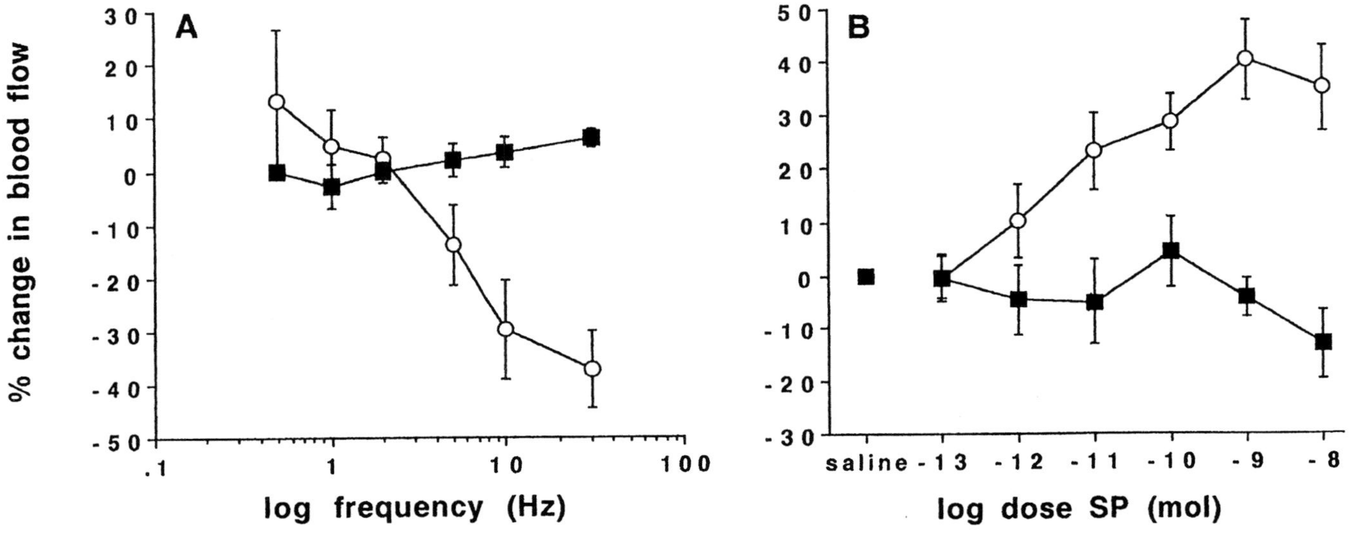

FIGURE 3. Nerve-mediated vasoconstrictor responses (A) and substance P-mediated (SP) vasodilator responses (B) in rat knees injected with Freunds complete adjuvant 1 week previously. In normal knees (-○-) electrical stimulation (15 V, 1 ms) produced frequency-dependent vasoconstriction which was abolished in the arthritic knees (-■-). Topical application of SP produced dose-related vasodilatation in normal knees (-○-) which was absent in arthritic knees (-■-). Change in blood flow is given as percentage change in laser Doppler signal from control. (Reproduced from McDougall, J. J. et al., *Neurosci. Lett.*, 174, 127, 1994. With permission.)

VIII. SUMMARY

In summary, we have outlined a number of different mechanisms by which sensory neuropeptides may contribute to arthritis (see Figure 4). The normal rat and human synovium are richly innervated by small unmyelinated nerve fibers containing these peptides. Apart from their important role in conveying nociception to the central nervous system, the peptides may also be released from peripheral terminals of the sensory neurons upon nerve stimulation or by noxious stimuli. Once released, these neuropeptides, particularly SP, could exert potent inflammatory actions in the joint either by releasing proinflammatory mediators from mast cells and macrophages, or by a direct action on the joint microvasculature to increase permeability of the blood vessels. CGRP, which is often coreleased with SP, as a consequence of its potent vasodilator action in the synovium could potentiate the inflammatory reactions produced. The involvement of the sympathetic innervation of the joint in the inflammatory process is uncertain and is more likely to be of minor significance compared to the peripheral sensory neurons, particularly in carrageenan-induced acute joint inflammation. A bidirectional communication also exists between sensory nerves on one hand and inflammatory/immune cells on the other, sustained by humoral factors released from both sides. This cross-communication has the capacity to develop a positive feedback cycle which could play an important role in the genesis and maintenance of inflammation.

It is surmised that the transitory nature of neuropeptide action on the microvasculature of normal joints is an important self-regulatory mechanism for the animal; it contributes to a short, effective healing process during and after injury, and at the same time the potential for damage by prolonged action of neuropeptides is avoided. However, in acutely inflamed joints it appears that the short-lasting effects of neuropeptides become sustained and have deleterious actions in the joint. We postulate that in aberrant situations, such as that depicted by carrageenan-induced acute inflammation in the joint, sensory neuropeptides contribute to a positive feedback sequence of events producing the inflammatory response: (1) injection of carrageenan into the knee initiates an inflammatory response and causes release of neuropeptides from sensory nerve endings; (2) these substances reduce sympathetic vasoconstrictor influence on the vasculature, thus enhancing vasodilatation and plasma extravasation by these neuropeptides, allowing more cells and mediators to be accumulated at the site from the systemic circulation; (3) the accumulated inflammatory mediators cause further inflammation and release more neuropeptides from sensory nerve endings. The consequence of these events is an ongoing inflammatory condition in the joint.

The manifestation of arthritis involves complex interactions of a large number of endogenous mediators in the affected joint. Among these, it is becoming clear that sensory neuropeptides have an important role in the inflammatory process and evidence is accumulating to suggest that in acutely inflamed joints, these neuropeptides may perpetuate the vicious circle of inflammation via their enhanced efficacies on NK_1 and NK_2 receptors. It may be premature at this stage to suggest that curtailment of neurokinin receptor activities can provide relief of arthritic symptoms. Nevertheless, in view of the key role sensory neuropeptides play in the inflammatory process, the possibility of producing antiarthritic drugs based on the reduction of neuropeptide activity in the disease certainly warrants further investigation.

ACKNOWLEDGMENT

The authors thank the following organizations that have supported their work: the Arthritis and Rheumatism Council, the Wellcome Trust, the MacFeat Bequest of the University of Glasgow, and the University and Polytechnic Grant Council to the Chinese University of Hong Kong. Figures 1 and 2 involved the contribution of M.C.S. Wong and Figure 3 of J.J. McDougall and S.M. Karimian. The authors are indebted to Fujisawa Pharmaceuticals Ltd. and Sanofi Recherche for their gifts of FK 888 and SR 48968, respectively.

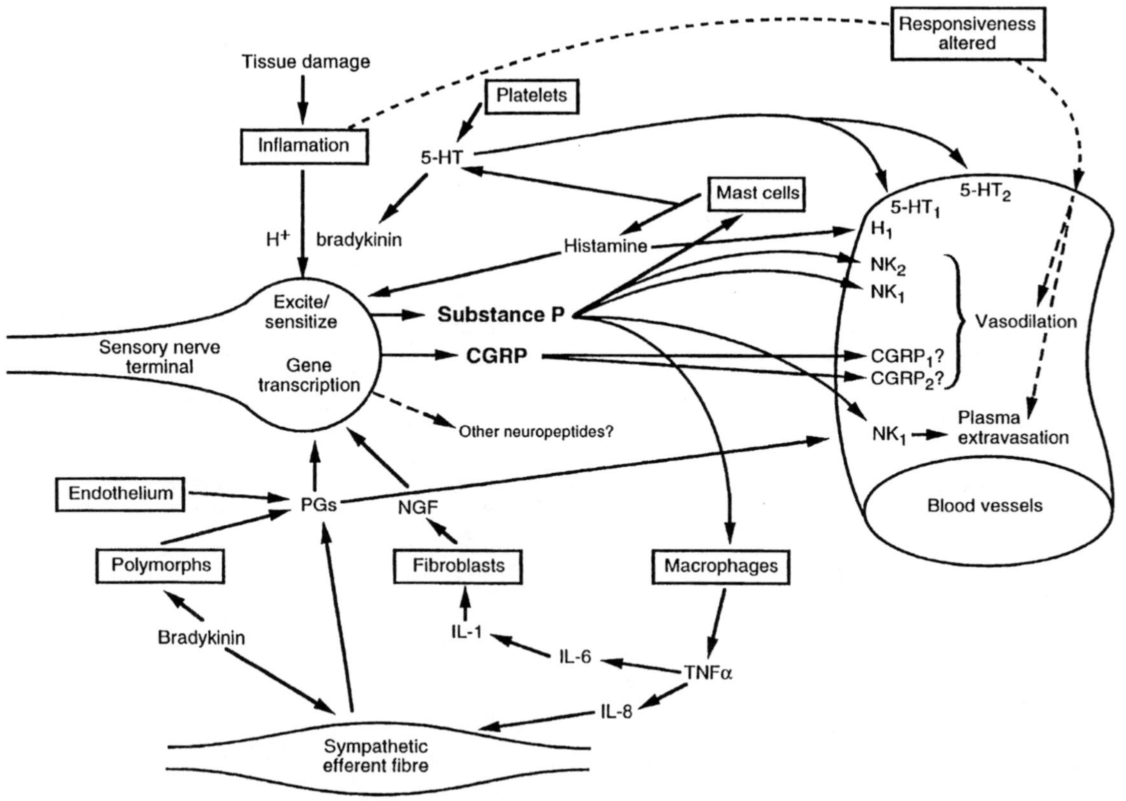

FIGURE 4. Summary of some of the interactions between peripheral sensory neurons, sympathetic neurons, the products of immune cells, and microvessels during inflammation. Not all mediators are shown; note that some of the illustrated examples are well established while others are more speculative. CGRP, calcitonin gene-related peptide; 5-HT, 5-hydroxytryptamine; IL-1, IL-6, IL-8, interleukins 1, 6, 8; NGF, nerve growth factor; PGs, prostaglandins; TNFα, tumor necrosis factor-α. (Modified from Dray A. and Bevan S., *Trends Pharmacol. Sci.*, 14, 287, 1993.)

REFERENCES

1. Lipsky, P. E., Davis, L. S., Cush, J. J., and Oppenheimer-Marks, N., The role of cytokines in the pathogenesis of rheumatoid arthritis, *Springer Semin. Immunopathol.*, 11, 123, 1989.
2. Arend, W. and Koopman, W. J., Molecular biology and the immunopathogenesis of rheumatoid arthritis, *J. Cell Biochem. Suppl.*, 15E, 141, 1991.
3. Gay, S. and Koopman, W. J., Immunopathology of rheumatoid arthritis, *Curr. Rheumatol.*, 1, 8, 1989.
4. Fassbender, H. G., Histomorphologic basis of articular cartilage destruction in rheumatoid arthritis, *Collagen Relat. Res.*, 3, 141, 1983.
5. Gay, S. and Gay, R. E., Cellular basis and oncogene expression of rheumatoid joint destruction, *Rheumatol. Int.*, 9, 105, 1989.
6. Baker, G. H. B., Life events before the onset of rheumatoid arthritis, *Psychother. Psychosom.*, 38, 173, 1982.
7. Thompson, M. and Baywater, E. G. L., Unilateral rheumatoid arthritis follows hemiplegia, *Ann. Rheum. Dis.*, 21, 320, 1962.
8. Levine, J. D., Clarke, R., Devor, M., Helms, C., and Moskowitz, M. A., Intraneuronal substance P contributes to the severity of experimental arthritis, *Science*, 266, 547, 1984.
9. Lewis, T., Experiments relating to cutaneous hyperalgesia and its spread through somatic nerves, *Clin. Sci.*, 2, 373, 1936.
10. Samuel E. P., The autonomic and somatic innervation of the articular capsule, *Anat. Rec.*, 113, 84, 1952.
11. Boyd, I. A., The histological structure of the receptors in the knee joint of the cat correlated with their physiological response, *J. Physiol.*, 124, 476, 1954.
12. Ferrell, W. R., Articular proprioception and nociception, *Rheumatol. Rev.*, 1, 161, 1992.
13. Langford, L. A. and Schmidt, R. F., Afferent and efferent axons in the medial and posterior articular nerves of the cat, *Anat. Rec.*, 206, 71, 1983.
14. Kidd, B. L., Mapp, P. I., Blake, D. R., Gibson, S. J., and Polak, J. M., Neurogenic influences in arthritis, *Ann. Rheum. Dis.*, 49, 649, 1990.
15. Freeman, M. A. R. and Wyke, B., The innervation of the knee joint. An anatomical and histological study in the cat, *J. Anat.*, 101, 505, 1967.
16. Mapp, P. I., Kidd, B. L., Gibson, S. J., Terry, J. M., Revell, P. A., Ibrahim, N. B. N., Blake, D. R., and Polak, J. M., Substance P-, calcitonin gene-related peptide- and C-flanking peptide of neuropeptide Y-immunoreactive nerve fibers are present in normal synovium but depleted in patients with rheumatoid arthritis, *Neuroscience*, 37, 143, 1990.
17. Grondblad, M., Konntinen, Y. T., Korkala, O., Liesi, P., Hukkanen, M., and Polak, J., Neuropeptides in the synovium of patients with rheumatoid arthritis and osteoarthritis, *J. Rheumatol.*, 15, 1807, 1988.
18. Hanesch, U., Heppelmann, B., and Schmidt, R. F., Acute monoarthritis of the cat's knee joint alters the proportion of CGRP-immunoreactive articular afferents, *Neuropeptides*, 26 (Suppl. 1), 57, 1994.
19. Hanesch, U., Pfrommer, U., Grubb, B. D., and Schaible, H.-G., Acute and chronic phases of unilateral inflammation in rat's ankle joint are associated with an increase in the proportion of calcitonin gene-related peptide-immunoreactive dorsal root ganglion cells, *Eur. J. Neurosci.*, 5, 154, 1993.
20. Lembeck, F., Donnerer, J., and Colpaert, F. C., Increase of substance P in primary afferent nerves during chronic pain, *Neuropeptides*, 1, 175, 1981.
21. Minami, M., Kuraishi, Y., Kawamura, M., Yamaguchi, T., Masu, Y., Nakanishi, S., and Satoh, M., Enhancement of preprotachykinin A gene expression by adjuvant-induced inflammation in the rat spinal cord: possible involvement of substance P-containing spinal neurons in nociception, *Neurosci. Lett.*, 98, 105, 1989.
22. Noguchi, K., Mortia, Y., Kiyama, H., Ono, K., and Tohyama, M., A noxious stimulus induces preprotachykinin-A gene expression in the rat dorsal root ganglion: a quantitative study using in situ hybridization histochemistry, *Mol. Brain Res.*, 4, 31, 1988.
23. Lygren, I., Ostensen, M., Burhol, P. G., and Husby, G., Gastrointestinal peptides in serum and synovial fluid from patients with inflammatory joint disease, *Ann. Rheum. Dis.*, 45, 637, 1986.
24. Devillier, P., Weill, B., Renoux, M., Menkes, C., and Pradelles, P., Elevated levels of tachykinin-like immunoreactivity in joint fluids from patients with inflammatory diseases, *N. Engl. J. Med.*, 314, 1323, 1981.
25. Marshall, K. W., Chiu, B., and Inman, R. D., Substance P and arthritis: analysis of plasma and synovial fluids, *Arthritis Rheum.*, 33, 87, 1989.
26. Marabini, S., Matucci-Cerinic, M., Geppetti, P., Del Bianco, E., Marchesoni, A., Tosi, S., Cagnoni, M., and Partsch, G., Substance P and somatostatin levels in rheumatoid arthritis, osteoarthritis, and psoriatic arthritis synovial fluid, *Ann. N.Y. Acad. Sci.*, 632, 435, 1991.
27. Larsson, J., Ekblom, A., Henrikson, K., Lundeburg, T., and Theodorsson, E., Immunoreactive tachykinins, calcitonin gene-related peptide and neuropeptide Y in human synovial fluid from inflamed knee joints, *Neurosci. Lett.*, 100, 326, 1989.

28. Larsson, J., Ekblom, A., Henrikson, K., Lundeburg, T., and Theodorsson, E., Concentration of substance P, neurokinin A, calcitonin gene-related peptide, neuropeptide Y and vasoactive intestinal polypeptide in synovial fluid from knee joints in patients suffering from rheumatoid arthritis, *Scand. J. Rheumatol.*, 20, 326, 1991.
29. Lam, F. Y. and Ferrell, W. R., Effects of interactions of naturally-occurring neuropeptides on blood flow in the rat knee joint, *Br. J. Pharmacol.*, 108, 694, 1993.
30. Brain, S. D. and Williams, T. J., Substance P regulates the vasodilator activity of calcitonin gene-related peptide, *Nature*, 335, 73, 1988.
31. Bileviciute, I., Lundberg, T., Ekblom, A., and Theodorsson, E., Bilateral changes of substance P-, neurokinin A-, calcitonin gene-related peptide- and neuropeptide Y-like immunoreactivity in rat knee joint synovial fluid during acute monoarthritis, *Neurosci. Lett.*, 153, 37, 1993.
32. Lewis, T., *The Blood Vessels of Human Skin and Their Responses*, Shaw, London, 1927.
33. Jancso, N., Jancso-Gabor, A., and Szolcsányi J., Direct evidence for neurogenic inflammation and its prevention by denervation and pretreatment with capsaicin, *Br. J. Pharmacol.*, 31, 138, 1967.
34. Ferrell, W. R. and Russell, N. J. W., Extravasation in the knee induced by antidromic stimulation of articular C fiber afferents of the anaesthetised cat, *J. Physiol. (London)*, 379, 407, 1986.
35. Ferrell, W. R. and Cant, R., Vasodilatation of articular blood vessels induced by antidromic electrical stimulation of joint C fibers, in *Fine Afferent Nerve Fibres and Pain*, Schmidt, R. F., Schaible, H. G., and Vahle-Hinz, C., Ed., VCH Verlagsgesellschaft, Weinheim, FRG, 1987, 187.
36. Kenins, P., Identification of unmyelinated sensory nerves which evoke plasma extravasation in response to antidromic stimulation, *Neurosci. Lett.*, 25, 137, 1981.
37. Linde, B., Chisolm, G., and Rosell, S., The influence of sympathetic activity and histamine on the blood-tissue exchange of solutes in canine adipose tissue, *Acta Physiol. Scand.*, 92, 145, 1974.
38. Engel, D., The influence of sympathetic nervous system on capillary permeability, *J. Physiol. (London)*, 99, 161, 1941.
39. Coderre, T. J., Basbaum, A. I., and Levine, J. D., Neural control of vascular permeability: interactions between primary afferents, mast cells, and sympathetic efferents, *J. Neurophysiol.*, 62, 48, 1989.
40. Fitzgerald, M., Arthritis and the nervous system, *Trends Neurosci.*, 12, 86, 1989.
41. Neugebauer, V. and Schaible, H.-G., Peripheral and spinal components of the sensitization of spinal neurons during an acute experimental arthritis, *Agents Actions*, 25, 234, 1988.
42. Hylden, J. L. K., Nahin, R. L., Traub, R. J., and Dubner, R., Expansion of receptive fields of spinal lamina I projection neurons in rats with unilateral adjuvant-induced inflammation: the contribution of dorsal horn mechanisms, *Pain*, 37, 229, 1989.
43. Sluka, K. A. and Westlund, K. N., Centrally administered non-NMDA but not NMDA antagonists block peripheral knee joint inflammation, *Pain*, 55, 217, 1993.
44. Sluka, K. A., Willis, W. D., and Westlund, K. N., Joint inflammation and hyperalgesia are reduced by spinal bicuculine, *Neuroreport*, 5, 109, 1993.
45. Rees, H., Sluka, K. A., Westlund, K. N., and Willis W.D., Do dorsal root reflexes augment peripheral inflammation?, *Neuroreport*, 5, 821, 1994.
46. De Castro Costa, M., De Sutter, P., Gybels, J., and van Hees, J., Adjuvant-induced arthritis in rats: a possible animal model of chronic pain, *Pain*, 10, 173, 1981.
47. Dardick, S. J., Basbaum, A. I., and Levine, J. D., The contribution of pain to disability in experimentally induced arthritis, *Arthritis Rheum.*, 29, 1017, 1986.
48. Landis, C. A., Robinson, C. R., and Levine, J. D., Sleep fragmentation in the arthritic rat, *Pain*, 34, 93, 1988.
49. Schaible, H.-G. and Grubb, B.D., Afferent and spinal mechanisms of joint pain, *Pain*, 55, 5, 1993.
50. Gibson, S. J., Polak, J. M., Bloom, S. R., and Wall, P. D., The distribution of nine peptides in rat spinal cord with special emphasis on the substantia gelatinosa and on the area around the central canal (lamina X), *J. Comp. Neurol.*, 201, 65, 1981.
51. Schaible, H.-G., Heppelmann, B., Craig, A. D., and Schmidt, R. F., Spinal termination of primary afferents of the cat's knee joint, in *Processing of Sensory Information in the Superficial Dorsal Horn of the Spinal Cord*, Cervero, F., Bennett, G. J., and Headley, P. M., Eds., Plenum, New York, 1989, 89.
52. Yaksh, T. L., Jessell, T. M., Gamse, R., Mudge, A. W., and Leeman S. E., Intrathecal morphine inhibits substance P release from mammalian spinal cord in vivo, *Nature*, 286, 155, 1980.
53. Duggan, A. W., Morton, C. R., Zhao, Z. Q., and Hendry, I. A., Noxious heating of the skin releases immunoreactive substance P in the substantia gelatinosa of the cat: a study with antibody microprobes, *Brain Res.*, 403, 345, 1987.
54. Pernow, B., Substance P, *Pharmacol. Rev.*, 35, 85, 1983.
55. Saria, A., Gamse, R., Petermann, J., Fischer, J. A., Theodorsson-Norheim, E., and Lundburg, J. M., Simultaneous release of several tachykinins and calcitonin gene-related peptide from rat spinal cord slices, *Neurosci. Lett.*, 63, 310, 1986.
56. Woolf, C. and Wiesenfeld-Hallin, Z., Substance P and calcitonin gene-related peptide synergistically modulate the gain of the nociceptive flexor withdrawal reflex in the rat, *Neurosci. Lett.*, 66, 226, 1986.

57. Henry, J. L., Effects of substance P on functionally identified units in cat spinal cord, *Brain Res.*, 114, 439, 1976.
58. Randic, M., and Miletic, V., Effect of substance P in cat dorsal horn neurons activated by noxious stimuli, *Brain Res.*, 128, 164, 1977.
59. Sugiura, Y., Terui, N., Hosoya, Y., and Kohno, K., Distribution of unmyelinated primary afferent fibers in the dosal horn, in *Processing of Sensory Information in the Superficial Dorsal Horn of the Spinal Cord*, Cervero, F., Bennett, G. J., and Headley, P. M., Eds., Plenum, New York, 1989, 15.
60. Buck, S. H. and Burcher, E., The tachykinins: a family of peptides with a brood of 'receptors', *Trends Pharmacol. Sci.*, 7, 65, 1986.
61. Quirion, R., Multiple tachykinin receptors, *Trends Neurosci.*, 8, 183, 1985.
62. Stucky, C. L., Galeazza, M. T., and Seybold, V. S., Time-dependent changes in Bolton-Hunter-labeled I-substance P binding in rat spinal cord following unilateral adjuvant-induced peripheral inflammation, *Neuroscience*, 57, 297, 1993.
63. Garry, M. G. and Hargreaves, K. M., Enhanced release of immunoreactive CGRP and SP from spinal dorsal horn slices occurs during carrageenan inflammation, *Brain Res.*, 582, 139, 1992.
64. Schaible, H.-G., Jarrott, B., Hope, P. J., Lang, C. W., and Duggan, A. W., Release of immunoreactive substance P in the spinal cord during development of acute arthritis in the knee joint of the cat: a study with antibody microprobes, *Brain Res.*, 529, 214, 1990.
65. Inoue, A., Takeda, R., Fukuyshu, T., Nakata, Y., and Segawa, T., Agonist-induced substance P receptor down-regulation in rat central nervous system, *Pharmacol. Res.*, 5, 795, 1988.
66. Holland, L. N., Goldstein, B. D., and Aronstam, R. S., Substance P receptor desensitization in the dorsal horn: possible involvement of receptor-G protein complexes, *Brain Res.*, 600, 89, 1993.
67. Schaible, H.-G. and Schmidt, R. F., Responses of fine medial articular nerve afferents to passive movements of knee joint, *J. Neurophysiol.*, 54, 1109, 1985.
68. Schaible, H.-G. and Schmidt, R. F., Effects of an experimental arthritis on the sensory properties of fine articular afferent units, *J. Neurophysiol.*, 49, 1118, 1983.
69. Grubb, B. D., Birrell, G. J., McQueen, D. S., and Iggo, A., The role of PGE2 in the sensitization of mechanoreceptors in normal and inflamed ankle joints of the rat, *Exp. Brain Res.*, 84, 383, 1991.
70. Guilbaud, G., Iggo, A., and Tegner, R., Sensory receptors in ankle joint capsules of normal and arthritic rats, *Exp. Brain Res.*, 58, 29, 1985.
71. Lembeck, F. and Holzer, P., Substance P as a neurogenic mediator of antidromic vasodilatation and neurogenic plasma extravasation, *Naunyn-Schmiedeberg's Arch. Pharmacol.*, 310, 175, 1979.
72. Foreman, J. C., Peptides and neurogenic inflammation, *Br. Med. Bull.*, 43, 386, 1987.
73. Yaksh, T. L., Bailly, J., Roddy, D. R., and Harty, G. J., Peripheral release of substance P from primary afferents, in *Proceedings Vth World Congress on Pain*, Dubner R., Gebhart, G. F., and Bond, M. R., Eds., Elsevier, Amsterdam, 1988, 51.
74. Holzer, P., Local effector functions of capsaicin-sensitive sensory nerve endings: involvement of tachykinins, calcitonin gene-related peptide and other neuropeptides, *Neuroscience*, 24, 139, 1988.
75. Hughes, S. R. and Brain, S. D., Effect of a calcitonin gene-related peptide (CGRP) antagonist (CGRP8-37) on blood flow and the potentiation of inflammatory oedema induced by CGRP and capsaicin in skin, *Br. J. Pharmacol.*, 104, 738, 1992.
76. Lam, F. Y. and Ferrell, W. R., Acute inflammation in the rat knee joint attenuates sympathetic vasoconstriction but enhances neuropeptide-mediated vasodilatation assessed by laser Doppler perfusion imaging, *Neuroscience*, 52, 443, 1993.
77. Lam, F. Y. and Ferrell, W. R., Capsaicin suppresses substance P-induced inflammation in the rat, *Neurosci. Lett.*, 105, 155, 1989.
78. Lam, F. Y. and Ferrell, W. R., Inhibition of carrageenan-induced inflammation in the rat knee joint by substance P antagonist, *Ann. Rheum. Dis.*, 48, 928, 1989.
79. Lam, F. Y. and Ferrell, W. R., Mediators of substance P-induced inflammation in the rat knee joint, *Agents Actions*, 31, 298, 1990.
80. Scott, D. T., Lam, F. Y., and Ferrell, W. R., Time course of substance P-induced protein extravasation in the rat knee joint measured by micro-turbidimetry, *Neurosci. Lett.*, 129, 74, 1991.
81. Scott, D. T., Lam, F. Y., and Ferrell, W. R., Acute inflammation enhances substance P-induced plasma protein extravasation in the rat knee joint, *Regul. Pept.*, 39, 227, 1992.
82. Lotz, M., Carson, D. A., and Vaught, J. L., Substance P activation of rheumatoid synoviocytes: neural pathway in pathogenesis of arthritis, *Science*, 235, 893, 1987.
83. Kimball, E. S., Perisco, F. J., and Vaught, J. L., Substance P, neurokinin A and neurokinin B induce generation of IL-1-like activity in P388D1 cells, *J. Immunol.*, 141, 3564, 1988.
84. Neveu, P. J. and Le Moal, M., Physiological basis for neuroimmunomodulation, *Fundam. Clin. Pharmacol.*, 4, 281, 1990.
85. Courtwright, L. J. and Kuzell, W. C., Sparing effect of neurological deficit and trauma on the course of adjuvant arthritis in the rat, *Ann. Rheum. Dis.*, 24, 360, 1965.

86. Cambridge, H. and Brain, S. D., Calcitonin gene-related peptide increases blood flow and potentiates plasma protein extravasation in the rat knee joint, *Br. J. Pharmacol.*, 106, 746, 1992.
87. Ferrell, W. R., Lam, F. Y., and Montgomery, I., Differences in the axon composition of nerves supplying the rat knee joint following intra-articular injection of capsaicin, *Neurosci. Lett.*, 141, 259, 1992.
88. Colpaert, F. C., Donnere, J., and Lembeck, F., Effects of capsaicin on inflammation and on the substance P content of nervous tissue in rats with adjuvant arthritis, *Life Sci.*, 32, 1827, 1983.
89. Levine, J. D., Clark, R., Devor, M., Helms, C., Moskowitz, M. A., and Basbaum, A. I., Intraneuronal substance P contributes to the severity of experimental arthritis, *Science*, 226, 547, 1984.
90. Lam, F. Y. and Wong, M. C. S., unpublished data, 1994.
91. Lam, F. Y. and Ferrell, W. R., Specific neurokinin receptors mediate plasma extravasation in the rat knee joint, *Br. J. Pharmacol.*, 103, 1263, 1991.
92. Fujii, T., Murai, M., Morimoto, H., Maeda, M., Yamaoka, M., Hagiwara, D., Miyake, H., Ikari, N., and Matsuo, M., Pharmacological profile of a high affinity dipeptide NK_1 receptor antagonist, FK888, *Br. J. Pharmacol.*, 107, 785, 1992.
93. Lam, F. Y. and Ferrell, W. R., CGRP modulates nerve-mediated vasoconstriction of the rat knee joint blood vessels, *Ann. N.Y. Acad. Sci.*, 657, 519, 1991.
94. Basbaum, A. I. and Levine, J. D., The contribution of the nervous system to inflammation and inflammatory disease, *Can. J. Physiol. Pharmacol.*, 69, 647, 1991.
95. Gray, E. and Ferrell, W. R., Acute joint inflammation alters adrenoceptor profile of synovial blood vessels in the knee joints of rabbits, *Ann. Rheum. Dis.*, 51, 1129, 1992.
96. Lam, F. Y. and Ferrell, W. R., Neurogenic component of different models of acute inflammation in the rat knee joint, *Ann. Rheum. Dis.*, 50, 747, 1991.
97. Najafipour, H. and Ferrell, W. R., Role of prostaglandins in regulation of blood flow and modulation of sympathetic vasoconstriction in normal and acutely inflamed rabbit knee joints, *Exp. Physiol.*, 79, 93, 1994.
98. Andrew, P. V., Helme, R. D., and Thomas, K. L., NK_1 receptor mediation of neurogenic plasma extravasation in rat skin, *Br. J. Pharmacol.*, 97, 1232, 1989.
99. Emonds-Alt, X., Advenier, C., Vilain, P., Goulaouic, P., Proietto, V., Van Broeck, D., Naline, E., Neliat, G., Le Fur, G., and Le Breliere, J. C., Pharmacological profile of SR48968, a potent non-peptide antagonist of the neurokinin A (NK_2) receptor, *Neuropeptides*, 22, 21, 1992.
100. McDougall, J. J., Karimian, S. M., and Ferrell, W. R., Alteration of substance P-mediated vasodilatation and sympathetic vasoconstriction in the rat knee joint by adjuvant-induced inflammation, *Neurosci. Lett.*, 174, 127, 1994.
101. Konttinen, Y. T., Rees, R., Hukkanen, M., Grönblad, M., Tolvanen, E., Gibson, S. J., Polak J. M., and Brewerton, D. A., Nerves in inflammatory synovium: immunohistochemical observations on the adjuvant arthritic rat model, *J. Rheumatol.*, 17, 1586, 1990.
102. Walsh, D. A., Salmon, M., Mapp, P. I., Wharton, J., Garrett, N., Blake, D. R., and Polak, J. M., Microvascular substance P binding to normal and inflamed rat and human synovium, *J. Pharmacol. Exp. Ther.*, 267, 951, 1993.

Chapter 18

SENSORY NEUROPEPTIDES IN THE SKIN

Susan D. Brain

CONTENTS

0-8493-7646-7/96/$0.00+$.50

I. INTRODUCTION

Cutaneous sensory nerves contain and release neuropeptides, of which the best known are substance P and calcitonin gene-related peptide (CGRP). Both of these peptides are readily detected in skin and first observed at an early stage (17 weeks) in the developing human fetus.[1] The afferent role of sensory nerves in mediating nociceptive stimuli (e.g., pain and itch) is well established and neuropeptides are involved in these responses. Cutaneous innervation enables the skin to pass sensory information concerning both the external and internal environments to higher nervous centers. The nerve endings, especially of Aδ-fibers and C-fibers, have been studied extensively using electrophysiological techniques and the specialization of different afferent nerve endings in skin has been reviewed elsewhere.[2,3] This chapter concentrates on the efferent activation, most commonly as a consequence of antidromic stimulation via axon reflexes, of sensory nerves. The release of neuropeptides leads to profound effects in skin, as a consequence of potent vasoactive properties and immunological activities of the neuropeptides. The vasodilatation and edema formation that results from antidromic stimulation of sensory nerves is most usually used to define "neurogenic inflammation".

II. HISTORICAL ASPECTS

A. EARLY STUDIES

The suggestion that nerves that originate from the dorsal root ganglion release a vasodilator in peripheral tissues was made in the late 1880s by Goltz and Stricker. This was confirmed by Bayliss (1901), who demonstrated that antidromic stimulation of peripheral sensory nerves results in cutaneous vasodilatation.[4] The results indicate that the sensory nerve terminals, in addition to transmitting orthodromically to the central nervous system, are capable of transmitting motor information antidromically to the peripheral tissue which the nerve innervates. In related studies Bruce (1913) demonstrated that topical application of mustard oil stimulated an acute inflammatory response, which was not observed when the sensory nerve supply to the skin had degenerated.[5]

B. THE USE OF CAPSAICIN

The use of capsaicin has enabled us to learn more about these nerves and the contribution of a neurogenic component to inflammation in the skin.[6,7] The majority of nerves involved in neurogenic inflammation in skin appear capsaicin-sensitive. Capsaicin acts selectively on most sensory nerves to stimulate pain and neurogenic inflammation. This response is not observed in denervated skin.[6] Subsequent application of capsaicin leads to desensitization of the fibers to stimuli and, depending on the treatment regime, sensory nerve degeneration (see Chapter 23).

C. THOMAS LEWIS AND THE TRIPLE RESPONSE

Thomas Lewis (1927), taking results from his own studies in addition to the earlier studies, established the concept of "the triple response" to skin injury. This comprises wheal, local erythema, and flare.[8,9] The triple response is clearly observed in human skin in response to many pin-point injuries (e.g., insect bites) and in response to intradermal injection of mediators (e.g., histamine[8] and bradykinin[10]). The flare (see Figure 1), which is a sensory nerve-mediated axon-reflex flare, can spread for up to several centimeters around the injection site. It is inhibited by prior treatment over several days with topical capsaicin to deplete sensory nerves.[11-13] Lewis suggested that injury to the skin leads to stimulation of sensory nerves, resulting in transmission of impulses to the spinal cord, as well as antidromic stimulation of connecting nerve fibers (axon reflexes) which innervate adjacent skin. Antidromic stimulation of these nerve fibers leads to the release of a vasodilator to mediate the flare response. The

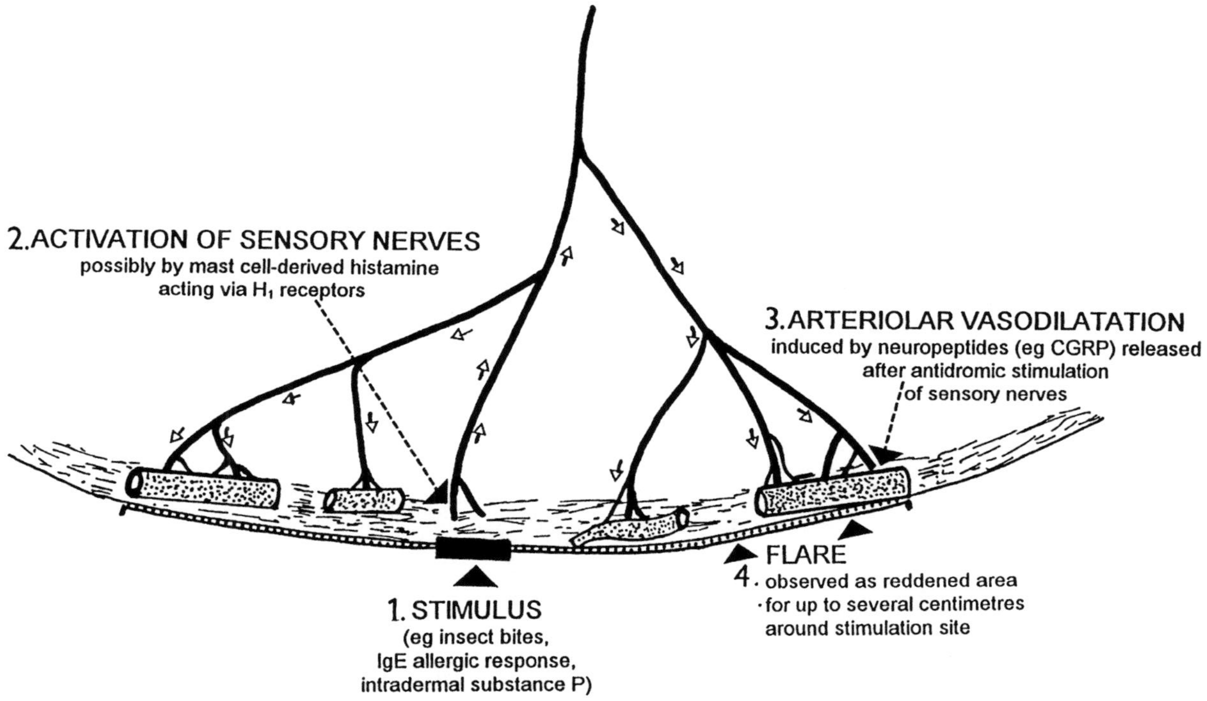

FIGURE 1. The axon-reflex flare in human skin.

possible identity of the vasodilator released from nerves to mediate the flare is discussed in Section III.D.

Activation of cutaneous mast cells and histamine release plays a central role in mediating the axon-reflex flare in human skin. There is good evidence that most agents (e.g., substance P, vasoactive intestinal peptide [VIP], and neurotensin, see Section IV) that stimulate an axon-reflex flare in skin do so as a consequence of their ability to stimulate mast cell degranulation. Histamine then plays an essential role by acting via H_1 receptors on afferent nerve terminals. This is probably why topical antihistamines play a beneficial role in the treatment of certain skin conditions in man.

The role, or even necessity, of an axon-reflex flare is uncertain, but it is generally assumed to be a signaling system which enables adjacent skin sites to be prepared for changes in the immediate environment. The axon-reflex flare induced by some agents (e.g., endothelin and nicotine) is not totally inhibited by systemic H_1 antagonists.[14,15] Thus, it is possible that certain agents can act independently of histamine, by stimulating receptors situated on afferent nerve terminals. Intradermal capsaicin, on initial application, directly activates C-fiber nerves in skin to stimulate a flare which is histamine independent.[16]

III. THE LOCALIZATION AND VASOACTIVE PROPERTIES OF NEUROPEPTIDES IN HUMAN SKIN

A. SUBSTANCE P AND THE NEUROKININS

The neurokinins, which include substance P, neurokinin A, and neurokinin B, have identical C-terminal amino acid structures. Neurokinin A and substance P are encoded by the same gene[17] and the localization of substance P and neurokinin A is similar.[18] They are found in perivascular nerves, where there is a close correlation with mast cells.[19,20] Substance P is also found in free nerve endings in the epidermis, in Meissners corpuscles of digital skin,[19,21] and close to sweat glands.[21,22] It is now established that neurokinin B is not found in skin.

Intradermal injection of substance P into human skin induces a triple response.[8] The flare component of the response to substance P is inhibited by H_1 receptor antagonists.[23] This indicates that the flare is initiated as a consequence of mast cell histamine release, as supported by Foreman and colleagues, who investigated structure–activity relationships of substance P-like peptides in human skin and showed that the basic N-terminal region of substance P is necessary for stimulating mast cell activation and the axon-reflex flare.[24] Substance P also releases histamine from human isolated skin mast cells.[25] There is little evidence to suggest that substance P directly stimulates nociceptors, although intradermal injection of substance P is often associated with itch. Substance P acts partly via histamine release and partly via activation of neurokinin NK_1 receptors on postcapillary venule endothelial cells to increase microvascular permeability and thus stimulate wheal (edema) formation. The ability of neurokinins A and B to stimulate wheal formation is solely dependent on their common C-terminal amino acid structure to increase microvascular permeability via NK_1 receptors and they have a similar potency to substance P in inducing a wheal in human skin.[26,27] Interestingly, as they possess less basic N-terminals, they do not activate mast cells and thus do not induce a substantial axon-reflex flare in human skin.[26,27] Recently, it has been suggested that the antiallergic drug sodium cromoglycate acts as a tachykinin receptor antagonist in human skin.[28]

B. CGRP

The calcitonin gene encodes for CGRP, but alternative processing of the mRNA of the gene leads to the production of CGRP in nervous tissue, with little CGRP in the normal thyroid.[29,30] CGRP is one of the most abundant neuropeptides in skin.[31] It is commonly colocalized with either substance P,[31-33] somatostatin,[31] or found in some nerves alone (e.g., around sweat glands). CGRP is found with substance P in perivascular nerves and in nerves that terminate close to mast cells.[18,33] Evidence to date suggests that the α form of CGRP predominates in human skin.[34,35]

The intradermal injection of CGRP induces a local erythema which can be observed for several hours after injection of picomolar doses, but for shorter time periods with lower doses.[36] The erythema induced by CGRP is due to an increase in cutaneous blood flow and is visually similar to that observed after intradermal injection of vasodilator prostanoids, but the response to the latter is transient.[37,38] Higher doses of CGRP can cause a triple response; this may be of relevance in pathological conditions.[36,39] The intravenous administration of CGRP leads to facial flushing,[40] which highlights the potent peripheral microvascular vasodilating activity of CGRP.

C. OTHER NEUROPEPTIDES

The vasodilator neuropeptide vasoactive intestinal peptide (VIP) and related peptides (e.g., pituitary adenylate cyclase activity peptide, PACAP) are found in nerves which terminate more deeply in normal skin, than sensory nerves as well as in sensory nerves. The nerves mainly terminate on arterial vessels and on sweat glands.[19,32,41,42] VIP is also found in sympathetic cholinergic nerves. There is also some suggestion that the synthesis of VIP is upregulated in sensory nerves during inflammation. VIP, when injected intradermally, initiates a response that is a mixture of both substance P and CGRP. Initially a wheal and flare is observed. A local erythema response (due to increased blood flow) is then revealed as the axon-reflex flare response fades.[37,43] The flare response to VIP is inhibited by H_1 antihistamines, which is indicative that VIP activates human cutaneous mast cells.

Other neuropeptides that have been demonstrated to be present in human skin include somatostatin,[19,31,44] galanin,[44] neurotensin,[42] bombesin,[19] and opioids.[18,42] The immediate response of human skin to these neuropeptides after intradermal injection is not well documented. Neurotensin (structurally related to substance P),[13] somatostatin,[39] and opioids,[18] all stimulate a triple response due to activation of mast cells. Studies, mainly in other tissues, suggest that opioids and some other neuropeptides can act via receptors on sensory nerve terminals to prevent further neuropeptide release (see References 46 and 47 and Section IV.D) and that the neuropeptides have immunological actions (see Section V). Thus, the acute actions of these peptides observed after intradermal injection may not be their most important action in skin. It is also possible that other novel neuropeptides with significant biological activities remain undiscovered.

D. CANDIDATES FOR THE MEDIATOR OF THE FLARE RESPONSE

The flare (see Figure 1) is mediated by vasodilator substances released from capsaicin-sensitive nerves in human skin, as discussed in Section II.C. The nerve terminals are adjacent to cutaneous arterioles. The axon reflex can be initiated in several ways, as discussed in Section II.C, but the substance(s) that mediate the flare will always be vasodilator(s) released from, or resulting from, activation of sensory nerves. The identity of the vasodilator(s) is debatable. Candidates have included histamine, ATP, substance P, VIP, CGRP and there has been interest in the possibility that a combination of substance P and histamine could mediate the flare.[13,48] Substance P is not an active vasodilator when injected into skin at doses lower than those required to stimulate degranulation of mast cells.[49] Capsaicin activates sensory nerve terminals directly and stimulates pain and an axon-reflex flare but it has been demonstrated that H_1 antagonists had no effect on the capsaicin-induced flare.[16] This evidence suggests that substance P and histamine are not strong candidates and another candidate, VIP, is not usually localized in capsaicin-sensitive nerves.

Femtomole amounts of CGRP injected intradermally stimulate erythema for about 20 min, which is a similar time course to that of the flare.[36,49] Thus, low concentrations of CGRP released from periarteriolar nerves could mediate the flare. The possibility cannot be proven due to a lack of a CGRP receptor antagonist which can be used in human skin.

IV. VASOACTIVE ACTIVITY OF ENDOGENOUS NEUROPEPTIDES: STUDIES IN ANIMAL MODELS

The use of inflammatory models in animal species to investigate mechanisms has been essential for determining the importance of endogenously released neuropeptides in skin. Increases in blood flow and vascular permeability in the skin in response to the release of neuropeptides from capsaicin-sensitive nerves is observed after chemical or electrical stimulation (see References 46 and 47). Interestingly, an axon-reflex flare is not as readily demonstrated or measured in animal skin,[50] although there is evidence for an axon-reflex flare in the pig.[51]

A. NEUROKININS

Neurokinins, as in human skin, are potent mediators of increased microvascular permeability in the rat[52-55] and guinea pig skin,[56,57] but inactive when injected intradermally in rabbit skin.[54] Edema formation (plasma protein extravasation), which results from increased microvascular permeability of the postcapillary venules,[55] can be readily measured at multiple sites in skin by the extravascular accumulation of either dyes (which bind to plasma proteins) or radiolabeled albumin. Substance P acts partly via histamine release and partly via neurokinin NK_1 receptors on vascular endothelial cells to increase microvascular permeability in rat skin.[13,47,57] Neurokinins A and B, as in human skin, act solely via endothelial cell NK_1 receptors.[26,53,58,59] The effect of the NK_1 receptor antagonists CP 96234,[60] RP 67580,[61,62] and SR 140333[63] is to abolish edema formation induced in rat skin in response to electrical stimulation of the sensory saphenous nerve. This is strong evidence that substance P and/or neurokinin A are the major mediators of increased microvascular permeability released by sensory nerves. Evidence suggests that NK_1 receptor antagonists also act to significantly inhibit carrageenan-induced edema in the rat.[56,64] This suggests that edema induced by carrageenan contains a neurogenic component.

B. CGRP

CGRP is a potent mediator of increased blood flow in the cutaneous microvasculature of all species tested.[36] The C-terminal fragment of human αCGRP, $CGRP_{8-37}$, has been identified as a CGRP receptor antagonist,[65] which acts as a competitive antagonist of vascular CGRP receptors *in vitro*.[66] In *in vivo* studies $CGRP_{8-37}$ has been shown to possess selective antagonist activity in the rat,[67,68] and in rabbit skin.[69] The increase in blood flow in response to intradermal capsaicin (100 nmol/site) or capsaicin analogs (e.g., olvanil and resinferatoxin) can be totally inhibited by intradermal $CGRP_{8-37}$,[70] in rabbit skin. Substance P is often colocalized with CGRP in capsaicin-sensitive nerves. However, substance P (despite well-established potent vasoactive effects in rat skin[52]) is extremely weak when injected intradermally in rabbit skin[36,54] and a substance P receptor-selective NK_1 antagonist[71] has no effect on capsaicin-induced blood flow in rabbit skin.[70] Thus, CGRP appears to be the important vasodilator neuropeptide released from capsaicin-sensitive nerves in rabbit skin.

$CGRP_{8-37}$ has to be administered intravenously (i.v.) in the rat as intradermal $CGRP_{8-37}$ is proinflammatory.[72] $CGRP_{8-37}$ i.v. acts in a selective manner to antagonize the actions of intradermal CGRP in increasing blood flow in the rat cutaneous microvasculature. $CGRP_{8-37}$ also inhibits the increased blood flow induced by topical capsaicin and by mild short electrical stimulation of the saphenous nerve.[73,74] Thus, the results are compatible with the suggestion that CGRP is the principal vasodilator released after mild and short electrical stimulation of sensory nerves.

1. Interaction of CGRP with Neurokinins and Other Inflammatory Mediators

CGRP, as a consequence of its vasodilator activity, potentiates edema formation induced by mediators of increased microvascular permeability, such as substance P, in species that include the rat.[54,58] $CGRP_{8-37}$ causes a partial but significant inhibition of edema induced by electrical stimulation of the sensory saphenous nerve. This is in contrast to similar experimental protocols where NK_1 receptor antagonists totally inhibit edema formation induced by saphenous nerve stimulation.[60-63] These results suggest that substance P and/or neurokinin A are the major mediators of increased microvascular permeability and that the potentiating effect of endogenously released CGRP on increased microvascular permeability, which results from saphenous nerve stimulation, is not essential for the edema formation. However, an increase in blood flow is observed and this is concomitant with edema formation and not inhibited by an NK_1 receptor antagonist.[75] Indeed it is unlikely that substance P acts directly via NK_1 receptors to increase blood flow as the selective NK_1 receptor agonist GR73632 does not increase blood flow in rat skin, although it has potent edema-inducing effects, which are potentiated by CGRP.[64,76] Thus, a non-neurokinin vasodilator component is obviously involved in the resulting edema formation. The increased blood flow could be induced totally by CGRP, which it is not possible to antagonize in the present experiments. It should be remembered that $CGRP_{8-37}$ is a peptide antagonist that is poorly characterized, while CP 96345 and RP 67580 are nonpeptide structures, developed on the basis of their affinity and selectivity as NK_1 receptor antagonists.

Thus, CGRP, as a consequence of arteriolar vasodilatation, can potentiate edema formation induced by a wide range of mediators of increased microvascular permeability.[54] CGRP, although not chemotactic in the rabbit, can potentiate neutrophil accumulation induced by chemotactic agents and the cytokine interleukin-1.[77] This means CGRP can contribute to inflammation as a neurogenically derived component by potentiating the actions of a range of non-neuropeptide mediators.

The injection of substance P with CGRP into human skin leads to a vasoactive response due to the combined response of substance P and CGRP. However, the long-lasting erythema normally observed with CGRP is not seen. This is due to the action of proteases (e.g., tryptase and chymase) from substance P-activated mast cells in degrading and thus inactivating CGRP.[78] This could be an important mechanism by which CGRP and other peptides are metabolized in diseases that are associated with mast cell activation.

C. INVOLVEMENT OF NITRIC OXIDE

Nitric oxide synthase inhibitors have no effect on the vasodilator activity of CGRP in rat and rabbit skin.[79,80] Nitric oxide synthase inhibitors do inhibit edema formation when coinjected with substance P into rat skin, but this seems to be a secondary effect due to inhibition of basal blood flow.[81] Thus, evidence suggests that substance P and CGRP do not induce their responses via nitric oxide-dependent mechanisms in skin. In contrast, there is evidence to suggest that the release of neuropeptides from sensory nerves is under the influence of nitric oxide.

The nitric oxide synthase inhibitor L-N^G-arginine methyl ester (L-NAME) has no effect on the vasodilator response induced by CGRP, but significantly inhibits capsaicin-induced blood flow in rabbit skin.[80] The data indicate the novel finding that nitric oxide plays an essential role in the release of CGRP but not in the activity of CGRP in the cutaneous microvasculature. This is supported by the findings that L-NAME attenuates mustard oil-induced vasodilatation in rat skin[82] and neurogenic edema formation induced by electrical stimulation of the saphenous nerve is inhibited by two distinct inhibitors of nitric oxide synthase.[83]

D. INHIBITION OF NEUROPEPTIDE RELEASE BY AGENTS THAT ACT PRESYNAPTICALLY

Several agents are suggested to inhibit neuropeptide release by acting via receptors situated on sensory nerve terminals (for review, see Reference 84). The best evidence in skin comes from studies of the effect of opioids in the rat saphenous nerve, where neurogenic inflammation is inhibited in a naloxone-reversible manner. The increased blood flow induced by short electrical stimulation of the sensory saphenous nerve is inhibited, in addition to the neurogenic edema formation.[73] The inhibition of edema formation induced by electrical stimulation of the saphenous nerve has been studied in several laboratories. Morphine,[85,86] the μ selective agonist lofentanil,[87] and the enkephalin analog [D-Met^2, Pro^5]enkephalinamide[85,88] are all active in this model. It has recently been shown that selective κ agonists also inhibit neurogenic edema formation.[89] Galanin has also been shown to inhibit C-fiber-evoked plasma extravasation, acting partially via presynaptic receptors,[90] and there is evidence that 5-hydroxytryptamine receptor (5-$HT_{1B/D}$)-like agonists and somatostatin analogs have inhibitory effects.

E. INFLUENCE OF GROWTH FACTORS

Studies of dorsal root ganglion cells, *in vitro,* have revealed the reliance of immature sensory nerve cells on growth factors, especially nerve growth factor (NGF), for their normal development. It has been demonstrated, in an elegant *in vivo* study in the rat, that NGF is involved in mediating the increased content and transport of substance P and CGRP in sensory nerves innervating inflamed tissue, including skin.[91]

V. NEUROPEPTIDE EFFECTS ON THE CELLULAR AND IMMUNOLOGICAL COMPONENTS OF INFLAMMATION

There is evidence from a substantial amount of research that neuropeptides can have varying, sometimes extremely potent, effects on the cellular and immunological components of inflammation. The pathological significance of the multiple and complex actions of neuropeptides is poorly understood. The reader is referred to recent reviews.[92,93] The discussion in this section is restricted to results of studies in which substance P and CGRP are suggested to have effects that are directly relevant to skin.

A. SUBSTANCE P AND THE NEUROKININS

The effect of substance P in activating skin mast cells is well documented (see Section III.A). This leads to the release of mast cell amines and enzymes. It is now realized that mast cells are also capable of secreting multiple cytokines. Recently substance P has been found to induce tumor necrosis factor-α gene expression and secretion from murine mast cells.[94] In addition substance P stimulates the release of interleukin-1 and tumor necrosis factor-α from human monocytes.[95] Human mononuclear leukocytes are chemotactic to substance P[96] and neurokinins, as well as other neuropeptides, also have modulatory effects on lymphocyte function and immunomodulation.[93,97,98]

Substance P is a potent priming agent for neutrophils, making the neutrophils more responsive to subsequent stimulation by chemotactic factors.[99] Substance P can induce neutrophil accumulation in mouse skin as a consequence of mast cell degranulation, leading to the release of leukotriene B_4,[100] but the activation of vascular endothelial cells is also involved in substance P-induced neutrophil accumulation in the skin. Recent evidence[101] suggests that substance P, VIP, and CGRP all induce neutrophil accumulation after intradermal injection into human skin. It is suggested that the accumulation is due to the ability of these neuropeptides to stimulate upregulation of endothelial cell adhesion molecules (especially P and E selectin). These authors also provide evidence that substance P stimulates eosinophil accumulation.

Substance P stimulates human skin fibroblasts[102] and keratinocytes to proliferate.[103] It is now clear that the proliferative effects of substance P on fibroblasts are mediated via NK_1 receptors as effects are inhibited by NK_1 receptor antagonists.[104] There is an interesting study which suggests that the mitogenic effects of leukotriene B_4, which is found in psoriatic lesions, may be enhanced in the presence of substance P and VIP.[105]

B. CGRP

CGRP stimulates a brief protein synthesis-independent adhesion of human neutrophils to cultured endothelial cells.[106,107] However, it does not directly stimulate neutrophil accumulation when injected intradermally into rabbit skin.[77] By comparison it has been suggested that CGRP, as well as substance P and VIP, induce neutrophil accumulation after intradermal injection into human skin.[98]

CGRP binds to and inhibits proliferation of mouse T cells[108] and it is chemotactic for human T-lymphocytes.[109] CGRP has been suggested to have potentially important modulatory effects on Langerhans cells in that it inhibits the ability of these cells to present antigen.[110]

VI. EVIDENCE FOR THE INVOLVEMENT OF NEUROPEPTIDES IN INFLAMMATORY MODELS

A. THERMAL INJURY

An involvement of substance P in thermal injury is suggested from experiments carried out in the anesthetized rat. Immersion of the rat paw in water (43 to 60°C) is associated with edema formation[111] and an increase in firing of C-fiber nerves in the saphenous nerve.[112] The use of capsaicin in toxic doses to deplete a neurogenic component leads to a loss of both responses, suggesting the involvement of neuropeptides. Of interest is the observation that in man heat pain thresholds fall and skin heating is very painful for 24 to 48 h after topical capsaicin.[11] Thus, a link could well exist between the mechanisms by which capsaicin and heat activate sensory nerves. Furthermore, there is evidence that postocclusion reactive hyperemia in human forearm skin is mediated by a local reflex involving sensory nerves.[113]

B. NEUROPEPTIDES: A ROLE IN THE MAINTENANCE OF SKIN INTEGRITY AND IN WOUND HEALING

The chronic administration of capsaicin to rats leads to damaged skin, as observed by lesions and loss of fur, and is accompanied by scratching behavior.[114,115] This suggests that neuropeptides are involved in maintaining skin integrity. Further, the growth of sensory nerves is observed during healing of burn wounds in guinea pig skin.[116] The proliferation of various cell types is essential to wound healing; thus, the proliferative effects of substance P outlined in Section V.A is relevant. Senapati and co-workers have shown a depletion of substance P and CGRP after wounding in the rat; this may be due to an increased neuropeptide release leading to neuropeptide depletion.[117]

1. Importance of Sensory Nerves in Skin Flap Survival

Local intravenous injection of CGRP increased skin flap survival in the rat, while increasing blood flow.[118] In an immunocytochemical study of the reinnervation in mouse skin flaps, CGRP and substance P appeared in new nerves in the skin flaps, some days before neuropeptide Y and VIP.[119] These studies suggest that sensory nerves have an important efferent role in releasing vasodilator peptides and protecting against ischemia.

VII. EVIDENCE FOR THE INVOLVEMENT OF SENSORY NEUROPEPTIDES IN RAYNAUD'S DISEASE

Raynaud's disease patients exhibit a lack of reflex vasodilatation after exposure to various stresses, in particular the cold. This leads to reduced blood flow to the skin, with the development of abnormalities such as skin ulcers that can lead to necrosis in the chronic disease state. Raynaud's disease patients respond normally to CGRP when it is injected into the forearm or digits.[38,120,121] Intravenous injection of CGRP leads to a flushing of the hands in Raynaud's patients, possibly due to a denervation supersensitivity[122] and evidence suggests that there is a reduction in CGRP-immunoreactive nerves (and also other sensory neuropeptides) in digital skin in Raynaud's disease.[123] Thus, the lack of a vasodilator response in Raynaud's disease patients could be due to a lack of release of vasodilator CGRP. Intravenous CGRP effectively dilates the compromised cutaneous vasculature in severe Raynaud's phenomenon and this is associated with an increased healing of ulcers.[124,125]

VIII. EVIDENCE FOR THE INVOLVEMENT OF NEUROPEPTIDES IN HUMAN SKIN DISEASE

A. DETECTION AND MEASUREMENT OF NEUROPEPTIDE LEVELS IN SKIN DISEASE

The levels of neuropeptides that can be measured by radioimmunoassay in skin biopsy extracts is low; however, various, sometimes high, levels of CGRP, somatostatin, and neuropeptide Y can be measured in spontaneous blisters from conditions that include bullous pemphigoid and herpes zoster.[126] Increased levels of neuropeptides (CGRP and substance P) have been found in tissue taken from patients with nodular prurigo with little change noted in samples taken from patients with lichinfied eczema.[127] In a study of atopic dermatitis levels of VIP and substance P were raised[129] and in another study levels of neuropeptides were increased in atopic dermatitis, but not for urticaria.[130] One problem associated with measuring substances in skin extracts/fluids is that proteases are readily available to degrade peptides. Urticaria is a mast cell-dependent phenomenon in which neuropeptides are considered to be involved, as capsaicin inhibited acute effects in a model of cold urticaria,[130] and in which tryptase levels would be expected to be high. As a consequence, neuropeptides might be expected to be metabolized soon after their release from cutaneous nerves. Therefore, the measurement of neuropeptides in skin samples is probably not a good indicator of either their release or activity.

B. PSORIASIS

The possibility that neuropeptides play a role in psoriasis was initially suggested because of symmetry of the distribution of lesions which can occur, indicating a neural component.[131] Interestingly an association has been found between cutaneous mast cells, sensory nerves, and the patient's stress state.[132] Patients who were diagnosed as suffering from high stress appeared to have more severe symptoms and more VIP- and CGRP-reactive nerves. VIP and substance P levels have been shown to be elevated in lesional psoriatic skin.[133,134]

The most convincing evidence for a sensory nerve component is that topical capsaicin can act to reduce both the scaling and erythema[135] and also the itching[136] associated with disease. Further, a new treatment called peptide T is thought to be effective in psoriasis by acting via somatostatin-dependent mechanisms.[137]

IX. CLINICAL POTENTIAL

It is clear that sensory neuropeptides have the capacity to act as a damaging neurogenic component in inflammatory diseases in skin. Therefore, it would seem appropriate to conclude this chapter by briefly discussing the mechanisms by which it is proposed to treat neurogenic inflammation in the clinic.

A. CAPSAICIN

The flare is inhibited by capsaicin pretreatment in a model of an IgE cutaneous allergic response in man.[138] The local pretreatment with capsaicin attenuated the flare response to intradermal histamine and substance P, but caused an overall enhancement of the erythema responses in ultraviolet irradiation, contact dermatitis, and the tuberculin reaction.[139] It is possible that this is because modulatory neuropeptides (e.g., opioids and somatostatin) may normally be released from capsaicin-sensory nerves to modulate responses. Thus, the use of capsaicin to learn about the efferent function of neuropeptides in man has led to some difficulties in interpretation.

Topical capsaicin at low concentrations selectively reduces the feeling of heat pain; this is maintained by continuous application.[140] Capsaicin is a useful treatment in a number of skin conditions, especially in relieving pain. Skin conditions include herpes[141] and diabetic neuropathy,[142] and also psoriasis, as discussed in Section VIII.B). These positive results suggest that there is every incentive to develop capsaicin analogs that act to deplete the sensory nerve component in skin.

B. NEUROPEPTIDE ANTAGONISTS AND INHIBITORS OF NEUROPEPTIDE RELEASE

There is immense interest in the clinical efficacy of nonpeptide neurokinin antagonists. It will become clear in the next few years whether they will be useful in the treatment of skin conditions. Their potential is related to their efficacy as analgesics, in addition to possible anti-inflammatory actions. The development of antagonists for other neuropeptides is less well progressed. In particular it would be of interest to determine the efficacy of a CGRP antagonist.

An alternative approach to inhibiting neurogenic inflammation is to develop agents that act to inhibit the release of neuropeptides. This approach has the advantage that all the effects of the neuropeptides are inhibited rather than just those mediated via one receptor subtype. However, apart from the opioids, little is known about the clinical potential of this approach.

It is anticipated that events over the next few years will do much to clarify the role of neurogenic inflammation in skin disease.

REFERENCES

1. Terenghi, G., Sundaresan, M., Moscoso, G., and Polak, J. M., Neuropeptides and a neuronal marker in cutaneous innervation during human foetal development, *J. Comp. Neurol.*, 22, 595, 1993.
2. Lynn, B., Capsaicin: actions on C-fibre afferents that may be involved in itch, *Skin Pharmacol.*, 5, 9, 1992.
3. Winklemann, R. K., Cutaneous sensory nerves, *Semin. Dermatol.*, 7, 236, 1988.
4. Bayliss, W. M., On the origin from the spinal cord of the vasodilator fibres of the hind limb, and on the nature of these fibres, *J. Physiol.*, 26, 173, 1901.
5. Bruce, A. N., Vasodilator axon reflexes, *Q. J. Physiol.*, 6, 339, 1913.
6. Jancsó, N., Jancsó-Gábor, A., and Szolcsányi, J., Direct evidence of neurogenic inflammation and its prevention by denervation and by treatment with capsaicin, *Br. J. Pharmacol.*, 31, 138, 1967.

7. Jancsó, G., Kiraly, E., and Jancsó-Gábor, A., Pharmacologically induced selective degeneration of chemosensitive primary sensory neurons, *Nature,* 270, 741, 1977.
8. Lewis, T., *The Blood Vessels of the Human Skin and Their Responses,* Shaw, London, 1927.
9. Lewis, T., Observations upon reactions of vessels in human skin to cold, *Heart,* 15, 177, 1930.
10. Wallengren, J. and Hakanson, R., Effects of capsaicin, bradykinin and prostaglandin E_2 in human skin, *Br. J. Dermatol.,* 126, 111, 1992.
11. Carpenter, S. E. and Lynn, B., Vascular and sensory responses of human skin to mild injury after topical treatment with capsaicin, *Br. J. Pharmacol.,* 73, 755, 1981.
12. Bernstein, J. E., Capsaicin in dermatologic disease, *Semin. Dermatol.,* 7, 304, 1988.
13. Foreman, J. an Jordan, C., Histamine release and vascular changes induced by neuropeptides, *Agents Actions,* 13, 105, 1983.
14. Crossman, D. C., Brain, S. D., and Fuller, R., Potent vasoactive effects of endothelin in skin, *Am. J. Physiol.,* 70, 260, 1991.
15. Izumi, H. and Karita, K., Axon reflex flare evoked by nicotine in human skin, *Jpn. J. Physiol.,* 42, 721, 1992.
16. Barnes, P. J., Brown, M. J., Dollery, C. T., Fuller, R. W., Heavey, D. J., and Ind, P. W., Histamine is released from skin by substance P but does not act as the final vasodilator in the axon reflex, *Br. J. Pharmacol.,* 88, 741, 1986.
17. Nawa, H., Hirose, T., Takashima, H., Inayama, S., and Nakamishi, S., Nucleotide sequences of cloned cDNAs for two types of bovine brain substance P precursor, *Nature,* 306, 32, 1983.
18. Weihe, E. and Hartschuh, W., Multiple peptides in cutaneous nerves: Regulator under physiological conditions and a pathogenetic role in skin disease?, *Semin. Dermatol.,* 7, 284, 1988.
19. Bloom, S. R. and Polak, J. M., Regulatory peptides and the skin, *Clin. Exp. Dermatol.,* 8, 3, 1983.
20. Holzer, P., Peptidergic sensory neurons in the control of vascular functions: mechanisms and significance in the cutaneous and splanchnic vascular beds, *Rev. Physiol. Biochem. Pharmacol.,* 121, 49, 1992.
21. Dalsgaard, C. J., Johnsson, C. E., Hokfelt, T., and Cuello, A. C., Localization of substance P-like immunoreactive nerve fibres in the human digital skin, *Experientia,* 39, 1018, 1983.
22. Tanio, H., Vaalasti, A., and Rechardt, L., The distribution of substance P-, CGRP-, galanin- and ANP-like immunoreactive nerves in human sweat glands, *Histochem. J.,* 19, 375, 1987.
23. Hagermark, O., Hokfelt, T., and Pernow, B., Flare and itch induced by substance P in human skin, *J. Invest. Dermatol.,* 71, 233, 1980.
24. Foreman, J. C., Jordan, C. C., Oehme, P., and Renner, H., Structure-activity relationships for some substance P-related peptides that cause wheal and flare reactions in human skin, *J. Physiol.,* 335, 449, 1983.
25. Benyon, R. C., Church, M. K., and Lowman, M. A., Histamine release from human dispersed skin mast cells induced by substance P, *Br. J. Pharmacol.,* 90, 102, 1987.
26. Devillier, P., Regoli, D., Asseraf, A., Descurs, B., Marsac, J., and Renoux, M., Histamine release and local responses of rat and human skin to substance P and other mammalian tachykinins, *Pharmacology,* 32, 340, 1986.
27. Fuller, R. W., Conradson, T.-B., Dixon, C. M. S., Crossman, D. C., and Barnes, P. J., Sensory neuropeptide effects in human skin, *Br. J. Pharmacol.,* 92, 781, 1987.
28. Crossman, D. C., Dashwood, M. R., Taylor, G. W., Wellings, R., and Fuller, R. W., Sodium cromoglycate: evidence of tachykinin antagonist activity in the human skin, *J. Appl. Physiol.,* 75, 167, 1993.
29. Amara, S. G., Jonas, V., Rosenfeld, M. G., Ong, E. S., and Evans, R. M., Alternative RNA processing in calcitonin gene expression generates mRNAs encoding different polypeptide products, *Nature,* 298, 240, 1982.
30. Rosenfeld, M. G., Amara, S. G., and Evans, R. M., Alternative RNA processing: determining neuronal phenotype, *Science,* 225, 1315, 1984.
31. Gibbins, I. L., Wattchow, D., and Coventry, G., Two immunohistochemically identified populations of calcitonin gene-related peptide (CGRP)-immunoreactive axons in human skin, *Brain Res.,* 414, 143, 1987.
32. Wallengren, J., Ekman, R., and Sundler, F., Occurrence and distribution of neuropeptides in the human skin. An immunochemical and immunocytochemical study on normal skin and blister fluid from inflamed skin, *Acta Derm. Venereol.,* 66, 185, 1987.
33. Alving, K., Sundstrom, C., Matran, R., Panula, P., Hokfelt, T., and Lundberg, J. M., Association between histamine-containing mast cells and sensory nerves in the skin and airways of control and capsaicin-treated pigs, *Cell Tissue Res.,* 264, 529, 1991.
34. O'Halloran, D. and Bloom, S. R., Calcitonin gene-related peptide, *Br. Med. J.,* 302, 739, 1991.
35. Mulderry, R. K., Ghatei, M. A., Spokes, R. A., Jones, P. M., Pierson, A. M., Hamid, Q. A., Kanse, S., Amara, S. G., Burrin, J. M., Legon, S., Polak, J. M., and Bloom, S. R., Differential expression of αCGRP and βCGRP by primary sensory neurons of the rat, *Neuroscience,* 25, 195, 1988.
36. Brain, S. D., Williams, T. J., Tippins, J. R., Morris, H. R., and MacIntyre, I., Calcitonin gene-related peptide is a potent vasodilator, *Nature,* 313, 54, 1985.
37. Brain, S. D., Tippins, J. R., Morris, H. R., MacIntyre, I., and Williams, T. J., Potent vasodilator activity of calcitonin gene-related peptide in human skin, *J. Invest. Dermatol.,* 87, 533, 1986.

38. Brain, S. D., Petty, R. G., and Williams, T. J., Cutaneous blood flow responses in the forearm of Raynaud's patients induced by local cooling and intradermal injections of CGRP and histamine, *Br. J. Clin. Pharmacol.*, 30, 853, 1990.
39. Piotrowski, W. and Foreman, J. C., Some effects of calcitonin gene-related peptide in human skin and on histamine release, *Br. J. Dermatol.*, 114, 37, 1986.
40. Struthers, A. D., Brown, M. J., Beacham, J. L., Morris, H. R., MacIntyre, I., and Stevenson, J. C., The acute effect of human calcitonin gene-related peptide in man, *J. Endocrinol.*, 104 (Suppl.), 129, 1985.
41. Moller, K., Zhang, Y. Z., Hakanson, R., Luts, A., Sjolund, B., Uddman, R., and Sunder, F., Pituitary adenylate cyclase activating peptide is a sensory neuropeptide: immunocytochemical and immunochemical evidence, *Neuroscience*, 57, 725, 1993.
42. Hartschuh, W., Weihe, E., and Reinecke, M., Peptidergic (neurotensin, VIP, substance P) nerve fibres in the skin. Immunohistochemical evidence of an involvement of neuropeptides in nociception, pruritus and inflammation, *Br. J. Dermatol.*, 109 (Suppl. 25), 14, 1983.
43. Anand, A., Bloom, S. R., and McGregor, G. P., Topical capsaicin pretreatment inhibits axon reflex vasodilatation caused by somatostatin and vasoactive intestinal polypeptide in human skin, *Br. J. Pharmacol.*, 78, 665, 1983.
44. Johansson, O. and Vaalasti, A., Immunohistochemical evidence for the presence of somatostatin-containing sensory nerve fibres in skin, *Neurosci. Lett.*, 73, 225, 1987.
45. Johansson, O., Vaalasti, A., Tainio, H., and Ljungberg, A., Immunohistochemical evidence of galanin in sensory nerves of human digital skin, *Acta Physiol. Scand.*, 132, 261, 1988.
46. Maggi, C. A. and Meli, A., The sensory efferent function of capsaicin-sensitive sensory neurons, *Gen. Pharmacol.*, 19, 1, 1988.
47. Holzer, P., Local effector functions of capsaicin-sensitive sensory nerve endings: involvement of tachykinins, calcitonin gene-related peptide and other neuropeptides, *Neuroscience*, 24, 739, 1988.
48. Lembeck, F. and Gamse, R., Substance P in peripheral sensory processes, in *Substance P in the Nervous System*, Ciba Foundation Symp. G1, Pitman, London, 1982, 35.
49. Brain, S. D. and Edwardson, J. A., Neuropeptides and skin, in *Pharmacology of the Skin*, Vol. 1, *Handbook of Experimental Pharmacology*, Greaves, M. W. and Shuster, S., Eds., Springer-Verlag, Berlin, 1987, 89.
50. Lynn, B. and Shakhanbeh, J., Neurogenic inflammation in the skin of the rabbit, *Agents Actions*, 25, 228, 1988.
51. Janscó, G., Pierau, F. K., and Sann, H., Mustard oil-induced cutaneous inflammation in the pig, *Agents Actions*, 39, 31, 1993.
52. Lembeck, F. and Holzer, P., Substance P as neurogenic mediator of antidromic vasodilation and neurogenic plasma extravasation, *Naunyn-Schmiedeberg's Arch. Pharmacol.*, 310, 175, 1979.
53. Gamse, R. and Saria, A., Potentiation of tachykinin-induced plasma protein extravasation by calcitonin gene-related peptide, *Eur. J. Pharmacol.*, 114, 61, 1985.
54. Brain, S. D. and Williams, T. J., Inflammatory oedema induced by synergism between calcitonin gene-related peptide (CGRP) and mediators of increased vascular permeability, *Br. J. Pharmacol.*, 86, 855, 1985.
55. Kenins, P., Hurley, J. V., and Bell, C., The role of substance P in the axon reflex in the rat, *Br. J. Dermatol.*, 111, 551, 1984.
56. Nagahisa, A., Kanai, Y., Suga, O., Taniguchi, K., Lowe, J. A., and Hess, H. J., Anti-inflammatory and analgesic activity of a non-peptide substance P receptor antagonist, *Eur. J. Pharmacol.*, 217, 191, 1992.
57. Wilsoncroft, P., Euzger, H., and Brain, S. D., Effect of a neurokinin-1 (NK1) receptor antagonist on oedema formation induced by tachykinins, carrageenan and an allergic response in guinea-pig skin, *Neuropeptides*, 26, 405, 1994.
58. Brain, S. D. and Williams, T. J., Interactions between the tachykinins and calcitonin gene-related peptide lead to the modulation of oedema formation and blood flow in rat skin, *Br. J. Pharmacol.*, 97, 77, 1989.
59. Andrews, P. V., Thomas, K. L., and Helme, R. D., NK-1 receptor mediation of neurogenic plasma extravasation in rat skin, *Br. J. Pharmacol.*, 97, 1232, 1989.
60. Lembeck, F., Donnerer, J., Tsuchiya, M., and Nagahisa, A., The non-peptide tachykinin antagonist, CP-96,345, is a potent inhibitor of neurogenic inflammation, *Br. J. Pharmacol.*, 105, 527, 1991.
61. Garret, C., Caruette, A., Fardin, V., Moussaoui, S., Peyronel, J.-F., Blanchard, J.-C., and Laduron, P. M., Pharmacological properties of a potent and selective nonpeptide substance P antagonist, *Proc. Natl. Acad. Sci. U.S.A.*, 88, 10208, 1991.
62. Xu, J.-X., Dalsgaard, J.-C., Maggi, C. A., and Wiesenfeld-Hallin, Z., NK-1, but not NK-2, tachykinin receptors mediate plasma extravasation induced by antidromic C-fiber stimulation in rat hindpaw: demonstrated with the NK-1 antagonist CP-96,345 and the NK-2 antagonist Men 10207, *Neurosci. Lett.*, 139, 249, 1992.
63. Emonds-Alt, X., Doutremepuich, J.-D., Heaulme, M., Neliat, G., Santucci, V., Steinberg, R., Vilain, P., Bichon, D., Ducoux, J.-P., Proietto, V., Van Broeck, D., Soubrie, P., Le Fur, G., and Breliere, J.-C., In vitro and in vivo biological activities of SR140333, a novel non-peptide tachykinin NK_1 receptor antagonist, *Eur. J. Pharmacol.*, 250, 403, 1993.

64. Birch, P. J., Harrison, S. M., Hayes, A. G., Rogers, H., and Tyers, M. B., The non-peptide NK1 receptor antagonist CP-96,345, produces antinociceptive and anti-oedema effects in the rat, *Br. J. Pharmacol.*, 105, 508, 1992.
65. Chiba, T., Yamaguchi, A., Yamatani, T., Nakamura, A., Morishita, T., Inui, T., Fukase, M., Noda, T., and Fujita, T., Calcitonin gene-related peptide receptor antagonist human CGRP(8-37), *Am. J. Physiol.*, 256, E331, 1989.
66. Han, S. P., Naes, L., and Westfall, T. C., Inhibition of periarterial nerve stimulation-induced vasodilation of the mesenteric arterial bed by CGRP (8-37) and CGRP receptor desensitization, *Biochem. Biophys. Res. Commun.*, 168, 786, 1990.
67. Donoso, V. S., Fournier, A., St.-Pierre, S., and Huidobro-Toro, P. J., Pharmacological characterization of CGRP1 receptor subtype in the vascular system of the rat: studies with hCGRP fragments and analogs, *Peptides*, 11, 885, 1990.
68. Gardiner, S. M., Compton, A. M., Kemp, P. A., Bennett, T., Bose, C., Foulkes, R., and Hughes, B., Antagonistic effect of human αCGRP [8-37] on the in vivo regional haemodynamic actions of human αCGRP, *Biochem. Biophys. Res. Commun.*, 171, 938, 1990.
69. Hughes, S. R. and Brain, S. D., A calcitonin gene-related antagonist ($CGRP_{8-37}$) inhibits microvascular responses induced by CGRP and capsaicin in skin, *Br. J. Pharmacol.*, 104, 738, 1991.
70. Hughes, S. R., Buckley, T. L., and Brain, S. D., Olvanil: more potent than capsaicin at stimulating the efferent function of sensory nerves, *Eur. J. Pharmacol.*, 219, 481, 1992.
71. Beresford, I. J. M., Birch, P. J., Balsat, M. C., Rogers, H., Fernandez, L., and Hagan, R. M., Effect of the spirolactam NK-1 receptor antagonist, GR82334, on neurokinin- and electrical stimulation-induced oedema in the rat, *Br. J. Pharmacol.*, 102, 360P, 1991.
72. Brain, S. D., Cambridge, H., Hughes, S. R., and Wilsoncroft, P., Evidence that calcitonin gene-related peptide (CGRP) contributes to inflammation in the skin and joint, *Proc. Ann. N.Y. Acad. Sci.*, 657, 412, 1992.
73. Gamse, R. and Saria, A., Antidromic vasodilatation in the rat hindpaw measured by laser Doppler flowmetry: pharmacological modulation, *J. Auton. Nerv. Syst.*, 19, 105, 1987.
74. Escott, K. J. and Brain, S. D., Effect of a calcitonin gene-related peptide antagonist ($CGRP_{8-37}$) on skin vasodilatation and oedema induced by stimulation of the rat saphenous nerve, *Br. J. Pharmacol.*, 109, 539, 1993.
75. Shepheard, S. L., Cook, D. A., Williamson, D. J., Hurley, C. J., Hill, R. G., and Hargreaves, R. J., Inhibition of neurogenic plasma extravasation but not vasodilation in the hind limb of the rat by the non-peptide NK-1 receptor antagonist RP 67580, *Br. J. Pharmacol.*, 107, 150P, 1992.
76. Richards, K. J., Cambridge, H., and Brain, S. D., Comparative effects of a neurokinin-1 (NK-1) receptor agonist, GR73632, and CRP on microvascular tone and permeability in rat skin, *Neuropeptides*, 24 (Abstr.), 206, 1993.
77. Buckley, T. L., Brain, S. D., Collins, P. D., and Williams, T. J., Inflammatory oedema induced by interactions between interleukin-1 and the neuropeptide calcitonin gene-related peptide, *J. Immunol.*, 146, 3424, 1991.
78. Brain, S. D. and Wiliams, T. J., Substance P regulates the vasodilator activity of calcitonin gene-related peptide, *Nature*, 335, 73, 1988.
79. Ralevic, V., Khalil, Z., Dusting, G. J., and Helme, R. D., Nitric oxide and sensory nerves are involved in the vasodilator response to acetylcholine but not calcitonin gene-related peptide in rat skin microvasculature, *Br. J. Pharmacol.*, 106, 650, 1992.
80. Hughes, S. R. and Brain, S. D., Nitric oxide-dependent release of vasodilator quantities of calcitonin gene-related peptide from capsaicin-sensitive nerves in rabbit skin, *Br. J. Pharmacol.*, 111, 425, 1994.
81. Hughes, S. R., Williams, T. J., and Brain, S. D., Evidence that endogenous nitric oxide modulates oedema formation induced by substance P, *Eur. J. Pharmacol.*, 191, 481, 1990.
82. Lippe, I. T., Stabentheiner, A., and Holzer, P., Participation of nitric oxide in the mustard oil-induced neurogenic inflammation of the rat paw skin, *Eur. J. Pharmacol.*, 232, 113, 1993.
83. Kajekar, R., Moore, P. K., and Brain, S. D., Modulation of neurogenic oedema formation by nitric oxide synthase inhibitors in the rat, *Br. J. Pharmacol.*, 112, 223P, 1994.
84. Barnes, P. J., Belvisi, M. G., and Rogers, D. F., Modulation of neurogenic inflammation, *Trends Pharmacol. Sci.*, 11, 185, 1990.
85. Bartho, L. and Szolcsányi, J., Opiate agonists inhibit neurogenic plasma extravasation in the rat, *Eur. J. Pharmacol.*, 72, 101, 1981.
86. Smith, T. W. and Buchan, P., Peripheral opioid receptors located on the rat saphenous nerve, *Neuropeptides*, 5, 217, 1984.
87. Lembeck, F. and Donnerer, J., Opioid control of the function of primary afferent substance P fibres, *Eur. J. Pharmacol.*, 114, 241, 1985.
88. Lembeck, F., Donnerer, J., and Bartho, L., Inhibition of neurogenic vasodilatation and plasma extravasation by substance P antagonists, somatostatin and [D-Met^2,Pro^5]enkephalinamide, *Eur. J. Pharmacol.*, 85, 171, 1982.

89. Barber, A., μ- and κ-opioid receptor agonists produce peripheral inhibition of neurogenic plasma extravasation in rat skin, *Eur. J. Pharmacol.*, 236, 113, 1993.
90. Xu, J.-X., Hao, J. X., Wiesenfeld-Hallin, Z., Hakanson, R., Folkers, H., and Hokfelt, T., Spantide II, a novel tachykinin antagonist, and galanin, inhibit plasma extravasation induced by antidromic C-fiber stimulation in rat hind paw, *Neurocience*, 42, 731, 1991.
91. Donnerer, J., Schuligoi, R., and Stein, C., Increased content and transport of substance P and calcitonin gene-related peptide in sensory nerves innervating inflamed tissue: evidence for a regulatory function of nerve growth factor in vivo, *Neuroscience*, 49, 693, 1992.
92. Maggi, c. A., Patacchini, R., Rovero, P., and Giachetti, A., Tachykinin receptors and tachykinin receptor antagonists, *J. Auton. Pharmacol.*, 13, 23, 1993.
93. Pincelli, C., Fantini, F., and Giannetti, A., Neuropeptides and skin inflammation, *Dermatology*, 187, 153, 1993.
94. Ansel, J. C., Brown, J. R., Payan, D. G., and Brown, M. A., Substance P selectively activates TNF-alpha gene expression in murine mast cells, *J. Immunol.*, 150, 4478, 1993.
95. Lotz, M., Vaughan, J. H., and Carson, D. A., Effect of substance P and substance K on the growth of cultured ketatinocytes, *J. Invest. Dermatol.*, 241, 1218, 1988.
96. Ruff, M. R., Wahl, S. M., and Pert, C. B., Substance P-mediated chemotaxis of human monocytes, *Peptides*, 6 (Suppl. 2), 107, 1985.
97. McGillis, J. P., Organist, M. L., and Payan, D. G., Substance P and immunoregulation, *Fed. Proc.*, 46, 196, 1987.
98. Scicchitano, R., Biennenstock, J., and Stanisz, A. M., In vivo immunomodulation by the neuropeptide substance P, *Immunology*, 63, 733, 1988.
99. Perianin, A., Synderman, R., and Malfroy, B., Substance P primes human neutrophil activation: a mechanism for neurological regulation of inflammation, *Biochem. Biophys. Res. Commun.*, 161, 520, 1989.
100. Iwamoto, I., Tomoe, S., Tomioka, H., and Yoshida, S., Leukotriene B_4 mediates substance P-induced granulocyte infiltration into mouse skin: comparison with antigen-induced granulocyte infiltration, *J. Immunol.*, 151, 2116, 1993.
101. Smith, C. H., Barker, J. N., Morris, R. W., MacDonald, D. M., and Lee, T. H., Neuropeptides induce rapid expression of endothelial cell adhesion molecules and elicit granuloyctic infiltration in human skin, *J. Immunol.*, 151, 3274, 1993.
102. Nilsson, J., von Euler, A. M., and Dalsgaard, C. J., Stimulation of connective tissue cell growth by substance P and substance K, *Nature*, 315, 61, 1985.
103. Tanaka, T., Danno, K., Ikai, K., and Imamura, S., Effect of substance P and substance K on the growth of cultured keratinocytes, *J. Invest. Dermatol.*, 90, 399, 1988.
104. Morbidelli, L., Maggi, C. A., and Ziche, M., Effect of selective tachykinin receptor antagonists on the growth of human skin fibroblasts, *Neuropeptides*, 24, 335, 1993.
105. Rabier, M. J., Farber, E. M., and Wilkinson, D. I., Neuropeptides modulate leukotriene B_4 mitogenicity towards cultured human keratinocytes, *J. Invest. Dermatol.*, 100, 132, 1993.
106. Sung, C. P., Arleth, A. J., Aiyar, N., Bhatnager, P. K., Lysko, P. G., and Feuerstein, G., CGRP stimulates the adhesion of leukocytes to vascular endothelial cells, *Peptides*, 13, 429, 1992.
107. Zimmerman, B. J., Anderson, D. C., and Granger, D. N., Neuropeptide promote neutrophil adherence to endothelial cell monolayers, *Am. J. Physiol.*, 263, G678, 1992.
108. Umeda, Y., Takamiya, M., Yoshizaki, H., and Arisawa, M., Inhibition of mitogen-stimulated T-lymphocytes proliferation by calcitonin gene-related peptide, *Biochem. Biophys. Res. Commun.*, 154, 227, 1992.
109. Foster, C. A., Mandak, B., Kromer, E., and Rot, A., Calcitonin gene-related peptide is chemotactic for human T lymphocytes, *Ann. N.Y. Acad. Sci.*, 657, 397, 1992.
110. Hosoi, J., Murphy, G. F., Egan, C. L., Lerner, E. A., Grabbe, S., Asahina, A., and Granstein, R. D., Regulation of Langerhans cell function by nerves containing calcitonin gene-related peptide, *Nature*, 363, 159, 1993.
111. Saria, A., Substance P in sensory nerve fibres contributes to the development of oedema in the rat hind paw after thermal injury, *Br. J. Pharmacol.*, 82, 217, 1984.
112. Lynn, G., Ye, W., and Costell, B., The actions of capsaicin applied topically to the skin of the rat on C-fibre afferents, antidromic vasodilatation and substance P levels, *Br. J. Pharmacol.*, 107, 400, 1992.
113. Larkin, S. W. and Williams, T. J., Evidence for sensory nerve involvement in cutaneous reactive hyperemia in humans, *Cir. Res.*, 73, 147, 1993.
114. Maggi, C. A., Borsini, F., Santiccioli, P., Geppetti, P., Abelei, L., Evangelista, S., Manzini, S., Theodorsson-Norheim, E., Somma, V., Amenta, F., Bacciarelli, C., and Meli, A., Cutaneous lesions in capsaicin-pretreated rats: a trophic role of capsaicin-sensitive afferents?, *Naunyn-Schmiedeberg's Arch. Pharmacol.*, 336, 538, 1987.
115. Thomas, D. R., Dubner, R., and Ruda, M. A., Neonatal capsaicin treatment in rats results in scratching behavior with skin damage: potential model of non-painful dysesthesia, *Neurosci. Lett.*, 171, 101, 1994.
116. Kishimoto, S., The regeneration of substance P containing nerve fibres in the process of wound healing in the guinea pig skin, *J. Invest. Dermatol.*, 1, 219, 1983.

117. Senepati, A., Anand, P., McGregor, G. P., Ghatei, M. A., Thompson, R. P. H., and Bloom, S. R., Depletion of neuropeptides during wound healing in rat skin, *Neurosci. Lett.*, 71, 101, 1986.
118. Kjartansson, J. and Dalsgaard, C. J., Calcitonin gene-related peptide increases survival of a musculo cutaneous flap in the rat, *Eur. J. Pharmacol.*, 142, 355, 1987.
119. Karanth, S. S., Dhital, S., Springall, D. R., and Polak, J. M., Reinnervation and neuropeptides in mouse skin flaps, *J. Auton. Nerv. Syst.*, 31, 127, 1990.
120. Bunker, C. B., Foreman, J., and Dowd, P. M., Digital cutaneous vascular responses to histamine, compound 48/80 and neuropeptides in normal subjects and Raynaud's phenomenon, *J. Invest. Dermatol.*, 92(Abstr.), 409, 1989.
121. Shawket, S., Dickerson, C., Hazleman, B., and Brown, M. J., Selective suprasensitivity to calcitonin gene-related peptide in the hands in Raynaud's phenomenon, *Lancet*, 2, 1354, 1989.
122. Shawket, S. and Brown, M. W., Pathogenetic and therapeutic implications of the calcitonin gene-related peptide in the cardiovascular system, *Trends Cardiovasc. Med.*, 1, 211, 1991.
123. Terenghi, G., Bunker, C. B., Lium, Y. F., Springall, D. R., Cowen, T., Dowd, P. M., and Polak, J. M., Image analysis quantification of peptide-immunoreactive nerves in the skin of patients with Raynaud's phenomenon and systemic sclerosis, *J. Pathol.*, 164, 245, 1991.
124. Bunker, C. B., Reavley, C., O'Shaughnessy, D. J., and Dowd, P. M., Calcitonin gene-related peptide in treatment of severe peripheral vascular insufficiency in Raynaud's phenomenon, *Lancet*, 342, 80, 1993.
125. Shawket, S., Dickerson, C., Hazleman, B., and Brown, M. J., Prolonged effect of CGRP in Raynaud's patients: a double-blind randomised comparison with prostacyclin, *Br. J. Clin. Pharmacol.*, 32, 209, 1991.
126. Wallengren, J., Ekman, R., and Moller, H., Substance P and vasoactive intestinal peptide in bullous and inflammatory skin disease, *Acta Derm. Venereol.*, 66, 23, 1986.
127. Abadia Molina, F., Burrows, N. P., Jones, R. R., Terenghi, G., and Polak, J. M., Increased sensory neuropeptides in nodular prurigo: a quantitative immunohistochemical analysis, *Br. J. Dermatol.*, 127, 344, 1992.
128. Giannettie, A., Fantini, F., Cimitan, A., and Pincelli, C., Vasoactive intestinal polypeptide and substance P in the pathogenesis of atopic dermatitis, *Acta Derm. Venereol.*, 176, 90, 1992.
129. Tobin, D., Nabarro, G., Baart de la Faille, H., van Vloten, W. A., van der Putte, S. C., and Schuurman, H. J., Increased number of immunoreactive nerve fibers in atopic dermatitis, *J. Allergy Clin. Immunol.*, 90, 613, 1992.
130. Husz, S., Toth-Kasa, I., Obal, F., and Jansso, G., A possible pathomechanism of the idiopathetic cold contact urticaria, *Acta Physiol. Hung.*, 77, 209, 1991.
131. Farber, E. M., Nickologg, B. J., Recht, B., and Fraki, J. E., Stress, symmetry and psoriasis: possible role of neuropeptides, *J. Am. Acad. Dermatol.*, 14, 305, 1986.
132. Harvima, I. T., Viinamaki, H., Naukkarinen, A., Paukkonen, K., Neittaanmaki, H., Harvima, R. J., and Horsmanheimo, M., Association of cutaneous mast cells and sensory nerves with psychic stress in psoriasis, *Psychother. Psychosom.*, 60, 168, 1993.
133. Anand, P., Springall, D. R., Blank, M. A., Sellu, D., Polack, J. M., and Bloom, S. R., Neuropeptides in skin disease: increased VIP in eczema and psoriasis but not axillary hyperhidrosis, *Br. J. Dermatol.*, 124, 547, 1991.
134. Eedy, D. J., Johnston, C. F., Shaw, C., and Buchanan, K. D., Neuropeptides in psoriasis: an immunocytochemical and radioimmunoassay study, *J. Invest. Dermatol.*, 96, 434, 1991.
135. Bernstein, J. E., Parish, L. C., Rappaport, M., Rosenbaum, M. M., and Roenigk, H. H., Effects of topically applied capsaicin on moderate and severe psoriasis vulgaris, *J. Am. Acad. Dermatol.*, 15, 504, 1986.
136. Ellis, C. N., Berberian, B., Sulica, V. I., Dodd, W. A., Jarratt, M. T., Katz, H. I., Prawer, S., Krueger, G., Rex, I. H., and Wolf, J. E., A double-blind evaluation of topical capsaicin in pruritic psoriasis, *J. Am. Acad. Dermatol.*, 29, 438, 1993.
137. Johansson, O., Hilliges, M., Talme, T., Marcusson, J. A., and Wetterberg, L., Somatostatin immunoreactive cells in lesional psoriatic human skin during peptide T treatment, *Acta Derm. Venereol.*, 74, 106, 1994.
138. Lundblad, L., Lundberg, J. M., Anggard, A., and Zetterstom, O., Capsaicin pretreatment inhibits the flare component of the cutaneous allergic reaction in man, *Eur. J. Pharmacol.*, 113, 461, 1985.
139. Wallengren, J. and Moller, H., The effect of capsaicin on some experimental inflammations in human skin, *Acta Derm. Venereol.*, 66, 375, 1986.
140. Simone, D. A. and Ochoa, J., Early and late effects of prolonged topical capsaicin on cutaneous sensibility and neurogenic vasodilatation in humans, *Pain*, 47, 285, 1991.
141. Berstein, J. E., Capsaicin in dermatologic disease, *Semin. Dermatol.*, 7, 304, 1988.
142. Ross, D. R. and Varipapa, R. J., Treatment of painful diabetic neuropathy with topical capsaicin, *Semin. Dermatol.*, 321, 474, 1989.

Chapter 19

INTERACTION OF THE AFFERENT AND THE ENDOCRINE SYSTEMS

Rainer Amann

CONTENTS

I. INTRODUCTION

Communication between different specialized cells of a living organism is essential for survival. The task of transmitting information is performed by three major networks, the immune system, the nervous system, and the endocrine system. These work in concert to achieve the best possible adaptation of the organism to changing environmental influences. This chapter focuses on the afferent nervous system, its role as input for endocrine adaptive responses, and the possible paracrine function of part of the afferent system.

A. THE NERVOUS AND ENDOCRINE SYSTEMS

Neurons typically serve as point-to-point connections; electrical transmission along the nerve fiber results in release of transmitter(s) at the nerve ending. Secretory cells of the endocrine system, in contrast, secrete hormones into the circulation and thus regulate the function of receptive tissues. Both systems use different methods to guarantee that information reaches the addressee: every organ that has the appropriate receptor will respond to alterations in circulating hormone concentration, while in the nervous system both receptor distribution and anatomic connection determine the response.

Nevertheless, there is a remarkable similarity between these two systems. Neurons, as well as endocrine cells, activate their targets through secretion of mediators, which act via specific receptors; both cell types can generate electric potentials. Furthermore, several mediators that are used by endocrine cells also serve as neurotransmitters. Later in this chapter, it will be argued that the rather imprecise distinction between endocrine system and nervous system becomes even more difficult when considering the function of small-diameter primary afferent neurons.

B. INTERACTION BETWEEN THE AFFERENT NERVOUS AND THE ENDOCRINE SYSTEMS

The most obvious way in which the afferent nervous system interacts with the endocrine system is by providing the input for integrative centers within the central nervous system

0-8493-7646-7/96/$0.00+$.50

(CNS). In addition, there are reports which suggest that afferent neurons may influence hormone secretion also at the peripheral level by regulating the sensitivity of secretory cells for the central stimulus.

Apart from serving as an input pathway, part of the afferent nervous system has the capacity to secrete bioactive peptides from peripheral nerve branches. These peptides can spread from the site of release, and thus can influence tissue function in a paracrine manner. This suggests that locally secreted neuropeptides act together with hormones in regulating organ function.

II. AFFERENT NEURONS AS INPUT

Afferent neurons provide the input for a variety of adaptive endocrine responses. Apart from their well-known function as baroreceptors, they possibly also participate in monitoring plasma osmolarity[1] and glucose concentration.[2,3] In rodents, afferent neurons also seem involved in the hormonal regulation of reproductive function.[4,5] Several reviews on this topic with special reference to the participation of capsaicin-sensitive afferent neurons in neuroendocrine mechanisms have been published.[6-8] This chapter focuses therefore only on one aspect of this input function of primary afferents, on their role in mediating activation of the hypothalamic-pituitary-adrenal axis after noxious stimulation.

A. ADRENOCORTICOTROPHIC HORMONE (ACTH) RELEASE

Any change in environmental conditions that is perceived as potentially harmful will trigger a uniform stress response, characterized by activation of the hypothalamic-pituitary-adrenal axis. Naturally, tissue injury, and subsequent activation of afferents — perceived as pain — is a potent stimulant for the stress response. Corticotropin-releasing hormone (CRF), synthesized in neurons of the paraventricular nucleus of the hypothalamus, is secreted into the pituitary portal blood, and stimulates adrenocorticotropin (ACTH) release from the anterior lobe of the pituitary gland. ACTH, in turn, stimulates synthesis and release of glucocorticoids from the adrenal gland.

Afferent neurons that mediate pain sensations are C- and Aδ-afferent neurons, which enter the spinal cord, bifurcate, ascend and descend for one to three segments, and synapse on dorsal horn neurons. Second-order neurons, in turn, project via ascending pathways to thalamic and brainstem areas. The ventral hypothalamus receives input from these areas via the ventral tegmentum and medial forebrain bundle.[9]

It is known that in humans, "pain" is no uniform experience. As early as the beginning of this century, the suggestion was made that two groups of sensations could be separated: protopathic, C-fiber mediated, and epicritic sensations, mediated by myelinated afferents.[10] Aδ-fiber activation is associated with sharp, pricking pain, while unmyelinated C-fibers are thought to mediate long-lasting burning pain, characterized by lack of precise spatial and stimulus intensity discrimination.[11,12] An example of this type of sensation is pain originating from the viscera, which are mainly innervated by C-fiber afferents.[12]

Treatment of newborn rats with capsaicin causes life-long degeneration of a large part of C-fiber afferent neurons.[13] In most afferent nerves, capsaicin causes an about 50% reduction in the number of unmyelinated fibers, while moderately, if at all, affecting the number of Aδ-fibers.[14] Although capsaicin treatment results in loss of chemonociception, there is conflicting evidence as to whether mechanical or thermal nociception is affected. Possible reasons for the lack of consistent effects of capsaicin treatment on mechanical and thermal nociception have been discussed in detail by Holzer.[14]

In contrast to the moderate effect of capsaicin treatment on acute (nonchemical) nociceptive responses is the observation that in capsaicin-treated rats plasma ACTH fails to increase following somatosensory stimuli (e.g., surgery, electrical stimulation of the sciatic nerve). The absence of ACTH response was confined to somatosensory stimuli, whereas the response to CRF injection or restraint stress was not different from control rats.[15-17]

Obviously, there are several possible causes for the absence of an ACTH response to noxious stimuli in capsaicin-denervated rats, including possible CNS effects of capsaicin treatment (see later). One particularly intriguing speculation would be that the lack of C-fiber-mediated protopathic afferent input in capsaicin-denervated rats was causally related to the lack of ACTH response.

If the destruction of a subset of primary afferent neurons with capsaicin was preventing an ACTH response to noxious stimulation, it would be feasible to mimic the capsaicin effect by pharmacological blockade of their transmitters and thus eventually produce a type of "analgesia" that blocks that part of sensory input which triggers the hypothalamo-pituitary-adrenal axis. Current knowledge suggests that glutamate, several neurokinins, and calcitonin gene-related peptide (CGRP) are stored in capsaicin-sensitive afferents, and released at the spinal terminals following noxious stimulation (see Reference 18). However, at present, it is not known whether antagonists at the respective receptors can produce this type of "analgesia".

Several stimuli have so far been shown to cause an ACTH response by a "capsaicin-sensitive mechanism", i.e., they have been found to be less effective or not effective in rats treated with a neurotoxic dose of capsaicin. Interestingly, conditions that are usually not thought to be associated with afferent neuron activation, such as morphine withdrawal,[16] or, more recently, administration of interleukin-1β or prostaglandin E_2,[19] have been found to be significantly less effective to increase plasma ACTH in capsaicin-treated rats. The latter seems particularly intriguing, because it suggests that afferent neurons participate in the interleukin 1β-induced ACTH response, thus establishing a link between the immune and endocrine systems.

However, these assumptions depend to a large extent on the selectivity of capsaicin for primary afferent neurons. Only if the neurotoxicity of capsaicin is restricted to peripheral afferents, can conclusions, as discussed above, be drawn from experiments in capsaicin-treated animals. Given the importance of this issue, the arguments for and against capsaicin selectivity will be discussed separately later in this chapter.

III. PERIPHERAL MODULATION OF ENDOCRINE FUNCTION BY AFFERENT NEURONS

Most endocrine glands are innervated by capsaicin-sensitive primary afferent neurons, which have the capability to secrete neuropeptides from their peripheral endings.[20] Although little is known about the circumstances that lead to release of these peptides under physiological conditions, an increased local concentration of these peptides may have an effect on glandular function.

In the rat, the effect of the neurokinin substance P (SP) on the responsiveness of the adrenal medulla to acetylcholine has been thoroughly investigated. The rat adrenal medulla receives not only autonomic efferent innervation via preganglionic cholinergic fibers but also SP-containing afferent fibers. Combination of SP immunohistochemistry and retrograde tracing with Fast Blue injected into the adrenal medulla of the rat have shown that SP-containing sensory neurons in the dorsal root ganglia provide an ipsilateral innervation of the adrenal medulla.[21]

The release of acetylcholine brings about secretion of catecholamines by acting on nicotinic and muscarinic receptors on the adrenal chromaffin cells. *In vitro* experiments have shown that SP does not affect muscarine-induced catecholamine release, while it interferes with the effect of nicotine. The latter action was dose-dependent: at low concentrations (10^{-9} *M*), SP changed the time course of nicotine-induced catecholamine secretion by enhancing catecholamine secretion initially and inhibiting catecholamine secretion thereafter. At a higher concentration (10^{-5} *M*), SP inhibited total nicotinic catecholamine secretion.[22]

Experiments in conscious rats also indicate that capsaicin-sensitive mechanisms participate in adrenal catecholamine release. Cold stress, swimming stress, but not hypovolemic or immobilization stress, were found to be less efficient to increase plasma catecholamine concentration in capsaicin-treated rats than in controls. This deficit in capsaicin-treated rats may be explained by reduced local availability of SP in the adrenal medulla.[23]

IV. CAPSAICIN AS A TOOL TO DETERMINE THE ROLE OF AFFERENT NEURONS IN PHYSIOLOGICAL RESPONSES

The pharmacology of capsaicin has been reviewed in detail before;[14] therefore, the following will only deal with the neurotoxic effect of high-dose capsaicin treatment. In newborn rats, systemic administration of capsaicin leads to a life-long degeneration of small-diameter primary afferent neurons.[13] This property of capsaicin has been considered useful to determine the physiological function of these afferents, assuming that deficits that can be determined in capsaicin-treated rats can be ascribed to the loss of a subpopulation of neurons. However, selective neurotoxicity of capsaicin is essential for this line of argument.

There is evidence that treatment of rats in the early postnatal period causes permanent destruction of a subset of primary afferent neurons, which store neurokinins, CGRP, and several other neuropeptides (see Reference 14). Degenerating nerve terminals have been observed in the dorsal spinal cord and in the brainstem, and this was ascribed to capsaicin-induced degeneration of primary afferent nerve terminals.[13,24-26] Consequently the concentration of these neuropeptides is markedly decreased in peripheral tissues and substantia gelatinosa of capsaicin-treated rats.

Most studies obtained no evidence for a neurotoxic effect of capsaicin outside the primary afferent system: in areas that do no receive primary afferent projections (ventral spinal cord, CNS above the brainstem) SP is not reduced by capsaicin treatment.[27] Furthermore, in the CNS, capsaicin treatment of newborn rats has no effect on tissue levels of several neurotransmitters, which are not associated with primary afferent neurons.[28-30] Therefore, there was reason to assume that capsaicin selectively destroys primary afferent neurons. There were only few known examples of a nonselective action of capsaicin: Panerai et al. found that after capsaicin treatment, β-endorphin was permanently depleted from the hypothalamus;[31] Szolcsányi et al. showed that capsaicin causes ultrastructural changes in neurons of the preoptic area, which are thought to be involved in thermoregulation.[32] However, this effect was observed only when capsaicin was administered to adult rats.[30]

Since then, work by Ritter and Dinh[33-36] has raised doubts about capsaicin selectivity for the primary afferent system. Using a cupric silver stain to identify degenerating neurons, Ritter and Dinh showed that administration of capsaicin to 10-d-old rats causes widespread staining in brain areas that are not associated with primary afferent function.[33] Electron microscopy confirmed the presence of degenerating neuron terminals at sites where capsaicin-induced argyrophilia was observed.[34] Since a second treatment of rats with capsaicin 10 d or 3 months later failed to produce degeneration, the authors concluded that the first treatment had produced permanent destruction of susceptible neurons.[35]

The use of capsaicin treatment in adult rats in order to investigate the physiological function of small-diameter primary afferents is complicated by several factors. First, also after treatment of adult rats, degeneration has been found of neurons in several brain areas that are not associated with primary afferents.[33,36] Thus, this form of treatment offers no advantage over neonatal treatment as far as possible nonselective CNS actions are concerned. Second, the acute effect of capsaicin treatment may severely influence behavior and endocrine function for several days. Treatment of rats with capsaicin under ether anesthesia has been shown to result in elevated ACTH plasma concentration several days after treatment.[37]

Other effects of capsaicin treatment, related or not to neurotoxicity, such as reduction of brown adipose tissue,[38,39] or effects on adrenal glandular cell function,[40] may lead to long-term consequences in metabolism and homeostasis, further complicating the interpretation of results obtained from capsaicin-treated animals.

This is the case especially when capsaicin is used as a probe to determine the influence of capsaicin-sensitive afferents in complex behavior. Nevertheless, when evaluating the results obtained with capsaicin treatment in the investigation of ACTH regulation, the case for selectivity of capsaicin seems quite strong. It has been shown that in capsaicin-denervated rats, secretory capabilities of pituitary corticotroph cells, per se, are not reduced,[15,19] and basal ACTH plasma concentration was within the normal range. The fact that capsaicin-treated rats respond normally to emotional stress, e.g., restraint stress,[37] open field exposure,[16] or cage switch stress[19] provides additional indication that there is no profound capsaicin-induced change in hypothalamic-pituitary function.

In conclusion, the use of capsaicin as a tool for investigating the physiological role of small-diameter primary afferent neurons has brought considerable advances in knowledge. In the absence of better tools, capsaicin treatment is still regarded as a valid method to obtain an indication for afferent neuron involvement in physiological responses. However, in view of evidence for nonselective effects of capsaicin, the interpretation of results obtained with capsaicin treatment requires greater caution than was anticipated some years ago.

V. PARACRINE FUNCTION OF CAPSAICIN-SENSITIVE AFFERENT NEURONS?

Nervous transmission typically represents a fast method of point-to-point transmission of information, exemplified by the motor system. Electrical impulses that travel down an axon cause release of chemical mediators, which then act on a specialized postjunctional site via receptor-operated mechanisms.

Capsaicin-sensitive primary afferent neurons, on the other hand, are less suited for precise point-to-point transmission. The peripheral nerve endings branch out; neuropeptides are stored in vesicles within boutons along the course of these branches. Once secreted, the neuropeptides may diffuse quite a distance to act on their targets. Breakdown by peptidases limits their action.[20]

The morphological arrangement of transmitter storage along nerve fibers resembles that found in sympathetic efferents, where action potentials generated at the central endings pass the storage sites and initiate the secretion process. In contrast, neuropeptides of capsaicin-sensitive afferent neurons are thought to be liberated as a consequence of peripherally generated nerve activity. They enter the local circulation where detectable levels are found also in the absence of C-fiber stimulation (see, for example, References 41 to 43), suggesting basal mediator release from afferent neurons. In most organs, released neuropeptides cause change of smooth muscle tone and vascular permeability. This "local effector function"[20] of primary afferents, therefore, may have profound influence on organ function.

The attempt to answer the question about the adequate stimulus which causes activation of capsaicin-sensitive afferents leads to a more speculative approach. There are reports which suggest that certain stimuli that potently excite C-polymodal nociceptors cause no, or only very moderate, local mediator release from afferents. In the rat skin, mechanical stimulation, sufficient to excite most C-polymodal nociceptors, did not cause neurogenic vasodilation, although antidromic electrical stimulation of the cut saphenous nerve at C-fiber strength (but not when only exciting Aδ- and Aα,β-fibers) was clearly effective.[44] Furthermore, Treede[45] obtained evidence in human skin that activation of C-polymodal nociceptors with painful heat stimuli produces only a weak local flare reaction, which should typically follow local neuropeptide release. This was in contrast to the pronounced flare response that was obtained by histamine injections. Therefore, C-fiber afferents that convey mechanical and thermal nociceptive input have obviously only moderate local secretory capabilities. A similar situation seems to exist in many tissues with regard to bradykinin-sensitive C-fiber afferents. Although bradykinin very effectively causes excitation of C-fiber afferents, it fails to produce local neuropeptide release in some preparations.[46,47]

This raises the question as to the adequate stimulus which excites those fibers that *have* the capacity for local neuropeptide secretion. In the absence of a conclusive answer to that question, it is tempting to speculate that these afferents are more involved in regulation of local tissue function than in the transmission of afferent information.

Local neuropeptide secretion from afferents may be controlled by the rate of synthesis in the cell body. Tissue injury, for example, leads to increased neuropeptide synthesis via retrograde axonal transport of nerve growth factor, and thus probably results in elevated basal release of neuropeptides from afferent terminals.[48] On the other hand it seems conceivable that local neuropeptide secretion is not necessarily dependent on electrical excitation.[49] Thus, alteration in the local environment may increase neuropeptide secretion without activating centripetal impulse transmission at all. According to this speculation, this group of C-fibers may be regarded as a paracrine system, which serves the maintenance of organ function.

ACKNOWLEDGMENTS

The author would like to thank J. Donnerer for helpful discussions and valuable suggestions. Work done in the author's laboratory was supported by the *Fonds zur Förderung der wissenschaftlichen Forschung* (Grants P7884-M, P7676-M, and P09823-M).

REFERENCES

1. Stoppini, L., Bariy, F., Mathison, R., and Baertschi, J., Spinal substance P transmits bradykinin but not osmotic stimuli from hepatic portal vein, *Neuroscience*, 11, 903, 1984.
2. Amann, R. and Lembeck, F., Capsaicin sensitive afferent neurons from peripheral glucose receptors mediate the insulin-induced increase in adrenaline secretion, *Naunyn-Schmiedeberg's Arch. Pharmacol.*, 334, 71, 1986.
3. Donnerer, J., Reflex activation of the adrenal medulla during hypoglycemia and circulatory dysregulations is regulated by capsaicin-sensitive afferents, *Naunyn-Schmiedeberg's Arch. Pharmacol.*, 338, 282, 1988.
4. Traurig, H.H., Papka, R.E., Saria, A., and Lembeck, F., Substance P immunoreactivity in rat mammary nipple and the effect of capsaicin on lactation, *Naunyn-Schmiedeberg's Arch. Pharmacol.*, 328, 1, 1984.
5. Traurig, H.H., Papka, R.E., and Rush, M.E., Role of capsaicin-sensitive afferent nerves in the neuroendocrine copulation reflex, in *Substance P and Neurokinins*, Henry, J.L., Couture, R., Cuello, A.C., Petellier, G., Quirion, R., and Regoli, D., Eds., Springer, New York, 1987, 226.
6. Lembeck, F., The 1988 Ulf von Euler Lecture. Substance P: from extract to excitement, *Acta Physiol. Scand.*, 133, 435, 1988.
7. Donnerer, J., Amann, R., Skofitsch, G., and Lembeck, F., Substance P afferents regulate ACTH-corticosterone release, *Ann. N.Y. Acad. Sci.*, 632, 296, 1991.
8. Lembeck, F. and Donnerer, J., The wide spectrum of functions of capsaicin-sensitive afferents, in *Advances in Pain Research and Therapy*, Vol. 20, Sicuteri, F., Ed., Raven Press, New York, 1992, 45.
9. Gann, S., Dallman, M.F., and Engeland, W.C., Reflex control and modulation of ACTH and corticosteroids, in *Endocrine Physiology*, Vol. 3, McGann, S.M., Ed., University Park Press, Baltimore, 1981, 535.
10. Ranson, S.W., Unmyelinated nerve fibers as mediators of protopathic sensation, *Brain*, 38, 381, 1915.
11. Jessel, T.M. and Kelly, D.D., Pain and analgesia, in *Principles of Neural Science*, 3rd ed., Kandel, E.R., Schwartz J.H., and Jessel, T.M., Eds., Appelton and Lange, Norwalk, CT, 1991, 385.
12. Cervero, F., Visceral nociception: peripheral and central aspects of visceral nociceptive systems, *Philos. Trans. R. Soc. London Ser. B*, 308, 325, 1985.
13. Jancsó, G., Király, E., and Jancsó-Gábor, A., Pharmacologically induced selective degeneration of chemosensitive primary sensory neurones, *Nature*, 270, 741, 1977.
14. Holzer, P., Capsaicin: cellular targets, mechanism of action, and selectivity for thin sensory neurons, *Pharmacol. Rev.*, 43, 143, 1991.
15. Amann, R. and Lembeck F., Stress-induced ACTH release in capsaicin treated rats, *Br. J. Pharmacol.*, 90, 727, 1987.
16. Donnerer, J. and Lembeck, F., Neonatal capsaicin treatment of rats reduces ACTH secretion in response to peripheral neuronal stimuli but not to centrally acting stressors, *Br. J. Pharmacol.*, 94, 647, 1988.

17. Donnerer, J. and Lembeck, F., Different control of the adrenocorticotropin–corticosterone response and of prolactin secretion during cold stress, anaesthesia, surgery, and nicotine injection in the rat: involvement of capsaicin-sensitive sensory neurons, *Endocrinology*, 126, 921, 1990.
18. Otsuka, M. and Yoshioka, K., Neurotransmitter function of mammalian tachykinins, *Physiol. Rev.*, 73, 229, 1993.
19. Watanabe, T., Morimoto, A., Tan, N., Makisumi, T., Shimada, S.G., Nakamori, T., and Murakami, N., ACTH response induced in capsaicin- desensitized rats by intravenous injection of interleukin-1 or prostaglandin E, *J. Physiol. (London)*, 475, 139, 1994.
20. Holzer, P., Local effector functions of capsaicin-sensitive sensory nerve endings: involvement of tachykinins, calcitonin gene-related peptide and other neuropeptides, *Neuroscience*, 24, 739, 1988.
21. Zhou, X.F., Oldfield, B.J., and Livett, B.G., Substance-P-containing sensory neurons in the rat dorsal-root ganglia innervate the adrenal-medulla, *J. Auton. Nerv. Syst.*, 33, 247, 1991.
22. Zhou, X.F., Marley, P.D., and Livett, B.G., Substance-P modulates the time course of nicotinic but not muscarinic catecholamine secretion from perfused adrenal-glands of rat, *Br. J. Pharmacol.*, 104, 159, 1991.
23. Zhou, X.F. and Livett, B.G., Capsaicin-sensitive sensory neurons are involved in the plasma-catecholamine response of rats to selective stressors, *J. Physiol. (London)*, 433, 393, 1991.
24. Jancsó, G., Selective degeneration of chemosensitive primary sensory neurones induced by capsaicin: glial changes, *Cell Tissue Res.*, 195, 145, 1978.
25. Jancsó, G. and Király, E., Distribution of chemosensitive primary sensory afferents in the central nervous system of the rat, *J. Comp. Neurol.*, 190, 781, 1980.
26. Jancsó, G., and Király, E., Sensory neurotoxins: chemically induced selective destruction of primary sensory neurons, *Brain Res.*, 210, 83, 1981.
27. Gamse, R., Leeman, S.E., Holzer, P., and Lembeck, F., Differential effects of capsaicin on the content of somatostatin, substance P, and neurotensin in the nervous system of the rat, *Naunyn-Schmiedeberg's Arch. Pharmacol.*, 317, 140, 1981.
28. Jancsó, G., Hökfelt, T., Lundberg, J.M., Király, E., Halász, N., Nilsson, G., Terenius, L., Rehfeld, J., Steinbusch, H., Verhofstad, A.E.R., Said, S., and Brown, M., Immunohistochemical studies on the effect of capsaicin on spinal and medullary peptide and monoamine neurons using antisera to substance P, gastrin/CCK, somatostatin, VIP, enkephalin, neurotensin, and 5-hydroxytryptamine, *J. Neurocytol.*, 10, 963, 1981.
29. Singer, E.A., Sperk, G., and Schmid, R., Capsaicin does not change tissue levels of glutamic acid, its uptake, or release in the spinal cord, *J. Neurochem.*, 38, 1383, 1982.
30. Hajós, M., Svensson, K., Nissbrandt, H., Obál, F., and Carlsson, A., Effects of capsaicin on central monoaminergic mechanisms in the rat, *J. Neural Transm.*, 66, 221, 1986.
31. Panerai, A.E., Martini, A., Locatelli, V., and Mantegazza, P., Capsaicin decreases b-endorphin hypothalamic concentrations in the rat, *Pharmacol. Res. Commun.*, 15, 825, 1983.
32. Szolcsányi, J., Joó, F., and Jancsó-Gábor, A., Mitochondrial changes in preoptic neurones after capsaicin desensitization of the hypothalamic thermodetectors in rats, *Nature*, 229, 116, 1971.
33. Ritter, S. and Dinh, T.T., Age-related changes in capsaicin-induced degeneration in rat brain, *J. Comp. Neurol.*, 318, 103, 1992.
34. Dinh, T.T. and Ritter, S., Electron microscopic evidence of capsaicin-induced degeneration in rat brain and retina, *Soc. Neurosci. Abstr.*, 17, 113,1991.
35. Ritter, S. and Dinh, T.T., Capsaicin-induced neuronal degeneration in the brain and retina of preweanling rats, *J. Comp. Neurol.*, 296, 447, 1990.
36. Ritter, S. and Dinh, T.T., Capsaicin-induced neuronal degeneration: silver impregnation of cell bodies, axons, and terminals in the central nervous system of the adult rat, *J. Comp. Neurol.*, 271, 79, 1988.
37. Lembeck, F. and Amann, R., The influence of capsaicin-sensitive neurons on stress-induced release of ACTH, *Brain Res. Bull.*, 16, 541, 1986.
38. Cui, J.Y. and Himmshagen, J., Rapid but transient atrophy of brown adipose-tissue in capsaicin-desensitized rats, *Am. J. Physiol.*, 262, R562, 1992.
39. Cui, J.Y. and Himmshagen, J., Long-term decrease in body-fat and in brown adipose-tissue in capsaicin-desensitized rats, *Am. J. Physiol.*, 262, R568, 1992.
40. Peltohuikko, M., Dagerlind, A., Ceccatelli, S., and Hökfelt, T., The immediate-early genes c-fos and c-jun are differentially expressed in the rat adrenal-gland after capsaicin treatment, *Neurosci. Lett.*, 126, 163, 1991.
41. Saria, A., Martling, C.-R., Yan, Z., Theodorsson-Norheim, E., Gamse, R., and Lundberg, J.M., Release of multiple tachykinins from capsaicin-sensitive sensory nerves in the lung by bradykinin, histamine, dimethylphenyl piperazinum, and vagal nerve stimulation, *Am. Rev. Respir. Dis.*, 137, 1330, 1988.
42. Amann, R., Donnerer, J., and Lembeck, F., Capsaicin-induced stimulation of polymodal nociceptors is antagonized by ruthenium red independently of extracellular calcium, *Neuroscience*, 32, 255, 1989.
43. Holzer, P., Peskar, B.M., Peskar, B.A., and Amann, R., Release of calcitonin gene-related peptide induced by capsaicin in the vascularly perfused rat stomach, *Neurosci. Lett.*, 108, 195, 1990.
44. Lynn, B. and Cotsell, B., Blood flow increases in the skin of the anaesthetized rat that follow antidromic nerve stimulation and strong mechanical stimulation, *Neurosci. Lett.*, 137, 249, 1992.

45. Treede, R.-D., Vasodilator flare due to activation of superficial cutaneous afferents in humans: heat-sensitive versus histamine-sensitive fibers, *Neurosci. Lett.*, 141, 169, 1992.
46. Hua, X.-Y., Saria, A., Gamse, R., Theodorsson-Norheim, E., Brodin, E., and Lundberg, J.M., Capsaicin-induced release of multiple tachykinins (substance P, neurokinin A, eledoisin-like material) from the guinea-pig spinal cord and ureter, *Neuroscience*, 19, 313, 1986.
47. Andreeva, L. and Rang, H.P., Effect of bradykinin and prostaglandins on the release of calcitonin gene-related peptide-like immunoreactivity from the rat spinal cord in vitro, *Br. J. Pharmacol.*, 108, 185, 1993.
48. Donnerer, J., Schuligoi, R., and Stein, C., Increased content and transport of substance P and calcitonin gene-related peptide in sensory nerves innervating inflamed tissue: evidence for a regulatory function of nerve growth factor in vivo, *Neuroscience*, 49, 693, 1992.
49. Donnerer, J. and Amann, R., Capsaicin-evoked neuropeptide release is not dependent on membrane potential changes, *Neurosci. Lett.*, 117, 331, 1990.

Chapter 20

SENSORY NEUROPEPTIDES: MITOGENIC AND TROPHIC FUNCTIONS

Marina Ziche

CONTENTS

I. INTRODUCTION

Tissue homeostasis is the end result of complex physiological processes, a major aspect of which is the control of cell replication. Under physiological conditions the cells in a living organism exist in a dynamic equilibrium. In most tissues cell turnover is slow, with the majority of the cells viable and metabolically active but in a nonproliferative state. Many cells, however, retain the capacity to respond by division to extracellular signals such as hormones, antigens, and growth factors. Neuropeptides are increasingly recognized to act as cellular growth factors and their mechanisms of action are attracting considerable attention.[1]

The evidence that peripheral nerve lesions are paralleled by alterations in tissue growth and repair has been known for a long time.[2] In recent years this concept has received much support by the observations that: (1) the primary sensory neurons, which are selectively susceptible to the action of capsaicin[3-5] exert an "efferent" function in the periphery, mediated by the release of several peptides, including substance P (SP);[6-9] (2) following chemical deafferentation by capsaicin, experimental animals exhibit "spontaneous" lesions at cutaneous and corneal levels, trophic disorders of skin and cutaneous annexa, and decreased survival and regrowth of surgical cutaneous flaps;[10-12] (3) at a cellular level, tachykinins (TK) and particularly SP mediate a variety of responses relevant for tissue repair in response to noxious stimuli.[7,13,14] The work from different laboratories and from our group has strengthened the evidence that TK contribute to the maintainance of tissue trophism by exerting a direct action on distinct cellular types *in vitro* and *in vivo*.

II. GROWTH-PROMOTING EFFECT OF TACHYKININS

The repair of damaged vertebrate tissues requires the concerted action of complex systems and cellular processes[15,16] involving, in a distinct but integrated program, the activation of

0-8493-7646-7/96/$0.00+$.50

inflammatory cells and products and the growth and migration of connective tissue cells and capillaries into a "primed" extracellular matrix. The evidence that TK exert a proliferative effect will be discussed for the different cell types.

A. CONNECTIVE TISSUE CELLS

The skin receives a rich supply of primary sensory neurons, which are particularly abundant in SP and predominantly localized around dermal blood vessels.[17] Replication and migration of fibroblasts are necessary to favor healing of skin wounds *in vivo*.[18] Using a stabilized culture of human skin fibroblasts (HF) from healthy volunteers, it is possible to reproduce *in vitro* this condition by seeding sparse cells in the culture dishes and by depriving the cells of their optimal growth substratum, i.e., serum. HF actively proliferated when exposed to increasing concentrations of SP.[19] The cell number increased significantly after 48 h exposure to picomolar and nanomolar concentrations of the peptide. The extent of the proliferative response evoked by the maximal effective dose of SP, could be compared to that of a potent mitogen like basic fibroblast growth factor (bFGF) (Figure 1, panel A). The growth-promoting effect of SP on skin fibroblasts exibited a peculiar shape of the dose-response curve. At increasing concentrations of the peptide there was an apparent reduction in the extent of the proliferation. This effect occurred with no sign of impairement of cell viability or vitality. One possible interpretation, as shown for other growth-promoting agents,[20] is that, at higher concentrations of SP, other cell functions can be preferentially stimulated, like cytoskeleton reorganization and cell differentiation. Indeed, micromolar concentrations of SP have been shown to promote migration of HF.[21] Nilsson and co-workers[22] reported that SP promoted DNA synthesis in fibroblasts at micromolar concentration of the peptide and that neurokinin A (NKA) reproduced the proliferative effect of SP. This apparent discrepancy with our data is probably linked to the rapid degradation of SP by peptidases.[23] In fact, in the presence of the peptidase inhibitor thiorphan the extent of the proliferative response to picomolar concentration of SP was doubled.[19]

The mitogenic effect of TK has been reported also for connective tissue cells other than skin fibroblasts. Smooth muscle cells and synoviocytes exposed to SP replicate and activate their metabolic functions,[24-26] thus indicating the ability of TK to be involved in the repair of specialized territory.

B. ENDOTHELIAL CELLS

The vascular endothelium plays a pivotal role in the control of the vascular tone[27] and participates in a variety of cellular events, including the elaboration of growth factors and regulation of vascular growth.[28] The localization and the biological effects exerted by sensory neuropeptides suggest the existence of a direct effect of these peptides on vascular endothelium. SP is released by capasaicin-sensitive unmyelinated nerve fibers in peripheral organs, where they envelop the vasculature and SP is colocalized with calcitonin gene-related peptide (CGRP).[9] SP and CGRP has been proposed as the main mediators of neurogenic inflammation (see Reference 29 and the other chapters of this book). In regenerating cutaneous tissue and gastric mucosa, nerve fiber sproutings expressing an increased immunoreactivity for SP and CGRP have been described in close proximity of regenerating blood vessels.[13,30] Based on these observations, we have hypothesized that SP could contribute to the growth of microvascular vessels by directly stimulating endothelial cells.

The ability of endothelial cells to migrate and proliferate from preexisting vessels is instrumental to regenerate the capillary network.[31] By mimiking *in vitro* endothelial cell migration with a chemotactic chamber we could show a potent and dose-dependent effect of SP in promoting capillary endothelium mobilization.[32] Both SP and CGRP have been described to exert a proliferative effect on endothelial cells.[33,34] The ability of both neuropeptides to be mitogenic and to synergize in promoting cell growth on cultured endothelium,[35]

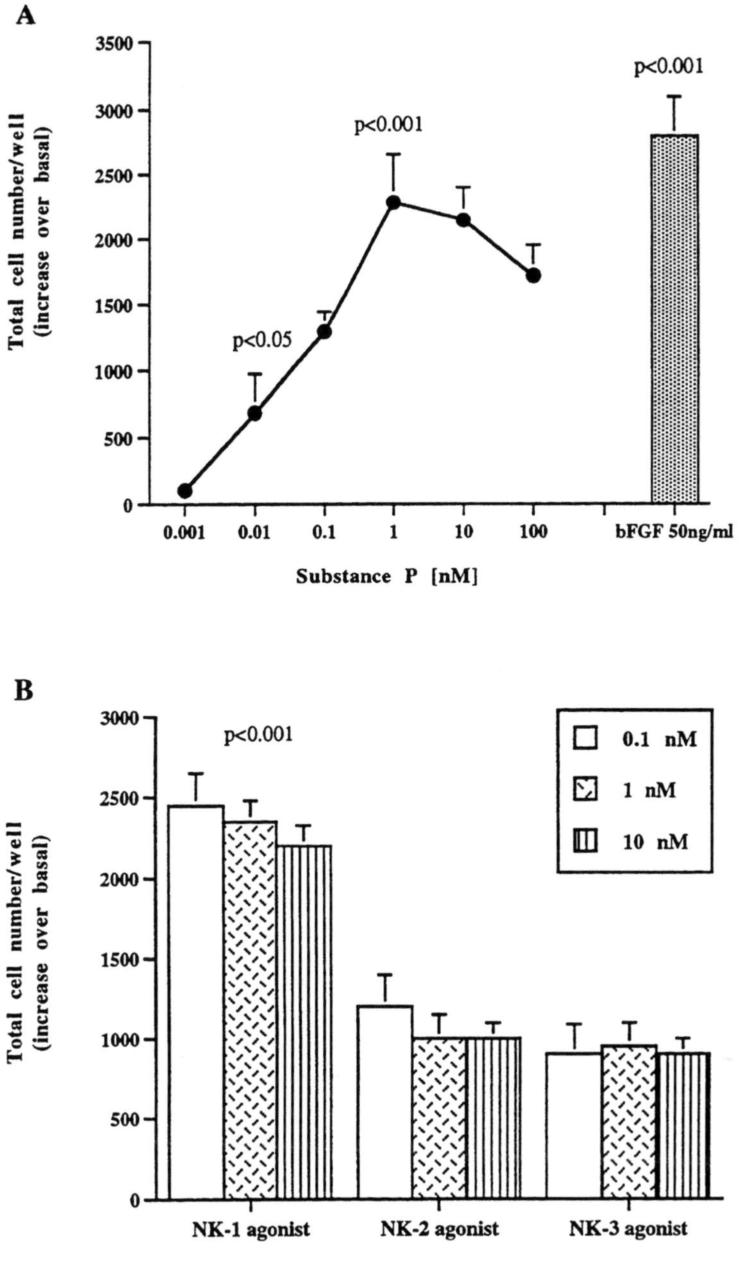

FIGURE 1. Effect of substance P (A) and selective NK_1, NK_2, and NK_3 receptor agonists (B) on human skin fibroblast proliferation. Subconfluent HF[19] from healthy volunteers were incubated with increasing concentrations of the peptides for 48 h in the absence of serum in 48 multiwell plates. The total cell number was evaluated microscopically after fixing and staining of the cells. The effect was compared to that obtained by exposing the cells to the maximal effective dose of the mitogen bFGF. Data (mean ± SEM) are expressed as increase of cell number/well over basal condition ($n = 6$). (NK_1 agonist, [Sar^9]-SP-sulfone; NK_2 agonist, [β-Ala^8]-NKA(4-10); NK_3 agonist, [$MePhe^7$]-NKB.)

underlines the direct role played by TK in promoting vascular growth *in vivo*. Indeed, reports from our group and from others have proved that SP induces angiogenesis *in vivo*, as will be discussed in a later section of this chapter.

C. IMMUNE SYSTEM

The role of the nervous system in modulating immunological and inflammatory responses has been supported by the identification of neuropeptide receptor on leukocytes and by the demonstration that these peptides can regulate leukocyte functions.[36] SP has been shown to be a potent modulator of the immune response *in vitro*[37,38] as well as *in vivo*.[39,40] It stimulates cell proliferation and protein synthesis by human peripheral T lymphocytes and by the B lymphoblastic cell line IM-9.[41] Mononuclear neutrophils of human and rodent origin have been described to migrate and release lysosomal enzymes in response to SP *in vitro* and *in vivo*.[42,43] Release of histamine by mast cells and eosinophils activation have also been reported.[44] Exposure of microvascular endothelium to SP activates the expression of specific adhesion molecules which trigger inflammatory cell infiltrate and leukocyte extravasation.[45]

III. TROPHIC FUNCTIONS OF TACHYKININS

Beside the presence of a balanced cell turnover, a major requirement for the maintenance of tissue trophism *in vivo* relies on the ability to restore tissue integrity following an injury. A rate-limiting step in this condition is the occurrence of a rapid and efficient neovascularization at the site of injury, where it controls the wound healing process.

A. ANGIOGENESIS

The healing process offers the appropriate condition to study the effects of different factors in maintaining and restoring tissue integrity. The process can be divided into distinct but interrelated stages, including migration and proliferation of connective tissue cells and blood vessels. In the healing process a major role is played by the rapid occurrence of angiogenesis in the injured tissue. This term indicates the sequence of events leading to the *de novo* formation of capillary vessels, which is crucial to support and to restore tissue integrity.[46] Neovascularization occurs in physiological processes such as embryo development, ovulation, and corpus lutheum formation, and healing processes including recovery of the heart from myocardial infarction. Although angiogenesis retains primarily a physiological significance in the healing process, its occurrence as part of an inflammatory process in physiologically avascular tissues, like the cartilage, can be detrimental and becomes a major pathological complication of chronic arthopaties. In most of the physiological and pathophysiological conditions listed the sensory neuropeptides are involved as well.[7]

In the adult tissue angiogenesis requires two major events: the presence at microvascular level of "factors" targeting the mobilization and proliferation of the capillary endothelium and the modification of the microenviroment to favor the organization and the morphogenesis of a new microvascular bed.[47] The localization of SP-containing fibers in close proximity to the postcapillary venules and the endothelium-dependent vasodilation elicited by SP have attracted our attention to study the effect of SP on isolated microvascular endothelial cells, the natural target for SP effect and for neovascular growth. Using cultured endothelial cells we were able to show that SP dose-dependently produced migration and proliferation of this cell type[32,33] (Figure 2, panel A). This effect could be reproduced *in vivo* in the avascular cornea of albino rabbits that received SP in slow-releasing polymer preparations to mimic a tonic release of the peptide[48,49] (Figure 3). Angiogenesis occurred *in vivo*, as well as *in vitro*, in a dose-dependent manner and the stable agonist for the NK_1 receptor was 10 times more potent than SP in promoting corneal neovascularization.[33] It is remarkable to note that the range of concentration that produced angiogenesis *in vivo* did not attract inflammatory cells in the tissue. Thus, *in vivo*, the effect could be ascribed to a direct action of SP on endothelial cells and not to the effect of intermediate products released by mast cells or immune cell activation. The ability of SP to synergize with interleukin-1 to promote neovascularization in subcutaneous sponge models, as reported by Fan and co-workers,[50] suggests that interactions exist between SP and cytokines released during chronic inflammatory process.

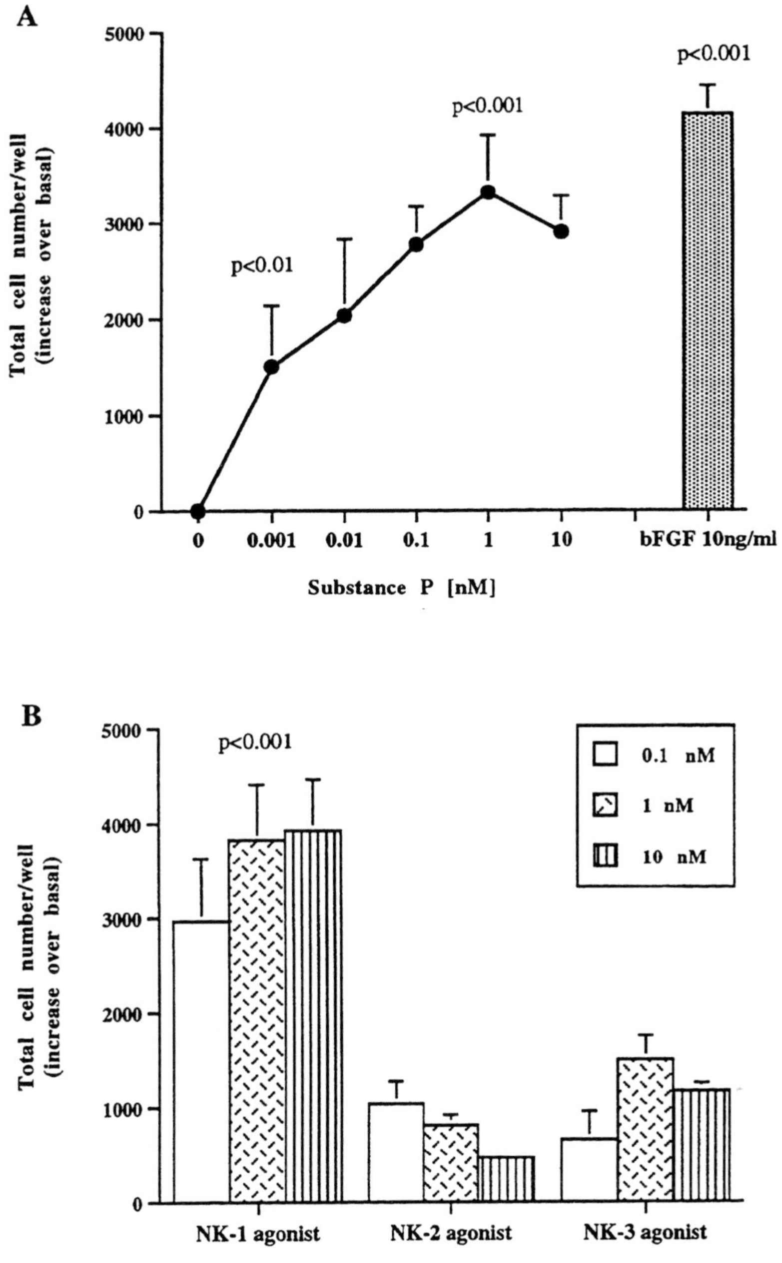

FIGURE 2. Effect of substance P (A) and selective NK_1, NK_2, and NK_3 receptor agonists (B) on postcapillary endothelial cell proliferation. Subconfluent postcapillary endothelial cells[53] were incubated with increasing concentrations of the peptides for 48 h in the absence of serum. Experimental conditions were as reported for Figure 1. Data (mean ± SEM) are expressed as increase of cell number/well over basal condition ($n = 6$). (NK_1 agonist, [Sar^9]-SP-sulfone; NK_2 agonist, [β-Ala^8]-NKA(4-10); NK_3 agonist, [$MePhe^7$]-NKB.)

SP is the main mediator of neurogenic inflammation (see Reference 29 and other chapters in this book). The vasorelaxant response produced by SP is endothelium-dependent[51] and it is linked to the production of nitric oxide (NO).[52] We could show that, although independent from the modification of blood flow, angiogenesis promoted by SP is dependent on the constitutive activation of NO synthase in vascular endothelium, leading to the generation of NO and cyclic GMP formation.[53,54]

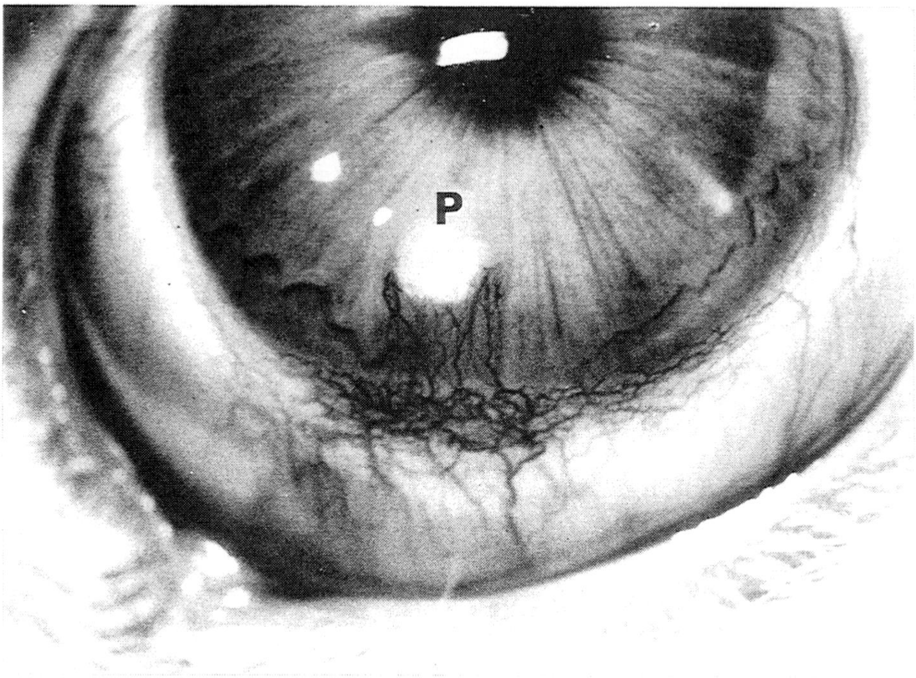

FIGURE 3. Substance P promotes angiogenesis *in vivo*. *In vivo* angiogenesis has been assessed by evaluating the neovascular response occurring in the normally avascular cornea of albino rabbits by a sterile surgical procedure.[33] The animals received the intracorneal implant of a pellet (P) containing 0.5 μg of [Sar^9]-SP-sulfone, a stable and selective agonist for the NK_1 receptor. The total amount of the agonist was slowly released by the polymer preparation of the pellet, to mimic the tonic release of neuropeptides occurring *in vivo*. The growth of capillaries from the preexisting limbal vessels started after 48 h from surgical implant. The pictures were taken at day 10 through a slit stereomicroscope (18× magnification).

B. WOUND HEALING

Experimental skin flaps provide a useful model of tissue healing, as well as of plastic and reconstructive surgical procedures. The sequential course of cutaneous nerve degeneration and regeneration and the distribution of reinnervating nerve fibers immunoreactive for neuropeptides have been extensively studied.[13] Results from these studies and from functional studies have indicated that the apperance of CGRP- and SP-containing fibers and the occurrence of pain sensation, hyperestesia, and itching are early features of tissue repair, suggesting an important role of sensory neuropeptide in tissue trophism.[13,55] These observations have been susbtantiated *in vivo* by the reports of Kjartansson and co-workers,[11] indicating that the depletion of tachykinins by capsaicin treatment significantly delayed the survival of critical skin flaps in the rat and that CGRP administration increased the survival of the skin flaps in untreated animals.

IV. TACHYKININ RECEPTORS AND GROWTH PROMOTION

Cell proliferation requires days to occur. A crucial issue in the understanding of the mechanism of action of TK in promoting cell growth *in vitro* and *in vivo*, is whether or not this is a primary effect of the peptide on the cell. An essential step for the characterization of biological events that take days to occur in response to a peptide whose half-life is of the order of seconds, is the availability of stable and selective receptor molecules. A susbtantial improvement in the understanding of the biological effects of TKs has been achieved by the development of synthetic peptides with agonistic activity for the three distinct receptor subtypes. Nonetheless the elucidation of the precise role(s) of TK in physiological and pathological processes has been hindered by the lack of high-affinity, receptor-selective, and stable antagonists.[56] The recent discovery of selective nonpeptide antagonists has provided a major advance towards this end.[57]

Using stable selective agonists for the three types of receptor, we have shown that the proliferative effect of SP on both fibroblasts and endothelial cells is mediate by the interaction with a NK_1 receptor subtype[19,33] (Figures 1 and 2, panels B). The proliferation elicited by SP on HF can be suppressed by selective antagonists for the NK_1 receptor without affecting the spontaneous or growth factor-induced cell replication.[58,59] The stable and selective agonist for the NK_1 receptor is 10 times more potent than SP in modulating capillary growth *in vivo*.[33,50] The efficacy of NK_1 receptor antagonist to block selectively SP effect was also evident in the angiogenesis model *in vivo*.[50]

V. SIGNAL TRANSDUCTION

More complex is the task to identify the transduction mechanism underlying the growth-promoting effect of SP coupled to NK_1 activation. Cell cycle control is complicated by the profusion of growth factors affecting on different messenger pathways. Both G protein-coupled receptors and tyrosine kinase-linked receptors can stimulate proliferation. The phosphoinositide-derived signals $InsP_3$ and DAG are common to both pathways, but there is conflicting evidence concerning their roles in promoting cell proliferation.[60]

The three tachykinin receptors have been shown to have seven putative transmembrane segments, to be coupled to G proteins, and to mediate the stimulation of phosphatidylinositol hydrolysis in many tissue preparations and cell types.[61] Intracellular signal transduction involved in the tachykinin receptors, however, seems to be more complex, because it has been shown that TK peptides stimulate cyclic AMP generation in some tissues but not in others, in which they may even inhibit cyclic AMP formation.[56,62] Furthermore, the precise characterization of the signal transduction underlying individual TK receptor has been limited in tissue or cell preparations because of the presence of multiple receptors and the lack of specific agonists and antagonists. Since the major part of these studies has been performed in transfected cells or in tumoral cell lines, there is relatively little or no information on the coupling mechanism in homogeneous population of mammalian cells that spontaneously express NK_1 receptors.

The NK_1 receptor mediates those biological actions encoded by the C-terminal sequence of TK, for which SP is a more potent agonist than NKA or NKB.[56]

Although the NK_1 receptor appears to be responsible for TK proliferation on both HF and endothelial cells, nevertheless the transduction mechanism appears to be different for the different cell type. In cultured fibroblasts inositolphosphate metabolism and calcium mobilization from intracellular stores are activated following exposure to SP.[63] In our experience this pathway is not activated on capillary endothelial cells by neither SP or the selective agonist for the NK_1 receptor [Sar^9]-SP-sulfone.[64] On the other hand, a recent report has indicated a transient cytosolic calcium mobilization following SP treatment on human venous endothelium.[65] Endothelial cells in culture have been reported to produce SP[66] and to undergo a rapid internalization of NK_1 receptor following SP exposure.[67] Both these conditions can reduce the availability of binding sites for the NK_1 receptor and thus impair the study of the rapid and transient second messenger activation. Furthermore, posttranslational modifications of the receptor, of its ligand binding recognition properties, or of transduction mechanisms are likely.

Evidence has been presented that multiple forms of the NK_1 receptor exist in mammalian tissues.[68,69] This is of particular interest because of the possibility that multiple forms of the NK_1 receptors can be generated from the NK_1 receptor gene. As an example Martin and co-workers[70] described the presence of an "atypical" NK_1 receptor when studying neonate rat brain purified microglia in culture. In particular, in addition to the full-length copy of the NK_1 receptor, the existence of C-terminally truncated "short" forms of the receptor has been described in both humans and rat.[68,69] However, it is also possible that other intracellular pathways might predominate in transducing NK_1 receptor activation on endothelial cells.

We have recently shown that following SP treatment cyclic GMP levels increase as a result of NO production and both NO and cyclic GMP mediate the proliferative effects of SP on the endothelium *in vivo* as well as *in vitro*.[53,54] Moreover, on endothelial cells CGRP has been decribed to promote proliferation via cyclic AMP formation but not InsP.[34]

VI. CONCLUSIONS

The nervous system collects information from the outer world by specific senses and from the interior milieu by somatic senses. This information is processed and stored in memory and affects various bodily functions through the efferent arm of the nervous system. The efferent chemical neuropeptide message is transported intraxonally to the site of action which imparts site specificity to the peripheral paracrine neuropeptide effects. The classical role of neuropeptides as fast-acting neurohumoral signalers has been recently challenged by the discovery that many neuropeptides are also growth factors stimulating slow-acting mitogenesis. TK take part to neurogenic inflammation and a direct participation of TK on wound healing has been clearly demonstrated. TK released from peripheral neurons in connection with tissue injury not only contribute to the inflammatory response but also can help other mitogens and cytokines to stimulate proliferation of surrounding connective tissue cells and thus initiate a healing response.

The identification of highly selective agonists and antagonists of peptide and nonpeptide nature, combined with the experimental information on the mitogenic and trophic functions of sensory neuropeptides, points to new perspectives to the clinical application of these agents. The role of TK in the angiogenesis process suggests that also "angiogenesis-dependent diseases" can become a target for the clinical application of stable and receptor-selective agonists and antagonists, thus contributing to the development of new therapeutical strategies to treat conditions like rheumatoid arthritis, diabetic retinopathy and microangiopathy, cancer growth, and diffusion.

ACKNOWLEDGMENTS

I wish to acknowledge the help over the years of Dr. Lucia Morbidelli, the contribution to the preparation of the manuscript by Dr. Astrid Parenti, and the financial contribution received from the Italian Association for Cancer Research and the National Research Council of Italy.

REFERENCES

1. Rosengurt, E., Neuropeptides as cellular growth factors: role of multiple signalling pathways, *Eur. J. Clin. Invest.*, 21, 123, 1988.
2. Lewis, T. and Marvin, H. M., Observations relating to vasodilatation arising from antidromic impulses to Herpes Zoster and trophic effects, *Heart*, 14, 27, 1927.
3. Jancsó G., Kiraly, E., and Jancsó-Gabor, A., Pharmacologically-induced selective degeneration of chemoreceptive primary sensory neuron, *Nature (London)*, 270, 741, 1977.
4. Nagy, J. I., Capsaicin: a chemical probe for sensory neuron mechanisms, *Handb. Psychopharmacol.*, 15, 185, 1982.

5. Buck, S. H. and Burks, T. S., The neuropharmacology of capsaicin: review of some recent observations, *Pharmacol. Rev.*, 38, 179, 1986.
6. Szolcsányi, J., Capsaicin-sensitive chemoceptive neural system with dual sensory-efferent function, in *Antidromic Vasodilatation and Neurogenic Inflammation*, Chahl, L. A., Szolcsányi, J., and Lembeck, F., Eds., Akademiai Kiado, Budapest, 1984, 26.
7. Lembeck, F., The 1988 Ulf Von Euler lecture. Substance P: from extract to excitement, *Acta Physiol. Scand.*, 133, 435, 1988.
8. Maggi, C. A. and Meli, A., The sensory-efferent function of capsaicin-sensitive sensory neurons, *Gen. Pharmacol.*, 19, 1, 1988.
9. Holzer, P., Local effector functions of capsaicin-sensitive sensory endings: involvement of tachykinin, calcitonin gene related peptide and other neuropeptides, *Neuroscience*, 24, 739, 1988.
10. Maggi, C. A., Borsini, F., Santicioli, P., Geppetti, P., Abelli, L., Evangelista, S., Manzini, S., Theodorsson Norheim, E., Somma, V., Amenta, F., Bacciarelli, C., and Meli, A., Cutaneous lesions in capsaicin pretreated rats, *Naunyn-Schiederberg's Arch. Pharmacol.*, 336, 538, 1987.
11. Kjartansson, J., Dalsgaard, C. J., and Jonsson, C. E., Decreased survival of experimental critical flaps in rats after sensory denervation with capsaicin, *Plast. Reconstr. Surg.*, 79, 218, 1987.
12. Fujita, S., Shimizu, T., Izumi, K., Fukuda, T., Sameshima, M., and Ohba, N., Capsaicin-induced neuroparalytic keratitis-like corneal changes in the mouse, *Exp. Eye Res.*, 38, 165, 1984.
13. Hermanson, A., Dalsgaard, C.-J., Bjorklund, H., and Lindblom, U., Sensory reinnervation and sensibility after superficial skin wounds in human patients, *Neurosci. Lett.*, 74, 377, 1987.
14. Pernow, B., Substance P, *Pharmacol. Rev.*, 35, 85, 1983.
15. Ross, R., Wound healing, *Sci. Am.*, 220, 40, 1969.
16. Howes, R. M. and Hoopes, J. E., Current concepts in wound healing, *Clin. Plast. Surg.*, 2, 173, 1977.
17. Karanth, S. S., Springall, D. R., Kuhn, D. M., Levene, M. M., and Polak, J. M., An immunocytochemical study of cutaneous innervation and the distribution of neuropeptides and protein gene product 9.5 in man and commonly employed laboratory animals, *Am. J. Anat.*, 191, 369, 1991.
18. Kingsnorth, A. N. and Slavin, J., Peptide growth factors and wound healing, *Br. J. Surg.*, 78, 1286, 1991.
19. Ziche, M., Morbidelli, L., Pacini, M., Dolara, P., and Maggi, C. A., NK-1 receptors mediate the proliferative response of human fibroblasts to tachykinins, *Br. J. Pharmacol.*, 100, 11, 1990.
20. Ahmed, A., Plevin, R., Shoaibi, M. A., Fountain, S. A., Ferriani, R. A., and Smith, S. K., Basic FGF activates phospholipase D in endothelial cells in the absence of inositol-lipid hydrolysis, *Am. J. Physiol.*, 266, C206, 1994.
21. Kähler, C. M., Sitte, B. A., Reinisch, N., and Wiedermann, C. J., Stimulation of the chemotactic migration of human fibroblasts by substance P, *Eur. J. Pharmacol.*, 249, 281, 1993.
22. Nilsson, J., Von Euler, A. M., and Dalsgaard, C. J., Stimulation of connective tissue cell growth by substance P and substance K, *Nature (London)*, 315, 61, 1985.
23. Patacchini, R., Maggi, C. A., Rovero, P., Regoli, D., and Drapeau, G., Effect of thiorphan on tachykinin-induced potentiation of nerve-mediated contraction of rat isolated vas deferens, *J. Pharmacol. Exp. Ther.*, 250, 678, 1989.
24. Nilsson, J., Sejerssen, T., Hultgartdh-Nilsson, A., and Dalsgaard, C.-J., DNA synthesis induced by the neuropeptide substance K correlates to the level of myc-gene transcript, *Biochem. Biophys. Res. Commun.*, 137, 167, 1986.
25. Payan, D. G., Receptor-mediated mitogenic effects of substance P on cultured smooth muscle cells, *Biochem. Biophys. Res. Commun.*, 130, 104, 1985.
26. Lotz, M., Carson, D. A., and Vaughan, J. H., Substance P activation of rheumatoid synoviocytes: neural pathway in pathogenesis of arthitis, *Science*, 235, 893, 1987.
27. Vane, J. R., Änggärd, E. E., and Botting, R. M., Regulatory functions of the vascular endothelium, *N. Engl. J. Med.*, 323, 27, 1990.
28. Dzau, V. J. and Gibbons, G. H., Vascular remodeling: mechanisms and implications, *J. Cardiovasc. Pharmacol.*, 21 (Suppl. 1), S1, 1993.
29. Lembek, F. and Holzer, P., Substance P as a neurogenic mediator of antidromic vasodilation and neurogenic plasma extravasation, *Naunyn-Schmiedeberg's Arch. Pharmacol.*, 310, 176, 1979.
30. Holzer, P., Lippe, I. T., Raybould, H. E., Pabst, M. A., Livingstone, E. H., Amann, R., Peskar, B. M., Peskar, B. A., Tachè, Y., and Guth, P. H., Role of peptidergic sensory neurons in gastric mucosal blood flow and protection, in *Substance P and Related Peptides*, Vol. 632, Leeman, S. E., Krause, J. E., and Lembeck, F., Eds., *Ann. N.Y. Acad. Sci.*, New York, 1991, 272.
31. Ausprunk, D. H. and Folman, J., Migration and proliferation of endothelial cells in preformed and newly formed blood vessels during tumor angiogenesis, *Microvasc. Res.*, 14, 53, 1977.
32. Ziche, M., Morbidelli, L., Geppetti, P., Maggi, C. A., and Dolara, P., Substance P induces migration of capillary endothelial cells: a novel NK-1 selective receptor mediated activity, *Life Sci.*, 48, PL7, 1991.
33. Ziche, M., Morbidelli, L., Pacini, M., Geppetti, P., Alessandri, G., and Maggi, C. A., Substance P stimulates neovascularization *in vivo* and proliferation of cultured endothelial cells, *Microvasc. Res.*, 40, 264, 1990.

34. Hægerstrand, A., Dalsgaard, C. J., Jonzon, B., Larsson, O., and Nilsson, J., Calcitonin gene-related peptide stimulates proliferation of human endothelial cells, *Proc. Natl. Acad. Sci. U.S.A.*, 87, 3299, 1990.
35. Villablanca, A. C., Murphy, C. J., and Reid, T. W., Growth-promoting effects of substance P on endothelial cells *in vitro*. Synergism with calcitonin gene-related peptide, insulin, and plasma factors, *Circ. Res.*, 75, 1113, 1994.
36. Payan, D. G., Neuropeptides and inflammation: the role of substance P, *Annu. Rev. Med.*, 40, 341, 1989.
37. Lotz, M., Vaughan, J. H., and Carson, D. A., Effect of neuropeptides on production of inflammatory cytokines by human monocytes, *Science*, 241, 1218, 1988.
38. Payan, B. G., Brewester, R., and Goetzl, E. J., Specific stimulation of human T lymphocytes by substance P, *J. Immunol.*, 131, 1613, 1983.
39. Scicchitano, R., Bienenstock, J., and Stanisz, A. M., *In vivo* immunomodulation by the neuropeptide substance P, *Immunology* , 63, 733, 1988.
40. Perretti, M., Ahluwalia, A., Flower, R. J., and Manzini, S., Endogenuous tachykinins play a role in IL-1-induced neutrophil accumulation: involvement of NK-1 receptors, *Immunology*, 80, 73, 1993.
41. Payan, B. G., Brewster, D. R., and Goetzl, E. J., Stereospecific receptors for substance P on cultured human IM-9 lymphoblasts, *J. Immunol.*, 133, 3260, 1984.
42. Serra, M. C., Bazzoni, F., Della Bianca, V., Greskowiak, M., and Rossi, F., Activation of human neutrophils by substance P, *J. Immunol.*, 141, 2118, 1988.
43. Helme, R. D., Eglezos, A., and Hosking, C. S., Substance P induces chemotaxis of neutrophils in normal and capsaicin-treated rats, *Immunol. Cell Biol.*, 65, 267, 1987.
44. Kroegel, C., Giembycz, M. A., and Barnes, P. J., Characterization of eosinophil cell activation by peptides. Differential effects of subatance P, mellittin and fMet-Leu-Phe, *J. Immunol.*, 145, 2581, 1990.
45. Smith, C. H., Barker, J. N. W. N., Morris, R. W., MacDonald, D. M., and Lee, T. H., Neuropeptides induce rapid expression of endothelial cell adhesion molecules and elicit granulocytic infiltration in human skin, *J. Immunol.*, 151, 3274, 1993.
46. Folkman, J., Angiogenesis: initiation and control, *Ann. N.Y. Acad. Sci.*, 401, 212, 1982.
47. Risau, W., Angiogenesis and endothelial cell function, *Arzneim. Forsch./Drug. Res.*, 44, 416, 1994.
48. Langer, R. and Folkman, J., Polymers for the sustained release of proteins and other macromolecules, *Nature (London)*, 263, 797, 1976.
49. Ziche, M., Jones, J., and Gullino, P. M., Role of prostaglandin E_1 and copper in angiogenesis, *J. Natl. Cancer Inst.*, 69, 475, 1982.
50. Fan, T.-P., Hu, D.-E., Guard, S., Gresham, G. A., and Watling, K. J., Stimulation of angiogenesis by substance P and interleukin-1 in the rat and its inhibition by NK-1 or interleukin-1 receptor antagonists, *Br. J. Pharmacol.*, 110, 43, 1993.
51. D'Orleans-Juste, P., Dion, S., Mizrahi, J., and Regoli, D., Effects of peptides and nonpeptides on isolated arterial smooth muscles: role of endothelium, *Eur. J. Pharmacol.*, 114, 9, 1985.
52. Hughes, S. R., Williams, T. J., and Brain, S. D., Evidence that endogenous nitric oxide modulates oedema formation induced by substance P, *Eur. J. Pharmacol.*, 191, 481, 1990.
53. Ziche, M., Morbidelli, L., Masini, E., Amerini, S., Granger, H. J., Maggi, C. A., Geppetti, P., and Ledda, F., Nitric oxide mediates angiogenesis *in vivo* and endothelial cell growth and migration *in vitro* promoted by substance P, *J. Clin. Invest.*, 94, 2036, 1994.
54. Zich, M., Morbidelli, L., Parenti, A., Amerini, S., Granger, H. J., and Maggi, C. A., Substance P increases cyclic GMP levels on coronary postcapillary venular endothelial cells, *Life Sci.*, 53, 1105, 1993.
55. Adeymo, O. and Wyburn, G. M., Innervation of skin grafts, *Transplant. Bull.*, 4, 152, 1957.
56. Maggi, C. A., Patacchini, R., Rovero, P., and Giachetti, A., Tachykinin receptors and tachykinin receptor antagonists, *J. Auton. Pharmacol.*, 13, 23, 1993.
57. Snider, R. M., Constantine, J. W., Lowe Iii, J. A., Longo, K. P., Lebel, W. S., Woody, H. A., Drodza, S. E., Desai, M. C., Vinick, F. J., Spencer, R. W., and Hess, H.-J., A potent nonpeptide antagonist of the substance P (NK-1) receptor, *Science*, 251, 435, 1991.
58. Morbidelli, L., Maggi, C. A., and Ziche, M., Effect of selective tachykinin receptor antagonists on the growth of cultured human skin fibroblasts, *Neuropeptides*, 24, 335, 1993.
59. Morbidelli, L., Parenti, A., Maggi, C. A., and Ziche, M., Basic fibroblast growth factor (bFGF) effect on cultured human skin fibroblasts is potentiated by substance P, *Neuropeptides*, 22, 45, 1992.
60. Berridge, M. J., Inositol triphosphate and calcium signalling, *Nature (London)*, 361, 315, 1993.
61. Guard, S. and Watson, S. P., Tachykinin receptor types: classification and membrane signalling mechanisms, *Neurochem. Int.*, 18, 149, 1991.
62. Mitsuhashi, M., Ohashi, Y., Shichijo, S., Christian, C., Sudduth-Klinger, J., Harrowe, G., and Payan, D. J., Multiple intracellular signalling pathways of the substance P receptor, *FASEB J.*, 6, A1562, 1992.
63. Ziche, M., Parenti, A., Morbidelli, L., Amerini, S., Ledda, F., Baldi, E., and Maggi, C. A., Effect of SP and selective NK-1 receptor agonist on inositolphosphates and cytosolic calcium levels in cultured human fibroblasts, *Neuropeptides*, 24, 231, 1993.
64. Ziche, M., unpublished data, 1995.

65. Greeno, E. W., Mantyh, P., Vercellotti, G. M., and Moldow, C. F., Functional neurokinin 1 receptors for substance P are expressed by human vascular endothelium, *J. Exp. Med.*, 177, 1269, 1993.
66. Ralevic, V., Milner, P., Hudlick, O., Kristek, F., and Burnstock, G., Substance P is released from the endothelium of normal and capsaicin-treated rat hind-limb vasculature, in vivo, by increased flow, *Circ. Res.*, 66, 1178, 1990.
67. Bowden, J. J., Garland, A. M., Baluk, P., Lefevre, P., Grady, E. F., Vigna, S. R., Bunnett, N. W., and MacDonald, D. M., Direct observation of substance P-induced internalization of neurokinin 1 (NK-1) receptors at sites of inflammation, *Proc. Natl. Acad. Sci. U.S.A.*, 91, 8964, 1994.
68. Fong, T.M., Anderson, S., Yu, H., Huang, R. R. C., and Strader, C. D., Differential activation of intracellular effector by two isoforms of human NK-1 receptor, *Mol. Pharmacol.*, 41, 24, 1992.
69. Kage, R., Leeman, S. E., and Boyd, N. D., Biochemical characterization of two different forms of the substance P receptor in rat submaxillary gland, *J. Neurochem.*, 60, 347, 1993.
70. Martin, F. C., Anton, P. A., Gornbein, J. A., Shanahan, F., and Merrill, J. E., Production of interleukin-1 in response to substance P: role for a nonclassical NK-1 receptor, *J. Neuroimmmunol.*, 42, 53, 1993.

Part IV. Clinical Studies

Chapter 21

HYPERALGESIA AND NEUROVASCULAR REACTIONS FOLLOWING CAPSAICIN APPLICATION TO HUMAN SKIN

Jordi Serra and José L. Ochoa

CONTENTS

I. INTRODUCTION

Nociceptors generate prominent sensory and vascular symptoms and signs during neurogenic inflammation. Indeed, C-nociceptors induce the pains and vasodilatation that characterize the specific human health disorder described under the term erythralgia or ABC syndrome. C-nociceptors also mediate, in one way or another, the pains and neurovascular changes induced experimentally by capsaicin. Like other neurotoxins that target specific neuronal receptors, capsaicin lends itself as a useful tool for experimental investigation of fine aspects of electrogenic and neurosecretory function of nociceptors. Moreover, like other neurotoxins applied to other fields, capsaicin, or its analogs, offers promise in the therapeutic control of nociceptor excitability.

We have chosen to lay out this chapter along the following sequence. We briefly introduce the structure and function of the human C-nociceptor, incorporating the newly described "silent" C afferent units. Next comes a discussion of heat hyperalgesia and mechanical hyperalgesia in the context of neurogenic vasodilatation, as induced by topical capsaicin in human volunteers. We then deal with the concept of cross-modality receptor threshold modulation, emerged from observations on experimental volunteers treated with capsaicin and patients with a particular variety of neuropathic pains.

The difficult and confused subject of "primary" vs. "secondary" hyperalgesia is approached at this point. Since common polymodal nociceptors in the sensitized state are not readily detectable substrates to account for the broad area of hyperalgesia caused by cutaneous injection of capsaicin, and given the historical availability of alternative mechanisms, the symptom has become fashioned as a secondary "centralized" anomaly. As a counterpoint, we refer to results of ongoing research on neurogenic inflammation induced by application of capsaicin to human skin, showing that the area of "secondary" hyperalgesia is most probably caused by dysfunction of primary nociceptors. We close by offering a reinterpretation of the cumulative evidence on capsaicin-induced cutaneous phenomena: all sensory and neurovascular effects of capsaicin on human skin would be consequent to afferent and neurosecretory actions of peripheral nociceptor units.

0-8493-7646-7/96/$0.00+$.50

II. STRUCTURE AND FUNCTION OF HUMAN C-NOCICEPTORS

The human unmyelinated C-nociceptor is better known than the larger Aδ-nociceptor, especially electrophysiologically, because it is much easier to access through microneurography.[1] The overall population of unmyelinated axons in human cutaneous nerves is three or four times more abundant than the large- and small-caliber myelinated populations put together.[2] At the fine structural level it remains unknown what percentage of unmyelinated axons in nerve trunks serving human skin is part of the nociceptor system, as opposed to being warm specific afferents, or part of the sympathetic efferent system. The size–frequency distribution of unmyelinated axons in human cutaneous nerves is unimodal, ranging between 0.2 and close to 3 μm, with a peak around 1.4 to 1.6.[2] Free unmyelinated nerve endings, almost certainly sensory in nature, penetrate the human epidermis from "penicillate" subepidermal formations.[3,4] Recent powerful confocal light microscopy studies have addressed the distribution of unmyelinated axons in the human skin and have revealed new vistas of these structures.[5]

In terms of physiological receptor-response characteristics, C-polymodal nociceptors are regularly excited by strong mechanical, heat, and chemical stimuli,[6] but respond unpredictably to low temperature. Selective intraneural microstimulation of C-nociceptors evokes a dull or a burning pain sensation, with long reaction time, that resists selective A-fiber block.[8] Selective microstimulation of A-fibers connected to low-threshold mechanoreceptors consistently evokes specific nonpainful sensations.[9]

The physiological capacity of C-nociceptors to release transmitters at their terminals is well established. One consequence is peripheral vasodilatation determined antidromically, which is confined to the cutaneous territory of the excited nerve bundles.[10] This phenomenon seems to be mediated by calcitonin gene-related peptide (see Chapter 18).

Fine structural criteria for pathology of human unmyelinated axons include the emergence of a subpopulation of immature axonal sprouts and the presence of characteristic denervated Schwann cell bands.[11-13] With natural aging, unmyelinated axons drop out and ultrastructural signs of pathology increase progressively.[11-13]

Pathological sensitization of human C-nociceptors in natural human disease causes spontaneous pain associated with heat hyperalgesia and temperature-dependent mechanical hyperalgesia. Over and above the sensory symptomatology, pathological hyperexcitability of C-polymodal nociceptors typically causes cutaneous vasodilatation.[14-16] A recent review on human nociceptors in health and disease is available.[17]

Most human cutaneous C-nociceptors that respond to strong natural mechanical stimuli also respond to heat and to chemical stimuli.[8] In the study by Ochoa and Torebjörk,[8] like in previous microneurographic studies in humans, the population of C-nociceptors was defined by either: (1) primary receptor stimulation with mechanical stimuli followed by elective use of other stimulus energies (heat, histamine injection, etc.) or (2) by searching with mechanical stimuli given at the monofocal projected field of dull or burning pain sensation evoked by intraneural microstimulation at threshold for sensation. These approaches do not cater for high-threshold afferent C units which might selectively respond to energies other than mechanical, or which may not respond at all under baseline circumstances. This proviso is proven real for visceral nociceptors, as shown in animal studies,[18-21] and for somatic C-nociceptors, as shown in monkey,[22,23] and also preliminarily in humans[24] using the electrical stimulus searching method for receptive fields of cutaneous nociceptors. In the context of neurogenic inflammation and capsaicin actions, these mechanically insensitive afferents are relevant because under baseline conditions they may respond exclusively to chemical stimuli and may only become sensitive to mechanical energy after chemical challenge. Thus, mechanically insensitive afferents are bound to contribute to mechanical hyperalgesia in skin injury, and they actually do (see below).

III. FLARE AND HYPERALGESIA INDUCED BY CAPSAICIN

It is known that capsaicin applied acutely to skin may cause hyperexcitability of the common C-polymodal nociceptor in animals[25-27] and in humans.[28,29] As it is known that burning pain is a specific sensation evoked by cutaneous C-polymodal nociceptors, it is inferred logically that the spontaneous burning pain induced by capsaicin in humans is due to spontaneous discharge in C-nociceptors from skin.

Matters become more complicated in the context of stimulus-induced hyperalgesia. If hyperalgesia is simply defined as "increased sensitiveness to pain",[30] capsaicin causes dose-dependent cutaneous hyperalgesia for mechanical and also for heat stimuli.[31] However, capsaicin does not cause detectable hyperalgesia for low-temperature stimuli. On the contrary, low temperature typically relieves both the spontaneous pain and the temperature-dependent mechanical hyperalgesia that capsaicin acutely induces.[31] While it is reasonable to envisage this hyperalgesic response to plural stimuli to be based on neurotoxin-determined hyperexcitability of plurally sensitive (polymodal) nociceptors, it is currently believed that this mechanism only explains the comparatively trivial "primary" hyperalgesia that by definition develops close to the site of capsaicin application.[32,33] The significantly broader area of hyperalgesia that surrounds the locus of a skin injury was historically attributed to either hyperexcitability of pain-signaling primary sensory units, or to secondary central nervous system repercussions. For Lewis,[34] all the hyperalgesia, which he called "nocifensor tenderness", was mediated by cutaneous nerve endings in "a state of hyperexcitability". These nerve fibers would respond with lowered threshold to plural stimulus energies and would even discharge spontaneously giving rise to "a very light spontaneous smarting".[34] Lewis found the "nocifensor tenderness" to extend over an area that often matched the area of associated vasodilatation. For Hardy et al.[35] the "secondary" hyperalgesia was due to "a central excitatory state . . .". Today's prevalent thinking is consistent with Hardy's, partly because the "secondary" area has been described as behaviorally different (would involve pure mechanical hyperalgesia assumed to be mediated by low-threshold tactile mechanoreceptor afferents).[32,36] Partly, the concept is also favored because, through microneurography, the common polymodal nociceptor has not been found to be sensitized in the area of secondary hyperalgesia induced by capsaicin in humans.[29,37] The flare that commonly associates with certain physical or chemical injuries to skin was not regarded by Hardy's school as significant enough in terms of neural mechanisms. Today the flare is typically taken by naked eye to extend over an area much smaller than the area of overall hyperalgesia.[34,38] Even by color thermography, the area of cutaneous hyperthermia may appear restricted relative to the area of abnormal sensation.

Later in this chapter it will be cautioned that the area of vasodilatation is decidedly underestimated both by naked eye and by color thermography,[39,40] and the hyperalgesia is polymodal, not purely mechanical.[41,42] It will also be shown that nociceptors other than the common polymodal nociceptor do become sensitized in the area of "secondary" hyperalgesia.[43] Thus, it will be necessary to reconcile the dissident new evidence with prevailing theory. Such theory rationalizes, through mechanisms remote to the nerve fiber itself, the "secondary" experimental hyperalgesia and much of the "brush induced", "dynamic", "A-fiber mediated"[44] expansive hyperalgesia (and spontaneous pains) expressed by "neuropathic" patients.

IV. CROSS-MODALITY RECEPTOR THRESHOLD MODULATION OF C-NOCICEPTORS

The mechanical hyperalgesia, as well as the spontaneous pain induced by capsaicin, are strikingly relieved through passive cooling of the symptomatic skin. The same happens in a particular variety of neuropathic painful conditions proven to be due to sensitization of primary nociceptors: the ABC syndrome.[14,15] This phenomenon has been theoretically

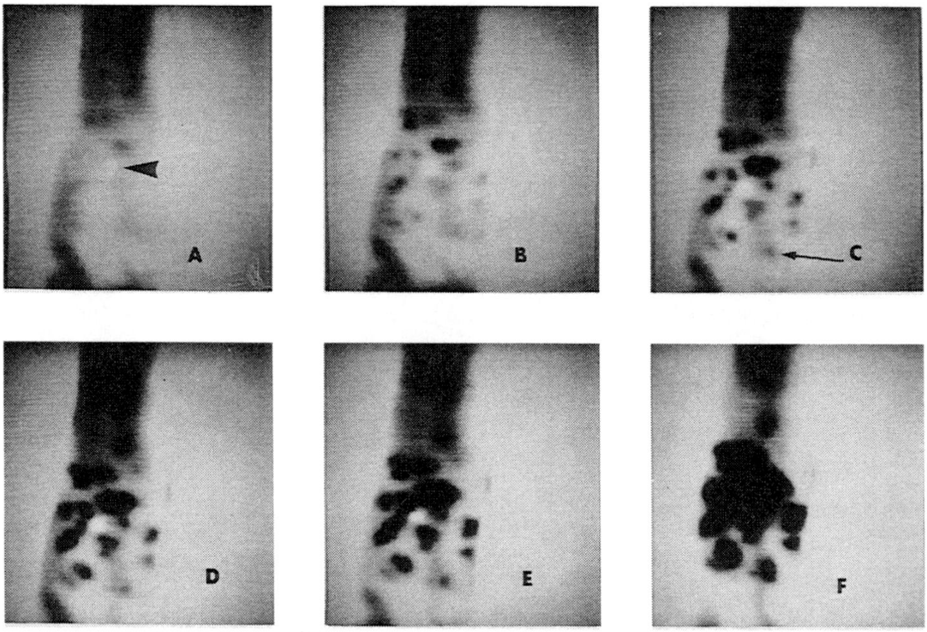

FIGURE 1. *Left:* Typical spatiotemporal development of the flare response following capsaicin injection (100 μg) in the anterior aspect of the forearm, as recorded using dynamic infrared telethermography (Mark V Flexitherm®). A monochrome display system was used (50 grades from black to white, black representing high temperature) with a range of 2.5°C. Discrete multifocal areas of increased skin temperature reflecting dilatation of cutaneous arterioles appear in a wide area. Distance between injection site (cold spot at arrowhead) and the most distant spot of increased temperature (arrow) was 10 cm in this case. The area of mechanical and heat hyperalgesia precisely matched the area of this broad neurovascular reaction, indicating that all three phenomena are most probably mediated at peripheral level (A, 5 sec; B, 30 sec; C, 1 min; D, 2 min; E, 3 min; and F, 5 min). *Right:* Schematic representation of the sequence.

explained through two possible mechanisms.[31] The hypothetical mechanism we have favored foresees that in the hyperalgesic state determined by sensitization of nociceptors, as induced by capsaicin or natural disease (ABC syndrome), it would be specifically the heat transducers of the polymodal excitable membrane that would be leaky and thus depolarize the membrane to, or near, threshold. Under those circumstances, a normally subthreshold contribution to depolarization through mechanical energy acting on transmembrane (stretch) transducers, would be enough to bring the membrane to threshold for firing, thus evoking a painful sensation at hyperalgesic level. In that state, the application of low temperature, the antithesis of a thermal stimulus, would subtract energy that might otherwise further open the leaky heat transducers. Thus, one energy (temperature) modulates threshold for another energy (mechanical) acting adequately on the same excitable membrane.[14,17] This hypothesis for cross-modality threshold modulation assumes that the excitable polymodal nociceptor membrane operates under the principle of cooperative threshold.

Alternatively, the relief of spontaneous pain and hyperalgesia through passive cooling might also be theoretically negotiated in terms of the "centralized" hypothesis for hyperalgesia as a symptom of experimental skin injury. Low temperature actually decreases afferent nociceptor activity: such activity would abnormally maintain the hypothetical secondary state of hyperexcitability in pain-signaling dorsal horn neurons. Those central neurons would now abnormally evoke pain when activated by primary low-threshold, Aβ, tactile units.[45] We favor the first alternative.

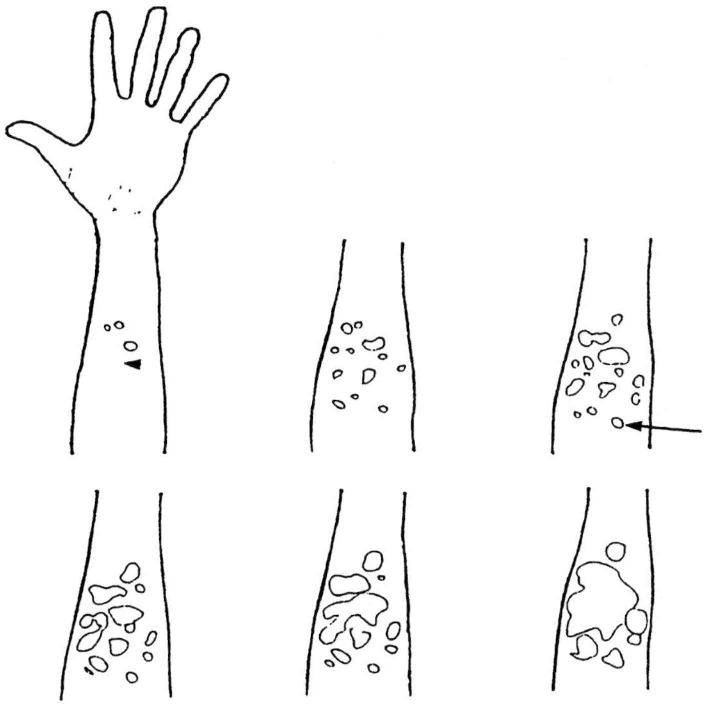

FIGURE 1. (Continued).

V. THE ANGRY BACKFIRING C-NOCICEPTOR (ABC) SYNDROME, OR ERYTHRALGIA

Lewis[46] described patients with erythralgia as expressing "a redness of the skin, associated with tenderness, the gentlest manipulation elicits pain, pain is also provoked by warming, cooling abolishes the pain . . . and the pain burns". The condition was termed the ABC syndrome (angry backfiring C-nociceptor syndrome)[14,15] when it was directly documented by microneurography that the patients have sensitized nociceptors accounting for both the sensory and antidromic vasomotor phenomena. This clinical condition is identical to the acutely induced anomalous state observed in human volunteers through experimental application of capsaicin to the skin.[31] The common features are dose-dependent rubor, spontaneous pain, and hyperalgesia. In volunteers treated with capsaicin, as well as in patients with the ABC syndrome, passive cooling abolishes spontaneous burning pain and mechanical hyperalgesia through hypothetical cross-modality threshold modulation. The beneficial effect of cooling persists after selective block of myelinated fibers and, therefore, could not be explainable through central gating exerted by cold specific input. Instead, the low-temperature status must directly rectify the excitability of sensitized peripheral nociceptors.

VI. PRIMARY AND SECONDARY HYPERALGESIA: SOME HISTORY AND RECENT EXPERIMENTAL DISCOVERIES USING CAPSAICIN

Today's belief that a peripheral injury may cause pain and hyperalgesia due to the development of long-term pathophysiological repercussions in the central neuraxis is somewhat programmed by clinical memories from the 1940s. The expansion of clinical "causalgic" symptoms, particularly the hyperalgesia, to a broad area, led Livingston[47] to postulate that the "internuncial pool" in the spinal cord develops an abnormal "irritable state". Evans[48] had hypothesized that "a prolonged bombardment of pain impulses sets up a vicious cycle of reflexes spreading through a pool of many neuronal connections upward, downward and even across the spinal cord . . . depending on the wide spread of the pool would detect the phenomena of pain and sympathetic disturbances observed a long distance from the injured area in the limb . . .". Insightful neurologists[49] were disturbed by the observation that in some "neuropathic" pain patients the sensory symptomatology may spread in ways that cannot be explained by peripheral nerve dysfunction: "in the painful states associated with hyperaesthesia and hyperpathia, the lesion in the periphery has induced abnormal functioning of the central nervous system, presumably at spinal level. This statement is based on the fact that tactile stimulation causes pain and the fact of the spread of pain and hypersensitivity beyond the original territory of the lesion".

In the context of human experimental cutaneous injury, the concept of centralization was originally conceived by Hardy et al.[35] to explain the area of "secondary" hyperalgesia that develops beyond the site of injury itself. Hardy et al. hypothesized "a central excitatory state in a network of internuncial neurons", in the same vein as the clinicians of his time, mentioned above, had. Apparent support for "centralization" eventually came from measurement of psychophysical reaction times in patients with "neuropathic" hyperalgesia: these were found to be shorter than expected for nociceptor mediation.[50] Further apparent support came from the realization that cutaneous hyperalgesia-allodynia may be abolished with selective block of large myelinated fibers, while nociceptors continue to conduct.[51,52] A painful experience if evoked by activation of tactile low-threshold mechanoreceptors, logically suggests that rostral to the peripheral nervous system there must be some anomalous transfer of non-noxious input towards the pain-signaling apparatus.

The concept that experimental "secondary" mechanical hyperalgesia is determined by low-threshold mechanoreceptor input to sensitized wide dynamic range neurons[29,32,37,44,53,54] has recently received apparent support from sophisticated human studies using microneurography. Their methodological goal has been to grasp an indirect measure of function of second-order dorsal horn neurons (presumed wide dynamic range convergent cells) before, during, and after experimental manipulations in the skin that elicit "secondary" hyperalgesia in human volunteers. This has been approached through combined microneurographic recordings and intraneural microstimulation: the combined strategy might enable testing of whether primary peripheral, as opposed to secondary central, anomalies modulate quality of sensation evoked by excitation of primary sensory units of identifiable types. These units are normally expected to evoke predictable sensory qualities when selectively microstimulated.[8,9,55] Should hyperalgesia be determined by secondary dysfunction of pain-signaling central convergent neurons, then selective microstimulation of non-nociceptor, low-threshold, tactile afferents from the region should be painful. In the setting of an acute transient peripheral injury, microstimulation would become painful throughout the period of expansion of the symptom hyperalgesia, when the experimentally induced "secondary" hyperalgesia encroaches the receptive field of the low-threshold mechanoreceptor. Under those experimental circumstances, intraneural microstimulation of non-nociceptive, probably mechanoreceptive, fibres innervating the hyperalgesic skin area gives rise to abnormal sensations of pain.[45] These results have been taken to provide the strongest support for the concept that the expanding secondary hyperalgesia following painful injury is the real consequence of sensitization of wide dynamic dorsal horn neurons in humans. These results remain to be confirmed, and equally valid hypothetical alternatives can explain behavioral events attributed to secondary sensitization of central neurons. It is pertinent to remind the reader that intraneural microstimulation of tactile afferents serving hyperalgesic skin in "neuropathic" patients with or without actual nerve injury elicits tactile sensations and not pain.[56] The strongest question against Torebjörk's interpretation of "secondary" hyperalgesia and sensitization of central neurons, as based upon experiments using capsaicin in volunteers and as extrapolated to "neuropathic pain patients", comes from recent studies by Serra et al.,[39-42] suggesting that the area of "secondary" hyperalgesia induced by capsaicin is caused by dysfunction of primary nociceptors.

The findings of Serra et al. supersede the prevailing "centralization" theory through establishing that traditional naked-eye assessment of the area of neurogenic vasodilatation has underestimated the phenomenon: its broad extent is sensitively revealed by quantitative thermography. Moreover, the area of heat hyperalgesia is coextensive with the area of mechanical hyperalgesia when the former is measured through suprathreshold magnitude estimation, rather than through the thresholds method. Past failure to identify sensitized polymodal nociceptors in the region of "secondary" hyperalgesia may be explainable through sensitization of the "silent" class of C-chemonociceptors, a possibility not tested previously. Evidence that this is probably the case has been raised recently.[43]

VII. POTENTIAL USE OF CAPSAICIN IN THERAPY

Chronic application of capsaicin to skin eventually desensitizes human nociceptors. A review of the literature on this subject, as well as a report on human experiments attempting to desensitize human cutaneous nociceptors through repeated application of a topical agent containing 0.75% capsaicin, was offered by Simone and Ochoa.[57] The use of capsaicin in human therapy is discussed in Chapter 23 of this volume.

REFERENCES

1. Torebjörk, H. E. and Ochoa, J. L., New method to identify nociceptor units innervating glabrous skin of the human hand, *Exp. Brain Res.*, 81, 509, 1990.
2. Ochoa, J. L. and Mair, W. G. P., The normal sural nerve in man. I. Ultrastructure and numbers of fibres and cells, *Acta Neuropathol.*, 13, 197, 1969.
3. Cauna, N., The free penicillate nerve endings of the human hairy skin, *J. Anat.*, 115, 277, 1973.
4. Ochoa, J. L., Peripheral unmyelinated units in man: structure, function, disorder and role in sensation, in *Advances in Pain Research and Therapy*, Vol. 6, Kruger, L. and Liebeskind, J. C., Eds., Raven Press, New York, 1984, 53.
5. Kennedy, W. R. and Wendelschafer-Crabb, G., The innervation of human epidermis, *J. Neurol. Sci.*, 115, 184, 1993.
6. Torebjörk, H. E. and Hallin, R. G., C-fibre units recorded from human sensory nerve fascicles in situ, *Acta Soc. Med. Ups.*, 75, 81, 1970.
7. Campero, M., Serra, J., and Ochoa, J. L., Response of human cutaneous C-polymodal nociceptors elicited by low-temperature stimulation, *J. Neurol.*, 241, S61, 1994.
8. Ochoa, J. L. and Torebjörk, H. E., Sensations evoked by intraneural microstimulation of C-nociceptor fibres in human skin nerves, *J. Physiol.*, 415, 583, 1989.
9. Ochoa, J. L. and Torebjörk, H. E., Sensation evoked by intraneural microstimulation of single mechanoreceptor units innervating the human hand, *J. Physiol.*, 342, 633, 1983.
10. Ochoa, J. L., Comstock, W. J., Marchettini, P., and Nizamuddin, G., Intrafascicular nerve stimulation elicits regional skin warming that matches the projected field of evoked pain, in *Fine Afferent Nerve Fibers and Pain*, Schmidt, R. F., Schaible, H. G., and Vahle-Hinz, C., Eds., VCH, Weinheim, 1987, 476.
11. Ochoa, J. L. and Mair, W. G. P., The normal sural nerve in man. II. Changes in axons and Schwann cells due to aging, *Acta Neuropathol.*, 13, 217, 1969.
12. Ochoa, J. L., Isoniacid neuropathy in man — quantitative electron microscope study, *Brain*, 93, 831, 1970.
13. Ochoa, J. L., Recognition of unmyelinated fiber disease: morphologic criteria, *Muscle Nerve*, 1, 375, 1978.
14. Ochoa, J. L., The newly recognized painful ABC syndrome: thermographic aspects, *Thermology*, 2, 65, 1986.
15. Cline, M. and Ochoa, J. L., Chronically sensitized C nociceptors in skin: patient with hyperalgesia, hyperpathia and spontaneous pain, *Soc. Neurosci. Abstr.*, 12, 331, 1986.
16. Cline, M. A., Ochoa, J. L., and Torebjörk, H. E., Chronic hyperalgesia and skin warming caused by sensitized C nociceptors, *Brain*, 112, 621, 1989.

17. Ochoa, J. L., Human nociceptors in health and disease, in *Neurobiology and Disease: Contributions from Neuroscience to Clinical Neurology. A Symposium in Honour of T. A. Sears*, Bostock, H., Kirkwood, P. A., and Pullen A. H., Eds., Cambridge University Press, London, 1996, 151.
18. McMahon, S. B., Neuronal and behavioural consequences of chemical inflammation of rat urinary bladder, *Agents Actions*, 25, 231, 1988.
19. Schaible, H.-G. and Schmidt, R. F., Effects of an experimental arthritis on the sensory properties of fine articular afferent units, *J. Neurophysiol.,*, 54, 1109, 1985.
20. Häbler, H.-J., Jänig, W., and Koltzenburg, M., Activation of unmyelinated afferent fibres by mechanical stimuli and inflammation of the urinary bladder in the cat, *J. Physiol.*, 425, 545, 1990.
21. Häbler, H.-J., Jänig, W., and Koltzenburg, M., Receptive properties of myelinated primary afferents innervating the inflammed urinary bladder of the cat, *J. Neurophysiol.*, 69, 395, 1993.
22. Meyer, R. A., Davis, K. D., Cohen, R. H., Treede, R.-D., and Campbell, J. N., Mechanically insensitive afferents (MIAs) in cutaneous nerves of monkey, *Brain Res.*, 561, 252, 1991.
23. Davis, K. D., Meyer, R. A., and Campbell, J. N., Chemosensitivity and sensitization of nociceptive afferents that innervate the hairy skin of monkey, *J. Neurophysiol.*, 69, 1071, 1993.
24. Handwerker, H. O., Schmidt, R., Forster, C., Schmelz, M., Traversa, R., and Torebjörk, H.-E., Microneurographic assessment of sensitive and insensitive C-fibers in a human skin nerve, *Soc. Neurosci. Abstr.*, 19, 1404, 1993.
25. Forster, R. and Ramage, A. G., The action of some chemical irritants on somatosensory receptors of the cat, *Neuropharmacology*, 20, 191, 1981.
26. Kenins, P., Response of single nerve fibres to capsaicin applied to the skin, *Neurosci. Lett.*, 29, 83, 1982.
27. Szolcsányi, J., Capsaicin-sensitive chemoceptive neural system with dual sensory-efferent function, in *Antidromic Vasodilatation and Neurogenic Inflammation*, Chahl, L. A., Szolcsányi J., and Lembeck F., Eds., Akademiai Kiádo, Budapest, 1984.
28. Konietzny, F. and Hensel, H., The effect of capsaicin on the response characteristics of human C-polymodal nociceptors, *J. Therm. Biol.*, 8, 213, 1983.
29. LaMotte, R. H., Lundberg, L. E. R., and Torebjörk, H.-E., Pain, hyperalgesia and activity in nociceptive C units in humans after intradermal injection of capsaicin, *J. Neurophysiol.*, 448, 749, 1992.
30. Head, H., On disturbances of sensation with especial reference to the pain of visceral disease, *Brain*, 16, 1, 1893.
31. Culp, W. J., Ochoa, J. L., Cline, M. A., and Dotson, R., Heat and mechanical hyperalgesia induced by capsaicin: cross modality threshold modulation in human C nociceptors, *Brain*, 112, 1317, 1989.
32. LaMotte, R. H., Shain, C. N., Simone, D. A., and Tsai, E. F. P., Neurogenic hyperalgesia: psychophysical studies of underlying mechanisms, *J. Neurophysiol.*, 66, 190, 1991.
33. LaMotte, R. H., Neurophysiological mechanisms of cutaneous secondary hyperalgesia in the primate, in *Hyperalgesia and Allodynia*, Willis, W. D. J., Ed., Raven Press, New York, 1992, 175.
34. Lewis, T., *Pain*, Macmillan, London, 1942.
35. Hardy, J. D., Wolff, H. G., and Goodell, H., *Pain Sensations and Reactions*, Williams & Wilkins, Baltimore, 1952.
36. Raja, S. N., Campbell, J. N., and Meyer, R. A., Evidence for different mechanisms of primary and secondary hyperalgesia following heat injury to the glabrous skin, *Brain*, 107, 1179, 1984.
37. Baumann, T. K., Simone, D. A., Shain, C. N., and LaMotte, R. H., Neurogenic hyperalgesia: the search for the primary cutaneous afferent fibers that contribute to capsaicin-induced pain hyperalgesia, *J. Neurophysiol.*, 66, 212, 1991.
38. Simone, D. A., Baumann, T. K., and LaMotte, R. H., Dose-dependent pain and mechanical hyperalgesia in humans after intradermal injection of capsaicin, *Pain*, 38, 99, 1989.
39. Serra, J., Campero, M., and Ochoa, J. L., Mechanisms of neurogenic flare in human skin, *J. Neurol.*, 241, S34, 1994.
40. Serra, J., Campero, M., and Ochoa, J. L., Common peripheral mechanism for neurogenic flare and hyperalgesia (capsaicin) in human skin, *Muscle Nerve Suppl.*, 1, S250, 1994.
41. Serra, J., Campero, M., and Ochoa, J. L., "Secondary" hyperalgesia (capsaicin) mediated by C-nociceptors, *Soc. Neurosci. Abstr.*, 20, 965, 1993.
42. Serra, J., Campero, M., and Ochoa, J. L., Flare and hyperalgesia following intradermal capsaicin injection in human skin, *J. Physiol.*, submitted.
43. Serra, J., Campero, M., and Ochoa, J., Sensitization of "silent" C-nociceptors in areas of secondary hyperalgesia (SH) in humans, *Neurology*, 45, A36, 1992.
44. Ochoa, J. L. and Yarnitsky, D., Mechanical hyperalgesias in neuropathic pain patients: dynamic and static subtypes, *Ann. Neurol.*, 33, 465, 1993.
45. Torebjörk, H. E., Lundberg, L. E. R., and LaMotte, R. H., Central changes in processing of mechanoreceptive input in capsaicin-induced secondary hyperalgesia in humans, *J. Physiol.*, 448, 765, 1992.
46. Lewis, T., *Vascular Disorders of Limb Described for Practioners and Students*, Macmillan, New York, 1936.
47. Livingston, W. K., *Pain Mechanisms*, Macmillan, New York, 1947, 209.

48. Evans, J. A., Reflex sympathetic dystrophy, *Surg. Clin. North Am.*, 26, 780, 1946.
49. Loh, L. and Nathan, P. W., Painful peripheral states and sympathetic blocks, *J. Neurol. Neurosurg. Psychiatry*, 41, 664, 1978.
50. Lindblom, U. and Verrillo, R. T., Sensory functions in chronic neuralgia, *J. Neurol. Neurosurg. Psychiatry*, 42, 422, 1978.
51. Torebjörk, H. E. and Hallin, R. G., Microneurographic studies of peripheral pain mechanisms in man, in *Advances in Pain Research and Therapy*, Vol. 3, Bonica, J. J., Liebeskind, J. C., Albe-Fessard D. G., Eds., Raven Press, New York, 1979, 121.
52. Campbell, J. N., Raja, S. N., Meyer, R. A., and Mackinnon, S. E., Myelinated afferents signal the hyperalgesia associated with nerve injury, *Pain*, 32, 89, 1988.
53. Simone, D. A., Oh, U., Sorkin, L. S., Owens, C., Chung, J. M., LaMotte, R. H., and Willis, W. D., Neurogenic hyperalgesia: central neural correlates in responses to spinothalamic tract neurons, *J. Neurophysiol.*, 66, 228, 1991.
54. Torebjörk, H. E., Human microneurography and intraneural microstimulation in the study of neuropathic pain, *Muscle Nerve*, 16, 1063, 1993.
55. Burke, D., Microneurography, impulse conduction and paresthesias, *Muscle Nerve*, 16, 1025, 1993.
56. Dotson, R., Ochoa, J. L., Cline, M., Marchettini, P., and Yarnitsky, D., Intraneural microstimulation of low-threshold mechanoreceptors in patients with causalgia/RSD/SMP, *Soc. Neurosci. Abstr.*, 18, 290, 1992.
57. Simone, D. A. and Ochoa, J. L., Early and late effects of prolonged topical capsaicin on cutaneous sensibility and neurogenic vasodilatation in humans, *Pain*, 47, 285, 1991.

Chapter 22

CLINICAL STUDIES ON THE AIRWAY EFFECTS OF SENSORY NEUROPEPTIDES

Guy F. Joos

CONTENTS

I. INTRODUCTION

The presence of substance P (SP) and neurokinin A (NKA) in the upper and lower human airway has been documented by radioimmunoassay and immunocytochemistry.[1-4] However, in comparison to rodents, the SP innervation of the human airways is rather scarce.[5] Nerve fibers containing SP-immunoreactivity (SP-IR) have been described beneath and within the airway epithelium, around blood vessels and submucosal glands, within the bronchial smooth muscle layer, and around local tracheobronchial ganglion cells. These nerve fibers are found in and around bronchi, bronchioles, and more distal airways, occasionally extending into the alveoli.[1]

A vast number of experimental studies, performed mainly in rodents, have demonstrated various airway effects of SP and NKA. These include smooth muscle contraction, submucosal gland secretion, vasodilatation, increase in vascular permeability, stimulation of cholinergic nerves, stimulation of mast cells, stimulation of B- and T-lymphocytes, stimulation of macrophages, chemoattraction of eosinophils and neutrophils, and the vascular adhesion of neutrophils. In view of their potential airway effects, the sensory neuropeptides have been implicated in the pathogenesis of various airway diseases, including asthma, chronic bronchitis, and rhinitis.

In this chapter the knowledge that has been gained from clinical studies on human airways is reviewed. The arguments that favor a possible role for SP and NKA in the pathogenesis of asthma are discussed, and, where appropriate, reference is also made to animal studies.

0-8493-7646-7/96/$0.00+$.50

II. ENHANCED EXPRESSION OF SENSORY NEUROPEPTIDES IN ASTHMA

Although SP-IR nerves invariably are not found in bronchial biopsies from patients with mild asthma,[6] others have reported that in tissue obtained at autopsy, after lobectomy, and at bronchoscopy, both the number and the length of SP-IR nerve fibers was increased in airways of patients with asthma, when compared to airways from those without asthma.[7]

SP has been measured in the bronchoalveolar lavage (BAL) fluid. Nieber et al.[8] examined six atopic subjects with grass pollen allergy and six nonallergic, healthy volunteers. A significantly larger amount of SP was found in the atopic compared to the nonallergic subjects. After intrasegmental provocation with allergen, a significant increase in BAL SP levels was observed in the allergic subjects. SP has now also been measured in sputum induced by inhalation of hypertonic saline: SP-IR was detected in patients with asthma and chronic bronchitis, but not in healthy subjects.[9]

III. CLINICAL STUDIES ON THE BRONCHOCONSTRICTOR EFFECT OF SENSORY NEUROPEPTIDES

A. SP AND NKA CONTRACT HUMAN AIRWAYS

The *in vitro* contractile effect of SP and NKA has been studied extensively. SP contracts human bronchi and bronchioli,[10,11] being less potent than histamine or acetylcholine.[12,13] NKA is a more potent constrictor of human bronchi than SP and was reported to be two to three orders of magnitude more potent than histamine or acetylcholine.[13] In contrast to its effect in guinea pigs, NKB had no contractile effect on human airways.[13] Noncholinergic pathways might be more important in the smaller airways: NKA and SP were found to contract small airways to a larger extent and at lower concentrations compared to the large airways.[14]

The *in vivo* bronchoconstrictor effect of SP and NKA, administered by inhalation or intravenous infusion, has been reported by several groups, as summarized in Table 1. In these studies NKA was found to be a more potent bronchoconstrictor than SP and asthmatics were found to be hyperresponsive to SP and NKA.

Intravenous infusion of SP (0.2 to 3.3 pmol/kg/min) in four normal and two asthmatic subjects produced a pronounced fall in diastolic blood pressure (8.5 ± 2.9 mmHg) and an increase in heart rate. At low infusion rates a small fall in $\dot{V}p30$ was observed (123 ± 45 to 111 ± 47 l/min) ($\dot{V}p30$ = flow at 30% of vital capacity during a partial flow-volume maneuver).[15] Comparing the effect of intravenous SP and NKA, Evans et al. reported that, in six normal subjects, SP was more potent than NKA in raising skin temperature and heart rate. In this study intravenous SP caused bronchodilatation, while NKA up to 64 pmol/kg/min caused a maximal fall in $\dot{V}p30$ of 79%.[16]

We have studied the effect of inhaled SP and NKA on airway caliber in normal and asthmatic airways.[17] Inhalation of SP and NKA by six healthy, nonsmoking subjects did not cause a significant change in specific airways conductance (sGaw). Inhalation of SP, up to 10^{-6} mol/ml, caused no change in sGaw in asthmatics, a finding also reported by Fuller et al.[15] After inhalation of NKA by the asthmatics, a concentration-dependent bronchoconstriction was observed, with a mean decrease in sGaw of 48%, at a concentration of 5×10^{-7} mol/ml. The bronchoconstriction occurred rapidly and sGaw returned to the baseline within 30 min in most patients. All patients reported chest tightness. No cough and no change in heart rate or blood pressure was noted.[17] In a group of 19 asthmatics, we found a weak but nonsignificant correlation between the bronchial responsiveness to methacholine and neurokinin A, suggesting that different mechanisms are involved in the bronchoconstrictor action of these two agents.[18] In a recent study we determined the degree of reproducibility of the bronchial response to NKA in a group of 10 mild asthmatics who were on inhaled

TABLE 1
Contractile Effect of Tachykinins: Studies on Human Airways *In Vivo*

Author	Agonist	Mode of Administration	Lung Function Parameter Studied	Comments
Joos et al.[17,18,25,26]	SP/NKA, normal/asthma	Aerosol	sGaw	SP had no effect in normal and asthmatic subjects. NKA caused a dose-dependent bronchoconstriction in asthmatic subjects.
Fuller et al.[15]	SP, normal/asthma	i.v./aerosol	$\dot{V}$p30/sGaw	
Evans et al.[16]	SP, NKA, normal	i.v.	$\dot{V}$p30	
Crimi et al.[20,27]	SP, asthma	Aerosol	FEV_1	SP caused bronchoconstriction in moderate asthmatic patients.
Nakai et al.[24]	SP, asthmatic children	Aerosol	FEV_1/$\dot{V}$50	
Cheung et al.[21–23]	NKA, normal	Aerosol	$\dot{V}$40p	No change in bronchial responsiveness to methacholine 24 h after NKA.
	NKA, asthma	Aerosol	FEV_1/$\dot{V}$40p	
	SP, asthma	Aerosol	FEV_1	Inhaled SP caused a drop in diastolic blood pressure.

Note: sGaw, specific airways conductance; $\dot{V}_{50}$, expiratory flow at 50% of vital capacity; FEV_1, forced expiratory volume in 1 s; $\dot{V}$40p, expiratory flow from partial expiratory flow-volume curve; $\dot{V}$p30, airflow at 70% of vital capacity measured from total lung capacity after a forced partial expiratory flow maneuver.

sympathomimetics only. In 9 of the 10 asthmatics the PC_{35} sGaw NKA (the provoking concentration of NKA causing a 35% decrease in sGaw) performed with an interval of at least 1 week, was reproducible, differing less than 0.5 log units from each other.[19]

The *in vivo* bronchoconstrictor effect of inhaled sensory neuropeptides in man has been confirmed and extended by other groups. Studying patients with a more pronounced bronchial responsiveness, Crimi et al. demonstrated that inhaled SP is also able to cause bronchoconstriction in asthmatic patients.[20] High doses of SP, up to 8 mg/ml, caused a significant decrease in the forced expiratory volume in one second (FEV_1) but also a significant fall in blood pressure and increase in heart rate.[21] Using an efficient nebulizer system and a sensitive measure of airway caliber, Cheung et al. found that NKA caused bronchoconstriction not only in asthmatics but also in normal persons, the asthmatics being more sensitive than the normal subjects.[22,23] Aerosolized SP was also shown to cause bronchoconstriction in children with asthma, an effect that was dependent on the severity of the asthma.[24]

B. SP AND NKA ARE INDIRECT BRONCHOCONSTRICTOR AGENTS

To study the mechanism of NKA-induced bronchoconstriction in asthmatics, the effect of nedocromil sodium and the anticholinergic agent oxitropium bromide on NKA-induced bronchoconstriction in mild asthmatics has been studied. Nedocromil sodium prevented the bronchoconstriction caused by NKA[25] and SP.[20] Oxitropium bromide offered some protection against the bronchoconstrictor effect of NKA, an effect that was evident in 4 out of 11 asthmatics.[26] In seven moderate asthmatic patients ipratropium bromide caused a small but significant rightward shift in the dose-response curve to SP.[27] The protective effect offered by nedocromil sodium against tachykinin-induced bronchoconstriction in asthmatics indicates that, unlike histamine and methacholine, SP and NKA cause bronchoconstriction by an indirect mechanism. This could arise from an effect on inflammatory cells (e.g., mast cells) and/or nerves. An alternative explanation is that nedocromil sodium and sodium cromoglycate act as a tachykinin antagonist, a suggestion that was raised from a study with sodium cromoglycate in human skin.[28]

In experimental animals tachykinins are able to cause acetylcholine release from postganglionic cholinergic airway nerve endings.[29-31] In normal human airways sensory neuropeptides do not enhance the effects of cholinergic neural stimulation,[31,32] except in the presence of K^+ channel blockade.[32] Indeed, in human isolated airways, NKA, but not SP, potentiated the contractile response to cholinergic neural stimulation in the presence of K^+ channel blockade with 4-aminopyridine.[32]

Histamine release by SP has been documented in rat lung *in vitro*[33] and *in vivo*.[34] The protective effect of nedocromil sodium against NKA-induced bronchoconstriction in asthmatics suggests a possible involvement of mast cells in the human airway response towards tachykinins. Although SP released histamine in the human skin and from dispersed human skin mast cells, it did not cause histamine release from human lung mast cells that were obtained from patients undergoing pneumonectomy for lung cancer.[33,35] However, a recent report suggests that SP can release histamine from mast cells recovered from BAL fluid.[36] Moreover, mast cells recovered from BAL fluid of asthmatic patients are more sensitive to SP.[37] Nevertheless, the lack of effect of pretreatment with astemizole and terfenadine, two potent and specific H_1-antagonists, on NKA-induced bronchoconstriction in asthmatics would tend to suggest that histamine does not play a major role in NKA-induced bronchoconstriction in man.[27,38]

C. NEUTRAL ENDOPEPTIDASE IS INVOLVED IN THE BRONCHOCONSTRICTOR RESPONSE TO TACHYKININS IN MAN

The role of neutral endopeptidase (NEP) in the airways and its relation to nonadrenergic, noncholinergic (NANC) airway innervation has been mainly studied in animal airways.[39,40] NEP has been localized in the lung of different animal species, both by biochemical studies (i.e., measurement of enzyme activity) and by immunohistochemistry. It was shown that the *in vitro* contractile response of human bronchi to NKA, SP, and the nonmammalian tachykinins eledoisin and physalaemin was potentiated by phosphoramidon.[41] Moreover, phosphoramidon was shown to potentiate not only the SP-induced contraction, but it also increased and prolonged the contraction induced by capsaicin.[42] Thus, NEP regulates the contractile effect of SP and of endogenous substances, probably tachykinins, which are released from capsaicin-sensitive nerves.

The effect of an inhaled inhibitor of NEP on the bronchoconstrictor effect of NKA in man has recently been studied.[22,23] Cheung et al. demonstrated that the inhalation of thiorphan (0.5 ml of 2.5 mg/ml) 10 min before NKA challenge enhanced the bronchoconstrictor effect of NKA both in normal persons[22] and in asthmatic subjects.[23] Interestingly, the leftward shift of the NKA dose-response curve was not significantly different in the asthmatic compared to nonasthmatic subjects. In neither group did thiorphan have any effect on baseline lung function nor on the bronchial responsiveness to methacholine. Using another inhibitor of NEP, phosphoramidon, Crimi et al.[43] have confirmed these findings. They found that the inhalation of phosphoramidon by six asthmatics had no effect on baseline airway tone, but caused a significant leftward shift of the dose-response curve to inhaled NKA.

These two *in vivo* studies offer functional proof that NEP is involved in the *in vivo* breakdown of inhaled NKA in man, confirming findings obtained on isolated human airways.[41,42] As inhibition of NEP offers a similar shift of the dose-response curve both in normal and nonsmoking stable, mild to moderate asthmatics, it can be argued that the functional activity of NEP is not significantly decreased in stable asthmatics. Indeed it has been hypothesized that the epithelial damage frequently observed in asthmatic airways could result in decreased NEP activity, enhancing neurogenic airway inflammation. Moreover, as neither thiorphan nor phosphoramidon influenced baseline lung function, it can be assumed that tachykinins (or another peptide susceptible to breakdown by NEP), unlike histamine, leukotrienes, or acetylcholine, are not involved in the maintenance of basal airway tone.[44]

The studies by Cheung et al. and Crimi et al. would therefore tend to argue against an important role for endogenously released tachykinins in the pathogenesis of asthma. Obviously several caveats remain in the interpretation and generalization of these findings. Higher doses of potent NEP inhibitors might be necessary to unmask subtle quantitative differences in NEP activity between normal subjects and mild asthmatics. Moreover, it is not known whether a more pronounced loss of NEP activity occurs in more severe asthma. Since it has been shown that viral infections and exposure to occupational agents such as TDI result in loss of NEP activity, this mechanism may be important in well-defined subsets of asthma or during exacerbations.[44]

IV. CLINICAL STUDIES ON THE EFFECTS OF SENSORY NEUROPEPTIDES IN THE NOSE

The effects of exogenously administered SP, NKA and calcitonin gene-related peptide (CGRP) have also been studied in the nose. Topical administration of SP and NKA increased nasal airway resistance in a dose-dependent way, SP being more potent than NKA and methacholine.[45,46] The increase in nasal airway resistance was more pronounced in patients with allergic rhinitis, compared to control subjects.[45] Furthermore, in patients with allergic rhinitis, nasal challenge with SP or NKA increased the amount of protein and albumin, which was taken as evidence in favor of plasma protein extravasation.[46] SP also increased the percentage of neutrophils recovered from nasal lavage.[46,47] In contrast, application of CGRP to the nasal mucosa did not stimulate glandular or plasma protein extravasation, which is in line with animal studies showing that CGRP does not affect baseline plasma protein extravasation.[48]

As for the lower airways, neutral endopeptidase was found to be involved in the upper airway response to exogenously administered SP: oral administration of the NEP inhibitor acetorphan potentiated the effects of SP on nasal airway resistance. As in the patients with asthma, no difference was observed between the normal subjects and the subjects with allergic rhinitis.[49] Moreover, angiotensin-converting enzyme also seems to be involved in the breakdown of exogenously administered SP in allergic rhinitis patients.[49]

V. OTHER AIRWAY EFFECTS OF SENSORY NEUROPEPTIDES IN MAN

A. MUCUS SECRETION

Receptors for SP have been demonstrated by autoradiographic labeling on human airway submucosal glands.[50] SP was found to be more potent than NKA in inducing mucus secretion in human bronchi.[51] Capsaicin has also been shown to stimulate mucus secretion in surgically resected human bronchi *in vitro*, an effect that can be blocked by morphine.[52] However, at the present time no clinical studies on this important airway effect are available.

B. COUGH

In guinea pigs exogenously administered SP and the NEP inhibitor phosphoramidon cause cough.[53] However, in clinical studies in normal and asthmatic persons, no cough was observed after inhalation of SP, NKA, or thiorphan.[17,20,22,23] Sensory neuropeptides may, however, be involved in the cough response to stimuli such as citric acid.[54,55]

C. VASCULAR PERMEABILITY

Neurogenic inflammation has been described in the airways of rodents. Cigarette smoke and light mechanical or local chemical irritation by ether, formalin, bradykinin, or capsaicin also induced an increase in vascular permeability. Pretreatment with capsaicin, destroying nonmyelinated sensory airway nerve fibers, reduced the Evans blue extravasation to nerve stimulation. It is difficult to approach this problem in human airways. Some studies have been done on the nose (*vide supra*), but data on human lower airways are lacking. However, from studies on receptor localization we know that the SP binding sites in the human respiratory tract are predominantly located on the microvessels: in studies by Springall and Polak, a high silver-grain density representing iodinated Bolton-Hunter SP binding predominantly to the subepithelial microvessels, was shown.[5] A direct measurement of vascular permeability changes in human lower airways is not available at present.

D. PROINFLAMMATORY EFFECTS

The first evidence for a direct effect of SP on lymphocytes was the demonstration that nanomolar concentrations of SP alone could stimulate human peripheral blood lymphocyte proliferation and could enhance the response to the T-cell mitogens phytohemaglutinin A (PHA) and concanavalin A (con A).[56] SP is also able to stimulate macrophage, neutrophil, and eosinophil function. SP has a potent chemotactic effect on human monocytes. The production of cytokines by macrophages is also influenced by SP: nanomolar concentrations stimulate the secretion of interleukin (IL)-1, tumor necrosis factor-α, and IL-6.[57] In contrast, SP in concentrations up to 10^{-4} *M* was not able to stimulate the release of TxB_2 from bronchoalveolar macrophages derived from both normal and asthmatic subjects.[58] SP is also a chemoattractant for neutrophils, an effect that probably occurs through a subset of Pertussis toxin-sensitive G proteins not coupled to classical phospholipases.[59] Recently SP was found to induce in human dermis a rapid influx of neutrophils and eosinophils, effects that occurred in parallel with translocation of P-selectin and a significant upregulation of E-selectin, suggesting that SP can induce endothelial adhesion molecule expression.[60] Numao and Agrawal[61] demonstrated that eosinophils from allergic and normal subjects differed with regard to their chemotactic response to SP. SP alone was not chemotactic for eosinophils, whereas the chemotactic response to PAF and LTB_4 of eosinophils derived from asthmatic but not normal subjects was enhanced by pretreatment with nanomolar quantities of SP.[61] As it has been described for isolated mast cells, high concentrations of SP were also found to induce degranulation of eosinophils.[62]

VI. CLINICAL STUDIES ON THE RELEASE OF SP INTO AIRWAYS

SP and NKA are thus present in human airway nerves and, when exogenously applied, have potent excitatory actions on both upper and lower human airways. NEP is present in the airways and involved in the breakdown of exogenously administered SP. One of the important questions is whether sensory nerves indeed release SP and NKA in various airway disorders such as asthma and rhinitis. One approach to this problem has been to measure the amount of SP released into nasal or BAL fluid. Increased amounts of SP have been recovered from nasal and BAL fluid after the administration of allergen in the nose or through the bronchoscope.[8,63] These studies suggest that sensory neuropeptides can indeed be released from sensory nerves into the airways during an allergic reaction.

Another approach has been the use of the neurotoxin capsaicin. When inhaled capsaicin causes cough. It also causes a mild, transient bronchoconstriction which is largely due to a parasympathetic reflex.[64,65] In the guinea pig sensory neuropeptides are involved in these airway responses; evidence for such a mechanism in human airways is, however, lacking at the present time.

Intranasal application of capsaicin has been used to study the role of the "capsaicin-sensitive nerves". Capsaicin caused nasal irritation and burning pain, sneezing, nasal congestion, and secretion both in normal subjects and in subjects with allergic and vasomotor

TABLE 2
Tachykinin Receptors in Various Animal Models

- Nonadrenergic, noncholinergic (NANC) bronchoconstriction in guinea pigs: NK_2 >> NK_1 (Maggi et al.,[76] Hirayama et al.[77])
- Plasma extravasation in rodents: NK_1 (Abelli et al.,[78] Sakamoto et al.[79])
- Antigen-induced plasma extravasation in guinea pigs: NK_1 (Bertrand et al.[80])
- Increased tracheal vascular permeability by hypertonic saline in rats: NK_1 (Piedimonte et al.[81])
- Hyperpnea-induced bronchoconstriction in guinea pigs: NK_1 and NK_2 (Solway et al.[82])
- Citric acid-induced bronchoconstriction and cough: NK_2 (Satoh et al.,[54] Advenier et al.[55])
- Indirect bronchoconstriction and airway mast cell activation by SP and NKA in the F344 rat: NK_1; direct bronchoconstriction in the BDE rat: NK_2 (Joos et al.[83])

rhinitis.[66-69] The effect of capsaicin was more pronounced in patients with vasomotor rhinitis.[67,68] Capsaicin also induced a contralateral secretory response. As pretreatment with a locally and systemically administered muscarinic receptor antagonist almost completely blocked the secretory response to capsaicin, the involvement of cholinergic parasympathetic reflexes has been suggested.[67] Although capsaicin did enhance TAMEesterase, a presumable marker of glandular secretion,[68] it did not increase albumin or total protein content in nasal lavage fluid.[68,69] So, in contrast to SP, capsaicin had no measurable effect on the vascular permeability in the nose.

However, effects other than vascular permeability may be more important. For instance in a study on 16 adults with chronic nonallergic rhinitis, Lacroix et al. found an enhanced vascular response to capsaicin compared to normal subjects.[70] Both nasal vascular responses and subjective discomfort were markedly reduced after the fifth application of capsaicin. Interestingly, these changes in nasal response to capsaicin were accompanied by a 50% reduction in the content of CGRP in nasal biopsies.[70]

VII. EVALUATION OF TACHYKININ RECEPTOR ANTAGONISTS IN HUMAN AIRWAYS

The NK_1 and NK_2 tachykinin receptors have been found to be involved in the various airway effects of the tachykinins.[71] Both the human NK_1 receptor and NK_2 receptor have been cloned from human lung and trachea.[72-75]

A. ANIMAL MODELS

From studies on animal airways, NK_2 receptors and, to a lesser extent, NK_1 receptors have been shown to be involved in bronchoconstriction, whereas NK_1 receptors were found to be involved in mucus secretion, microvascular leakage, vasodilation, and most of the effects on inflammatory cells. Tachykinin antagonists have been tested in various animal models (Table 2). In F344 rats we showed that the specific and potent NK_1 receptor antagonists RP 67580 and CP-96,345 reduced the bronchoconstrictor effect and the airway mast cell activation caused by NKA, whereas a NK_2 receptor antagonist only affected the direct bronchoconstrictor effect of NKA.[83] FK224 is a recently described cyclopeptide tachykinin antagonist, which inhibits NK_1 and NK_2 receptor-mediated airway responses in the guinea pig.[77,84,85] Again, in our animal model, this compound was able to block both the bronchoconstriction and the airway mast cell activation induced by NKA in F344 rats.[86]

B. HUMAN STUDIES

The human airway tachykinin receptor involved in bronchoconstriction has, at the present time, been characterized *in vitro* only. Naline et al. demonstrated that the NK_2 receptor-selective agonist [Nle^{10}]NKA(4-10) was a potent contractor of isolated human bronchi, whereas agonists selective for the NK_1 and NK_3 receptor were almost inactive.[87] The recently described potent and specific nonpeptide receptor antagonists have been evaluated on isolated human bronchi. Advenier et al.[88] showed that the potent and specific NK_2 receptor-antagonist, SR 48968, displayed competitive antagonism for the contractions induced by Nle^{10}NKA(4-10). No effect on acetylcholine-, histamine-, KCl-, or $PGF_2\alpha$-induced contractions was seen. SR 48968 also shifted the dose-response curve for SP to the right. The specific NK_1 receptor antagonist CP-96,345 had no effect.[88] These *in vitro* results suggest that the contraction induced by SP and NKA in large, normal airways is mediated by NK_2 receptors, probably of the NK_{2A} subtype.[89,90] However, recent evidence has been presented that NK_1 receptors are also involved in the contraction of normal human airways.[91,92]

At the present time two tachykinin receptor antagonists have been evaluated in clinical trials: FK224, a cyclic peptide tachykinin antagonist for NK_1 and NK_2 receptors, and CP-99,994, a nonpeptide NK_1 tachykinin receptor antagonist.

FK224, 4 mg given by metered-dose inhaler, was shown to inhibit bradykinin-induced bronchoconstriction and cough in nine asthmatics.[93] We have recently studied the effect of inhaled FK224 on NKA-induced bronchoconstriction in 10 mild asthmatics.[19] On day 1 baseline lung function and PC_{20} methacholine (provocative concentration causing a 20% decrease in FEV_1) were determined. On days 2 and 3, doubling concentrations of NKA (3.3×10^{-9} to 1.0×10^{-6} mol/ml) were administered via a Wiesbadener-Doppel inhalator, with intervals of 10 min. On both days NKA caused a concentration-dependent decrease in sGaw. Mean (S.E.M.) log PC_{35} sGaw NKA (moles per milliliter) was –6.61 (0.10) on day 2 and –6.57 (0.14) on day 3 (NS). On days 4 and 5 FK224 (4 mg) or placebo were administered by a metered-dose inhaler, 30 min before NKA challenge in a double-blind, crossover manner. The study medication was well tolerated. FK224 had no significant effect on baseline FEV_1 and sGaw, measured 15 and 30 min later. After placebo and FK224, NKA caused a comparable concentration-dependent bronchoconstriction. The mean (S.E.M.) log PC_{35} sGaw NKA (moles per milliliter) was –6.04 (0.18) after placebo and –6.19 (0.23) after FK224 (NS). So, inhaled FK224 had no effect on baseline lung function and offered no protection against NKA-induced bronchoconstriction in this group of mildly asthmatic patients.[19] FK224, 4 mg via metered-dose inhaler q.i.d., administered over 4 weeks, was studied in patients with mild to moderate asthma. Compared to placebo no beneficial effects of FK224 on symptoms and lung function were found.[94] However, as FK224 was not able to antagonize the airway effects of NKA in asthmatics, the study by Lunde et al.[94] does not reject a possible role for sensory neuropeptides in asthma.

Recently, a preliminary report described the effect of the nonpeptide tachykinin NK_1 receptor antagonist CP-99,994, given intravenously (250 μg/kg). Compared to placebo, it had no effect on bronchoconstriction and cough induced by hypertonic saline in 14 male asthmatic subjects.[95] However, it remains to be proven whether this compound, at the given dose, is able to block the airway responses induced by inhaled SP or NKA. Clinical trials with other potent nonpeptide tachykinin antagonists, such as SR 48968, are needed to further elucidate the role of sensory neuropeptides in the pathogenesis of asthma.

VIII. CONCLUSIONS: FUTURE DIRECTIONS

There is now convincing evidence for the presence of SP and NKA in human airway nerves. Studies on autopsy tissue, BAL, and sputum suggest that SP may be present in increased amounts in the asthmatic airway. The intranasal or endobronchial instillation of allergen can release SP in the airways. SP and NKA cause bronchoconstriction and increase nasal resistance. The major enzyme responsible for the degradation of the tachykinins, the neutral endopeptidase (NEP), is present in the airways and is involved in the breakdown of exogenously administered SP and NKA. Other, less well documented, airway effects of SP and NKA include mucus secretion, vasodilation, and plasma extravasation, as well as the chemoattraction and stimulation of the various cells presumed to be involved in the allergic airway inflammation.

Hence, SP and NKA fulfill two of the three criteria that a presumed mediator of inflammation has to meet: their presence and release in the airways and their ability to mimic various features of the disease (asthma, rhinitis). Clinical studies with the potent and selective tachykinin antagonists in various models and forms of asthma and rhinitis will allow us to further define the role of the sensory neuropeptides SP and NKA in the pathogenesis of inflammatory airway disorders.

REFERENCES

1. Lundberg, J. M., Hökfelt, T., Martling, C.-R., Saria, A., and Cuello, C., Substance P immunoreactive sensory nerves in the lower respiratory tract of various mammals including man, *Cell Tissue Res.*, 235, 251, 1984.
2. Martling, C.-R., Theodorsson-Norheim, E., and Lundberg, J. M., Occurrence and effects of multiple tachykinins; Substance P, neurokinin A and neuropeptide K in human lower airways, *Life Sci.*, 40, 1633, 1987.
3. Luts, A., Uddman, R., Alm, P., Basterna, J., and Sundler, F., Peptide-containing nerve fibers in human airways: distribution and coexistence pattern, *Int. Arch. Allergy Immunol.*, 101, 52, 1993.
4. Baraniuk, J. N., Lundgren, J. S., Okayama, M., Goff, J., Mullol, J., Merida, M., Shelhamer, J. H., and Kaliner, M. A., Substance P and neurokinin A in human nasal mucosa, *Am. J. Respir. Cell. Mol. Biol.*, 228, 4, 1991.
5. Springall, D. R. and Polak, J. M., Neuropeptides in the lower airways investigated by modern microscopy, in *Neuropeptides in Respiratory Medicine*, Kaliner, M. A., Barnes, P. J., Kunkel, G. H. H., and Baraniuk, J. N., Eds., Marcel Dekker, New York, 1994, chap. 3.
6. Howarth, P. H., Djukanovic, R., Wilson, J. W., Holgate, S. T., Springall, D. R., and Polak, J. M., Mucosal nerves in endobronchial biopsies in asthma and non-asthma, *Int. Arch. Allergy Appl. Immunol.*, 94, 330, 1991.
7. Ollerenshaw, S. L., Jarvis, D., Sullivan, C. E., and Woolcock, A. J., Substance P immunoreactive nerves in airways from asthmatics and nonasthmatics, *Eur. Respir. J.*, 4, 673, 1991.
8. Nieber, K., Baumgarten, C. R., Rathsack, R., Furkert, J., Oehme, P., and Kunkel, G., Substance P and β-endorphin-like immunoreactivity in lavage fluid of subjects with and without allergic asthma, *J. Allergy Clin. Immunol.*, 90, 646, 1992.
9. Tomaki, M., Ichinose, M., Nakajima, N., Miura, M., Yamauchi, H., Inoue, H., and Shirato, K., Elevated Substance P concentration in sputum after hypertonic saline inhalation in asthma and chronic bronchitis patients, *Am. Rev. Respir. Dis.*, 147 (Abstr.), A478, 1993.
10. Lundberg, J. M., Martling, C.-R., and Saria, A., Substance P and capsaicin induced contraction of human bronchi, *Acta Physiol. Scand.*, 119, 49, 1983.
11. Finney, M. J. B., Karlsson, J.-A., and Persson, C. G. A., Effects of bronchoconstrictors and bronchodilators on a novel human small airway preparation, *Br. J. Pharmacol.*, 85, 29, 1985.
12. Martling, C.-R., Theodorsson-Norheim, E., and Lundberg, J. M., Occurrence and effects of multiple tachykinins; Substance P, neurokinin A and neuropeptide K in human lower airways, *Life Sci.*, 40, 1633, 1987.
13. Advenier, C., Naline, E., Drapeau, G., and Regoli, D., Relative potencies of neurokinins in guinea pig trachea and human bronchus, *Eur. J. Pharmacol.*, 139, 133, 1987.
14. Frossard, N. and Barnes, P. J., Effect of tachykinins in small human airways, *Neuropeptides*, 19, 157, 1991.
15. Fuller, R. W., Maxwell, D. L., Dixon, C. M. S., McGregor, G. P., Barnes, V. F., Bloom, S. R., and Barnes, P. J., Effect of Substance P on cardiovascular and respiratory function in human subjects, *J. Appl. Physiol.*, 62, 1473, 1987.
16. Evans, T. W., Dixon, C. M., Clarke, B., Conradson, T.-B., and Barnes, P. J., Comparison of neurokinin A and Substance P on cardiovascular and airway function in man, *Br. J. Clin. Pharmacol.*, 25, 273, 1988.
17. Joos, G., Pauwels, R., and Van der Straeten, M., The effect of inhaled Substance P and neurokinin A on the airways of normal and asthmatic subjects, *Thorax*, 42, 779, 1987.
18. Joos, G. F., The role of sensory neuropeptides in the pathogenesis of asthma, *Clin. Exp. Allergy*, 19 (Suppl. 1), 9, 1989.
19. Joos, G. F., Van Schoor, J., Kips, J. C., and Pauwels, R. A., The effect of inhaled FK224, a NK-1 and NK-2 receptor antagonist on neurokinin A-induced bronchoconstriction in asthmatics, *Am. J. Respir. Crit. Care Med.*, 149 (Abstr.), A890, 1994.
20. Crimi, N., Palermo, F., Oliveri, R., Palermo, B., Vancheri, C., Polosa, R., and Mistretta, A., Effect of nedocromil on bronchospasm induced by inhalation of Substance P in asthmatic subjects, *Clin. Allergy*, 18, 375, 1988.

21. Cheung, D., van der Veen, H., den Hartigh, J., Dijkman, J. H., and Sterk, P. J., Effects of inhaled Substance P on airway responsiveness to methacholine in asthmatic subjects in vivo, *J. Appl. Physiol.*, 77, 1325, 1994.
22. Cheung, D., Bel, E. H., Den Hartigh, J., Dijkman, J. H., and Sterk, P. J., The effect of an inhaled neutral endopeptidase inhibitor, thiorphan, on airway responses to neurokinin A in normal humans in vivo, *Am. Rev. Respir. Dis.*, 145, 1275, 1992.
23. Cheung, D., Timmers, M. C., Zwinderman, A. H., Den Hartigh, J., Dijkman, J. H., and Sterk, P. J., Neutral endopeptidase activity and airway hyperresponsiveness to neurokinin A in asthmatic subjects in vivo, *Am. Rev. Respir. Dis.*, 148, 1467, 1993.
24. Nakai, J., Iukura, Y., Akimoto, K., and Shiraki, K., Substance P-induced cutaneous and bronchial reactions in children with bronchial asthma, *Ann. Allergy*, 66, 155, 1991.
25. Joos, G., Pauwels, R., and Van der Straeten, M., The effect of nedocromil sodium on the bronchoconstrictor effect of neurokinin A in asthmatics, *J. Allergy Clin. Immunol.*, 83, 663, 1989.
26. Joos, G. F., Pauwels, R. A., and Van Der Straeten, M. E., The effect of oxitropiumbromide on neurokinin A-induced bronchoconstriction in asthmatics, *Pulmon. Pharmacol.*, 1, 41, 1988.
27. Crimi, N., Palermo, F., Oliveri, R., Palermo, B., Vancheri, C., Polosa, R., and Mistretta, A., Influence of antihistamine (astemizole) and anticholinergic drugs (ipratropium bromide) on bronchoconstriction induced by Substance P, *Ann. Allergy*, 65, 115, 1990.
28. Crossman, D. C., Dashwood, M. R., Taylor, G. W., Wellings, R., and Fuller, R. W., Sodium cromoglycate — evidence of tachykinin antagonist activity in the human skin, *J. Appl. Physiol.*, 75, 167, 1993.
29. Tanaka, D. T. and Grunstein, M. M., Mechanisms of Substance P induced contraction of rabbit airway smooth muscle, *J. Appl. Physiol.*, 57, 1551, 1984.
30. Hall, A. K., Barnes, P. J., Meldrum, L. A., and Maclagan, J., Facilitation by tachykinins of neurotransmission in guinea-pig pulmonary parasympathetic nerves, *Br. J. Pharmacol.*, 97, 274, 1989.
31. Belvisi, M. G., Patacchini, R., Barnes, P. J., and Maggi, C. A., Facilitatory effects of selective agonists for tachykinin receptors on cholinergic neurotransmission: evidence for species differences, *Br. J. Pharmacol.*, 111, 103, 1994.
32. Black, J. L., Johnson, P. R. A., Alouan, L., and Armour, C. L., Neurokinin, A with K^+ channel blockade potentiates contraction to electrical stimulation in human bronchus, *Eur. J. Pharmacol.*, 180, 311, 1990.
33. Ali, H., Leung, K. B. P., Pearce, F. L., Hayes, A., and Foreman, J. C., Comparison of the histamine releasing action of Substance P on mast cells and basophils from different species and tissues, *Int. Arch. Allergy Appl. Immunol.*, 79, 121, 1983.
34. Joos, G. and Pauwels, R., The in vivo effect of tachykinins on airway mast cells of the rat, *Am. Rev. Respir. Dis.*, 148, 922, 1993.
35. Lawrence, I. D., Warner, J. A., Cohan, V. L., Hubbard, W. C., Kagey-Sobotka, A., and Lichtenstein, L. M., Purification and characterization of human skin mast cells. Evidence for human mast cell heterogeneity, *J. Immunol.*, 139, 3062, 1989.
36. Heaney, L. G., Cross, L. J. M., Stanford, C. F., and Ennis, M., Stimulation of bronchoalveolar lavage mast cells by Substance P, *Eur. Respir. J.*, 6 (Suppl. 17), 435s (Abstr.), 1993.
37. Cross, L. J. M., Heaney, L. G., Stanford, C. F., and Ennis, M., Mast cell and basophil stimulation by neuropeptides in allergic asthma, *Allergy Clin. Immunol. News Suppl.*, 2 (Abstr.), A1036, 1994.
38. Crimi, N., Oliveri, R., Polosa, R., Palermo, F., and Mistretta, A., The effect of oral terfenadine on neurokinin-A induced bronchoconstriction, *J. Allergy Clin. Immunol.*, 91, 1096, 1993.
39. Nadel, J. A., Neutral endopeptidase modulates neurogenic inflammation, *Eur. Respir. J.*, 4, 745, 1991.
40. Borson, D. B., Roles of neutral endopeptidase in airways, *Am. J. Physiol.*, 260, L212, 1991.
41. Black, J. L., Johnson, P. R. A., and Armour, C. L., Potentiation of the contractile effects of neuropeptides in human bronchus by an enkephalinase inhibitor, *Pulmon. Pharmacol.*, 1, 21, 1988.
42. Honda, I., Kohrogi, H., Yamaguchi, T., Ando, M., and Araki, S., Enkephalinase inhibitor potentiates Substance P- and capsaicin-induced bronchial smooth muscle contractions in humans, *Am. Rev. Respir. Dis.*, 143, 1416, 1991.
43. Crimi, N., Palermo, F., Oliveri, R., Raccuglia, D. R., Pulverenti, G., and Mistretta, A., Inhibition of neutral endopeptidase potentiates bronchoconstriction induced by neurokinin A in asthmatic patients, *Clin. Exp. Allergy*, 24, 115, 1994.
44. Joos, G., Kips, J., and Pauwels, R., A role for neutral endopeptidase in asthma?, *Clin. Exp. Allergy*, 24, 91, 1994.
45. Devillier, P., Dessanges, J. F., Rakotosihanaka, F., Ghaem, A., Boushey, H. A., Lockhart, A., and Marsac, J., Nasal response to Substance P and methacholine in subjects with and without allergic rhinitis, *Eur. Respir. J.*,1, 356, 1988.
46. Braunstein, G., Fajac, I., Lacronique, J., and Frossard, N., Clinical and inflammatory responses to exogenous tachykinins in allergic rhinitis, *Am. Rev. Respir. Dis.*, 144, 630, 1991.
47. Braunstein, G., Buvry, A., Lacronique, J., Desjardins, N., and Frossard, N., Do nasal mast cells release histamine on stimulation with Substance P in allergic rhinitis?, *Clin. Exp. Allergy*, 24, 922, 1994.

48. Guarnaccia, S., Baraniuk, J. N., Bellanti, J., and Duina, M., Calcitonin gene-related peptide nasal provocation in humans, *Ann. Allergy*, 72, 515, 1994.
49. Lurie, A., Nadel, J. A., Roisman, G., Siney, H., and Dusser, D. J., Role of neutral endopeptidase and kininase II on Substance P-induced increase in nasal obstruction in patients with allergic rhinitis, *Am. J. Respir. Crit. Care Med.*, 149, 113, 1994.
50. Carstairs, J. R. and Barnes, P. J., Autoradiographic mapping of Substance P receptors in human lung, *Eur. J. Pharmacol.*, 127, 295, 1986.
51. Rogers, D. F., Aursudkjii, B., and Barnes, P. J., Effects of tachykinins on mucus secretion on human bronchi in vitro, *Eur. J. Pharmacol.*, 174, 283, 1989.
52. Rogers, D. F. and Barnes, P. J., Opioid inhibition of neurally mediated mucus secretion in human bronchi, *Lancet*, ii, 930, 1989.
53. Ujiie, Y., Sekizawa, K., Aikawa, T., and Sasaki, H., Evidence for Substance P as an endogenous substance causing cough in guinea pigs, *Am. Rev. Respir. Dis.*, 148, 1628, 1993.
54. Satoh, H., Lou, Y. P., Lee, L. Y., and Lundberg, J. M., Inhibitory effects of capsazepine and the NK_2 antagonist SR 48968 on bronchoconstriction evoked by sensory nerve stimulation in guinea-pigs, *Acta Physiol. Scand.*, 146, 535, 1992.
55. Advenier, C., Girard, V., Naline, E., Vilain, P., and Edmonds-Alt, X., Antitussive effect of SR 48968, a non-peptide tachykinin NK_2 receptor antagonist, *Eur. J. Pharmacol.*, 250, 169, 1993.
56. Payan, D. G., Brewster, D. R., and Goetzl, E. J., Specific stimulation of human T-lymphocytes by Substance P, *J. Immunol.*, 131, 1613, 1983.
57. McGillis, J. P., Mitsuhashi, M., and Payan, D. G., Immunomodulation by tachykinin neuropeptides, *Ann. N.Y. Acad. Sci.*, 594, 85, 1990.
58. Pujol, J.-L., Bousquet, J., Grenier, J., Michel, F., Godard, P., Chanez, P., De Vos, C., De Paulet, A. C., and Michel, F.-B., Substance P activation of bronchoalveolar macrophages from asthmatic patients and normal subjects, *Clin. Exp. Allergy*, 19, 625, 1989.
59. Haines, K. A., Kolasinski, S. L., Cronstein, B. N., Reibman, J., Gold, L. I., and Weissmann, G., Chemoattraction of neutrophils by Substance P and transforming growth factor-β1 is inadequately explained by current models of lipid remodeling, *J. Immunol.*, 151, 1491, 1993.
60. Smith, C. H., Barker, J. N. W. N., Morris, R. W., McDonald, D. M., and Lee, T. H., Neuropeptides induce rapid expression of endothelial cell adhesion molecules and elicit granulocytic infiltration in human skin, *J. Immunol.*, 151, 3274, 1993.
61. Numao, T. and Agrawal, D. K., Neuropeptides modulate human eosinophil chemotaxis, *J. Immunol.*, 149, 3309, 1992.
62. Kroegel, C., Giembycz, M. A., and Barnes, P. J., Characterization of eosinophil activation by peptides. Differential effects of Substance P, mellitin, and f-Met-leu-Phe, *J. Immunol.*, 145, 2581, 1990.
63. Mosimann, B. L., White, M. V., Hohman, R. J., Goldrich, M. S., Kaulbach, H. C., and Kaliner, M. A., Substance P, calcitonin gene-related peptide, and vasoactive intestinal peptide increase in nasal secretions after allergen challenge in atopic patients, *J. Allergy Clin. Immunol.*, 92, 95, 1993.
64. Fuller, R. W., Dixon, C. M. S., and Barnes, P. J., Bronchoconstrictor response to inhaled capsaicin in man, *J. Appl. Physiol.*, 58, 1080, 1985.
65. Midgren, B., Hansson, L., Karlsson, J.-A., Simonsson, B. G., and Persson, C. G. A., Capsaicin-induced cough in humans, *Am. Rev. Respir. Dis.*, 146, 347, 1992.
66. Geppetti, P., Fusco, B. M., Marabini, S., Maggi, C. A., Fanciullacci, M., and Sicuteri, F., Secretion, pain and sneezing induced by application of capsaicin to the nasal mucosa in man, *Br. J. Pharmacol.*, 93, 503, 1988.
67. Stjärne, P., Lundblad, L., Lundberg, J. M., and Anggard, A., Capsaicin and nicotine-sensitive afferent neurones and nasal secretion in healthy human volunteers and in patients with vasomotor rhinitis, *Br. J. Pharmacol.*, 96, 693, 1989.
68. Bascom, R., Kagey-Sobotka, A., and Proud, D., Effect of intranasal capsaicin on symptoms and mediator release, *J. Pharmacol. Exp. Ther.*, 259, 1323, 1991.
69. Rajakulasingam, K., Polosa, R., Lau, L. C. K., Church, M. K., Holgate, S. T., and Howarth, P. H., Nasal effects of bradykinin and capsaicin: influence on plasma protein leakage and role of sensory neurons, *J. Appl. Physiol.*, 72, 1418, 1992.
70. Lacroix, J. S., Buvelot, J. M., Polla, P. S., and Lundberg, J. M., Improvement of symptoms of non-allergic chronic rhinitis by local treatment with capsaicin, *Clin. Exp. Allergy*, 21, 595, 1991.
71. Maggi, C. A., Tachykinin receptors and airway pathophysiology, *Eur. Respir. J.*, 6, 735, 1993.
72. Nakanishi, S., Mammalian tachykinin receptors, *Annu. Rev. Neurosci.*, 14, 123, 1991.
73. Hopkins, B., Powell, S. J., Danks, P., Brigges, I., and Graham, A., Isolation and characterization of the human lung NK-1 receptor cDNA, *Biochem. Biophys. Res. Commun.*, 180, 1110, 1991.
74. Gerard, N. P., Eddy, R. L., Shows, T. B., and Gerard, C., The human neurokinin A (substance K) receptor. Molecular cloning of the gene, chromosome localization, and isolation of cDNA from tracheal and gastric tissues, *J. Biol. Chem.*, 265, 20455, 1990.

75. Graham, A., Hopkins, B., Powell, S. J., Danks, P., and Briggs, I., Isolation and characterization of the human lung NK-2 receptor gene using rapid amplification of cDNA end, *Biochem. Biophys. Res. Commun.*, 177, 8, 1991.
76. Maggi, C. A., Patacchini, R., Rovero, P., and Santicioli, P., Tachykinin receptors and noncholinergic bronchoconstriction in the guinea-pig isolated bronchi, *Am. Rev. Respir. Dis.*, 144, 363, 1991.
77. Hirayama, Y., Lei, Y.-H., Barnes, P. J., and Rogers, D. F., Effects of two novel tachykinin antagonists, FK224 and FK888, on neurogenic airway plasma exudation, bronchoconstriction and systemic hypotension in guinea pigs, *Br. J. Pharmacol.*, 108, 844, 1993.
78. Abelli, L., Maggi, C. A., and Rovero, P., Del Bianco, E., Regoli, D., Drapeau, G., and Giachetti, A., Effect of synthetic tachykinin analogues on airway microvascular leakage in rats and guinea-pigs: evidence for the involvement of NK-1 receptors, *J. Auton. Pharmacol.*, 11, 267, 1991.
79. Sakamoto, T., Barnes, P. J., and Chung, K. F., Effect of CP-96,345, a non-peptide NK_1 receptor antagonist, against Substance P-, bradykinin- and allergen-induced airway microvascular leakage and bronchoconstriction in the guinea-pig, *Eur. J. Pharmacol.*, 231, 31, 1993.
80. Bertrand, C., Geppetti, P., Baker, J., Yamawaki, I., and Nadel, J. A., Role of neurogenic inflammation in antigen-induced vascular extravasation in guinea-pig trachea, *J. Immunol.*, 150, 1479, 1993.
81. Piedimonte, G., Bertrand, C., Geppetti, P., Snider, M., Desai, M. C., and Nadel, J. A., A new NK_1 receptor antagonist (CP-99,994) prevents the increase in tracheal vascular permeability produced by hypertonic saline, *J. Pharmacol. Exp. Ther.*, 266, 270, 1993.
82. Solway, J., Kao, B. M., Jordan, J. E., Gitter, B., Rodger, I. W., Howbert, J. J., Alger, L. E., Necheles, J., Leff, A. R., and Garland, A., Tachykinin receptor antagonists inhibit hyperpnea-induced bronchoconstriction in guinea-pigs, *J. Clin. Invest.*, 92, 315, 1993.
83. Joos, G. F., Kips, J. C., and Pauwels, R. A., In vivo characterization of the tachykinin receptors involved in the direct and indirect bronchoconstrictor effect of tachykinins in the rat, *Am. J. Respir. Crit. Care Med.*, 149, 1160, 1994.
84. Morimoto, H., Murai, M., Maeda, Y., Yamaoka, M., Nishikawa, M., Kiyotoh, S., and Fujii, T., FK224, a novel cyclopeptide Substance P antagonist with NK_1 and NK_2 receptor selectivity, *J. Pharmacol. Exp. Ther.*, 262, 398, 1992.
85. Murai, M., Morimoto, H., Maeda, Y., Kiyotoh, S., Nishikawa, M., and Fujii, T., Effects of FK224, a novel compound NK_1 and NK_2 receptor antagonist, on airway constriction and airway edema induced by neurokinins and sensory nerve stimulation in guinea pigs, *J. Pharmacol. Exp. Ther.*, 262, 403, 1992.
86. Joos, G. F., Kips, J. C., Lefebvre, R., and Pauwels, R. A., The effect of FK224, a NK_1 and NK_2 receptor antagonist, on the in vivo and in vitro airway effects of neurokinin A in the rat, *Eur. Respir. J.*, 6 (Abstr.), 264s, 1993.
87. Naline, E., Devillier, P., Drapeau, G., Toty, L., Bakdach, H., Regoli, D., and Advenier, C., Characterization of neurokinin effects and receptor selectivity in human isolated bronchi, *Am. Rev. Respir. Dis.*, 140, 679, 1989.
88. Advenier, C., Naline, E., Toty, L., Bakdach, H., Emonds-Alt, X., Vilain, P., Brelière, J.-C., and Le Fur, G., Effects on the isolated human bronchus of SR 48968, a potent and selective nonpeptide antagonist of the neurokinin A (NK_2) receptors, *Am. Rev. Respir. Dis.*, 146, 1177, 1992.
89. Ellis, J. L., Undem, B. J., Kays, J. S., Ghanekar, S. V., Barthlow, H. G., and Buckner, C. K., Pharmacological examination of receptors mediating contractile responses to tachykinins in airways isolated from human, guinea pig and hamster, *J. Pharmacol. Exp. Ther.*, 267, 95, 1993.
90. Astolfi, M., Treggiari, S., Giachetti, A., Meini, S., Maggi, C. A., and Manzini, S., Characterization of the tachykinin NK_2 receptor in the human bronchus: influence of amastatin-sensitive metabolic pathways, *Br. J. Pharmacol.*, 111, 570, 1994.
91. Chitano, P., Di Blasi, P., Lucchini, R. E., Calabro, F., Saetta, M., Maestrelli, P., Fabbri, L. M., and Mapp, C. E., The effects of toluene diisocyanate and of capsaicin on human bronchial smooth muscle in vitro, *Eur. J. Pharmacol. Environ. Toxicol. Pharmacol.*, 270, 167, 1994.
92. Molimard, M., Naline, E., Regoli, D., Edmonds-Alt, X., and Advenier, C., Contractile effects of Substance P on the human isolated bronchus, *Eur. Respir. J.*, 7 (Suppl. 18), 248s (Abstr.), 1994.
93. Ichinose, M., Nakajima, N., Takahaschi, T., Yamauchi, H., Inoue, H., and Takishima, T., Protection against bradykinin-induced bronchoconstriction in asthmatic patients by neurokinin receptor antagonist, *Lancet*, 340, 1248, 1992.
94. Lunde, H., Hedner, J., and Svedmyr, N., Lack of efficacy of 4 weeks treatment with the neurokinin receptor antagonist FK224 in mild to moderate asthma, *Eur. Respir. J.*, 7 (Suppl. 18), 151s (Abstr.), 1994.
95. Fahy, J. V., Wong, H. H., Geppetti, P., Nadel, J. A., and Boushey, H. A., Effect of an NK_1 receptor antagonist (CP-99,994) on hypertonic saline-induced bronchoconstriction and cough in asthmatic subjects, *Am. J. Respir. Crit. Care Med.*, 149, A1057, 1994.

Chapter 23

CAPSAICIN AS A DRUG

Pierangelo Geppetti

CONTENTS

I. INTRODUCTION

Pungent extracts from the plants of the genus *Capsicum* have been used for centuries in traditional medicine. Most of the applications derived from the empirical use of these extracts were based on the counterirritant action of these remedies. It was only with the discovery by N. Jancsó and co-workers of the neuropharmacological properties of capsaicin and its selective action on a subpopulation of primary sensory neurons that capsaicin and related compounds obtained wide interest in human therapy. The functions and the possible pathophysiological roles of capsaicin-sensitive sensory neurons and the specific features of capsaicin pharmacology have been covered by previous[1,2] and more recent[3,4] review articles and by different sections of this book (see chapters by Szolcsányi, Bevan and Docherty, Dray, and Maggi). The aim of this chapter is to focus on the modalities of administration and the results obtained by the use of capsaicin in the treatment of human diseases, and to update the status of the clinical studies reported in previous review articles.[5-8]

II. CHEMISTRY, PHARMACOKINETICS, AND TOXICOLOGY

Capsaicin (*trans*-8-methyl-*N*-vanillyl-6-nonenamide) is a crystalline neutral principle, freely soluble in ethanol, ether, benzene, or chloroform, and insoluble in water. Over the counter (OTC) medications marketed in various countries for decades contain the active ingredients of capsicum extracts, including capsaicin. The pungency of these materials is due to capsaicin

0-8493-7646-7/96/$0.00+$.50

and to other vanillylamides of fatty acids. Bands, rubs, liniments, and creams are the most common pharmaceutical preparations containing *Capsicum* extracts. More recently, capsaicin itself has been approved for clinical use as a cream in the U.S., whereas applications for similar creams are currently under review by regulatory agencies in other countries.

Capsaicin and dihydrocapsaicin are readily absorbed by the gastrointestinal tract through a nonactive process in the jejunum via the portal system.[9] [^{3}H]-dihydrocapsaicin and capsaicin administed intragastrically in anesthetized rats are almost completely metabolized in the liver.[10] A certain dregree of biotransformation takes place in the intestinal lumen.[10] Approximately 80% of capsaicin or dihydrocapsaicin ingested orally in rats is found in the urine (8.7% as unchanged drug, the remaining as vanillylamine, vanillin, valillyl alcohol, and acid).[11] The proportion of free and glucuronide metabolites in the urine is 14.5 and 60.5% of the total dose, respectively.[11] Ten percent of the total dose is excreted in the feces unchanged, and hydrolyzing enzyme activity is found mainly in the liver.[11] After intravenous administration of capsaicin (2 mg/kg) in rats, it accumulates in the brain, spinal cord and liver.[12] Capsaicin and its congeners have been shown to inhibit the activity of NADH-coenzyme Q oxidoreductase of the mitochondrial respiratory chain.[13] As an alimentary additive, capsaicin is found in a large variety of foods[14] and analysis of the capsaicin content in the diet in different countries shows that the ingestion of 20 mg of capsaicin is not infrequent with a "hot" meal.[15] Therefore, pharmacokinetic studies mentioned above may be of interest in relation to the alimentary use of capsaicin. However, since the clinical utilization of capsaicin is mainly, if not exclusively, topical (skin creams and solutions or suspensions for vesical or nasal administration, etc.), these studies cannot give useful information about disposition of the drug following the use of the formulations for topical application. The extremely low amounts of capsaicin administered topically in human therapy (usually less than 500 μg per application) in comparison to the large amount of the drug used in the pharmacokinetic studies mentioned above, suggest that a negligible amount, if any, of capsaicin may diffuse from the site of application, and the attainment of significant systemic concentration of the drug is unlikely.

Various investigators have addressed the possibility that capsaicin and its analogs have some mutagenic effects, obtaining conflicting results. In short-term tests, capsaicin, vanillin, and chilli extracts were devoid of any mutagenic property.[16] A recent study proposed, but did not demonstrate, an association between alimentary use of chilli and the development of gastric cancer.[17] The ultrapotent capsaicin analog, resiniferatoxin (see Chapter 4), which has attracted attention for its therapeutic potential, was first considered a protein kinase C activator like the tumor-promoting phorbol esters. However, further studies have dissociated the proinflammatory effect of resiniferatoxin from the protein kinase C activation and the tumor-promoting activity of phorbol esters.[18] The mechanisms by which capsaicin exerts various actions on primary sensory neurons (stimulatory, desensitizing, and toxic effects) have been described in detail in previous reviews and in other parts of this book. Of interest for the present description is the possibility that the therapeutic goal sought with the clinical use of capsaicin, e.g., desensitization of some primary afferent neurons[19] may be obtained through a transient or permanent toxic effect exerted by capsaicin on these nerves. However, the precise mechanism by which repeated capsaicin application to the human skin causes its beneficial action in different diseases, and, in particular, whether this action is somehow related to the toxic effect of capsaicin on primary sensory neurons, has not been clarified.

III. CAPSAICIN IN HUMAN THERAPY: GENERAL FEATURES

OTC preparations containing capsaicin or capsicum extracts have gained success mostly because of their counterirritant effect. The principle that a moderate painful stimulus that causes a mild vasomotor and inflammatory response might counteract a preexistent painful condition, is not, however, specific for any irritant stimulus. The success of capsaicin and congeners as counterirritants is probably due to the fact that these compounds cause acute effects (pain and reddening of the skin) without any macroscopic visible damage of the tissue. In fact, capsaicin even at high doses does not produce any blistering. Recent acquisitions in the understanding of the physiopharmacology of primary sensory neurons have given the basis for a completely different use of medications containing capsaicin. Most clinical studies in which capsaicin was applied to human tissue were based on the ability of this drug to desensitize a subpopulation of primary sensory neurons, thus making them no longer sensitive to the action of capsaicin itself and other agents usually able to affect their function. This goal, clearly different from counterirritation, has been generally attained by repeated application of small doses of the drug to the same cutaneus or mucosal region. Similar to the use of capsaicin as counterirritant, utilization of capsaicin to desensitize primary afferents relies on topical use of the drug on defined cutaneous areas or mucosal surfaces.

In the last 10 years a number of clinical studies have documented the effects of capsaicin in different human diseases. Reduction of pain has been the major, but not the exclusive, purpose of capsaicin therapy, given that diseases in which pain is a negligible symptom, and inflammation plays a major role, have been successfully treated with capsaicin as well. Although many papers reporting the beneficial effect of capsaicin treatment were case report studies in which only one or few patients were treated, several studies enrolled a much larger number of patients and adopted a double-blind, placebo-controlled design. However, a common bias remains in clinical trials in which the effect of capsaicin was compared to that of placebo: even creams containing a low concentration of capsaicin (0.025%) caused stinging and burning sensations at the first applications, whereas the vehicles were, often, devoid of such effects. Therefore, the possibility of conducting fully placebo-controlled studies, and to completely blind the patient has been hampered by the fact that the active drug was recognizable by its pungent and burning effects. The strategy to reduce pain caused by capsaicin-containing preparations[6,20] by adding local anesthetics is hindered by the lack of complete abolition of the pungency, resulting from the difficulty of applying topical products precisely to the same areas and by differences in absorption kinetics, particularly at the cutaneous level. Difficulties in conducting clinical studies with capsaicin and in the interpretation of the results obtained with these studies are underlined by a recent investigation in which several trials were analyzed using meta-analytic methods.[21] Determining the odds ratio as the main outcome measure, results indicated that capsaicin cream was effective in providing pain relief in ostheoarthritis and in psoriasis, whereas less convincing results were obtained in postherpetic neuralgia.[21]

IV. DOUBLE-BLIND, PLACEBO-CONTROLLED TRIALS

A. POSTHERPETIC NEURALGIA

Clinical trials or case report studies in which capsaicin or capsicum extracts have been used to alleviate symptoms of different diseases have found little access to scientific literature in the past. It was only in the last 10 years that this field of investigation has obtained a much wider interest. In particular, extensive clinical trials have been carried out for postherpetic neuralgia, postmastectomy pain, and painful diabetic neuropathy. Severe and chronic pain or intractable itch not infrequently follow Herpes Zoster infection, particularly in elderly patients. Treatment of these conditions is poor and essentially symptomatic, requiring potent analgesics and tranquilizers.

An initial open study[22] reported that the application of 0.025% capsaicin cream three to five times daily for 4 weeks to the skin area affected by this segmental neuralgia resulted in a substantial relief of pain in 9 of the 14 patients treated, whereas 2 patients dropped out. This observation was followed by the publication of case report studies[23,24] describing similar

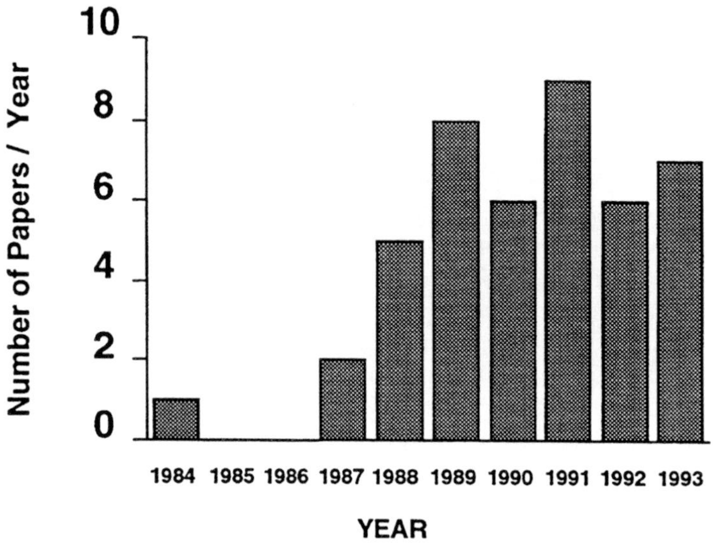

FIGURE 1. Number of papers published between 1984 and 1993 regarding the use of capsaicin in human therapy. The papers reported in the diagram include case reports, open-labeled studies, and double-blind, placebo-controlled studies.

positive results in patients unsuccessfully treated with narcotic analgesics, carbamazepine and amitriptyline. The beneficial effect of capsaicin treatment was confirmed by a double-blind study in which almost 80% of the 32 patients, with a history of severe intractable postherpetic neuralgia of 12 months or longer, experienced relief of their pain after repeated application of 0.075% capsaicin cream.[25] Another randomized double-blind study, however, did not show any effect superior to placebo.[26] Finally, an open nonrandomized study investigated the efficacy, time course of action, and predictors of capsaicin (0.025%) applied for 8 weeks in 39 patients with chronic postherpetic neuralgia.[27] Five patients dropped out because of subjective postcapsaicin burning, 13 patients showed no effect, and 19 patients reported moderate to excellent pain relief. The follow-up investigation, 10 to 12 months after the study, showed that 13 of the 18 responder patients were still improved.

B. PAINFUL DIABETIC NEURALGIA

Pain is commonly associated with diabetic neuropathy, a condition that is not infrequently refractory to common analgesics. Limited benefit may be obtained by the association of phenytoin, carbamazepine, phenothiazines, and other drugs, all therapies that have the disadvantage of important side effects and problems related to drug interactions. The observations of the beneficial effect of capsaicin in different painful conditions prompted the study of this medication in painful diabetic neuropathy. After a first report[28] of two cases successfully treated with 0.025% capsaicin cream, larger studies were initiated. A placebo-controlled study involving 58 patients did not show any benefit of 0.075% capsaicin cream application (four times daily for 4 weeks) as compared to the vehicle.[29] However, a multicenter, double-blind, vehicle-controlled study in which more than 250 patients were enrolled showed opposite results.[30] In this investigation, 69.5% of the patients treated with capsaicin (0.075%, four times daily for 8 weeks) showed improvement by the physician's global evaluation scale vs. 53.4% with vehicle. A similar significant difference between capsaicin and placebo was obtained considering decrease in pain intensity and amelioration in pain relief. Improvement in daily activity and quality of the patient's life accompanied the improvement of pain.[30]

C. POSTMASTECTOMY PAIN

The postmastectomy pain syndrome is a disorder that affects 4 to 13% of patients who undergo mastectomy. An initial open study demonstrated that capsaicin (0.025%) application to the painful area four times daily for 4 weeks caused a significant pain relief in 12 of 14 patients.[31] A subsequent randomized, double-blind, placebo-controlled trial conducted by the same investigators[32] using 0.075% capsaicin did not show any significant difference in the visual analog scale for steady pain between placebo and capsaicin. However, a significant difference in jabbing pain, category pain severity scales, and in overall pain relief in favor of capsaicin was observed. An independent group[33] confirmed the beneficial effect of 0.025% capsaicin applied three times daily for 2 months in 21 patients affected by the postmastectomy pain syndrome.

D. PSORIASIS

The belief that alteration in skin microcirculation (vasodilation, elongation and convolution of papillary vessel, exudate, and inflammatory cell recruitment in the papilla) plays a major role in the pathogenesis of psoriasis, and the observation that capsaicin blocks axon-reflex vasodilatation produced by a variety of erythematogenic substances, led to investigations of whether capsaicin treatment might be of any benefit in psoriasis. Under a double-blind, placebo-controlled design, 44 patients with symmetrically distributed psoriasis were treated with 0.01 and 0.025% capsaicin cream or placebo (six and four times daily for 6 weeks).[34] A significant improvement was found in the capsaicin-treated group. Patients with psoriasis (98) treated with capsaicin (0.025%, four times a day for 6 weeks) showed improvement in pruritus relief and global evaluation as compared to patients (99) treated with the vehicle.[35] The beneficial effect of capsaicin (0.025%) on psoriasis vulgaris was found also in a subsequent open trial involving 10 patients.[36]

E. CLUSTER HEADACHE

Cluster headache is a unilateral painful disorder of the head characterized by attacks of severe pain of 15 to 180 min in duration, occurring usually one to eight times a day, conjunctival reddening, lacrimation, nasal obstruction and rhinorrhea, miosis, edema of the eyelid ipsilateral to the pain accompanying the attack. The effect of the application of capsaicin (300 μg) to the nasal mucosa of both nostrils once a day for 5 days was studied in an open-label study involving 20 cluster headache patients affected by the episodic or chronic form of the disease.[37] The number of cluster headache attacks per 10 days of observation dropped by about 70% following capsaicin treatment. In a subsequent study[38] the same investigators treated 26 patients suffering from episodic cluster headache with 300 μg of capsaicin (once a day for 5 days) applied into the nasal mucosa of the nostril ipsilateral to the attack, whereas another group of patients (25 subjects) received the same dose of capsaicin into the contralateral nostril. None of the subjects receiving capsaicin in the contralateral side showed changes in the number of attacks, whereas a remarkable reduction in the number of pain attacks was observed in the patients treated on the side ipsilateral to the occurrence of the attack. Similar results were obtained in patients affected by the chronic form of cluster headache.[38] In a randomized, double-blind, placebo-controlled study capsaicin cream (0.025%) or placebo (3% camphor) were applied twice a day for 7 days in the ipsilateral nostril of episodic or chronic cluster headache patients.[39] A significant reduction in number and severity of the pain attacks was observed in the capsaicin-treated group, but the episodic cluster headache patients appeared to benefit more than the patients affected by the chronic form.

F. PAINFUL OSTEOARTHRITIS OF THE HANDS

Twenty-one patients suffering from rheumatoid arthritis or primary osteoarthritis with painful involvement of the hands were investigated with a double-blind, placebo-controlled design.[40] Patients received 0.075% capsaicin cream (four times a day for 4 weeks) or vehicle according to a randomization schedule. No benefit of capsaicin treatment was observed in rheumatoid arthritis patients, whereas patients suffering from osteoarthritis showed a significant reduction in tenderness scores, pain, and swelling.

V. OPEN-LABEL STUDIES AND CASE REPORTS

A. PERENNIAL RHINITIS (NONALLERGIC CHRONIC RHINITIS)

Perennial rhinitis is the term that describes a condition possessing most of the symptoms of allergic rhinitis, but without an allergic basis. The pathophysiology of this disease remains unclear and satisfactory therapy is not yet available. A first study reported favorable results with topical treatment with capsaicin in a group of nine patients affected by perennial rhinitis.[41] Two independent open-label studies[42,43] reported improvement of the symptoms of perennial rhinitis following repeated nasal application of capsaicin. A nasal spray of capsaicin (3.3 mmol) once weekly for 5 weeks reduced all symptoms (obstruction, rhinorrhea, sneezing) of rhinitis in 17 patients.[43] Similar results were obtained in 20 patients suffering from perennial rhinitis treated intranasally with capsaicin following a different protocol.[42] In this case, 15 μg of capsaicin in 100 μl solution were sprayed into the nostril of the patients three times a day for 3 days. Reduction in symptom score was observed until 30 days from the beginning of the capsaicin treatment.

B. NEUROGENIC BLADDER DYSFUNCTION

Intravesical instillation of capsaicin (0.1 to 10 μ*M*) in five patients with hypersensitive disorders of the lower urinary tract caused disappearance or marked attenuation of their symptoms: no effect, however, was produced in three patients by the instillation of the capsaicin vehicle.[44] Further evidence for the beneficial effect of capsaicin instillation in this pathological condition was obtained by the same investigators.[45] Three patients with neurogenic bladder dysfunction were treated with 100 ml containing 1 to 2 mmol/l capsaicin, thus improving their intractable detrusor hyperreflexia.[46] The same authors showed that intravesical instillation of capsaicin improved bladder function in an additional 12 patients affected from detrusor hyperreflexia.[47]

C. TRIGEMINAL NEURALGIA

Intense paroxysmal pain, confined to the area innervated by one or more of the branches of the trigeminal nerve, is the specific feature of trigeminal neuralgia. An open-label study[48] showed that application (three times daily for several days) of approximately 1 g of a 0.5% capsaicin cream to the cutaneous area corresponding to the affected trigeminal branch relieved pain in 12 patients suffering from trigeminal neuralgia. Of the 10 patients who were responsive, 4 had relapses of pain. These patients were treated successfully a second time with capsaicin.

D. OTHERS

Severe intractable stump pain may arise at the site of amputation. In a case of stump pain after bilateral below-knee amputation resistant to conventional therapies, treatment with 0.025% capsaicin cream completely relieved the pain.[49] However, this finding was not confirmed in a subsequent case report study in which 0.075% capsaicin cream was used.[50] Reflex sympathetic dystrophy is a syndrome of burning pain, hyperalgesia, allodynia, vasomotor, and dystrophic changes. A complete, although temporary, resolution of the syndrome was obtained in a case of intractable reflex sympathetic dystrophy with a 3-week treatment with 0.025% capsaicin cream.

Pruritus caused by different pathologies has been treated successfully with capsaicin. Pruritus caused by prurigo nodularis (three patients), chronic prurigo (two patients), and neurodermatitis circumscripta (two patients) was markedly reduced by a capsaicin cream.[51] Similar positive results against itch were obtained in notalgia parestetica.[52] Pruritus is a common symptom in patients undergoing hemodialysis. An open-label trial and a small double-blind study reported some efficacy of capsaicin cream (0.025%) in reducing the itch that follows hemodialysis.[53] Capsaicin application to the forearm reduced the itch in three patients affected by cold urticaria.[54] Application of topical capsaicin to the affected skin areas (0.025%) relieved pain caused by a malignant melanoma of the skin,[55] and hyperpathia in a patient suffering from Guillain-Barré syndrome.[56] Tenderness and burning of the vulva caused by vulvar vestibulitis were relieved by the application of 0.025% capsaicin cream in two of four treated patients.[57] In a case of apocrine chromhidrosis, the side of the face treated with capsaicin (0.025%) cream showed no discoloration of the sweat gland secretion, whereas no effect was observed in the side treated with capsaicin vehicle.[58]

VI. CONCLUSIONS

Evidence accumulated in the last 10 years suggests that capsaicin may enter the therapeutic armamentarium to relieve pain, itch, and inflammation in several pathological conditions. The neuropharmacological basis of the clinical use of capsaicin relies on the remarkable selectivity of this drug to act on a subpopulation of primary nociceptive neurons, and on the ability to desensitize these neurons after repeated administration. Drawbacks in the use of capsaicin in human therapy are its property to stimulate sensory neurons, thus causing stinging and burning sensation, particularly at the first applications of the drug, and the fact that its utilization is exclusively confined to topical preparations. Difficulties in conducting fully blind clinical trials, and, therefore, in the interpretation of the results, also reflects the stimulatory action of capsaicin. The possibility of dissociating the stimulatory from the desensitizing action of capsaicin, to obtain a drug, that while ablating the function of capsaicin-sensitive neurons, does not excite them, although theoretically possible,[59] has not yet been achieved. There is some evidence suggesting that the ultrapotent capsaicin analog, resiniferatoxin, is more potent in desensitizing than in exciting sensory nerves.[60] Ongoing clinical trials with this compound will disclose, together with its possible efficacy, whether this drug has a less prominent stimulatory effect and a more pronounced desensitizing action than capsaicin.

ACKNOWLEDGMENTS

Work performed in the author's laboratory was supported by grants from MURST (60%, Florence, Italy) and Consiglio Nazionale delle Ricerche (Rome, Italy), Progetto Bilaterale.

REFERENCES

1. Maggi, C. and Meli, A., The sensory-efferent function of capsaicin-sensitive sensory neurons, *Gen. Pharmacol.*, 19, 1, 1988.
2. Holzer, P., Local effector functions of capsaicin-sensitive sensory nerve endings: involvement of tachykinins, calcitonin gene-related peptide and other neuropeptides, *Neuroscience*, 24, 739, 1988.
3. Holzer, P., Capsaicin: cellular targets, mechanisms of actions, and selectivity for thin sensory neurons, *Pharmacol. Rev.*, 43, 143, 1991.
4. Maggi, C., The pharmacology of the efferent function of sensory nerves, *J. Auton. Pharmacol.*, 11, 173, 1991.
5. Lynn, B., Capsaicin: actions on nociceptive C-fibres and therapeutic potential, *Pain*, 14, 61, 1990.

6. Jancsó, G. and Lynn, B., Possible use of capsaicin in pain therapy, *Clin. J. Pain*, 3, 123, 1987.
7. Carter, R. B., Topical capsaicin in the treatment of cutaneous disorders, *Drug Dev. Res.*, 22, 109, 1991.
8. Maggi, C. A., Capsaicin and primary afferent neurons: from basic science to human therapy, *J. Auton. Nerv. Syst.*, 33, 1, 1991.
9. Kawada, T., Suzuki, T., Takahashi, M., and Iwai, K., Gastrointestinal absorption and metabolism of capsaicin and dihydrocapsaicin in rats, *Toxicol. Appl. Pharmacol.*, 72, 449, 1984.
10. Donnerer, J., Amann, R., Shuligoi, R., and Lembeck, F., Absorption and metabolism of capsaicinoids following intragastric administration in rats, *Naunyn-Schmiedeberg's Arch. Pharmacol.*, 342, 357, 1990.
11. Kawada, T. and Iwai, K., In vivo and in vitro metabolism of dihydrocapsaicin, a pungent principle of hot pepper, in rats, *Agric. Biol. Chem.*, 49, 441, 1985.
12. Saria, A., Lembeck, F., and Skofitsch, G., Determination of capsaicin in tissues and separation of capsaicin analogs by high performance liquid chromatography, *J. Chromatogr.*, 208, 41, 1981.
13. Shimomura, Y., Kawada, T., and Suzuki, M., Capsaicin and its analogs inhibit the activity of NADH-coenzyme Q oxidoreductase of the mitocrondrial respiratory chain, *Arch. Biochem. Biophys.*, 270, 573, 1989.
14. Edwards, S. J., Colquhoun, E. Q., and Clark, M. G., Levels of pungent principles in chilli sauces and capsicum fruit in Australia, *Food Aust.*, 42, 432, 1990.
15. Limlomwongse, L., Shaittauchawong, C., and Tongyai, S., Effect of capsaicin on gastric acid secretion and mucosal blood flow, *J. Nutr.*, 109, 773, 1979.
16. Nagabhushan, M. and Bhide, S. V., Mutagenicity of chili extract and capsaicin in short-term tests, *Environ. Mutagen.*, 7, 881, 1985.
17. Lopezcarrillo, L., Avila, M. H., and Dubrow, R., Chilli pepper consumption and gastric cancer in Mexico — a case-control study, *Am. J. Epidemiol.*, 139, 263, 1994.
18. Szallasi, A. and Blumberg, P. M., Effect of resiniferatoxin pretreatment on the inflammatory response to phorbol-12-myristate-13-acetate in mouse strains with different susceptibilies to phorbol ester tumor promotion, *Carcinogenesis*, 11, 583, 1990.
19. Craft, R. M. and Porreca, F., Therapeutic potential of capsaicin-like molecules. Treatment parameters of desensitization to capsaicin, *Life Sci.*, 51, 1767, 1992.
20. Watson, C. P. N., Evans, R. J., and Watt, V. R., Postherpetic neuralgia and topical capsaicin, *Pain*, 33, 333, 1988.
21. Zhang, W. Y. and Po, A. L. W., The effectiveness of topically applied capsaicin. A meta-analysis, *Eur. J. Clin. Pharmacol.*, 46, 517, 1994.
22. Bernstein, J. E., Bickers, D. R., Dahl, M. V., and Roshal, J. Y., Treatment of chronic postherpetic neuralgia with topical capsaicin, *J. Am. Acad. Dermatol.*, 17, 93, 1987.
23. Bucci, F. A., Gabriels, C. F., and Krohel, G. B., Successful treatment of postherpetic neuralgia with capsaicin, *Am. J. Ophthalmol.*, 106, 758, 1988.
24. Don, P. C., Topical capsaicin for treatment of neuralgia associated with herpes zoster infection, *J. Am. Acad. Dermatol.*, 18, 1135, 1988.
25. Bernstein, J. E., Korman, N. J., Bickers, D. R., Dahl, M. V., and Millikan, L. E., Topical capsaicin treatment of chronic postherpetic neuralgia, *J. Am. Acad. Dermatol.*, 21, 265, 1989.
26. Drake, H. F., Harries, A. J., Gamester, R. E., and Justins, D., Randomized double-blind study of topical capsaicin for treatment of postherpetic neuralgia, *Pain*, 5, S58, 1990.
27. Peikert, A., Hentrich, M., and Ochs, G., Topical 0.025% capsaicin in chronic postherpetic neuralgia: efficacy, predictors of response and long-term course, *J. Neurol.*, 238, 452, 1991.
28. Ross, D. R. and Varipapa, R. J., Treatment of painful diabetic neuropathy with topical capsaicin, *N. Engl. J. Med.*, 321, 474, 1989.
29. Chad, D. A., Aronin, N., Lundstrom, R., McKeon, P., Ross, D., Molitch, M., Schipper, H. M., Stall, G., Dyess, E., and Tarsy, D., Does capsaicin relieve the pain of diabetic neuropathy?, *Pain*, 42, 387, 1990.
30. The Capsaicin Study Group. Treatment of painful diabetic neuropathy with topical capsaicin, *Arch. Intern. Med.*, 151, 2225, 1991.
31. Watson, C. P., Evans, R. J., and Watt, V. R., The postmastectomy pain syndrome and the effect of topical capsaicin, *Pain*, 38, 177, 1989.
32. Watson, C. P. N. and Evans, R. J., The postmastectomy pain syndrome and topical capsaicin — a randomized trial, *Pain*, 51, 375, 1992.
33. Dini, D., Bertelli, G., Gozza, A., and Forno, G. G., Treatment of the postmastectomy pain syndrome with topical capsaicin, *Pain*, 54, 223, 1993.
34. Bernstein, J. E., Parish, L. C., Rapaport, M., Rosenbaum, M. M., and Roenigk, H. H., Effects of topically applied capsaicin on moderate and severe psoriasis vulgaris, *J. Am. Acad. Dermatol.*, 15, 504, 1989.
35. Ellis, C. N., Berberian, B., Sulica, V. I., Dodd, W. A., Jarratt, M. T., Katz, H. I., Prawer, S., Kueger, G., Rex, I. H., and Wolf, J. E., A double-blind evaluation of topical capsaicin in pruritic psoriasis, *J. Am. Acad. Dermatol.*, 29, 438, 1993.
36. Kurkuoglu, N. and Alaybei, F., Topical capsaicin for psoriasis, *Br. J. Dermatol.*, 123, 549, 1990.

37. Sicuteri, F., Fusco, B. M., Marabini, S., Campagnolo, V., Maggi, C. A., Geppetti, P., and Fanciullacci, M., Beneficial effect of capsaicin application to the nasal mucosa in cluster headache, *Clin. J. Pain*, 5, 49, 1989.
38. Fusco, B. M., Marabini, S., Maggi, C. A., and Geppetti, P., Preventive effect of repeated nasal applications of capsaicin in cluster headache, *Pain*, 59, 321, 1994.
39. Marks, D. R., Rapoport, A., Padla, D., Weeks, R., Rosum, R., Sheftell, F., and Arrowsmith, F., A double-blind placebo-controlled trial of intranasal capsaicin for cluster headache, *Cephalalgia*, 13, 114, 1993.
40. McCarthy, G. M. and McCarthy, D. J., Effect of topical capsaicin in the therapy of painful ostheoarthitis of the hands, *J. Rheumatol.*, 19, 604, 1992.
41. Wolf, G., Neue Aspekte zur Pathogenese und Therapie der hyperreflektorischen Rhinopatie, *Laryng. Rhinol. Otol.*, 67, 438, 1988.
42. Marabini, S., Ciabatti, P. G., Polli, G., Fusco, B. M., and Geppetti, P., Beneficial effects of intranasal applications of capsaicin in patients with vasomotor rhinitis, *Eur. Arch. Otorhinolaryngol.*, 248, 191, 1991.
43. Lacroix, J. S., Buvelot, J. M., Polla, B. S., and Lundberg, J. M., Improvement of symptoms of non-allergic chronic rhinitis by local treatment with capsaicin, *Clin. Exp. Allergy*, 21, 595, 1991.
44. Maggi, C. A., Barbanti, G., Santicioli, P., Beneforti, P., Misuri, D., Meli, A., and Turini, D., Cystometric evidence that capsaicin-sensitive nerves modulate the afferent branch of micturition reflex in humans, *J. Urol.*, 142, 150, 1989.
45. Barbanti, G., Maggi, C. A., Beneforti, P., Baroldi, P., and Turini, D., Relief of pain following intravesical capsaicin in patients with hypersensitive disorders of the lower urinary tract, *Br. J. Urol.*, 71, 686, 1993.
46. Flower, C. J., Jeweks, D., McDonald, W. I., Lynn, B., and de Groat, W. C., Intravesical capsaicin for neurogenic bladder dysfunction, *Lancet*, 339, 1239, 1992.
47. Fowler, C. J., Beck, R. O., Gerrard, S., Betts, C. D., and Fowler, C. G., Intravesical capsaicin for treatment of detrusor hyperreflexia, *J. Neurol. Neurosurg. Psychiatry*, 57, 169, 1994.
48. Fusco, B. M. and Alessandri, M., Analgesic effect of capsaicin in idiopathic trigeminal neuralgia, *Anesth. Analg.*, 74, 375, 1992.
49. Reyner, H. C., Atkins, R. C., and Westerman, R. A., Relief of local stump pain by capsaicin cream, *Lancet*, ii, 1276, 1989.
50. Weintraub, M., Golik, A., and Rubio, A., Capsaicin for treatment of post-traumatic amputation stump pain, *Lancet*, 336, 1003, 1990.
51. Tupker, R. A., Coenraads, P. J., and van der Meer, J. B., Treatment of prurigo nodularis, chronic prurigo and neurodermatitis circumscripta with topical capsaicin, *Acta Derm. Venereol.*, 72, 463, 1992.
52. Wallegren, J., Treatment of nostalgia paresthetica with topical capsaicin, *J. Am. Acad. Dermatol.*, 24, 286, 1991.
53. Breneman, D. L., Cardone, J. S., Blusmack, R. F., Lather, R. M., Searle, E. A., and Pollack, V. E., Topical capsaicin for treatment of hemodialysis-related pruritus, *J. Am. Acad. Dermatol.*, 26, 91, 1992.
54. Toth Kasa, I., Katona, M., Obal, J. F., Husz, S., and Jancsó, G., Pathological reactions of human skin: involvement of sensory nerves, in *Antidromic Vasodilatation and Neurogenic Inflammation*, Chahl, L. A., Lembeck, F., and Szolcsányi, J., Eds., Akademiai Kaido, Budapest, 1984, 317.
55. Wist, E. and Eisberg, T., Topical capsaicin in treatment of hyperalgesia, allodynia and dysestetic pain caused by malignant infiltration of the skin, *Acta Oncol.*, 32, 343, 1993.
56. Morgenlander, J. C., Hurwitz, J. B., and Massey, E., Capsaicin for the treatment of pain in Guillain-Barré syndrome, *Ann. Neurol.*, 28, 199, 1990.
57. Friedrich, E. G., Therapeutic studies in vulvar vestibulitis, *J. Reprod. Med.*, 33, 514, 1988.
58. Marks, J. G., Treatment of apocrine chromihidrosis with topical capsaicin, *J. Am. Acad. Dermatol.*, 21, 418, 1989.
59. Szolcsányi, J. and Jancsó-Gábor, A., Sensory effect of capsaicin congeners. II. Importance of chemical structure and pungency in desensitizing activity of capsaicin type compounds, *Arzneim. Forsch.*, 25, 33, 1975.
60. Szallasi, A. and Blumberg, P. M., Mechanisms and therapeutic potential of vanilloids (capsaicin-like molecules), *Life Sci.*, 24, 123, 1993.

Chapter 24

NONPEPTIDE NK_1 ANTAGONISTS: FROM DISCOVERY TO THE CLINIC

John A. Lowe, III, R. Michael Snider, and David B. MacLean

CONTENTS

The discovery of potent and highly selective nonpeptide NK_1 antagonists has ushered in a new chapter in our understanding of the role of SP and its preferred NK_1 receptor in human physiology and pathophysiology. As discussed throughout this volume, these antagonists have proven conclusively that SP is a partial or complete mediator of neurogenic inflammation in virtually all tissue beds that are innervated by small-diameter, unmyelinated C-fibers. Systemic administration has also permitted evaluation in peripheral disease models, including inflammatory and immune diseases of the gut. Initial studies on the role of NK_1 receptors in the spinal cord or brainstem have been hampered by limited efficacy at higher doses, species selectivity of specific antagonists, and nonspecific effects, e.g., on calcium channels. Overall, however, results to date suggest that NK_1 antagonism may have attenuating effects on windup or hyperalgesic pain, associated sensory motor reflexes, or, as in the bladder, heightened sensory motor reflex facilitation secondary to peripheral inflammation. Emesis, a complex nocifensor adaptation to gastrointestinal irritation, is also inhibited by NK_1 antagonism. Clinical investigations are just beginning in order to determine in what circumstances, e.g., for preemptive treatment of inflammation or pain, or for what indications SP alone is a major mediator that will allow detection of a potential therapeutic benefit of these antagonists. Regardless of their therapeutic potential, however, these and related NK_1 antagonists will greatly expand our knowledge of the role of peptidic sensory neurotransmitters in human physiology.

0-8493-7646-7/96/$0.00+$.50

I. INTRODUCTION

The mammalian tachykinin peptide substance P (SP) is widely distributed throughout the peripheral and central nervous systems. Synthesized in cell bodies of cranial and dorsal root sensory ganglia and transported directionally, predominantly in small, unmyelinated C-fibers, to the central and peripheral terminals of those sensory neurons, SP has been the most intensively studied sensory neuropeptide. It is also the most fundamental example of Dale's hypothesis, which predicted that one neurotransmitter might mediate both the peripheral effector and central sensory functions of sensory neurons. Dale's hypothesis, of course, did not predict the multiplicity of transmitters that may be synthesized within and coreleased from a single sensory neuron.

Lembeck and his Graz colleagues have amplified the nocifensor role of SP-containing C-fibers, emphasizing the teleological relationship of their peripheral effector functions with a corresponding central reflex. The peripheral effector function probably represents a primitive, rapid, active sentinel function of these neurons while the simultaneously initiated autonomic sensory motor and sensory higher central nervous system (CNS) reflex arcs represent subsequent evolutionary adaptation of these nocifensor reflexes. The best understood examples of such paired effector sensory functions include, of course, SP-induced neurogenic inflammation, inflammatory and immune functions, and the simultaneous, at least partial, mediation of inflammatory pain. Other more recently studied examples include peripheral inflammatory effects in the damaged bladder and simultaneous alteration of micturition reflexes, vascular and inflammatory responses to ingested ethanol, and, perhaps, the simultaneous mediation of the emetic reflex, and possible involvement in a range of cardiovascular and respiratory peripheral effector functions in response to noxious stimuli and the simultaneous triggering of sensory autonomic reflexes. It is ironic that although many of these complex peripheral effector and central reflex functions of SP no doubt evolved as essential protective mechanisms for the organism in response to noxious stimuli, they have now become the target for therapeutic intervention due to their role in inflammation. With the availability of potent nonpeptide antagonists that can be delivered both systemically or regionally, the next decade should hopefully fully clarify the role of SP and its preferred neurokinin receptor, the NK_1 receptor, in the pathogenesis of these symptoms and in the pathogenesis of these symptoms associated with acute or chronic tissue irritation/injury and inflammatory conditions.

II. DISCOVERY AND OVERVIEW OF THE BIOCHEMISTRY OF NONPEPTIDE NK_1 ANTAGONISTS

The initial development of nonpeptide SP antagonists has been followed by a proliferation of novel structural types. The first nonpeptide SP antagonist, CP-96,345 (Figure 1), discovered by optimization of a lead structure from file screening, shows potent and selective neurokinin (NK_1) antagonist activity *in vitro*, with K_d = 0.6 n*M* in bovine caudate and IC_{50} > 1000 n*M* at NK_2 and NK_3 sites in hamster urinary bladder and guinea pig cortex, respectively.[1] *In vivo*, CP-96,345 shows an ED_{50} value of 6.3 mg/kg for inhibition of mustard oil-induced edema in rat foot and an ED_{50} value of 10.7 mg/kg p.o. for inhibition of acetic acid-induced writhing in mouse.[2] Subsequent studies revealed that CP-96,345 also binds to the calcium L-channel at a site occupied by verapamil, with an IC_{50} value of 27 n*M*.[3] To circumvent possible side effects in the clinic, the structure was modified to produce CP-99,994 (Figure 1), which shows potent and selective NK_1 antagonism *in vitro* (K_d = 0.1 ± 0.01 n*M* in human IM-9 cells) and weak affinity for the L-type calcium channel (IC_{50} = 3010 n*M* in rat heart).[4] *In vivo*, CP-99,994 blocks Evans blue dye extravasation induced by capsaicin with an ID_{50} value of 4 mg/kg p.o., consistent with the anti-inflammatory effect seen with CP-96,345.

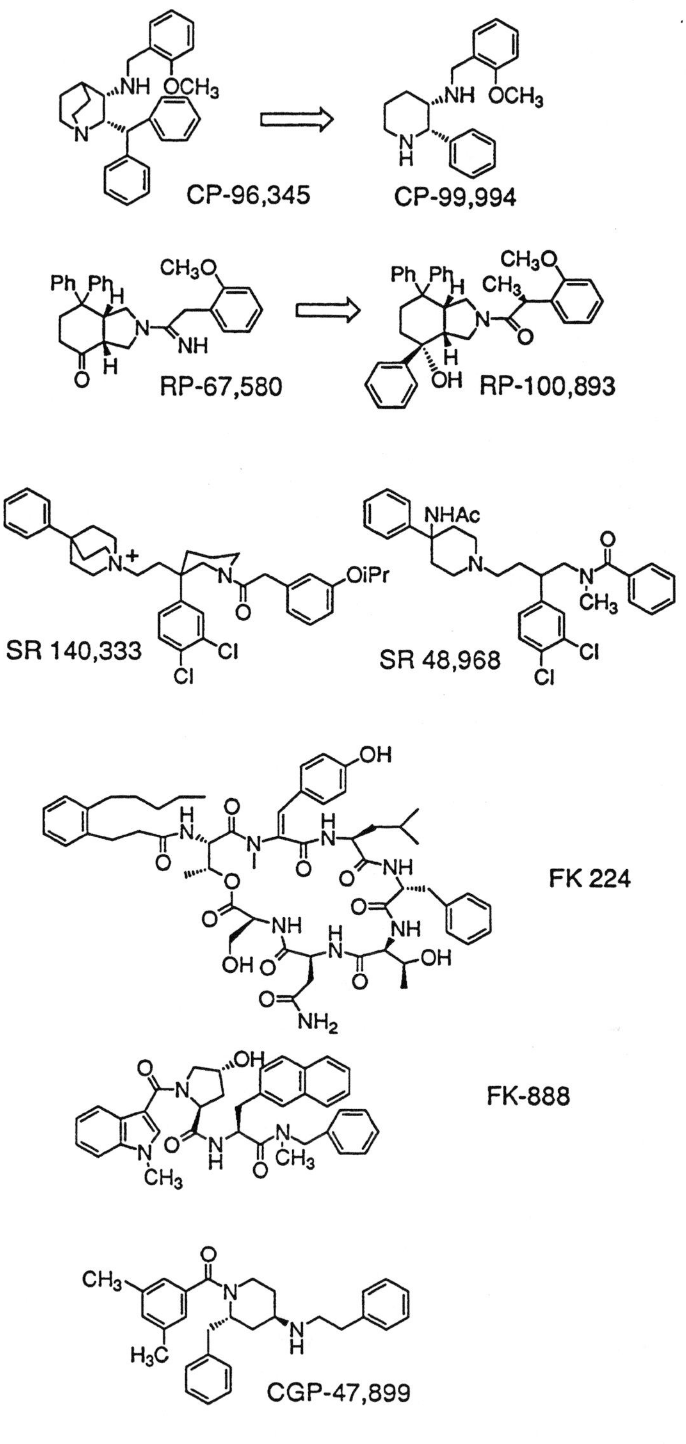

FIGURE 1. Nonpeptide NK_1 antagonists.

Another of the early compounds, which was subsequently modified to permit clinical study, is RP-67,580 (Figure 1), which shows an IC_{50} value of 4.17 n*M* for the rat NK_1 receptor, but has much lower affinity for the human NK_1 receptor.[5] RP 67,580 has also been used extensively for pharmacological characterization of SP, for example, in investigating the role of SP in neurogenic inflammation connected with animal models of headache (see below).[6] To improve affinity for the human NK_1 receptor, RP 67,580 was modified by addition of a 2-methoxyphenyl group, resulting in RPR 100,893[7] (Figure 1). In human IM-9 cells, RPR 100,893 shows an IC_{50} value of 54 n*M*, which increases to 12 n*M* after a 4-h incubation.

Starting from a common lead structure, the potent NK_1 receptor antagonist SR-140,333[8] (Figure 1, K_i = 0.02 and 0.01 n*M* at the NK_1 receptor in rat and human, respectively), and the potent NK_2 receptor antagonist SR-48,968[9] (Figure 1) were developed. SR-140,333 in particular shows promising anti-inflammatory activity, with a total reversal of heat-induced facilitation of the tail-flick response in rats at 0.06 mg/kg i.p.[10]

The cyclic peptidic antagonist, FK 224[11] (Figure 1) was discovered through screening of fermentation broths followed by chemical modification. FK224 affords a mixed spectrum of NK_1 (IC_{50} = 37 n*M*) and NK_2 (IC_{50} = 72 n*M*) antagonism. A different route, rational design based on a peptidic lead compound, was used to arrive at the nonpeptide FK 888[12] (Figure 1). FK-888 shows an IC_{50} value at the guinea pig NK_1 receptor of 6.9 n*M* and a pA_2 value of 9.29 in guinea pig ileum, and is active by the oral route in specifically blocking SP-induced airway edema in the guinea pig. Another entry in the clinical picture for tachykinin antagonists is CGP-47,899,[13] which was discovered by chemical modification of a lead structure detected by file screening (Figure 1). Thus, while many important SP pharmacology studies were carried out using the early compounds CP-96,345 and RP-67,580, the broad structural diversity of later compounds suggests the potential for unexpected discoveries in the clinic.

III. PHARMACOLOGICAL ISSUES IMPACTING THE CLINICAL PICTURE

A. SPECIES SELECTIVITY

Relating results of animal studies with NK_1 antagonists to the clinical situation may be complicated by the issue of species selectivity. For example, both CP-96,345 and CP-99,994 show 30- to 100-fold greater affinity for the guinea pig as compared to the rat NK_1 receptor, and thus doses of CP-96,345 or CP-99,994 required for full activity in the guinea pig or hamster are 1/30 to 1/100 of those required in the mouse or rat. The high doses required in rodent studies, in the range of 3 to 10 mg/kg s.c. or i.v., which result in plasma or local concentrations above 100 n*M*, open the possibility that activity in rodent models is associated with nonspecific effects due to binding with calcium channels or as yet other unidentified receptors.

B. NONSPECIFIC ACTIVITY

The demonstration that high doses of CP-96,345 show nonspecific blockade of smooth muscle contractility and reductions in blood pressure in the rat and dog suggested that these effects were not due to blockade of the NK_1 receptor. The equal activity of CP-96,345 and its 2*R*,3*R* enantiomer CP-96,344, which has negligible affinity for the NK_1 receptor, in animal models of inflammation, such as the rat foot edema test, support this idea.[14] Similar nonspecific effects of racemic RP-67,580 were demonstrated in antinociceptive assays in rodents in doses up to 30 mg/kg. Schmidt et al.[3] and Guard et al.[15] subsequently demonstrated that both CP-96,345 and CP-96,344 inhibit radiolabeled diltiazem binding to rat brain or muscle with K_I values between 25 and 120 n*M*. Rupniak et al.[16] demonstrated that RP-67,580 inhibits ^{3}[H]-diltiazem binding to rat cardiac muscle membranes with an IC_{50} value of 298 n*M*. This binding to the L-type channel was functionally significant, as demonstrated by depressed high-threshold calcium currents in rat cortex. Finally, racemic CP-96,345 has been shown to block sodium channels at concentrations above 10 μ*M*.[17] Thus, studies in animals using doses above 1 mg/kg that do not also include the inactive enantiomer as a negative control should be interpreted with caution.

TABLE 1
Sites of Action for NK_1 Antagonists

Site or Organ System	Effector Function	Afferent Function or Reflex
All	Neurogenic inflammation	Pain or hyperalgesia
Meninges	Plasma extravasation/vasodilation	Headache
Skin	Erythema/wheal	Pruritus/pain
Nose/Lung	Congestion/mucus secretion/bronchoconstriction	Cough/dyspnea/ventilatory drive
Gastrointestinal	Inflammation	Distention/injury
	Blood/Flow/ion transport	Pain/nausea/emesis
Bladder	Inflammation	Incontinence

C. BRAIN TO PLASMA RATIO

Partitioning of an NK_1 antagonist between the peripheral circulation and brain may also influence dose requirements and confound interpretation of dose response studies. Most of the reported NK_1 antagonists are indeed lipophilic and are expected to partition well into the CNS. However, dialyzable concentrations of CP-99,994 or, to our knowledge, other NK_1 antagonists have not been performed and free concentrations in the extracellular compartment within the CNS have not been determined. Thus, although the brain may act as a lipophilic sink for these compounds, the available concentrations at extracellular sites interacting with the NK_1 receptor are not known.

IV. STUDIES IN CLINICALLY RELEVANT DISEASE MODELS

As discussed in the introduction, consideration of the potential roles of SP and the NK_1 receptor can be viewed mechanistically and conceptually as those effects mediated by the release of SP from the peripheral terminals of sensory neurons (effector functions) and the CNS or reflex activity following the stimulation of peripheral cranial sensory neurons (afferent functions). Examples of conditions likely to be mediated by such NK_1-dependent processes, and thus amenable to treatment with an NK_1 antagonist, are summarized in Table 1.

A. NEUROGENIC INFLAMMATION

Nagahisa and co-workers have demonstrated the inhibitory activity of CP-96,345 on Evans blue dye extravasation in the guinea pig on local injections of SP, or mustard oil application.[2] Moussaoui and co-workers, using the more rat-selective RP-67,580, similarly demonstrated the postsynaptic site of SP action in peripheral tissue.[18] RP-67,580 almost totally abolished the Evans blue extravasation in the hind paw of the rat by antidromic stimulation, intravenous injection of SP, and topical application of xylene. There was no difference in dose response (ID_{50} = 0.026 mg/kg i.v.) or percent inhibition for these three diverse stimuli, all presumably acting through the same postsynaptic NK_1 receptor-mediated mechanism. However, direct tissue injury, which almost certainly releases tissue mediators in addition to stimulating sensory nerves, may not be as definitively inhibited as experimental manipulations such as mustard oil or xylene, which at low doses selectively stimulate sensory nerve neuropeptide release independently of more gross tissue injury.

B. NEUROGENIC INFLAMMATION IN SPECIFIC TISSUES

There has been a proliferation of reports assessing experimental neurogenic inflammation in a variety of tissues in which the role of capsaicin-sensitive afferent nerves had previously been demonstrated and in which therapeutic potential for tachykinin antagonists was previously hypothesized. A recurrent issue in each of these body systems is whether abrogation of the specific neurogenic response can be observed with a more clinically relevant challenge, e.g., tissue injury or adogenic stimulation.

1. Meninges

It has been previously hypothesized that neurogenic inflammation in the cranial distribution of the trigeminal nerve may trigger migraine. Two groups have reported the effects of RP-67,580 on neurogenic plasma extravasation in the meninges and dura mater of the rat or guinea pig. Following electrical stimulation of the trigeminal ganglion in rats, RP-67,580, with an ID_{50} of 0.6 μg/kg, almost totally inhibited the extravasation of radiolabeled serum albumin into the isolateral dura.[6] The NK_1 receptor-inactive enantiomer RP-68,651 was 400-fold less active. Extravasation was less completely inhibited into other tissues within the trigeminal distribution, e.g., conjunctiva, eyelid, and lip. In another study, Moussaoui et al.[19] similarly demonstrated that RP-67,580 inhibited radiolabeled albumin extravasation into the dura whether stimulated by capsaicin (ID_{50} 35 μg/kg) or exogenously administered SP (2.5 μg/kg). CP-96,345 was also effective at 100 and 1000 μg/kg. For both compounds, the dose response of inhibition of i.v. capsaicin-evoked extravasation into the conjunctiva and skin was shifted to the right, suggesting that tissue penetration at these extradural sites of the antagonist may be poorer or that other mediators may partially mask the inhibitory effect. In both studies, the selective 5-hydroxytryptamine receptor agonist sumatriptan was used as a positive control. Sumatriptan inhibited both capsaicin and trigeminal nerve stimulation of extravasation within the dura, but had no effect at extracranial sites in the skin or bladder. These encouraging findings suggest that neurogenic extravasation in the trigeminal distribution is NK_1 mediated. The efficacy of an NK_1 antagonist relative to sumatriptan remains to be determined in man.

2. Respiratory System

Neuropeptide sensory nerves have been implicated in the pathogenesis of vasomotor and allergic rhinitis and asthma in humans.[20] Increased SP, calcitonin gene-related peptide (CGRP), and other neuropeptides have been detected in nasal lavage after allergen challenge.[21] SP has been implicated in the local vessel dilation and obstruction in allergic rhinitis, and in associated fluid and cellular exudation following nasal SP challenge. Braunstein et al.[22] postulated, however, that the albumin and cellular leakage after neurokinin challenge may be mediated by the N-terminal rather than the C-terminal sequence of SP, which is common to the family of neurokinins. Okamoto et al.[23] have demonstrated increased cytokine mRNA in human nasomucosal biopsies following intranasal SP administration. In the rat, intranasal SP increased nasal blood flow measured by trapping of radionuclide-labeled microspheres, an effect abolished by CP-99,994 (2 μg/kg i.v.).[24] Capsaicin (25 μg/kg) caused a three- to fourfold increase in blood flow, an effect that was only partially reduced by CP-99,994 (average 1.5-fold increase in flow). The inactive enantiomer CP-100,263 had no effect. These studies again suggest that SP, through the NK_1 receptor, mediates most, but not all, of the capsaicin-stimulated increase in blood flow. These findings in the nasal mucosa are distinguished from those in the trachea and lung, in which almost all capsaicin-induced vasodilation is apparently mediated by the NK_1 receptor.

3. Trachea and Lungs

Nonpeptide NK_1 antagonists have confirmed the crucial role of the NK_1 receptor in neurogenic inflammation in the airways in various animal species. For example, CP-96,345 was shown to inhibit airway plasma leakage induced by cigarette smoke,[25] vagal nerve stimulation,[26] and capsaicin or SP.[27] Similarly, CP-96,345 selectively inhibited the bronchoconstriction elicited by exogenous SP in guinea pigs.[28] Importantly, however, capsaicin-induced bronchospasm was blocked by NK_2 receptor antagonism and unaffected by CP-96,345, suggesting that both NK_1 and NK_2 mechanisms might participate in airway narrowing, by different mechanisms.[29] The combined NK_1/NK_2 antagonist FK224 inhibited both the bronchoconstriction and plasma leakage in guinea pigs induced by capsaicin, SP, or neurokinin A (NKA), while not affecting responses elicited by acetylcholine or histamine.[30] In an early clinical trial, FK224 was shown to block the decrease in specific airway conductance induced by inhaled aerosolized bradykinin.[31] In addition, the increased vascular permeability in the trachea and bronchi following inhalation of hypertonic saline has been shown to be blocked by CP-99,994.[32]

Whether less selective inflammatory insults are also inhibited by NK_1 antagonists is of course critically relevant to their potential use in asthma or other airway disease. Although antigen-induced clearance of radiolabeled solute was reduced in rats pretreated with capsaicin,[33] prior capsaicin denervation did not affect an increase in airway resistance in guinea pigs in response to ovalbumin aerosol challenge.[34] Bronchial alveolar lavage (BAL) levels of prostaglandins, thromboxanes, and histamine were all elevated in this challenge model. Similarly, NK_1 antagonists have no effect on PAF or histamine-induced increases in airway resistance. However, Bertrand et al.[35] reported that the Evans blue dye extravasation induced by ovalbumin inhalation in guinea pigs was inhibited by up to 75% by CP-96,345 with no further inhibition by the bradykinin antagonist Hoe-140. They did not, however, report on antigen-induced changes in total pulmonary resistance, which are perhaps more clinically relevant.

4. Gastrointestinal Tract

Tachykinins and neurokinin receptors are distributed throughout the GI tract. Evidence suggests that SP-containing sensory nerves may be involved in modulating inflammatory bowel disease and other gut disorders. SP receptors are dramatically upregulated in perivascular and lymphoid tissue in colonic biopsies removed from patients with inflammatory bowel disease (IBD), while NK_2 receptors are unaltered.[36] Increased SP levels are also induced in trichinella-infected rats, an increase not observed in rats pretreated with capsaicin.[37] Treatment of infected rats with betamethazone attenuated both the inflammatory response and increase in SP, consistent with previous observations on adrenocorticotropic hormone-adrenal regulation of SP in the sensory vagus nerve.[38] Interleukin-1 (IL-1) may be one of the cytokines mediating the increase in SP or NK_1 receptors.[39] Whether the apparent inflammatory stimulation of C-fiber expression of SP and associated increased expression of the NK_1 receptor is an epiphenomenon that is induced to support the immune/inflammatory reaction or whether these increases play an initiating path of physiologic function is unclear. Furthermore, whether associated abdominal pain or increased bowel motility are directly mediated by the NK_1 receptor within the spinal cord and/or brainstem is unknown, but IBD will clearly be one of the first therapeutic targets in upcoming clinical trials with nonpeptide NK_1 antagonists.

SP and NK_1 receptors also play a role in ion transport and may mediate increases in secretion or bowel motility in response to ingested toxins[40]. A recent exciting finding by Leeman's group[41] demonstrated that pretreatment of rats with the NK_1 antagonist CP-96,345 blocked the diarrhea and hypersecretion associated with bowel perfusion with *Clostridium difficile* toxin. However, treatment after perfusion of the toxin was ineffectual, making a potential therapeutic role of an NK_1 antagonist in pseudomembranous colitis or related toxin-associated diarrhea uncertain.

5. Arthritis

Several lines of converging evidence also implicate capsaicin-sensitive sensory nerves and NK_1 receptors in experimental or spontaneous arthritis. Clinically, arthritic involvement is decreased in the denervated limb, either experimentally or by illness such as stroke. Capsaicin treatment, either before or after the induction of experimental arthritis, reduces inflammatory involvement. The reduction in diameter of the inflamed paw or joint is only partial (20%), occurs within 24 h, and is otherwise nonprogressive. This suggests that capsaicin may decrease the neurogenic inflammatory component of the global inflammatory response, without affecting the other components of that response.[42] SP transport within the sciatic nerve is increased in rats with ipsilateral adjuvant arthritis.[43] Increased SP immunoreactivity within nerve fibers and immunoreactive content correlate with the severity of arthritic involvement. Finally, indirect evidence of a pathophysiological role for SP has been suggested by its effects on increasing collagenase and prostaglandin secretion from synoviocytes[44] and cytokine production from macrophages.

Although nonpeptide NK_1 antagonists block carrageenan-induced extravasation into the knee joint and SP, but not NKA or CGRP, increased blood flow, there is no evidence that they affect the arthritic process. Collagen-induced arthritis in the guinea pig is unaffected by treatment with NK_1 antagonists.[58] Thus, the blockade of the neurogenic inflammatory response associated with arthritis — even if mediated by SP — may not alter the overall inflammatory and immune process associated with clinical arthritis.

6. Other Immune Responses

In addition to its effect on the neurogenic inflammatory component that may be important in the initiation and subsequent vascular support of the immune response, neurokinins have been implicated in the cellular and cytokine components of the immune response. SP may directly stimulate neutrophil chemotaxis, an effect mediated by its C-terminal, and not its N-terminal, tetrapeptide sequence,[45] which has also been implicated in macrophage and neutrophil activation.[46] Although these effects were demonstrated on neutrophils *in vitro*, SP may also stimulate granulocyte infiltration via activation of mast cells, an effect perhaps partially mediated by leukotriene B_4.[47] Both SP and NKA have shown equal activity in reducing the release of IL-1, tumor necrosis factor, and IL-6 from human blood monocytes *in vitro*.[48] This suggests an important role for both NK_1 and NK_2 receptors in this process.

The presence of NK_1 receptors has been demonstrated on T lymphocytes, whereas purified human B lymphocytes showed minimal binding of labeled SP.[49] NKA and $SP_{(4-11)}$ were equipotent with SP in inhibiting binding of [^{3}H]-SP. *In vitro*, SP stimulates IL-2 synthesis,[50] or enhances its synthesis when lymphocytes are first activated by phytohemagglutinin or concanavalin A.[51] These stimulatory effects of SP on IL-2 production or costimulation with IL-1 or other activators occur in the subnanomolar range. Rameshwar et al.[50] report that the effect of SP on IL-2 production, which peaks at 10^{-13} to 10^{-11} *M*, is blocked by CP-96,345.

7. Skin

Neurogenic inflammation was first described in the skin and it is thus appropriate to consider the application of NK_1 antagonists in dermatologic disorders. Increased SP immunoreactivity has been described in the skin of patients with atopic dermatitis and psoriasis.[52] Topical capsaicin has also been studied in psoriasis and shown to reduce overall severity scores of scaling, thickness, erythema, and pruritis.[53] Resolution of a psoriatic lesion has been noted after regional denervation.[54] NK_1 antagonists might also be appropriate for more clearly neurogenic stimuli, such as the erythema following ultraviolet irradiation, or following bee stings.

8. Bladder

The bladder has proven an excellent model to study the complex interactions of efferent and afferent functions of sensory neurokinins following experimental neurogenic or inflammatory stimulation, as reviewed in detail by Maggi.[55] Nonpeptide NK_1 antagonists have demonstrated that protein extravasation in the bladder, whether induced by capsaicin or by more potent stimulants such as xylene or cyclophosphamide, is completely or partially NK_1-mediated. Whereas CP-96,345 was completely effective in blocking capsaicin-induced extravasation, only combination with the bradykinin antagonist Hoe-140 inhibited the protein extravasation following xylene.[56] In contrast, however, a combination of RP-67,580 and Hoe-140 did not have any further additive effect on the inhibition of cyclophosphamide-mediated extravasation. Thus, the effect of NK_1 antagonism on bladder inflammation may be very stimulus-specific.

At the site of the bladder wall, NK_2 antagonists inhibit reflex contractions, but intrathecal or systemically administered NK_1 antagonists block inflammation-induced reductions in the volume-dependent generation of the micturition reflex.[57] Although the hypothesis remains to be tested in humans, NK_1 antagonism either alone or in combination with NK_2 or cholinergic antagonists may play a significant role in certain forms of inflammatory bladder or incontinence syndromes.

V. CONCLUSION

The modulatory roles of SP and other neuropeptides will continue to pose a challenge for the development of their receptor antagonists for specific clinical conditions. The structural diversity of the antagonists that have been discovered to date provides a rich array of clinical entities to probe the therapeutic potential of tachykinin antagonism, while keeping in mind their limitations, e.g., their species selectivity or possible nonspecific effects. Some of the most exciting potential for clinical effects may be in the area of headache pain such as migraine, neurogenic inflammation in the airways, inflammatory bowel disease, and urinary incontinence. Indeed, the initial clinical trial with FK224 already indicates the considerable potential in this field, which is poised for many exciting clinical findings.

REFERENCES

1. Snider, R. M., Constantine, J. W., Lowe, J. A., III, Longo, K. P., Lebel, W. S., Woody, H. A., Drozda, S. E., Desai, M. C., Vinick, F. J., Spencer, R. W., and Hess, H.-J., A potent nonpeptide antagonist of the substance P (NK1) receptor, *Science*, 251, 435, 1991.
2. Nagahisa, A., Kamano, K., Lowe, J. A., III, Suga, O., Tsuchiya, M., and Hess, H.-J., Antiinflammatory and analgesic activity of a non-peptide substance P receptor antagonist, *Eur. J. Pharmacol.*, 217, 191, 1992.
3. Schmidt, A. W., McLean, S., and Heym, J., The substance P receptor antagonist CP-96,345 interacts with Ca^{++} channels, *Eur. J. Pharmacol.*, 219, 491, 1992.
4. McLean, S., Ganong, A., Seymour, P. A., Snider, R. M., Desai, M. C., Rosen, T., Bryce, D. K., Longo, K. P., Reynolds, L. S., Robinson, G., Schmidt, A. W., Siok, C., and Heym, J., Pharmacology of CP-99,994: a nonpeptide antagonist of the tachykinin neurokinin-1 receptor, *J. Pharmacol. Exp. Ther.*, 267, 472, 1993.
5. Garret, C., Carruette, A., Fardin, V., Moussaoui, S., Peyronel, J.-F., Blanchard, J.-C., and Laduron, P. M., Pharmacological properties of a potent and selective nonpeptide substance P antagonist, *Proc. Natl. Acad. Sci. U.S.A.*, 88, 10208, 1991.
6. Shepheard, S., Williamson, D. J., Hill, R. G., and Hargreaves, R. J., The non-peptide $neurokinin_1$ receptor antagonist, RP 67580, blocks neurogenic plasma extravasation in the dura mater of rats, *Br. J. Pharmacol.*, 108, 11, 1993.
7. Fardin, V., Carruette, A., Menager, J., Bock, M., Flamand, O., Foucault, F., Heuillet, E., Moussaoui, S. M., Tabart, M., Peyronel, J. F., and Garret, C., In vitro pharmacological profile of RPR 100893, a novel nonpeptide antagonist of the human NK1 receptor, *Neuropeptides*, 26, 34, 1994.
8. Emonds-Alt, X., Doutremepuich, J. D., Jung, M., Proietto, E., Santucci, V., Van Broeck, D., Vilain, P., Soubrie, Ph., LeFur, G., and Breliere, J. C., SR 140,333, a non-peptide antagonist of substance-P (NK1) receptor, *Neuropeptides*, 24, 231, C18, 1993.
9. Emonds-Alt, X., Vilain, P., Goulaouic, P., Proietto, V., Van Broeck, D., Advenier, C., Naline, E., Neliat, G., LeFur, G., and Breliere, J. C., A potent and selective non-peptide antagonist of the neurokinin A (NK2) receptor, *Life Sci.*, 50, PL101, 1992.
10. Jung, M., Calassi, R., Maruani, J., Barnouin, M. C., Souilhac, J., Poncelet, M., Gueudet, C., Emonds-Alt, X., Soubrie, P., Breliere, J. C., and LeFur, G., Neuropharmacological characterization of SR 140,333, a nonpeptide antagonist of NK1 receptors, *Neuropharmacology*, 33, 167, 1994.
11. Hashimoto, M., Hayashi, K., Murai, M., Fujii, T., Nishikawa, M., Kitoyoh, S., Okuhara, M., Kohsaka, M., and Imanaka, H., WS9326A, a novel tachykinin antagonist isolated from *Streptomyces violaceusniger* no. 9326, *J. Antibiot.*, 45, 1064, 1992.
12. Hagiwara, D., Miyake, H., Igari, N., Morimoto, H., Murai, M., Fujii, T., and Matsuo, M., Design of a novel dipeptide substance P antagonist FK888, *Regul. Peptides Suppl.*, 1, S66, 1992.

13. Schilling, W., Bittiger, H., Brugger, F., Criscione, L., Hauser, K., Ofner, S., Olpe, H. R., Vassout, A., and Veenstra, S., Approaches Towards the Design and Synthesis of Nonpeptidic Substance P Antagonists. XIIth Int. Symp. Medicinal Chemistry, Basel, Switzerland, ML-11.3, 1992.
14. Nagahisa, A., Asai, R., Kanai, Y., Murase, A., Tsuchiya-Nakagaki, M., Nakagaki, T., Shieh, T. C., and Taniguchi, K., Nonspecific activity of (±)-CP-96,345 in models of pain and inflammation, *Br. J. Pharmacol.*, 107, 273, 1992.
15. Guard, S., Boyle, S. J., Tang, K., Watling, K. J., McKnight, A. T., and Woodruff, G. N., The interaction of the NK_1 receptor antagonist CP-96,345 with L-type calcium channels and its functional consequences, *Br. J. Pharmacol.*, 110, 385, 1993.
16. Rupniak, N. M. J., Boyce, S., Williams, A. R., Cook, G., Longmore, J., Seabrook, G. R., Caeser, M., Iversen, S. D., and Hill, R. G., Antinociceptive activity of NK1 receptor antagonists: non-specific effects of racemic RP67580, *Br. J. Pharmacol.*, 110, 1607, 1993.
17. Caeser, M., Seabrook, G. R., and Kemp, J. A., Block of voltage-dependent sodium currents by the substance P receptor antagonist (±)-CP-96,345 in neurones cultured from rat cortex, *Br. J. Pharmacol.*, 109, 918, 1993.
18. Moussaoui, S. M., Philippe, L., Le Prado, N., and Garret, C., A non-peptide NK1-receptor antagonist, RP 67580, inhibits neurogenic inflammation postsynaptically, *Br. J. Pharmacol.*, 109, 259, 1993.
19. Moussaoui, S. M., Morimoto, H., Maeda, Y., Kiyotoh, S., Nishikawa, M., and Fujii, T., Inhibition of neurogenic inflammation in the meninges by a non-peptide NK1 receptor antagonist, RP 67580, *Eur. J. Pharmacol.*, 238, 421, 1993.
20. Lowe, J. A., III and Snider, R. M., The role of tachykinins in pulmonary disease, *Annu. Rep. Med. Chem.*, 28, 99, 1993.
21. Mosimann, B. L., White, M. V., Hohman, R. J., Goldrich, M. S., Kaulbach, H. C., and Kaliner, M. A., Substance P calcitonin gene-related peptide, and vasoactive intestinal peptide increase in nasal secretions after allergen challenge in atopic patients, *J. Allergy Clin. Immunol.*, 92, 95, 1993.
22. Braunstein, G., Fajac, I., Lacronique, J., and Frossard, N., Clinical and inflammatory responses to exogenous tachykinins in allergic rhinitis, *Am. Rev. Respir. Dis.*, 144, 630, 1991.
23. Okamoto, Y., Shirotori, K., Kudo, K., Ishikawa, K., Ito, E., Togawa, K., and Saito, I., Cytokine expression after the topical administration of substance P to human nasal mucosa, *J. Immunol.*, 151, 4391, 1993.
24. Piedimonte, G., Hoffman, J. I. E., Husseini, W. K., Bertrand, C., Snider, R. M., Desai, M. C., Petersson, G., and Nadel, J. A., Neurogenic vasodilation in the rat nasal mucosa involves neurokinin tachykinin receptors, *J. Pharmacol. Exp. Ther.*, 265, 36, 1993.
25. Delay-Goyet, P. and Lundberg, J. M., Cigarette smoke-induced airway oedema is blocked by the NK1 antagonist, CP-96,345, *Eur. J. Pharmacol.*, 203, 157, 1991.
26. Lei, Y.-H., Barnes, P. J., and Rogers, D. F., Inhibition of neurogenic plasma exudation in guinea-pig airways by CP-96,345, a new non-peptide NK1 receptor antagonist, *Br. J. Pharmacol.*, 105, 261, 1992.
27. Eglezos, A., Guiliani, S., Viti, G., and Maggi, C. A., Direct evidence that capsaicin-induced plasma protein extravasation is mediated through tachykinin NK1 receptors, *Eur. J. Pharmacol.*, 209, 277, 1991.
28. Griesbacher, T., Donnerer, J., Legat, F. J., and Lembeck, F., CP-96,345, a non-peptide antagonist of substance P: II. Actions on substance P-induced hypotension and bronchoconstriction, and on depressor reflexes in mammals, *Naunyn-Schmiedeberg's Arch. Pharmacol.*, 346, 323, 1992.
29. Ballati, L., Evangelista, S., Maggi, C. A., and Manzini, S., Effects of selective tachykinin receptor antagonists on capsacin- and tachykinin-induced bronchospasm in anaesthetized guinea-pigs, *Eur. J. Pharmacol.*, 214, 215, 1992.
30. Murai, M., Morimoto, H., Maeda, Y., Kiyotoh, S., Nishikawa, M., and Fujii, T., Effects of FK224, a novel compound NK1 and NK2 receptor antagonist, on airway constriction and airway edema induced by neurokinins and sensory nerve stimulation in guinea pigs, *J. Pharmacol. Exp. Ther.*, 262, 403, 1992.
31. Ichinose, M., Nakajima, N., Takahashi, T., Yamauchi, H., Inoue, H., and Takishima, T., Protection against bradykinin-induced bronchoconstriction in asthmatic patients by neurokinin receptor antagonist, *Lancet*, 340, 1248, 1992.
32. Piedmonte, G., Bertrand, C., Geppetti, P., Snider, R. M., Desai, M. C., and Nadel, J. A., A new NK1 receptor antagonist (CP-99,994) prevents the increase in tracheal vascular permeability produced by hypertonic saline, *J. Pharmacol. Exp. Ther.*, 266, 270, 1993.
33. Sestini, P., Dolovich, M., Vancheri, D., Stead, R. H., Marshall, J. S., Perdue, M., Gauldie, J., and Bienenstock, J., Antigen-induced lung solute clearance in rats is dependent on capsaicin-sensitive nerves, *Am. Rev. Respir. Dis.*, 1989, 401, 1989.
34. Ingenito, E. P., Pliss, L. B., Martins, M. A., and Ingram, R. H., Jr, Effects of capsaicin on mechanical, cellular, and mediator responses to antigen in sensitized guinea pigs, *Am. Rev. Respir. Dis.*, 143, 572, 1991.
35. Bertrand, C., Nadel, J. A., Yamawaki, I., and Geppetti, P., Role of kinins in the vascular extravasation evoked by antigen and mediated by tachykinins in guinea pig trachea, *J. Immunol.*, 151, 4902, 1993.

36. Mantyh, C. R., Gates, T. S., Zimmerman, R. P., Welton, M. L., Passaro, E. P., Jr., Vigna, S. R., Magio, J. E., Kruger, L., and Mantyh, P. W., Receptor binding sites for substance P, but not substance K or neuromedin K, are expressed in high concentrations by arterioles, venules, and lymph nodules in surgical specimens obtained from patients with ulcerative colitis and Crohn disease, *Proc. Natl. Acad. Sci. U.S.A.*, 85, 3235, 1988.
37. Swain, M. G., Agro, A., Blennerhassett, P., Stanisz, A., and Collins, S. M., Increased levels of substance P in the myenteric plexus of trichinella-infected rats, *Gastroenterology*, 102, 1913, 1992.
38. MacLean, D. B., Adrenocorticotropin-adrenal regulation of transported substance P in the vagus nerve of the rat, *Endocrinology*, 121, 1540, 1987.
39. Hurst, S. M., Stanisz, A. M., Sharkey, K. A., and Collins, S. M., Interleukin 1-beta-induced increase in substance P in rat myenteric plexus, *Gastroenterology*, 105, 1754, 1993.
40. Jodal, M., Holmgren, S., Lundgren, O., and Sjoqvist, A., Involvement of the myenteric plexus in the cholera toxin-induced net fluid secretion in the rat small intestine, *Gastroenterology*, 105, 1286, 1993.
41. Pothoulakis, C., Castagliulo, I., LaMont, J. T., Jaffer, A., O'Keane, J. C., Snider, R. M., and Leeman, S. E., CP-96,345, a substance P antagonist, inhibits rat intestinal responses to Clostridium difficile toxin A but not cholera toxin, *Proc. Natl. Acad. Sci. U.S.A.*, 91, 947, 1994.
42. Colpaert, F. C., Donnerer, J., and Lembeck, F., Effects of capsaicin on inflammation and on the substance P content of nervous tissues in rats with adjuvant arthritis, *Life Sci.*, 32, 1827, 1983.
43. Donnerer, J., Schuligoi, R., and Stein, C., Increased content and transport of substance P and calcitonin gene-related peptide in sensory nerves innervating inflamed tissue: evidence for a regulatory function of nerve growth factor in vivo, *Neuroscience*, 49, 693, 1992.
44. Lotz, M., Vaughan, J. H., and Carson, D. A., Effect of neuropeptides on production of inflammatory cytokines by human monocytes, *Science*, 241, 1218, 1988.
45. Kolasinski, S. L., Haines, K. A., Siegel, E. L., Cronstein, B. N., and Abramson, S. B., A somatostatin analog as a selective antagonist of neutrophil activation by substance P, *Arthritis Rheum.*, 35, 369, 1992.
46. Serra, M. C., Bazzoni, F., Bianca, V. D., Greskowiak, M., and Rossi, F., Activation of human neutrophils by substance P: effect on oxidative metabolism, exocytosis, cytosolic Ca^{2+} concentration and inositol phosphate formation, *J. Immunol.*, 141, 2118, 1988.
47. Iwamoto, I., Tomoe, S., Tomioka, H., and Yoshida, S., Leukotriene B4 mediates substance P-induced granulocyte infiltration into mouse skin, *J. Immunol.*, 151, 2116, 1993.
48. Guerne, P. A., Carson, D. A., and Lotz, M., IL-6 production by human articular chondrocytes. Modulation of its synthesis by cytokines, growth factors, and hormones in vitro, *J. Immunol.*, 144, 499, 1990.
49. Payan, D. G., Brewster, D. R., Missirian-Bastian, A., and Goetzl, E. J., Substance P recognition by a subset of human T lymphocytes, *J. Clin. Invest.*, 74, 1532, 1984.
50. Rameshwar, P., Gascon, P., and Ganea, D., Stimulation of IL-2 production in murine lymphocytes by substance P and related tachykinins, *J. Immunol.*, 151, 2484, 1993.
51. Calvo, C.-F., Chavanel, G., and Senik, A., Substance P enhances IL-2 expression in activated human T cells, *J. Immunol.*, 148, 3498, 1992.
52. Tobin, D., Nabarro, G., Baart de la Faille, H., van Vloten, W. A., van der Putte, C. J., and Schuurman, H.-J., Increased number of immunoreactive nerve fibers in atopic dermatitis, *J. Allergy Clin. Immunol.*, 90, 613, 1992.
53. Ellis, C. N., Berberian, B., Sulica, V. I., Dodd, W. A., Jarratt, M. T., Katz, H. I., Prawer, S., Krueger, G., Rex, I. H., Jr., and Wolf, J. E., A double-blind evaluation of topical capsaicin in pruritic psoriasis, *J. Am. Acad. Dermatol.*, 29, 438, 1993.
54. Raychaudhuri, S. P. and Farber, E. M., Are sensory nerves essential for the development of psoriatic lesions?, *J. Am. Acad. Dermatol.*, 28, 488, 1993.
55. Maggi, C. A., Patacchini, R., Rovero, P., and Giachetti, A., Tachykinin receptors and tachykinin receptor antagonists, *J. Auton. Pharmacol.*, 13, 23, 1993.
56. Giuliani, S., Santicioli, P., Lippe, I. T., Lecci, A., and Maggi, C. A., Effect of bradykinin and tachykinin receptor antagonist on xylene-induced cystitis in rats, *J. Urol.*, 150, 1014, 1993.
57. Maggi, C. A., Giuliani, S., Ballati, L., Lecci, A., Manzini, S., Patachhini, R., Renzetti, R., Rovero, P., Quartara, L., and Giachetti, A., In vivo evidence for tachykininergic transmission using a new NK-2 receptor-selective antagonist, MEN 10,376, *J. Pharmacol. Exp. Ther.*, 257, 1172, 1991.
58. Nagahisa, A., personal communication.

Index

INDEX

A

B

C

D

E

M

N

O

P

R

T

U

V

W

X

Z